Christoph Schmitz
(ed.)

Handbook of **Aluminium Recycling**

2nd Edition

Mechanical Preparation I Metallurgical Processing I Heat Treatment

Vulkan Verlag

edition heat processing

Bibliographic information published by Deutsche Nationalbibliothek

Deutsche Nationalbibliothek lists this publication
in the Deutsche Nationalbibliografie;
detailed bibliographic data is available on the Internet at

www.dnb.de

ISBN 978-3-8027-2970-6

Please register for your interactive eBook:
www.di-verlag.de/de/Service/eBook-Registrierung

V7KM-WDSG-VDGZ-MWPE

Huyssenallee 52-56, D-45128 Essen, Fed. Rep. of Germany
Tel.: +49 (0)201 8 20 02-0, Internet: www.vulkan-verlag.de

Project management: Stephan Schalm, s.schalm@vulkan-verlag.de
Proofreading: Loren Mark Hamersley
Editor: Annamaria Frömgen

Production: Nilofar Mokhtarzada
Cover design: Daniel Klunkert
Cover photo: © Jasper Gesellschaft für Energiewirtschaft und Kybernetik mbH, Geseke, Germany
Print: Druckerei Chmielorz GmbH, Germany

Preface

Established processes do not change very rapidly. Therefore, the procedures and the related equipment design, as described in the first issue, are still common practises. However, public interest is focused more and more on requirements related to the "carbon footprint". Since our industry uses a substantial amount of fossil energy the reduction of CO_2 emissions is a key issue. This refers to a better energy balance and a more sophisticated combustion control. Therefore, the chapters describing burner technology have been expanded accordingly with special attention given to oxy-fuel technology. But also the use of the internal energy of plastic waste present in the collected scrap, to improve the energy balance, is covered to some extent. Since burner operation goes in hand with process control the chapters covering this subject have been expanded particularly with the description of the control of the unburned components in the flue gas.

The generation of waste is becoming more and more a problem since space available for landfill is very limited and the impact of hazardous contents in such waste must be considered. Salt slag makes up the majority of waste in a melting process. Therefore, a chapter with a detailed description of a process for salt and aluminum recovery from such salt slag waste was now added to the book.

Besides the efforts in regard to the carbon footprint and the recycling of waste, quality requirements by customers are becoming more stringent. Since heat treatment is one of the processes to improve the quality, a separate chapter focuses on this technology.

I trust that the new edition of the "Handbook of Aluminium Recycling" will help to combine the efforts of operators, designers and project engineers for a better understanding of the specific problems between the partners. This will enable them to upgrade the processes beyond the technologies described in this book for enhancement of the economy of the operations and to further reduce any negative impacts on the environment.

I am grateful for the support to complete the new edition. Once again, the companies producing aluminium process equipment provided me with valuable information and my co-authors contributed up-to-date additions to the book. Many thanks go to the Vulkan-Verlag team and specifically Annamaria Frömgen who carefully read through the old and new script. Her thorough proofreading eliminated many inaccuracies. Furthermore, I very much appreciate the proofreading efforts of Loren Mark Hamersley who ensured that the English expressions and wording are more easily understood. Last but not least, I thank Jana Maria Esser for converting the black and white figures into nice color pictures and graphs which added color to the book and thus helped that some of the graphs became easier to comprehend.

Bad Münstereifel, 2014

Christoph Schmitz

Authors

Dipl.-Ing. Josef Domagala
Engtra – Engineering & Trade Services
Naabstraße 2
D-40699 Erkrath
Email: j.domagala@engtra.de
Part III, chapter 3.3, 3.3.1-3, 3.3.5

Petra Haag
Economist
Hermann-Josef-Straße 1
D-53902 Bad Münstereifel
Email: hppv.haag@t-online.de
Part IV

Dr. Thomas Niehoff
Head of Industry Segment Nonferrous & Mining
Linde AG
Carl-von-Linde-Str. 25
D-85716 Unterschleißheim
Email: thomas.niehoff@linde-gas.com
Part III, chapter 3.3.4, 3.3.6

Dipl.-Ing. Christoph Schmitz
Aluserve Projektberatung Aluminium
Engelsbergweg 2
D-53902 Bad Münstereifel
Email: christophschmitz@gmx.de
Editor and main author

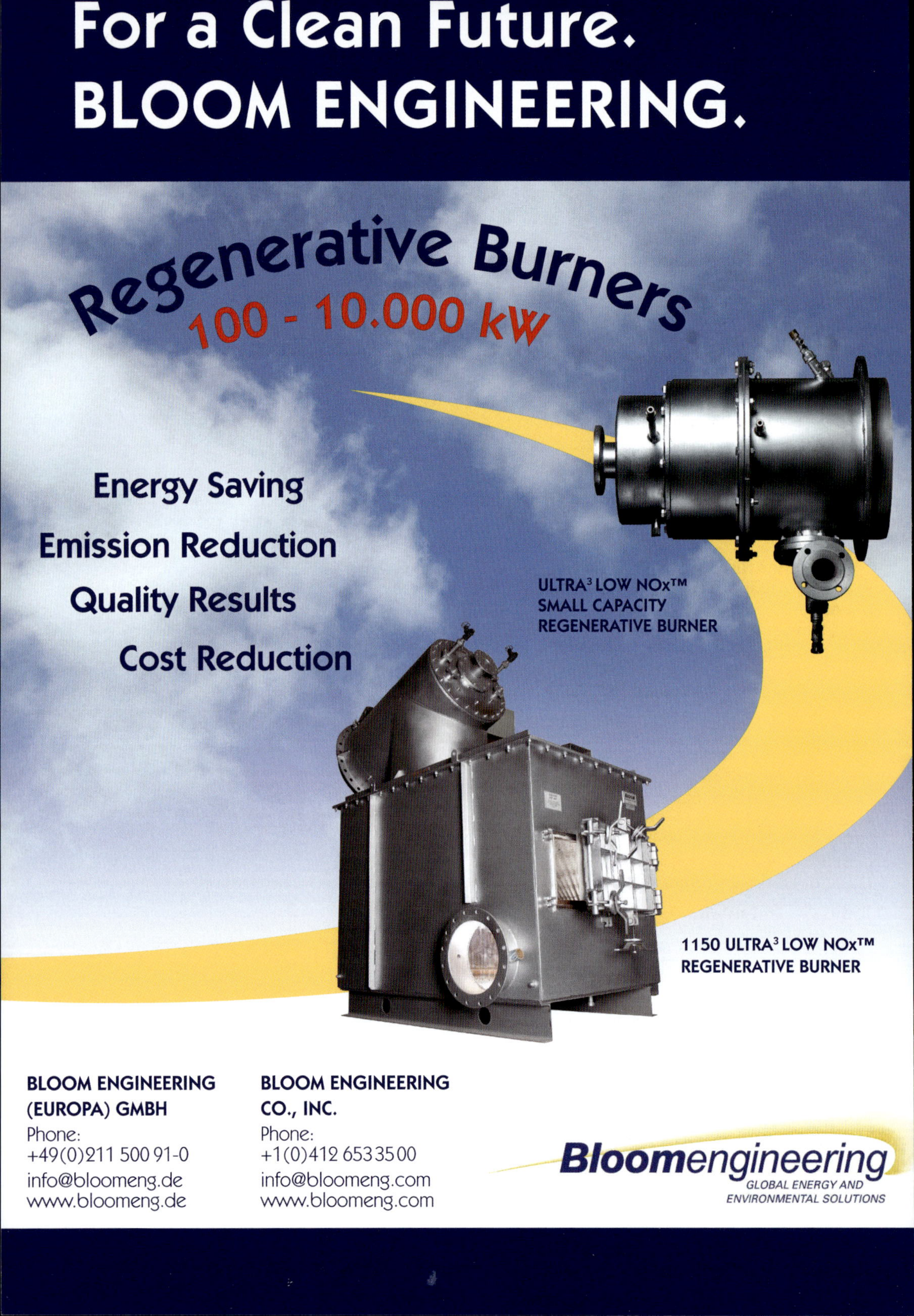
For a Clean Future.
BLOOM ENGINEERING.
Regenerative Burners
100 - 10.000 kW
Energy Saving
Emission Reduction
Quality Results
Cost Reduction
ULTRA³ LOW NOx™
SMALL CAPACITY
REGENERATIVE BURNER
1150 ULTRA³ LOW NOx™
REGENERATIVE BURNER
BLOOM ENGINEERING
(EUROPA) GMBH
Phone:
+49(0)211 500 91-0
info@bloomeng.de
www.bloomeng.de
BLOOM ENGINEERING
CO., INC.
Phone:
+1(0)412 653 35 00
info@bloomeng.com
www.bloomeng.com
Bloomengineering
GLOBAL ENERGY AND
ENVIRONMENTAL SOLUTIONS

Contents

DANIELI OLIVOTTO FERRE'
DANIELI
olivotto
FERRE
ADVANCED SOLUTIONS FOR ALUMINIUM FIELD REQUIREMENTS
C.so Unione Sovietica 612 int. 20, 10135 Torino - Italy
Phone +39 011 2633 111 - Fax +39 011 2633 552
E-mail info.dfo@danieli.it - sales@dfo.danieli.it
www.danielicentrocombustion.it
part of DANIELI Centro Combustion S.p.A.

Part I
Fundamentals

1. Introduction

Christoph Schmitz

During the huge bang of an exploding star as supernova in the outer region of our galaxis alumina came into existence together with the other heavy elements which astronomers call metals. The litter of the explosion formed our solar system. At some stage in the genesis of our planet heavier metals such as iron concentrated in the center of the planet while the lighter elements such as magnesium and, of course aluminium, formed the earth's crust. Aluminium has reacted with silicon and oxygen and this chemical compound forms the basis of most of the stones on earth.

Aluminium is the most common metal in the earth's crust and, therefore, provides an almost inexhaustible raw material source (**Table 1.1**). After oxygen and silicon it is the third common element and its content amounts to 8 % in the continental crust and 6.8 % in the entire earth's crust. But aluminium is a very young metal. Its industrial use dates back only 120 years. But modern life is unimaginable without aluminium. How could air transportation and space technology be realized without using aluminium. The small label attached to the Pioneer 10 space probe is made of aluminium and may inform other intelligent forms of life that may be found in the outer space some time about mankind on our blue planet.

Table 1.1: Occurrence of elements on earth

	Occurence in %	
	Total earth	Earth crust
Others	<1	<1
Aluminium	1.1	8.0
Ca	1.1	2.4
Na	<1	2.1
K	<1	2.3
S	1.9	<1
Ni	2.4	<1
Mg	13.0	4.0
Si	15.0	28.0
O	30.0	46.0
Fe	35.0	6.0

Many applications in various technologies would be at least very difficult without aluminium. It may be regarded as coincidence that technologies came to life together with the use of aluminium in air and road traffic, civil engineering, packing, power transmission and many other technologies that make life easier.

Although the industrial use of aluminium metal dates back only 120 years human beings have used aluminium when they started to produce pottery some 7,000 years ago. The clays that our forefathers have used contained essentially a silicate of aluminium and magnesium. The baking furnaces for the pottery may have also been the reason to find out about the processing of other metals. Metals such as gold, silver, iron, copper, lead and tin have a long history in the human society. Aluminium was not discovered as metal due to very stable compounds with oxygen and silicon. It was also unknown that aluminium in non-metallic form was an essential component of many gem stones. Only 150 years ago the patent was filed to extract aluminium on industrial scale. A few years before some small quantities of aluminium were obtained at very high price and for this reason it was called "silver out of clay". Economical value was not obtained until the Fifties of the 19th century as a result of some fortunate occasions that took place simultaneously. In Heidelberg the famous German chemist, Robert Bunsen, succeeded the electrolytic separation of metallic aluminium. In France the nephew of Napoleon III, Henry St. Claire - Deville, started the first production of aluminium with the target to produce strong but light armor for the soldiers.

In 1855 he could present the first ingot of the "silver out of clay". But all he was able to produce before he retired were some military helmets. The real breaking point came when the electrolysis of the alumina-cryolite melt was invented. The process was patented almost simultaneously, but independently, in 1886. Héroult patented the process in France while Hall patented a method with minor difference in America. The patents coincided with the developments in the generation of electric power, which made commercial fabrication technically feasible. Additionally, Bayer developed the extraction of alumina from bauxite. The first electrolytic cells operated with 600 A per cell. Continuous improvements in cell design over the last 100 years increased the scale up to 300,000 A and the appearance of cells has changed dramatically. However, the fundamental production process did not undergo any change.

The production of aluminium is very frequently linked to excessive use of valuable energy. The discussion is very often just highlighting only one side of the coin. So even in the packing industry the light weight of aluminium saves energy when transporting goods from one side of the world to the other. Another example is the high frequency train schedule with very short acceleration and deceleration. This definitely requires light vehicles in order to save energy. And it is certainly not a coincidence that the designers of automobiles use a steadily increasing content of aluminium.

No doubt, the production of primary aluminium is energy-intensive. Smelters are therefore built in regions with ample energy resources. It is obvious that electrical energy cannot be transported over a very large distance. It would, for instance, be quite a technical challenge to construct a power line system from the Middle East to Europe and this apart from the inevitable power losses. Import of primary aluminium can be regarded as import of energy. The electric power used for the smelter is available for other utilization.

The "green conscience" is, however not limited to energy savings. Aluminium is a metal predestined for recycling. The alloys used for the various applications can be recycled with only little loss in quality and at comparatively low energy cost. It is, however, still amazing that only approximately 30 % of the aluminium produced is available for recycling. Some material is lost forever. This will for instance be the content used for the steel industry for improving steel alloys. Some aluminium is lost in the domestic waste since there is still no economical technology available to recover the compound material with extremely thin layers of aluminium. Some material is available for recycling after lengthy periods of time, partly due to its durability, for instance in the construction industry. Maybe there is also a lack in collecting efficiency. Today aluminium scrap is a valuable commodity that is traded worldwide.

The technologies for recycling have been invented over the past centuries. In very early days processes have been developed to refine the collected scrap with the efficiency growing as energy cost increased, availability of scrap easy to process decreased and environmental regulation became more and more stringent. Process engineers did have visions how the ideal plant concept should be. This dates back to the mid of last century. As it is with visions, some fail because of economical shortcoming, some of them due to the very conservative attitude of producers and their process engineers. It is true that all operators want the most modern and efficient technology. But who wants to be the first one to try? Who wants to take the economical risk? So the standard expression is: "Yes, very interesting. But where can we see such a plant?" Some technologies have been introduced in spite of such hesitation. This is, for instance, the technology of the closed well furnace as well as the tiltable rotary drum furnace. Many improvements have been made in detail design, too. So, if one looks at today's design of reverberatory furnaces, the manufacturer can only be defined by insiders. Still, all these petty steps help to improve the technologies to meet today's challenge with regard to the more and more stringent environmental requirements, the propelling energy cost and the increasing shortage of raw material. Some technologies that are presented at meetings and conferences presenting a revolutionary new technology appear after having a closer look at remake of designs given up years ago or useful for very specific or limited application only. Real development can only be achieved by good and trustful cooperation between designers and operators. It is of no help if operators install equipment which they cannot operate properly due to lack of knowledge and the efforts of the designers will be in vain if they do not have the indispensable feedback from plant operation.

This book will give an overview of the status of the aluminium recycling technologies and intends to focus on the engineer in production, the designers of equipment, the plant operators and the owners and, last not but least, students of our trade to obtain their own view and evaluation of the processes. Additionally, it may be of interest to the reader to be aware where primary metal comes from, how it is produced and how the used aluminium comes back and to process it for reuse. The form of the recycled aluminium and the impact on the recycling technologies help to understand the process requirements involved. Also the knowledge of the very specific characteristics and the behavior of the metal while passing through the various process steps help to understand and perhaps to select the proper plant equipment.

The very common and basic technologies and designs are portrayed in some detail, while very special technologies used in particular but not so common applications are touched only. Since the processes for liquid metal are the key for a successful plant operation, furnaces play a substantial role in the book with the consequence that a large section will refer to all types of furnaces that have been developed from the very basic hearth furnace over the centuries.

So, dear reader, feel free to glance through the book page per page or study the chapters of particular interest to you and cross over to chapters explaining some basics in more detail or as it may be helpful when, for instance defining a new plant concept, go from type of material and its characteristics to select the best suited equipment for the purpose and finally design the entire plant. The book is focussed on the processing of aluminium in the recycling industry. Details of the aluminium metallurgy are only covered to the extent that the process requirements are understood. For more detail please refer to the amply available literature. This also refers to heat processing or other design routines and construction material features.

1.1 Aluminium characteristics

Aluminium is a white silvery shining metal of high electrical conductivity. It is characterized by the dense spherical packing leading to good ductility. Chemically, aluminium shows the characteristics of a high reactive base metal with a high affinity to hydrogen in the liquid phase and a strong tendency to form aluminium oxide compounds. It is resistant to the environment only because it forms a dense oxide skin that seals the metal against the air and provides a good resistance of the surface to chemical attacks. The density of aluminium is 2.7, the melting point 991 K and the boiling point 2,603 K.

Low specific gravity

The density of 2.7 g/cm^3 is only one third of the specific gravity of steel. The ratio is even better compared to heavy metals. This offers advantages where low weight is of importance, i. e. in vehicles or in components that undergo frequent movement. The reduction of mass forces during acceleration and movement of transportable equipment results in lower energy requirements. In stationary structures less weight of the construction often results in less heavy foundations. Together with the good resistance to weather and chemical attack this all leads to less cost for operation and maintenance. The low specific density is also indispensable for the construction of airplanes and spacecrafts.

Good strength

There is good selection of standardized aluminium alloys with different strength characteristics. The engineer is able to select the alloy best suited for his design. Compared with other materials good strength is also a remarkable advantage of aluminium. In combination with the other characteristics good solutions to technical applications can be found.

Good resistance to chemical attacks, weather and seawater

Particularly virgin aluminium and copper-free alloys are resistant to numerous chemicals. Besides the favorable characteristic with regard to strength and light weight, copper-free alloys are used with growing demand in the chemical industry, the construction of vehicles in the shipbuilding industry and in many other applications. Good weather resistance is important for aluminium application if using aluminium wire for power lines or other application in buildings. Even after many years the roofing, cladding and extruded shapes are still in perfect condition. The utilization for buildings, in the chemical industry, packing of food, shipbuilding and other applications can even be improved by selecting proper alloy composition or by surface protection.

Good shaping capability

This behavior permits production of shapes, tubes, plate and other shapes with difficult cross section. Furthermore, aluminium can be processed to very thin foils or wall thickness, an advantage for other utilization.

Good machining behavior

Particularly alloys for automatic machining permit short manufacturing time due to high permissible cutting speed.

Good suitability for welding and adhesive joining

All techniques for welding and adhesive joining of components can be applied and gain more and more importance throughout the industry.

Wide range of special surface treatment

Processes are available to obtain decorative effects of the surface and to offer efficient and durable surface protection.

High electric conductivity

The high electric conductivity of virgin aluminium and AlMgSi-alloys permits a wide range of application in electrical installations and equipment.

High thermal conductivity

This is one of the most important characteristics utilized for alloys applied in the automotive industry for pistons, cylinders and cylinder heads and for heat-exchangers.

Good optical characteristics

The characteristic to reflect light and other electro-magnetic rays (that includes heat) result in low absorption. Surfaces can be treated to maintain the metallic shining surface or to vary refection and absorption in wide ranges.

Table 1.2: Characteristics of aluminium in comparison to copper and steel

Qualities		Unit	Aluminium (Al 95.5)	Copper (Cu)	Steel (St 60-2)
Tensile Strength	R_m	(N/mm^2)	360	500	650
	$R_{0.2}$	(N/mm^2)	320	430	520
Spec. Gravity		(g/cm^3)	2.7	8.92	7.85
Conductivity		($m/\Omega mm^2$)	35	56	
E-modulus		(N/mm^2)	$0.7*10^5$	$1.2*10^5$	$2.1*10^5$
Melting Point		(°C)	658	1083	1539

Magnetic neutrality and behavior after nuclear attack

Aluminium is para-magnetic and has a very low half life value after nuclear attack.

No concern with regard to health

Aluminium and its alloys are not toxic, easy to sterilize and to clean and do fulfil all hygienic and toxic requirements.

Some physical characteristics in comparison to other metals are summarized in **Table 1.2**.

1.2 Aluminium applications

For the discussion of aluminium recycling technologies aluminium statistics may not be the ultimate challenge. But still, it is interesting since the use of aluminium and the percentage of aluminium in various applications give some indication of what the recycling plant has to process. **Fig. 1.1** shows the shares for the various industrial sectors which use aluminium alloys. The percentages refer to the total quantities but do not differentiate between primary and secondary aluminium. There are certainly sectors such as the steel production which use only secondary aluminium since this aluminium is added to the steel to improve characteristics of steel. Other industries use only primary aluminium but mostly it is realistic to assume that combination of primary and secondary aluminium will find their way to the end-user.

The secondary aluminium industry represents an important share for the supply of aluminium to the various end-users. This may be shown by the consumption figures for Germany. In 1998 the total aluminium consumption amounted to approx. 2.6 million tons. The primary production in Germany was only approx. 0.62 million tons while the secondary production was close to 0.5 million tons. The quantity for the secondary production does not include the closed loop production of the different down stream production plants. Fig. 1.1 also shows that the areas transport, packing and civil construction consume the highest quantities of aluminium.

With a share of 38 % of the total aluminium consumption, transport (**Fig. 1.2**) is the most important customer for the secondary aluminium industry. In fact, comparing the other industries, close to 90 % of its production goes to the automotive industry as casting alloys while the other share is provided by wrought aluminium alloys made of primary aluminium as shapes and plates. Remarkable growth rates have been experienced particularly in the automotive industry and even more growth can be expected in future. Starting with casting alloys for engine casings and gear box housings as well as for small components today also the vehicle frame is made of extruded aluminium and the back rest in the car is designed as die-casting element. In railway vehicles the outer shell is made of aluminium as well as some carriage structures. It is common knowledge that aircraft and,

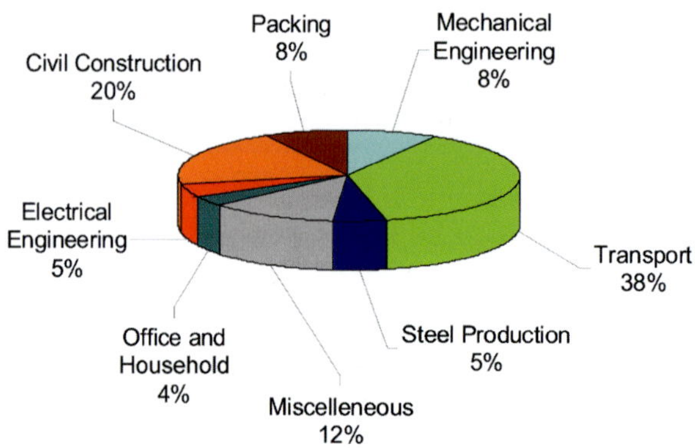

Fig. 1.1: End use of aluminium

Fig. 1.2: High speed train

to a much smaller extent, spacecraft require aluminium alloys. To a certain extent the dependency of the recycling industry on the automotive industry is a weak point. As soon as the production of cars slows down the secondary aluminium industry gets into trouble.

Civil construction or construction industry is the next large customer. In this section mostly shapes or sheets are utilized. These are in general produced from primary aluminium. But as supplier of scrap this industry is very important for the recycler.

The situation is similar to the packing industry with the exemption of used beverage containers (UBC). This material forms an almost closed loop circuit in the USA. But in Europe the bales of UBC are added to the batch to make up the alloy. The alloys produced by the secondary aluminium industry are mostly cast alloys which are not used to make up the starting material for producing beverage containers. The packing industry uses large quantities aluminium foil, sheets for food trays and very thin aluminium foil for lamination of other materials. At the end of its life the material ends up generally in the household waste. No real economical solution has been found yet to recover the aluminium from there. So it has to be regarded as lost to the industry.

A similar case is the material supplied to the steel industry. Here the aluminium is used to kill the steel which means suppressing the formation of gas in the steel batch in order to reduce the brittleness of the steel and to improve its resistance to low temperatures. The total of the aluminium added to the steel is definitely lost. And so is certainly also the aluminium in satellites which remain in an earth orbit and will glow away once their life is terminated.

The small quantity recycled from office and household use is quite unimportant for the industry.

Other recycling production is generally not counted in the statistics. In general, this is the closed loop recycling of the downstream industry. Using this feed material they will certainly produce the wrought alloy they require for their product. Extrusion plants, for example, recycle up to 15 % of their total production.

Still, considering all these facts and also the impact of the worldwide trading activities for scrap, it appears that a high percentage of aluminium disappears somewhere and the question remains whether the collection of aluminum scrap could still be improved.

1.3 Aluminium alloys

1.3.1 System of aluminium alloys

Primary aluminium, which means pure aluminium as produced by the primary smelters, is used for very special applications. Its characteristics depend very much on a series of factors. In particular, the addition of alloying elements to the aluminium or already present in the batch play an important role to improve the features. Except for the high purity alloy Al 99.99, only alloys containing other elements are used in technical applications. Even the characteristics of primary aluminium Al 99 and Al 99.9 are determined by certain contents of iron (Fe) or silicon (Si). Main alloying elements are copper (Cu), silicon (Si), magnesium (Mg), zinc (Zn), and manganese (Mn). Small quantities are added or are already present as contaminations. These are iron (Fe), chromium (Cr) and titanium (Ti). Some special alloys contain nickel (Ni), cobalt (Co), silver (Ag), lithium (Li), zirconium (Zr), tin (Sn), lead (Pb), vanadin (V), bismuth (Bi), calcium (Ca), and cadmium (Cd). Trace elements, such as strontium (Sr), beryllium (Be), boron (B), sodium (Na), strontium (Sr) or antimony (Sb), can also be found. Thus, almost all of the metals of the classification of elements could be present in aluminium alloys, some as alloying element, some of them as unwanted contamination. Looking to the increasing recycling, multiple turnaround of metal may result in culmination of some of these elements.

Primary aluminium shows limited mechanical characteristics. To improve the strength alloying elements have to be added to the melt. Also castability and resistance to chemical attack and weather as well as good behavior with changing temperatures can be improved by alloying. Increase of mechanical strength can be achieved by age-hardening. This also requires definite alloying elements.

There are two major groups of alloys.

Wrought alloys are used as basic material for mechanical forming such as extrusion, rolling, deep-drawing and forging. These alloys have a comparatively small quantity of alloying elements.

Good castability, particularly with regard to modern die-casting machines and considering the very complex structure of castings for the automotive industry, low shrinkage during solidification of the cast and good thermal stability are essential for casting alloys.

Other alloys with a very low addition of alloying elements are used as deoxidation agent for the steel industry, aluminium fashion jewellery and electrical conductors.

For standardization of aluminium and its alloys different systems are used, depending on country, international organization or manufacturer. Some of the brand names for alloys have been developed from manufacturer's company name or typical application or fantasy names or just numbers. The most easy-to-remember designation seems to be the denotation of the German DIN 1712 and DIN 1725 as well as the ISO-system. They provide an identification according to the alloying components and the approximate content of these elements. Very common is also the AA numbering system, comprising a 4 digit system for wrought alloys, whereby the first digit describes the main alloying component:

- 1xxx clean primary aluminium
- 2xxx alloy with Cu as main alloying element
- 3xxx alloy with Mn as main alloying element
- 4xxx alloy with Si as main alloying element
- 5xxx alloy with Mg as main alloying element, in combination with Cr or Mn
- 6xxx alloy of the system AlMgSi or AlSiMg
- 7xxx alloy with Zn as main alloying element
- 8xxx special alloys

Table 1.3: Typical analysis for wrought and casting alloys

Material		Content of alloying components in %. balance aluminium										Applications
		Si	Fe	Cu	Mn	Mg	Cr	Ni	Zn	Pb	Ti	
Wrought alloys												
Al99.5	1050	0.25	0.35	0.02					0.03		0.02	sheet, shapes, wire forgings, cans
AlCuMg2	2024	0.50	0.50	4.50	0.60	1.50	0.10		0.25		0.20	sheet, tubes,shapes forgings
AlMnCu	3003	0.60	0.70	0.20	1.00	1.20			0.10			sheet
AlMg2.5	5052	0.25	0.40	0.10	0.10	2.50	0.90		0.10			cans. body
AlMg4.5	5082	0.20	0.35	0.15	0.15	4.50	0.15		0.25		0.10	cans (lid)
AlMg4Mn	5086	0.40	0.40	0.10	0.70	5.00	0.25		0.20		0.15	cans (lid)
Casting alloys												
AlCu5Si3	223	3.00	1.10	5.00	0.60	0.50		0.50	2.50	0.30	0.15	engine and gearbox housings
AlSi9Cu3	226	9.00	1.20	3.00	0.30	0.30	1.20		0.15			sand casting, die-casting
Si12CuNiMg	260	12.00	0.70	1.00	0.20	1.00		1.00	0.20		0.20	engine pistons, simple castings
AlSi8Cu2	333	8.00	1.00	2.00	0.50	0.30		2.00	2.00	0.30	0.20	thin-walled die castings, pressure-tight

For casting alloys a 3 digit system is becoming very common. This originates from ALCOA standards. A logical system cannot be determined except that all 3 digit numbers refer to casting alloys.

Examples for some typical alloys are listed in **Table 1.3**.

1.3.2 Wrought aluminium alloys

The alloy AlCuMg was introduced shortly after the invention by Hall and Heroult and almost simultaneously with the start of industrial aluminium production. It was developed in the period 1906-1909 by WILMS of Dürener Metallwerke and produced under the name "Duraluminium". After expiration of the patent many other companies are producing this alloy using different brand names. Today's level of transport technology would be unthinkable without Duraluminium. In general the alloy contains 3.5 – 4.0 % copper and additional elements such as Mg, Si and Mn. AlCuMg is a metal with high mechanical strength that is mainly obtained by its very good age hardening behavior.

The AlMgSi is free of Cu and contains Si and Mg. Compared with AlCuMg the alloy has a better corrosion resistance as well as a higher electrical conductivity but less expansion. Its mechanical strength is almost identical to that of the alloy AlCuMg. Since the metal is free of copper it is more favorable for recycling. The alloy AlMgSi is normally used for producing electrical wire since the conductivity is good and is combined with the necessary mechanical strength.

The alloys AlZnMg and AlZnMgCu do have a very high deformation resistance. They can be age-hardened. It is used for highly-stressed components in vehicles and aircraft. Its weldability is also good.

The alloy group AlMg is characterized by very good corrosion resistance and good stress behavior and low sensitivity to thermal stress. It is therefore used for shipbuilding and fittings.

Alloys AlMgMn is also very corrosion-resistant. The addition of Mn reduces the generation of coarse grains during age-hardening. Its thermal strength is not as high as that of the other alloys but it is characterized by very good corrosion resistance in addition to the excellent deep-draw behavior. The alloy is used in the chemical industry and in the food industry and is a good material for beverage cans. For not so critical requirements the AlMn alloy is also used. It deep-draw behavior and its corrosion resistance are less than that for the AlMgMn alloys.

The alloys on the basis of AlMg and AlMn cannot be age-hardened.

1.3.3 Casting aluminium alloys

In order to increase the castability of aluminium alloys Si will be added to the aluminium. While wrought alloys receive only a low percentage of alloying elements, the portion of silicon added to obtain is quite high and may reach as much as 25 % for alloys used to manufacture car rims. Today pressure die-casting parts are produced with very delicate structures and low wall thickness. The aluminium must be able to fill all the fissures and must reach every small area within the casting. The metallurgical engineer can achieve this behavior only by addition of a substantial quantity of Si. Alloys for other castings, as may be used in mechanical engineering, may not be that critical, thus lower Si content will be sufficient. Besides the good to excellent castability, shrinkage during the cooling of the cast must be limited. High performance parts will require the addition of Cu.

The alloys of the group AlSi are best for casting and have a wide range of application. They have a good chemical resistance. At a silicon content of close to 13 % the alloy forms an eutecticum. This alloy is used for components which are exposed to high stress, mechanical shock and vibrations. Typical application is castings for engine pistons.

Having casting with variations of wall thickness within one single casting, AlSiMg may offer advantages. This type of alloy can be age-hardened and will thus receive high mechanical strength. The favorable casting behavior of the AlSi alloy is fully preserved.

Most common are alloys of the type AlSiCu. An example is the alloy 226 (G-AlSi9Cu3 according to DIN 17007) that most of the aluminium refiners tend to produce. It has 8 – 11 % Si and 2.0 – 3.5 % Cu. The alloy has good casting behavior, good machining characteristics and is quite resistant to chemical attack.

Casting alloys of the AlMg type are very resistant to seawater. They are easy to machine and decorative. Therefore, these alloys are used in marine applications as well as for fittings and parts for construction hardware.

Casting alloys are the main production for aluminium refiners. But it may also be helpful to be able to identify scrap as it is supplied to the plant yard from the application in its former life. The specific equipment requirements for processing these materials are described in part 2 of this book. For the metallurgical question please refer to the relevant literature.

1.4 Trends in aluminium production

In 2005 the world production of primary aluminium was close to 50 million tons. Comparing this figure with the annual production of 6,000 tons in 1900 one can be aware of the substantial growth rate experienced during the last 100 and some years. This growth rate has fluctuated in recent years but the overall expansion has exceeded that of most other metallurgical industries. In spite of the growing demand of the automotive industry, which substitutes other materials with aluminium to reduce the weight of the vehicle in order to save energy, a decline of the growth rate has to be expected since substitution in other applications has come to an end. The requirement to build plants at the most feasible size has caused the industry to cycle between overcapacity and undercapacity. The changes of the world economy intensify this effect even more.

In spite of constant or even growing demand for aluminium in industrialized countries, the spiralling energy cost forces the industry to close down older smelters that are either out of date or local cost factors make them unattractive. This is independent from the generation of safe nuclear power since the energy requirements of other consumers are also growing steadily. Thus new smelters, either to increase the world capacity or replace not economic plants in industrialized countries, will be built near the source of generation of inexpensive electrical energy.

The surrounding conditions of a smelter may also change. Growing population density in the area which had few inhabitants when the smelter was constructed some 40 or 50 years ago and also the environmental conscience may force the management to close the operation. Therefore, permanent closure of plants is becoming more prevalent at times of recession, while new plants are constructed elsewhere. Energy-rich countries will thus find a possibility to export their energy. An example may be the development of the primary aluminium industry in Germany. The total aluminium production in Germany as published by the Statistisches Bundesamt was 1,065,000 tons in the year 2011. This production was split into 612,000 tons of primary aluminium and close to 500,000 tons of secondary aluminium. Thus the secondary aluminium production contributed almost 50 % to the total aluminium production. However, this total production does not cover the demand of 2,450,000 tons per year subdivided into1,842,000 tons of wrought alloys and 605,000 tons casting alloys. The balance between production and demand had to be covered by imports. Unfortunately, the balance will increase. By the end of 2006 the total production of primary aluminium in Germany will be less than 400,000 t/a. For comparison, most of the smelters in regions with low cost of electrical energy have a capacity exceeding 500,000 t/a by far.

High cost for energy and low labor cost in combination with environmental problems are the reason to locate alumina refineries near the bauxite mining areas. Each ton of alumina produced

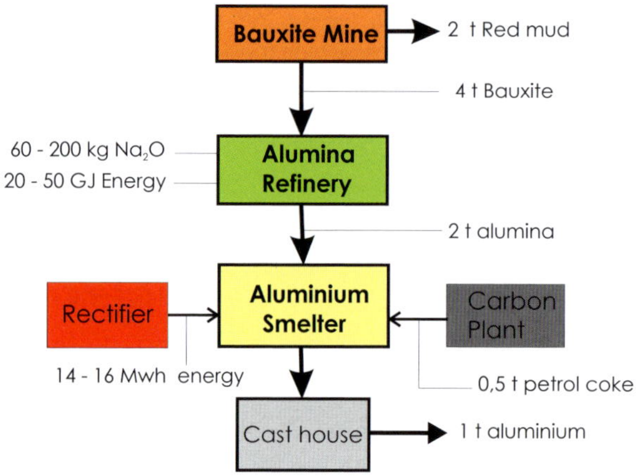

Fig. 1.3: Raw material and energy required for the production of primary aluminium

needs approximately 2 tons of bauxite (**Fig. 1.3**) resulting in 1 ton of waste as red mud. Thus, an alumina refinery with an annual production capacity of 1 million tons would generate 1 million tons of red mud (approx. 350,000 m^3). Area for this huge quantity is hardly available in industrialised countries considering the high prices for land. Dumping of this waste into the sea or into rivers, as practized in past years, is not accepted anymore. At the mining sites, which are generally located in unpopulated areas, the required space is available particularly if one considers that the removed bauxite provides free space.

There is a different situation to be observed looking at the secondary aluminium industry.

Over 12 million tons of old and new scrap were recycled in 2005. This figure does not even include the aluminium recycled as in-house scrap of rolling mills, extrusion plants and casting facilities. In relation to the primary aluminium production this amounts to 40 % or, in comparison with the total aluminium production, secondary aluminium production reaches a share of approximately 30 %. This percentage comprises certainly not the aluminium produced in that very year that is returned for recycling. The shapes and the alloys returned to the recycling industry reflect very much the use of the aluminium by the different industries. Although some products, such as beverage cans (UBC), may represent a closed loop recycling (with very short time to return UBC material and statistically back in the shelves, filled with beverage, in less than three months), quite a few materials have a long life span. Construction material used in buildings will be returned to the metallurgical industry within an average time of 30 years. Cars and other vehicles show an increase of aluminium content average to approximately 6 years service life. Thus, some of the aluminium produced long ago is still in use.

The secondary aluminium industry branches into two major lines. Aluminium colleted in defined format and alloy can be remelted without preparation and can be processed directly. Furthermore, a certain but not too large quantity of non-metallic contaminations can be tolerated. The material is just being remelted, thus production plants processing this type of material are consequently called remelter. They are mostly linked to rolling or extrusion plants. This simplifies the recycling of in-house scrap as well. The tendency of the production plants to proceed this way reduces the availability of clean or new scrap on the market.

A refiner process with not defined scrap is generally used to produce casting alloys. For this production all sorts of scrap and alloys can be utilized. But the metal must be refined. By proper mixing of available components and by removal of adhering non-metallic contaminations the correct alloy within the desired limitation of the metallurgical analysis is obtained.

Thus almost anything made of aluminium can be recycled repeatedly, not only used beverage cans (UBC), but plates and die molds, window frames, garden furniture, automotive components and – within limitations – aluminium foil are melted down and used to manufacture similar products again. The recycling of aluminium to produce secondary aluminium requires only a fraction of the energy as compared to primary metal and generates much less green house gas emissions. For this reason aluminium companies have invested in state-of-art secondary metal processing plants to recycle aluminium.

Refiners and remelters do not differ in product quality but in the type of alloy produced. Wrought alloys are suitable for rolling and extruding, while the refiners produce casting alloys. The starting materials, as used by the remelters, do have a very high content of aluminium and a narrowly defined content of alloying metals. Therefore, this type of scrap is called primary metal as well. Although refiners are able to use wrought alloys for adjusting the analysis of their product, no casting alloys can be added for producing wrought alloys. This is mainly due to their high content of copper and silicon.

The more and more stringent regulations for the protection of the environment require more and more sophisticated technologies for pollution control, the protection of the working environment and, last but not least, the improvement of heat efficiency. If we look at the ecological balance of both, the primary as well as the secondary aluminium industry will notice immediately the advantage of recycling.

No doubt the production of primary aluminium requires a considerable amount of energy. In contrast, recycling needs only 6 – 8 % of this energy. Although these figures are much higher, the 5 %, as frequently published, represent only a fraction of the energy required to produce primary aluminium which adds up to approx. 168,000 MJ (47 MW) for each ton of metal (refer also to section 5.1.2). Approximately 75 % of this is required for the fusion electrolysis, including the losses in generating the electric power. If hydropower is used, the energy balance becomes more favorable and the emission of green house gases is reduced. Roughly 20 % is required for the alumina refinery and the balance for transport and smaller consumers.

The secondary aluminium production requires only 12,000 MJ/t that is equal to 3.3 MW/t. This figure may be somewhat misleading if only melting of aluminium is considered. The components attached to the aluminium, such as oxides, have to be heated as well to a temperature above the melting point of aluminium and the inevitable losses need to be considered on top. Please refer also to section 5.1.2. However, the major portion is used for melting while the balance covers the energy requirements for holding and casting. One should also not forget the electrical power required for the pollution control. Mechanical preparation and salt slag recovery require a very much smaller fraction of the energy since not all of the scrap needs mechanical preparation and also not all of the melting involves salt.

Apart from the energy consumption, the production of aluminium generates, like most other industrial processes, a certain quantity of waste. Out of the primary aluminium production comes a total of 3,700 kg of waste per ton of aluminium. This is mainly red mud representing almost 60 % of the total mass. The balance is generated as waste of power generation. Contrary to that, the waste from the secondary production totals to less than 400 kg per ton of aluminium produced. Certainly this may be more if, for instance, dross is processed. But comparing the figure to the total secondary production, the average value of 400 kg/t is reached. This comprises slag, filter dust, non-recyclable waste from the mechanical preparation and also the waste caused by power generation.

Another factor to be considered is the atmospheric pollution. In the primary aluminium industry, approximately 85 % of the total of roughly 200 kg per ton of aluminium is caused by the combustion of fossil fuel of the power generation. Hydro-power, as used in quite a number of aluminium smelters i. e. in Canada, Norway, Brazil or Venezuela, is in a much better position. Process-oriented emissions are originating from chlorine, fluor or other process gases. In the primary production the major pollutant is hydro-fluoride appearing as HF. This is harmful to the environment and contributes to the green house gases as well.

During production of secondary aluminium the total of gaseous emissions reaches approximately 12 kg per ton of aluminium. They comprise to the extent of 80 % of combustion products of fossil fuel including carbon oxide (CO_2 and some CO), sulfuric oxide (SO_x), nitrogen oxide (NO_x) as well as dust. The balance is caused by the metallurgical process including the emissions due to the slag processing. You will find some more details in part 3, chapter 6.

When summarizing and comparing the environmental impact of the primary and the secondary aluminium industry the advantages of the secondary aluminium industry are obvious. As consequence, the focus should concentrate on recycling. Unfortunately, the choice is not that easy. A complete replacement of primary aluminium is not really possible. It is certainly true that there is no downgrading of an aluminium alloy during recycling. But not all of the alloys produced by the secondary aluminium industry can be used for all applications. It is not possible to obtain aluminium in a clean form by recycling. The alloying metals can only be removed to a very limited extent during the melting process. With an increasing rate of recycling more and more metal is in the recycling loop. Thus the content of alloying components will cumulate more and more and could eventually even exceed the tolerated limits. Most of the wrought alloys need fresh primary metal to obtain the requested characteristics. However, the recycling rate is presently still low enough for the downstream industry to produce their high quality alloys by using primary aluminium.

A target for the expansion of the secondary aluminium industry is difficult to define. Meeting the quality standards as well as reducing the propelling energy cost in combination with the need to fulfil the increasing requirements for environmental protection and all of that at economical operation calls for very thorough investigation of available technologies and careful selection of equipment going hand in hand with improvements of equipment design. A change in the structure of the secondary aluminium industry may also be expected to face today's challenge. Compared to the recycling procedures, as applied 50 years ago, a tendency to large plant capacities can be noted with more intensified mechanical preparation in separate plants.

The trend to high quality products and the ISO 9000ff requirements leads rolling and extrusion plants to produce their own slabs and billets based on primary aluminium ingots. Besides the better material quality, which reduces in many cases the quantity of reject material in downstream production steps, in-house rejects can be processed much easier. This measure does not result in additional growth rates for the aluminium production but for the process equipment.

Looking at the growth of aluminium used in transport and packing, increasing capacity of recycling plants can fairly be expected. It is assumed that the growth in primary aluminium production will be in the range of 4 % per year. Based on today's production level, this figure would mean that 2 or 3 primary smelters with a total capacity of more than 300,000 t/a must be constructed every year. This appears not to be very realistic particularly if one considers the necessary modernization or even complete reconstruction of out-of-date smelters. Thus, recycling will become even more important. However, before the recycling rate increases, improvement of infrastructure for collecting scrap is indispensable. It can be noticed already that most of the existing secondary aluminium plants are battling to get hold of the valuable commodity scrap. And it is a well-known economical law that prices react to the supply and demand situation. Consequently, in some areas companies have to face the decreasing demand at increasing cost for raw material. Thus, it is fair to assume that the secondary aluminium industry will relocate more and more to the sources of aluminium scrap i. e. follow the automotive industry.

2. Sources for the production of aluminium

Christoph Schmitz

Today, production of aluminium follows two lines: the primary aluminium production and the secondary aluminium production.

The primary aluminium industry gets aluminium from the natural resource bauxite, following the production line from mining via the intermediate product Al_2O_3, commonly called alumina, to the metallic aluminium.

The secondary aluminium industry uses, as raw material resource, the aluminium collected after its end of service life or obtained as by-product of a fabrication process. Wrought alloys are used to manufacture sheet metal as well as rolled or extruded shapes. Casting alloys are the input material for castings shaped into forms as required.

2.1 Aluminium ore

Thus, when you pick up a stone you hold most probably some aluminium in your hand. However, to separate it from the other compounds is difficult. In technical production the most utilized aluminium ore is bauxite. Although production from other ores is technically possible it is not (yet) economically feasible. Bauxite, which received its name from the today's ghost town in southern France, is the result of the weathering of aluminium silicates to the more chemically stable aluminium oxide. Consequently, it comprises an aggregation of aluminium hydroxides: Hydrargillite - Gibbsite (γ-$Al(OH)_3$), Bohemite (γ-AlOOH), Diaspor (α-AlOOH), iron oxide, silicate (quartz) and caolinite. The content and type of the aluminium hydroxide depend very much on the origin of the ore. Greek bauxite comprises almost entirely of diaspor with some bohemite, while Jamaica bauxite mainly contains hydrargillite. Obviously there is a relation between the geological age of the site and the type of bauxite. The weathering of the basic rock depends on a humid and warm climate as existing in a tropical environment. Time is also an important factor with the result that in the old bauxites only the very stable form of aluminium hydroxide, i. e. diaspor remains. The stable forms require higher temperature for processing than hydrargillite available in tropical bauxite (Western Africa, Australia, Jamaica and Brazil) which are, therefore, the preferred ores.

Due to its chemical characteristics, particularly the high affinity to chlorine and oxygen as well as the position in the electro-chemical series of elements, aluminium cannot be carbo-chemically processed. The required temperature of more than 2,000 °C is difficult to handle and limits the required process. Due to the chemical and physical condition of the system Al - O - C the technical realization of the standard metal recovery process is technically and economically not feasible and it appears that it will remain so in the near future.

2.2 Aluminium scrap

The raw material for secondary aluminium production is material that has been used already and is returned for further processing after the end of its life in the first application. It is certainly not the very pure aluminium that came out of the electrolysis plant. We have seen before that some foreign elements have been added to the metal to improve its characteristics. These additions cannot be removed in the secondary process – or at least to a very limited extent. It may also be contaminated by organic matter or with dirt. Some of the aluminium was firmly connected to steel or copper or brass or may also still carry bolts or other mechanical parts. Thus, the base or starting material for the secondary aluminium industry must be refined to be useful for a second life.

3. Alumina refinery (Bayer process)

Christoph Schmitz

Today aluminium is produced in a two step process. The first step comprises the alumina refinery where pure aluminium oxide (Al_2O_3) is extracted from the bauxite by means of the Bayer Process. The raw material is first digested with the aid of hot caustic liquor to achieve a solution. Next the aluminium hydrate particles crystallize and precipitate and finally calcinate to alumina. The principal process flow sheet is shown in **Fig. 3.1**.

3.1 Stockpiling of bauxite

Generally the refinery is not situated close to the bauxite mine. The criteria are good harbor facilities and, if possible, easy access to fossil fuel such as coal. Particularly for plants in Europe a blend of bauxite is processed. The material supplied is off-loaded from the bulk carrier (01) and stored in different stock piles (02). Depending on the production program, the bauxite is reclaimed by heavy stacker-reclaimer equipment and blended as required by feeding the material in the preset ratio to the process.

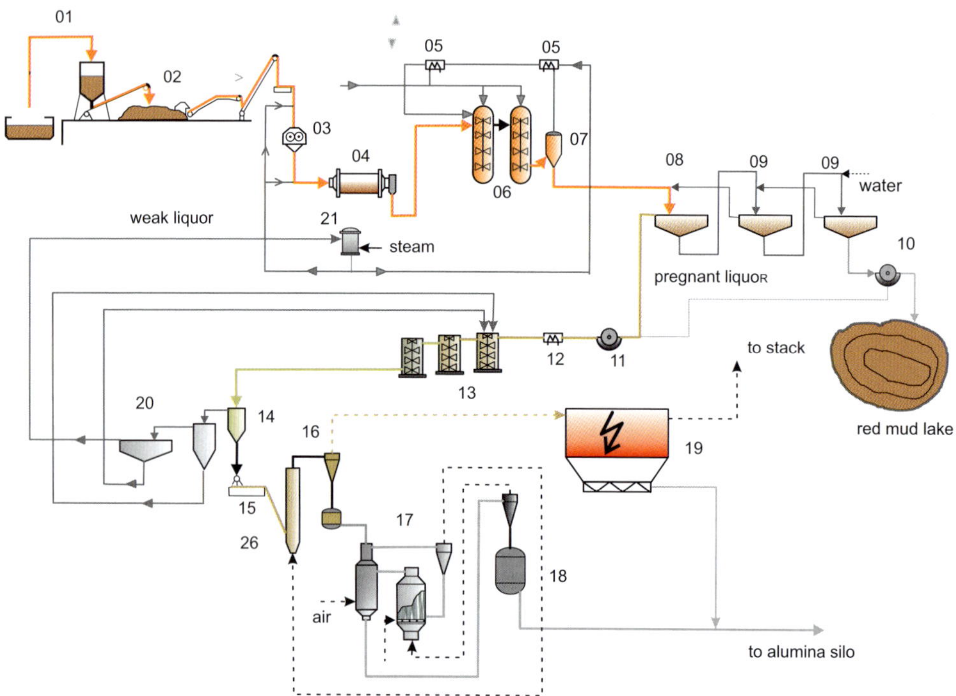

01 harbor facilities, 02 stockpile of bauxite, 03 pre-crusher, 04 grinding mill, 05 heat exchanger, 06 digester, 07 flash vessel, 08 red mud thickener, 09 red mud washer, 10 red mud filtration, 11 polishing filtration, 12 heat exchanger, 13 precipitators, 14 hydrate cyclone, 15 hydrate table filter, 16 drying stage, 17 calcination stage, 18 alumina cooler, 19 electro-static precipitator (ESP), 20 seed separation, 26 venturi feeder

Fig. 3.1: Principle flow sheet of an alumina refinery

3.2 Bauxite grinding

To provide satisfactory surface area for the reaction with the hot caustic liquor, bauxite arriving from the mine is first comminuted. The grain size distribution should provide sufficient specific surface area but should also permit easy settlement behavior when liquor and residuals are separated. In order to save energy, the bauxite will be pre-crushed (03) before it passes to a wet ball or rod mill (04). The choice between ball mill and rod mill depends on the grinding behavior of the material. Caustic is added in the crushing and grinding stage already to initiate good reaction. An additional portion of the caustic liquor is fed to the mill to obtain pumpable slurry with high solids content. Leaving the mill the slurry arrives at a mixing tank where the balance of the caustic liquor required for the digestion process is added.

3.3 Digestion

High pressure pumps now convey the bauxite/caustic to the digester section.

The extraction takes place according to the equation

$$Al(OH)_3 + NaOH \longleftrightarrow Na^+ + \left[Al(OH)_4\right]^- \tag{3.1}$$

This means that the aluminium hydroxide is dissolved in the liquid. The process requires a temperature above 100 °C since the solubility of the aluminium hydroxide increases not only with the concentration of Na_2O but also with temperature. Time, temperature, caustic concentration and residual time have to be balanced carefully. The parameters depend very much on the type of bauxite processed.

To be able to raise the temperature the process also requires sufficient pressure to avoid boiling, i. e. evaporation. Therefore, the digester section comprises a series of pressure vessels (06) called digesters. In the first stage the slurry is heated up by direct injection of steam and exhaust vapor from the downstream cooling section. After digestion the solution needs to be cooled. This takes place in the cooling section by gradually lowering the pressure. The evaporation heat cools the slurry and the resulting steam is directed to the heating section. Final cooling stage is the “flash vessel” (07). The aluminium hydroxide is now dissolved in the caustic liquor that is therefore called “pregnant liquor“.

3.4 Separation and filtration

On completion of the digestion, the liquor must be separated from the insoluble residue, called red mud, as well as purifying the liquor as much as possible. After cooling to a temperature below 100 °C the slurry flows to thickeners (08) to be clarified by settling, followed by additional thickeners (09) for washing the red mud. This is essential to remove Na_2O still adhering to the solid particles before it is directed to the red mud lake for landfill. The overflow from the washers (09) is used for dilution of the digester product in the primary thickener (08).

3.5 Precipitation

The pregnant liquor passes a polishing filtration (11) to remove traces of red mud and arrives at the precipitation section. This comprises a series of settling tanks with agitator. At a temperature between 75 °C and 55 °C the pregnant liquor becomes supersaturated and the aluminium hydroxide starts to crystallize to small grains. Thus, before passing to the precipitators (13), the liquor is cooled by means of a heat exchanger (12). The decomposition process is very slow and may not start spontaneously even if the aluminate is supersaturated. To support crystallization a

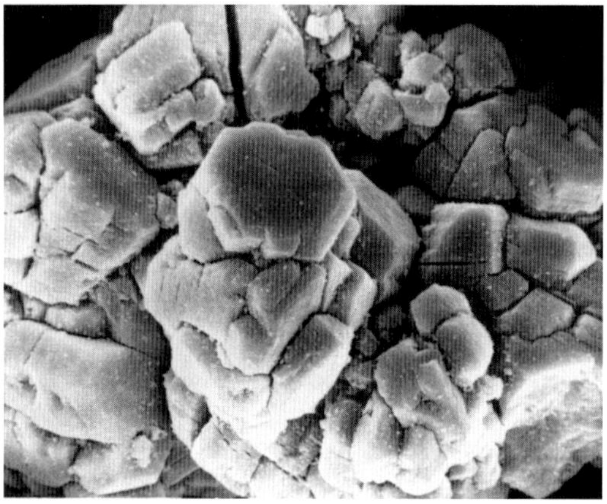

Fig. 3.2: Alumina grain (magnification 700 x)

large quantity of initial nucleus material must be added. This seed comprises already precipitated aluminium hydroxide that is separated from the overflow of the precipitating vessels and separated in thickener/settlers (20). Thus, up to 400 % of the total hydroxide production is recycled as seed and remains in the process in a closed loop. Crystal growth depends on the degree of supersaturation, temperature and time. But even with sufficient growth the individual crystals are very small. If the process parameters are carefully controlled, different particles agglomerate and finally cement to a larger grain (**Fig. 3.2**) whereby the microscopic fissures and gaps are filled with even smaller particles forming a stable structure. Fig. 3.2 shows an alumina grain with a 700x magnification after calcination. The fissures in the larger grains originate from the explosion like removal of crystal water during flash calcination. Since most of the hydroxide is used to provide seed for the crystal formation, only a small portion of the material leaving precipitation is used for production. Additional settlers (20) remove some of the coarser particles for recycling to the precipitators. The major portion is mixed with the pregnant liquor arriving from clarification to provide seeds for the crystallization. The caustic liquor overflow of the precipitator vessels is used for direct cooling of the hot pregnant liquor before entering precipitation and is then recycled to make up the slurry prior to digestion.

3.6 Calcination

The portion of precipitated aluminium hydroxide selected for production is sent to calcination. For the calcination today's aluminium refineries apply the fluidized bed technology although some elder plants still use the rotary kiln technology with an energy consumption of 400-500 MJ per ton while the modern fluidized bed technology requires approx. 50 to 60 % of this energy.

After washing and pre-drying on drum or table filters (15) the hydroxide is converted to alumina according the equation

$$2Al(OH)_3 \rightarrow Al_2O_3 + 3H_2O \tag{3.2}$$

The hydroxide is then passed to a ventury dryer (16) where the surface water is removed. From this diluted phase system the carrier gas is separated and the hydrate flows to the first stage of the flash reactor (17) after passing through a dense phase dryer. It is then sent to the second stage and finally as alumina to the dense phase cooler (18). Dust separated during the different process stages is removed by the electrostatic precipitator (19) and added to the product.

After the final stage of calcination, almost pure aluminum oxide as Al_2O_3 is obtained. It is a white, coarse sandy bulk material with a content of only 5 – 25 % of a - alumina and a grain size of 4 – 20 % <44 µm. Alumina is handled and transported to the smelter by bulk carrier and stored in large alumina silos.

3.7 Red mud disposal

The major environmental impact of an alumina refinery is the very high content of red mud. In former days the red mud was simply dumped into rivers or the nearby sea. Because of the inefficient washing of red mud it contained a substantial amount of Na_2O. Despite today's efficient washing process red mud is deposited as landfill. It presents no severe environmental hazard but it requires a large area for the red mud lake. An alumina refinery with a capacity of 1,000,000 t/a of alumina – resulting in 500,000 t/a of aluminium metal – produces annually more than 1,000,000 t of red mud on dry basis. Only a small quantity is used for other applications such as road construction or as additive for the cement industry. All processes to separate the metallic contents of the red mud (iron, titanium) are not yet economical.

The requirements for the disposal of red mud are one of the reasons for improving the recycling of aluminium.

4. Aluminium reduction

Christoph Schmitz

The product of the alumina refinery is the refined ore to be further processed in the aluminium smelter. Although there is a large field of applications most of the alumina is used to produce aluminium metal.

4.1 Hall-Heroult process

As mentioned above, its chemical and physical characteristics do not allow obtaining the metal by a carbo-chemical process. However, the alumina with its melting point of more than 2,000 °C has a high solubility in molten cryolite (Na_3AlF_6). This allows separating aluminium and oxygen at a temperature of 950 °C to 980 °C in an electrolytic process. Thus, knowing the work of Neville and Bunsen, Hall in the USA and Herault in France developed the electrolytic aluminium extraction process which in the honor of the two inventors is called the Hall-Heroult Process. Just for curiosity, both men were born in 1836 and died in 1914. Both applied for the patent in 1886 with a difference in time of only three months.

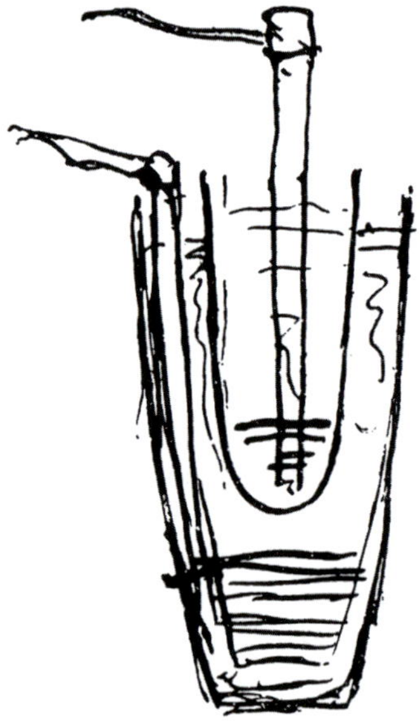

Fig. 4.1: Heroult's principle sketch to explain the fusion electrolysis of aluminium

The basic idea is quite simple as demonstrated by Heroult's sketch (**Fig. 4.1**). DC current is applied to the molten solution of 90 % cryolite and 10 % alumina having a temperature of 950 °C. The container accommodating the bath is shaped as the cathode while the anode is dipped into the bath from above (**Fig. 4.2**). Since the aluminium ions discharge their electrical load at the cathode the liquid metal collects at the bottom of the container. The technical concept of the first electrolytic plant was also very simple but the process became more and more sophisticated during the last 120 years in order to improve efficiency, i. e. lower energy consumption and better metal yield. The first industrial size aluminium electrolysis plants were located in Niagara Falls in the USA and the Rhine Falls in Switzerland.

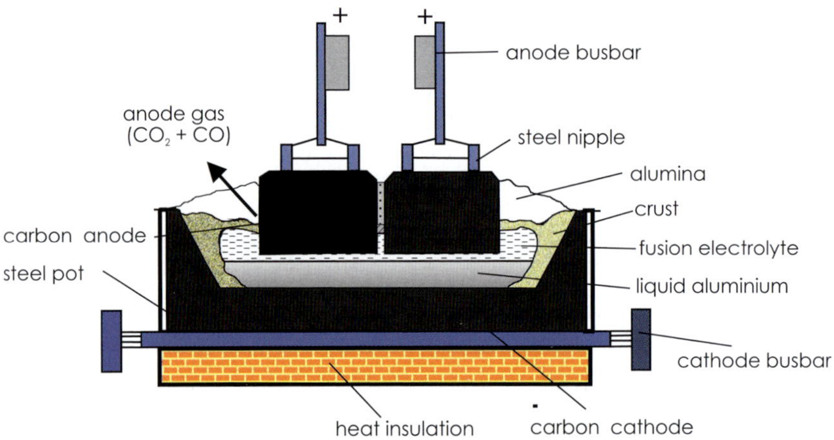

Fig. 4.2: Principle of a modern electrolytic pot for the fusion electrolysis

But even today the mechanism of metal separation, by considering the intermediate reaction with cryolite, is not fully understood. Generally it follows the equation

$$Al_2O_3 + \text{energy} \rightarrow 2Al + \frac{3}{2}O_2 \tag{4.1}$$

for the cathode reaction and

$$\frac{3}{2}O_2 + 2C \rightarrow CO_2 + CO + \text{energy} \tag{4.2}$$

for the anodic reaction. The equation shows that the aluminium ions are forced to pass their electrical charge to the cathode due to the electrical field and the energy introduced. The metal collects at the bottom of the electrolytic cell. The oxygen ions pass through the electrolyte to the anode and react with the carbon of the anode contributing some chemical energy to the energy balance of the process.

The theoretical energy requirement is 8.7 kWh/kg aluminium (enthalpy of reaction) or 9.3 kWh/kg if the energy required to heat up the bath is included. Due to the anodic reaction providing 3 kWh/kg, the theoretical total energy is lowered to 5.7 or 6.3 kWh/kg, respectively. The result is that approximately 0.5 kg of anode carbon is consumed per kg aluminium. Regarding the theoretical energy consumption for the reactions within the process, some losses, which are due to the electrical resistance of pot components such as bus bars, carbon, metal, electrolytic bath, anodes and anode rods, voltage drops in the electrolytic cell. Therefore, the total consumption amounts to 11.5 - 13.5 kWh/kg in industrial plants.

Although the principle of the electrolytic cell of today's design is still the same as in the days of Hall and Heroult, the technical concept is much more sophisticated. This is depicted in Fig. 4.2.

4.2 Pot lines

A modern electrolytic cell, called pot, consists of a large steel container that is lined with refractories. The inside (the cathode) consists of graphitized carbon blocks and is shaped to form a vessel for the alumina-cryolite bath. Electric power is supplied through iron bars attached to the cathode and connected to the bus bar system. A superstructure carries the anodic part that comprises two rows of individual pre-baked anode blocks which are lowered into the bath. The anodes are supported by the anode beam that can move vertically to maintain the inter-polar distance between the liquid aluminium and the anode as the anode burns off or the metal level rises due to metal separation or lowers when metal is tapped. The anode beam is electrically connected to the bus bar system by packages of flexible aluminium strip which allow the movement of the unit. Alumina is fed to the pot in frequent intervals by centrally arranged point feeders. The feeders are equipped with an impact hammer required to break the crust of the bath before alumina is charged in small portions to keep the alumina content at a constant level while metal is separated. Only a little cryolite is consumed and needs to be added from time to time. The reaction at the anode provides energy to hold the bath at a temperature of approximately 950 °C in order to maintain the liquid phase. The entire pot is covered with removable hoods to collect the gases generated – mainly CO_2 and some HF – which are then passed to the scrubbing system.

As anodes are consumed, used anodes have to be replaced by new ones. This is done by the pot tending crane equipped with the tools for changing the anodes, the holders for new and used anodes, the crust breaking hammers required to open the crust around the old anodes as well as a cryolite tank with feeder for adjusting the bath composition. From time to time the anode beam reaches its lowest position. It needs to be raised to the top and all anodes must be reset to the required inter-polar distance. The set of anodes is arranged in a way that only one anode must be replaced daily. That means that the tops of the anodes are at different levels. The anode comprises a pre-baked carbon block that is fixed to an aluminium rod for feeding the electric power. The rod is connected to the anode beam by special clamps and screws.

The sizing of an electrolytic cell is governed by Faraday's law. The quantity q in g/h of metal separated is proportional to the electric current I in A and time t in hours following the equation

$$q = c \cdot I \cdot t \tag{4.3}$$

c is the electrochemical equivalent of 0.335 g/Ah for aluminium. The equation shows the need to operate with high electric current to obtain an acceptable output of the electrolysis. Assuming an amperage of 100,000 A, the capacity of an electrolytic cell amounts to

$$Q = 0{,}335 \cdot 100{,}000 = 33{,}500 \text{ g/h or } 33.5 \text{ kg/h}$$

There are, however, some limitations to be considered for the design. The current density of the anodes is limited to max. 1 A/cm^2 and modern plants operate at even lower current densities. That means that an anode area of 100,000 cm^2 or 10 m^2 is required to produce the quantity of 33.5 kg of aluminium per hour. The resulting pot would be of remarkable size already but the amount of metal produced is by far too small for an economical plant size. The pots are sized to provide approx. 2,000 kg daily and to have one anode change per day. Consequently, a pot line consists of quite a number of pots arranged side by side (**Fig. 4.3**). Modern pot lines operate at 200 kA or more. It would be of advantage to increase the capacity by higher current. But high current induces a strong magnetic field that oscillates at low frequency since the required DC power has to be rectified from high voltage AC power. Both the oscillating strong magnetic field and the resulting bath movement become extremely difficult to handle with increasing amperage. On top of this, the homogeneous distribution of energy to the different anodes is a difficult task for pot design. Therefore, computer simulation for the pot behavior is indispensable before introducing a new design at very high investment cost.

Theoretically, a voltage of 2.8 V is required for the electrolytic de-composition of aluminium oxide that varies on bath composition. On account of the indispensable in energy transfer, modern pots operate at a voltage of 4 - 4.5 V according the equation

$$U = U_E + \sum R \cdot I \tag{4.4}$$

U is the pot voltage in V, UE the theoretical de-composition voltage in V and R the electrical resistance in W. It is obvious that the comparatively small resistance values of the various pot components create a significant voltage drop S R ◊ I when considering the high electric current.

A very important voltage drop is caused by the inter-polar distance that is the distance between carbon anode and liquid aluminium. This distance must be large enough to avoid short-circuiting between anode and metal that may be caused by fluctuations of the metal surface due to waves caused by the oscillating magnetic field. On the other hand, the inter-polar distance must be small in order to reduce the voltage drop, i. e. energy loss to the smallest possible level. It is the purpose of the sophisticated automatic pot control to optimize these values during all phases of operation.

Fig. 4.3 shows the flow sheet of a modern aluminium smelter. The electrolysis is shown as one pot only. Actually it is only a symbol for 300, 700 or even more pots. These are arranged in two rows installed in long buildings (**Fig. 4.4**) typical for aluminium smelters. Power is supplied to the pots by a bus bar system that is made of large size aluminium bars. The electric AC power, that is supplied

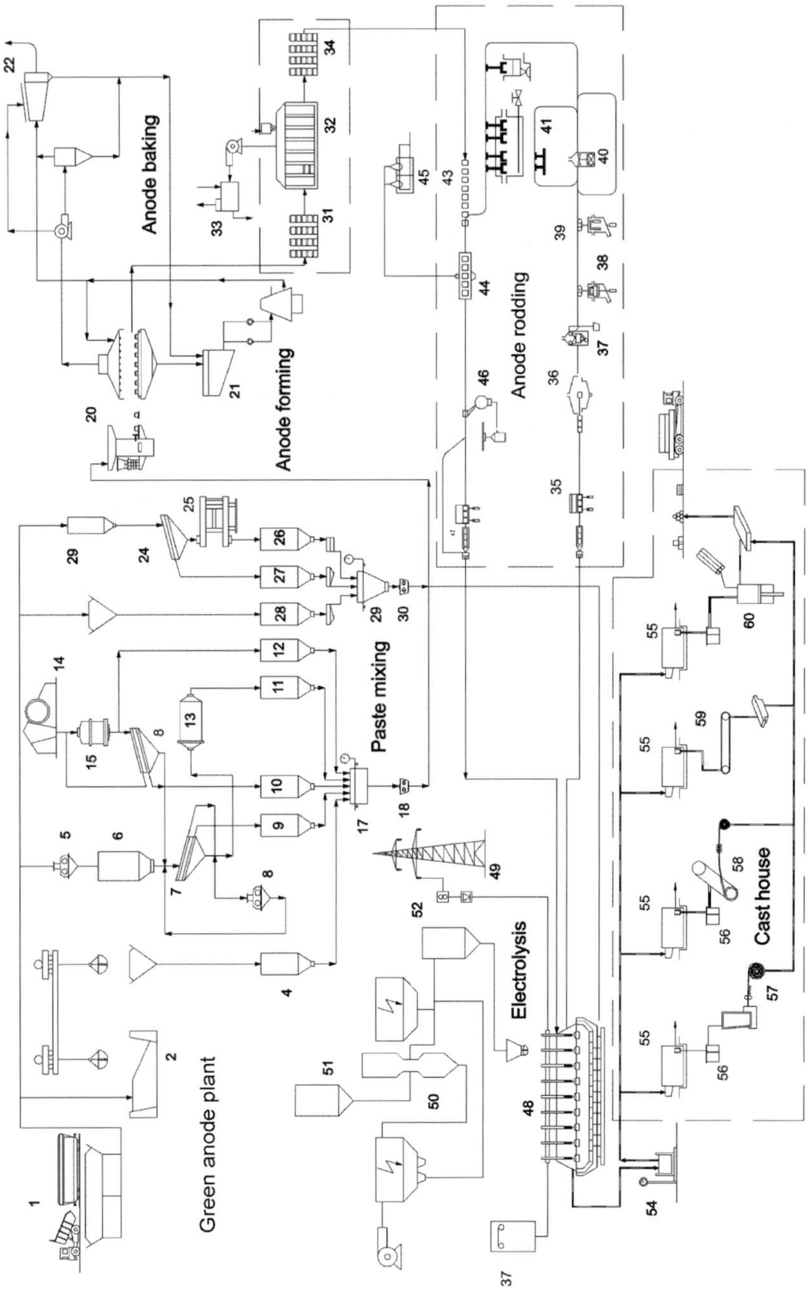

1 harbor facilities, 2 coke storage, 4 pitch silo, 5 primary crusher, 6 coke silo, 7 primary vibrating screen, 8 secondary crusher, 9, coke bin, 10 coke bin, 11 fines bin, 12returned butts, 17 mix scale, 18 mixer 20, anode forming machine, 21 anode cooling, 22 pitch fume scrubber, 28 packing material bin, 27 packing material fines fraction, packing material mixer, 31 green anode storage, 32 baking furnace, 33 waste gas scrubber, 34 baked anode storage, 35 used anode charging, 36 butts cleaner, 37 butts stripper, 38 thimble stripper, 40 and 41 rod repair loop, 43 mating station, 44 casting station, 46 spray station, 48 pot line, 49 power supply and rectifier, 50 pot line dry scrubber, 51 fresh alumina storage bin, 52 reacted alumina bin, product scale, 55 casting furnace, 56 in-line de-gassing, 58 rod casting machine, 59 ingot casting machine, 60 DC casting machine

Fig. 4.3: Flow sheet of a primary aluminium smelter

Fig. 4.4: View of a pot line

to the plant by high tension overhead power lines, is transformed to the required line voltage and rectified to DC power. Transformers and rectifiers are located at one end of the pot line buildings.

4.3 Cast house

The metal is tapped in frequent periods from the electrolytic cells by means of a vacuum vessel, then poured into a crucible and transported to the cast house (**Fig. 4.5**). In tiltable reverberatory (55) furnaces the temperature is adjusted and the necessary alloying components are added, if required. After passing a metal treatment system (56) to remove hydrogen, metallic contaminations caused by the electrolytic process (mainly Na) and mechanical contaminations, the metal is cast into shapes as required by customers. Depending on the product mix, DC-casting machines (60), ingot casting machines (59) or strip casters (58) are used. The cast house equipment is quite

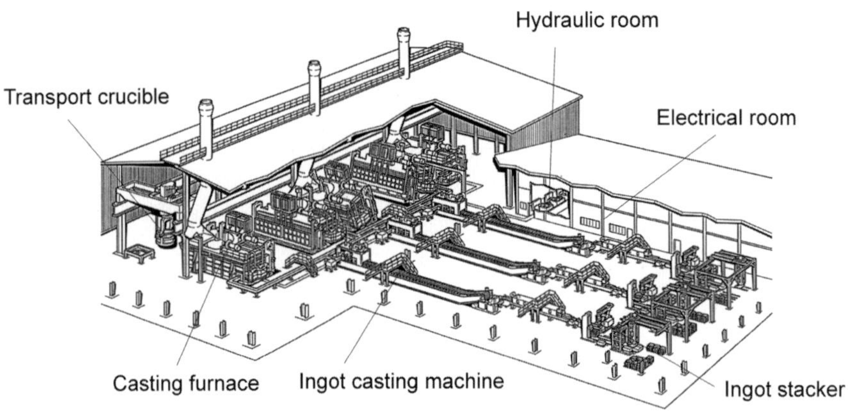

Fig. 4.5: Principle of a cast house for a primary aluminium smelter

similar to the units used in secondary smelters. It will be covered in the book sections in more detail. Since the capacity of a primary smelter is by far larger than that of a secondary smelter, the units are much larger in size, too. Furnaces have a bath capacity of 50 - 80 tons and ingot casting machines are sized for a production rate of 20 - 25 t/h. Although primary aluminium smelters supply traditionally extrusion and rolling ingots, a tendency to produce only 20 kg or 25 kg remelt ingots for the downstream aluminium industry can be observed. If extrusion ingots are produced, the cast house is equipped with a homogenizing plant that can be designed as batch-type unit or as continuous system. The decision, which type of is more favorable, depends largely on the specific requirements. Generally, it can be stated that a continuous plant is more favorable if there is little fluctuation in billet size and alloy while a batch-type furnace is more flexible with regard to size and alloy – and requires less investment, too. The homogenizing plant comprises the heating and holding section as well as the cooling section allowing operation at pre-defined cycles. Slab saw and billet saw complete the cast house equipment. Some special units may be installed to process the generated swarf from the saw and adjacent production plants. But this material should be passed onto the secondary aluminium industry that is better equipped for processing such material efficiently. Also dross skimmed from the cast house furnaces is generally sold to refining plants.

4.4 Carbon plant

Each kg of aluminium produced consumes approx. 0.5 kg of anode carbon. These anodes are made up of calcined petroleum coke – a by-product of the petrochemical industry – and pitch as binder. The material passes through various production steps before it is finally baked to form the strong anode block. The carbon plant comprises the sections green carbon plant, baking furnace and anode rodding plant.

It is easy to understand that the carbon anode plant is a large and important area of a primary aluminium smelter. Petrol coke is supplied as bulk material by ship or as it may be in some cases by railway to the plant (1).

To obtain the required density of 1.6 kg/dm^3 the recipe for the coke mixture is made up of a grain size distribution that is close to dense spherical packing. Consequently, the coke needs to be classified by screening into the various grain sizes (4 – 7). The quantities for the required grain size distribution are achieved by crushing and grinding (8, 15 and 13). The recycled anodes are crushed and classified to make up the coarse fraction (14, 15). All fractions are intimately mixed (30) and liquid coal tar pitch is added as binder. The result is a paste that is formed into green anode blocks on a vibrating forming machine (20). During forming the paste is shaped and compacted to the required size and density while the binder pitch fills the voids between the grains and penetrates into the pores of the coke.

The anode blocks obtained are called "green anodes". They are sent on to the baking furnace (32) where they are calcined at a temperature of approx. 1,200 °C. During this process the volatiles of mainly the pitch evaporate and the remaining carbon forms a rigid skeleton within the block. The baking furnace is built with refractory brick as ring kiln with a large number of pits to take the anode blocks in layers. One or more baking furnaces are accommodated in a long building. The firing is set step by step for different furnace sections thus configuring pre-heating, heating and cooling sections. Leaving the baking furnace the anode block is ready to be used in the pot.

But before this is possible it has to be attached to the anode rod. This comprises an aluminium bar having good electrical conductivity and – attached to it – the yoke. Welded to this steel structure are rows of pins which match exactly the holes in the anode block leaving a small gap. In the casting station this gap is filled with liquid cast iron that forms a sturdy connection between anode block and rod after solidification. Now the complete anode assembly is to be used in the pot. During the electrolytic process the anode burns down to a residue providing just enough carbon to cover pins and cast iron nipple. It is removed from the pot and recycled to the anode rodding shop. At first the used anode is removed. The stripped carbon is recycled to the green anode plant to make the coarse fraction after crushing and classifying. The next step for the rod is removal of the cast iron thimble. After inspection the rod is now ready again to mate with the carbon block.

The whole carbon plant is sized to process all anodes required for the pot lines. This could be more than 1,000 per day depending on the size of the smelter.

Besides the production buildings described, one important building is for alumina storage, petrol coke storage, storage for green and baked anodes required to provide sufficient buffer between green carbon plant, baking furnace and anode rodding. A large area is dedicated to the storage for aluminium produced.

Buildings for maintenance and administration complete the plant.

The typical plant layout of a primary aluminium smelter is characterized by the extended parallel buildings accommodating the electrolytic cells. The next largest building is the baking furnace for carbon anodes. All production buildings i. e. pot lines, cast house and carbon plant are arranged for optimizing the material flow.

5. Secondary aluminium

Christoph Schmitz

5.1 General consideration

In industrial countries infrastructure and processes for aluminum recycling have been developed over the last decades. To cope with the demand for secondary aluminium, plant capacities have grown steadily and the equipment became more and more sophisticated. In earlier days mainly such aluminum parts were used whereby the composites were only unavoidable attachments such as fastening bolts or functionally required copper, bronze, brass or zinc parts (e. g. bearings). Different types of furnaces were tried and improved to handle the various types of scrap. Many melters regarded their technology as the only solution for producing secondary aluminium alloys which may even be true for their specific product. The same applies for the casting equipment necessary to produce the shapes that customers want. Highly efficient equipment was developed to cope also with the more and more stringent environmental protection regulations. Considering the fairly substantial cost of personnel, a high degree of automation became essential for the overall economy of a plant. Today modern plants need an annual production capacity of 10,000 t or more to be able to operate economically and particularly to keep cost of manpower to a minimum. Due to the increasing amount of composite material, mechanical preparation became necessary to free the aluminum from adhering metals and plastics. This resulted also in the development of highly mechanized plants.

The secondary aluminium industry branches into two major lines. Aluminium colleted in defined format and alloy can be remelted without preparation and processed directly. Furthermore, it is possible to tolerate a certain but not too large quantity of non-metallic contaminations. The material is just being remelted, so production plants processing this type of material are consequently called remelter. They are mostly linked to rolling or extrusion plants. This simplifies the recycling of in-house scrap, too. The tendency of the production plants to proceed this way reduces the availability of clean or new scrap on the market.

A refiner process with not defined scrap is generally used to produce casting alloys. For this production all sorts of scrap and alloys can be utilized. But the metal must be refined. By proper mixing of available components and by removal of adhering non-metallic contaminations, the correct alloy within the desired limitation of the metallurgical analysis is obtained.

Today most of the available scrap – depending on the service life of the relevant product – is collected and processed in sophisticated and very efficient plants, operating with little manpower. To construct such plants a substantial amount of investment capital is needed. Careful consideration of the plant layout and the equipment involved is vital for the economical success of the plant. Special equipment, that may process a defined type of material, is certainly very efficient for the present situation but may represent a restriction when the market situation is different or the availability of raw material, i. e. aluminium scrap, changes.

5.1.1 Quality aspects

Almost anything made of aluminium can be recycled repeatedly, not only used beverage cans (UBC), but plates and die molds, window frames, garden furniture, automotive components and – within limitations – aluminium foil are melted down and used to manufacture similar products again. The recycling of aluminium to produce secondary aluminium requires only a fraction of the energy as compared to primary metal and generates much less green house gas emissions. Aluminium companies have invested in state-of-art secondary metal processing plants to recycle aluminium.

Refiners and remelters do not differ in product quality but in the type of alloy produced. Wrought alloys are suitable for rolling and extruding, while the refiners produce casting alloys. The starting materials, as used by the remelters, do have a very high content of aluminium and a much defined content of alloying metals. Therefore, this type of scrap is called primary metal as well. Although refiners are able to use wrought alloys for adjusting the analysis of their product, no casting alloys can be added for producing wrought alloys. This is mainly due to their high content of copper and silicon.

It requires substantial efforts to fulfil the quality requirements requested by clients. This needs thorough control of the scrap charged and efficient metal treatment to remove contaminations trapped in the melt. As soon as the metal is delivered to the plant it must be classified and stored at assigned areas in the scrapyard. This permits to compose the batch as defined by the batch calculation. Fast melting and rapid metal transfer are also factors for good metal quality because these procedures reduce the loss of valuable aluminium and alloying components as well as minimize the contamination by unwanted elements such as iron while turbulence-free transfer of metal reduces the generation and inclusion of metal oxides. Some of the unwanted elements, such as potassium, sodium or – at least sometimes – magnesium, can be removed or reduced by gas treatment in the casting furnace. Other elements, such as copper, iron, zinc, calcium, silicon and the like, cannot be removed. Their content must be limited by skilful blending of the melt with the other available scrap or – as in the worst case due to cost – by addition of primary metal.

The importance of one procedure in an operating plant is very often underestimated: cleaning of furnaces and launders. Not all of the alloying elements useful in a very specific alloy are wanted in another alloy. Thus, when changing alloys all of the previous alloy must be removed. Some of the metal may be left over as frozen material in launders or other transport equipment. However, most of it is contained in the dross cumulated at the furnace walls and the furnace bottom. It is indispensable to remove these scalings thoroughly. This is quite a tough job. Plant engineers should therefore have a close look at the design of the furnace they would like to install in their plant for easy and uncomplicated cleaning, i. e. descaling.

Even with utmost care sometimes a complete batch in the furnace is outside tolerances for the different elements in the alloy and no correction is possible by just adding other metal or primary aluminium. The batch has to go somewhere but processing it through the normal casting facility would block the operation of the plant for some time. On the other hand, the whole quantity must be stored in portions that allow adding the valuable material step by step to the production as soon as it is appropriate. For that reason, even the smallest plant should have a casting facility for large ingots, commonly called sows. In order to have a homogeneous quality for a certain quantity, the metal market requires a minimum of 20 t of aluminium originating from the same batch. This is also a factor to be considered when installing furnaces.

Proper de-gassing of the metal should be a standard procedure in the refining plant although one may argue that the metal will be remelted anyway. But it is commonly known that die-casters do not fancy metal treatment. Thus, entrapped gases, such as hydrogen, are not removed in the downstream casting and the metal may even pick up more hydrogen that finally results in blisters trapped in the casting. Thus, the responsibility for gas-free metal is with the refiner.

Quality is a very important factor for the production since poor or even lower quality may have a substantial effect on the end product. One should imagine the cost and the problems to call back thousand automobiles because one small but vital part made of aluminium fails or is unsafe.

Quality can only be assured by good quality equipment suitable for the specific process and careful working of the operators.

5.1.2 Energy requirements

The production of aluminium is very energy-intensive. The long way from bauxite to the aluminium alloy encompasses some very energy-consuming process steps. Recycling requires much less

Table 5.1: Energy required to produce 1 ton of primary aluminium

(MJ/t of produced primary aluminium)						
Energy source	**Alumina refining**	**Coke production**	**Pitch production**	**Anode production**	**Smelter**	**Casting**
Electricity	462	35	0	826	115330	1910
Natural gas	23336	763	20	696	752	2417
Distillate oil		327	8	149	20	698
Residual oil	243				5	698
LPG				149	8	465
Coal	729					
Gasoline					5	372
Carbon anodes					17325	
Subtotal	24770	1125	28	1820	133445	6560
Total						**167748**

energy. Only 6 to 8 % of the energy to produce primary aluminium is required for remelting and refining of aluminium in its repeated service life.

The raw material quantity and the energy requirement to produce primary aluminium is summarized in Fig. 1.3. The electrical energy indicated does not include the actual energy required to produce the electrical power.

The production of primary aluminium relies on the electrolytic process as outlined in section 4. One third of the average cost of aluminium is for the energy to make it. The average current consumption in an electrolytic cell is estimated to be in the range of 15 kWh/kg of aluminium. Modern smelters operate with an energy consumption of about 13 kWh per kg of aluminium. The energy consumption may decrease as old smelters are shut down or existing smelters are modernized and new smelters with modern technology are built.

The primary source of electrical power and the power company are very influential on the price of power and, therefore, on economics of aluminium production. Therefore, primary aluminium production facilities are concentrated in areas with low electricity cost.

Table 5.1 summarizes the energy requirements of the primary aluminium production process of alumina refining, anode manufacture, aluminium smelting and ingot casting. The conversion factors used for electricity (including line losses) were 7,600 kJ/kWh for primary aluminium smelting. This conversion factor includes also the efficiency of the power plant and the conversion of AC power to DC power as required for the electrolytic process. In normal industrial applications in other industries and also the secondary aluminium industry this conversion factor is not realistic. The most favorable technology requires 8,750 kJ/kWh while standard gas turbine plants or nuclear power plants need approximately 11,000 kJ/kW. Lignite-based power plants have a better efficiency and need only 8,500 kJ/kWh. Thus, using a conversion factor of 10,000 kJ/kWh seems to be quite realistic for the general industry. Hydro-power is the most favorable source for producing electrical energy and is, therefore, counted at its theoretical energy equivalence of 3,600 kJ/kWh with no pre-combustion impacts included. These impacts are not assumed for hydropower since water does not have an inherent energy value from which line losses etc. can be subtracted.

Besides the electrical energy consumed by the electrolysis, some energy is required for the cast-house and the production of the carbon anodes energy. This is partly thermal energy. Summarizing

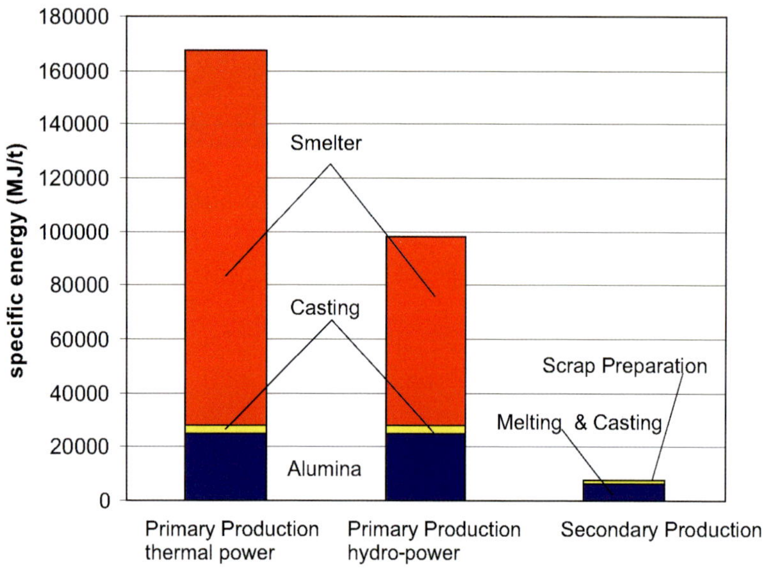

Fig 5.1: Specific energy consumption for the production of aluminium

the energy consumption per ton of aluminium produced, the total including all production steps comes to approximately 168,000 MJ/ton of aluminium (Table 5.1 and **Fig. 5.1**). Most of it, explicitly 83 %, is used for the electrolysis.

There is also a substantial heat requirement for the production of alumina. While most of the energy used in the aluminium electrolysis is electrical energy, the Bayer plant requires thermal energy provided by natural gas or coal. Some electrical energy is required for the various mechanical equipment such as crushing and grinding facilities or for pumping. The total energy used for alumina refining is 25,300 MJ/tons of primary alumina. This figure assumes a consumption of 1.9 tonne of alumina per ton of aluminium. The comparatively small energy requirement for bauxite mining is also included. Most of this energy has to be provided by thermal energy that comprises 96 % t of the total energy.

The total energy required per ton of aluminium comes to 168,000 MJ/ton of aluminium. The share of the electrical energy amounts to 118,563 MJ/ton. This is 70 % of the total energy and most of it is used for the smelter. Using the conversion figure indicated above this is equal to 118,563/7,600 = 15.6 kWh/ton of aluminium.

Considering the world aluminium production of 25 million tons per annum the total energy consumed by the world aluminium industry amounts to 2.96 x 109 GJ with a share of 3.90 x 102 GWh/a. Thus, the share of energy used by the world aluminium industry is, although quite high, only a minor fraction of the 3.4 x 1,022 GJ/a of the world's energy consumption.

Reusing aluminium by remelting and refining requires only 6 to 8 % of the original energy input of aluminium produced from bauxite. In Table 5.2, the energy requirements for producing 1 ton of secondary aluminium, including scrap preparation, refining, casting and slag treatment are approximately 11,200 MJ/t.

Table 5.2: Energy required to produce 1 ton of secondary aluminium

	(MJ/ton produced aluminium)		
	Specific energy consumption		
Energy source	Scrap preparation	Secondary melting/refining	Slag treatment
Electricity	2100	2580	800
Natural gas	900	3960	410
Light oil		40	
Heavy oil		40	
LPG		70	
Coal			
Diesel	200	30	30
Subtotal	3200	6720	1240
Total			**11160**

The figures listed in **Table 5.2** represent average values assuming that not all scrap is passed through mechanical preparation and also an average of 70 % of metal recovery is obtained meaning that only 30 % of the production tonnage will be the quantity of slag to be treated.

The actual energy consumption at the production plant may differ from the summary for the recycling in general as discussed above. The reason is that Table 5.2 compiles the general figures at an assumed breakdown for the variety of scrap generally processed. Furthermore, for the electricity the conversion factors as defined above are applied. The breakdown of energy consumption of the various production steps as shown in **Fig. 5.2** considers the capability of equipment and the individual process within a secondary aluminium plant. As it is common practise, the values

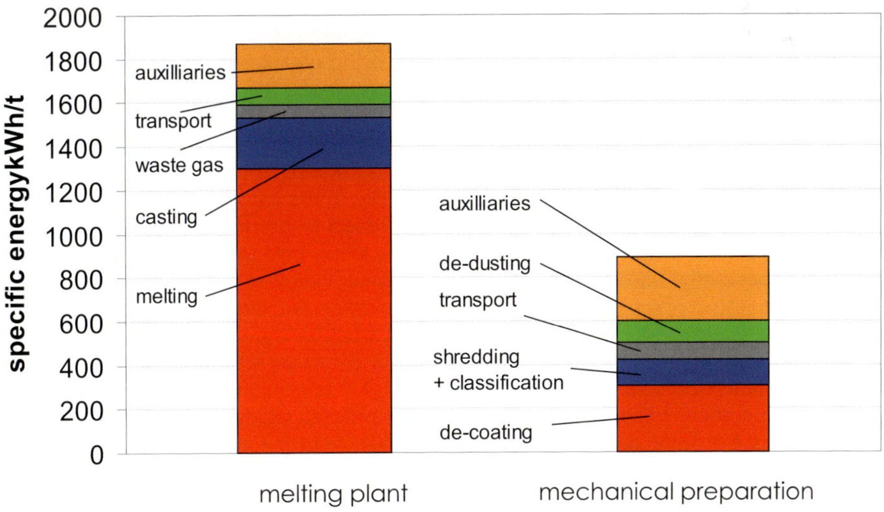

Fig. 5.2: Specific energy consumption for the production of secondary aluminium

are expressed in kWh no matter if natural gas or fuel oil is used. For electrical furnaces a similar definition applies. Most of the energy is used for melting. This is generally fossil energy provided as natural gas with some electrical energy for equipment such as blowers or hydraulic pumps. The net specific melting rate of a furnace is defined as the quantity of metal heated to the melting point, melting it and heating the liquid metal to 50 °C above the liquidus temperature within one hour. It is generally defined as tons per hour (t/h). Since this would define a different melting rate for every alloy, generally a liquid metal temperature of 720 °C is used as standard. This definition is valid for the period of the furnace cycle from the time the charging door is closed and the burners are switched on and continue to fire until the metal reaches the temperature as defined. All ancillary time, such as charging or skimming, is excluded. The net specific energy is defined as the energy used to obtain the net melting rate. It is therefore defined as kWh per ton melting rate and expressed as kWh/t. Although this expression is not correct in a mathematical sense, it is common practise to use it. Modern furnaces have a net specific energy consumption that is in the range of 600 kWh/t. This refers to the complete batch including, however, also the contamination. Assuming a recovery of 70 %, the net specific energy consumption of 1 ton of aluminium increases then to 600/0.7 = 857 kWh/t. There is also ancillary time to be considered where burners are firing, perhaps on low fire or heating up of the furnace after the door was open. Thus, energy required for melting is in the range of 1,300 kWh/t.

Metal transferred to the casting furnace must be kept at temperature or heated again. Also some alloying metal or clean scrap or material originating from no-go production must be melted. Thus, at least 200 – 300 kWh/t of aluminium is necessary. The energy required to operate the casting machine is very low and can be regarded as included in the energy for the casting furnace. But the launders for metal transfer must be heated prior to their use. A large blower supported by some smaller blowers is needed for the extraction and treatment of the waste gases in the filtration plant. This requires approximately 80 kWh/t of aluminium produced.

Also the cooling water system, compressors and other auxiliary equipment use energy. They are accounted for with 200 kWh/t. Vehicles within the plant area transport scrap, products and equipment. The subsequent fuel requirement is in the range of 80 kWh/t.

The consumption of the plant for mechanical preparation is summarized in a similar way. The shredder is very heavy equipment using most of the energy in the plant as electricity. The power requirement depends very much on the type of scrap to be processed. Wrought alloys returned as old building structures or window frames require approximately 50 kWh/t while castings require only a third of that energy. Considering also the conveying equipment, a power consumption of 50 kWh/t shall be used for comparison. Part of the process is the de-dusting plant. This is an important part of the system since a large quantity of non-metallic fines and also dirt are removed during shredding. Thus, large blower capacity is required which is expressed by the 60 kWh/t of aluminium processed. Some separating and classifying equipment is arranged downstream of the shredder requiring also electrical energy. Additionally, some material, such as used beverage containers (UBC), should be decoated and shavings need to be dried. Although this is usually done functionally in the melting plant, drying and decoating belongs to the section preparation of the scrap. The energy used for this process is thermal energy with some electrical energy for the drive units of the drying plant and the ancillary conveying and classification equipment. There is also electrical energy required for the circulation system for the heavy media plant and other auxiliary equipment. All these plant sections are summarized as auxiliaries and the subsequent energy requirements add up to approximately 120 kWh/t of aluminium processed. The total of the mechanical preparation is fairly low. Thus, this type of process would be of advantage but unfortunately not all of the scrap can be processed to the benefit of the melting plant.

Melting salt produces salt slag that can be recycled. This process comprises leaching and evaporation requiring thermal energy. Some electrical energy is required for crushing and screening as well as for circulation pumps. Some transport costs occur due to the logistics involved to supply the slag to the processing plant and return the recovered salt to the aluminium melting facilities. Compared to the other recycling processes, the energy requirements are comparatively low. Table 5.2 includes the energy needed recalculated to 1 ton of aluminium produced. Considering that approx. 300 kg of slag is produced for each ton of aluminium when processing and, furthermore, that only a fraction of the aluminium scrap needs processing with salt, the figures as listed in the table have to be regarded as very conservative.

5.2 Scrap processing

5.2.1 Mechanical preparation

To re-produce the different aluminum alloys the scrap has to be melted and cast into shapes as required by the market. To minimize the negative impact of the contaminations and compound materials these must be removed economically and environment-friendly. Although thermal processes are the key for aluminum recycling, preference is given to mechanical preparation and enrichment prior to the melting stage. Consequently, modern aluminum recycling comprises mechanical preparation that is followed by the metallurgical processing. The flow diagram of aluminium recycling is shown in **Fig. 5.3**.

5.2.2 Primary inspection

Scrap is collected from different sources such as fabrication plants, small scrap collectors, household waste deposits and individual scrap dealers. It reaches the preparation plant as bulk material after it is freed from loose contamination and waste as far as possible. Upon arrival, the very first step is qualified and thorough inspection of the material before it is sent to the stockpile for different types of scrap, i. e. different alloys. The statement that aluminium can be recycled without loss of quality refers to the specific alloys since alloying components cannot be removed. This means that only the original alloy can be produced although some quantity may be used for blending. Therefore, sorting into the type of material is essential. Different types of material require different processing. Clean scrap is bundled and goes directly to the melting plant. This is also the case with material having only little contamination. All other material has to undergo a particular preparation process.

5.2.3 Comminution

In order to be able to separate compound materials fixed to the aluminium, the scrap must be crushed into fairly small particles having a size smaller than 100 to 150 mm. Consequently, for this type of material the next process step is comminution. Specially designed shredders and crushers are applied. The de-dusting equipment removes the very fine fraction and dust already during crushing. The material now flows via belt conveyor to the secondary sorting section. First stage here is magnetic separation to remove steel and iron. This may also take place in two steps: An overhead magnetic separator picks larger lumps. Some aluminium may still adhere to these parts. In general this fraction is recycled to the crusher. A magnetic drum separator removes the finer iron parts. This fraction still contains some aluminium and other non-ferrous components, mainly Zn, and may be processed in a further stage providing efficiency permits. The crushed material now passes a vibrating screen or a screening drum to remove the fraction < 10 mm. It is somewhat more difficult to remove the stainless steel parts. This requires more and more attention because of the increasing quantity of space frames with stainless steel corner parts used in automobiles. Stainless steel is non-magnetic. However, the mechanical stress applied during crushing causes this material to become slightly magnetic and can, therefore, be removed by a magnetic separator with a very strong magnetic field. The next process step separates non-metallic material from the remaining metallic components comprising mainly aluminium. The material flow now reaches the eddy current separator. This classifier uses the principle where a fast rotating magnetic field induces an electric current in metallic particles which turns these particles into small temporary

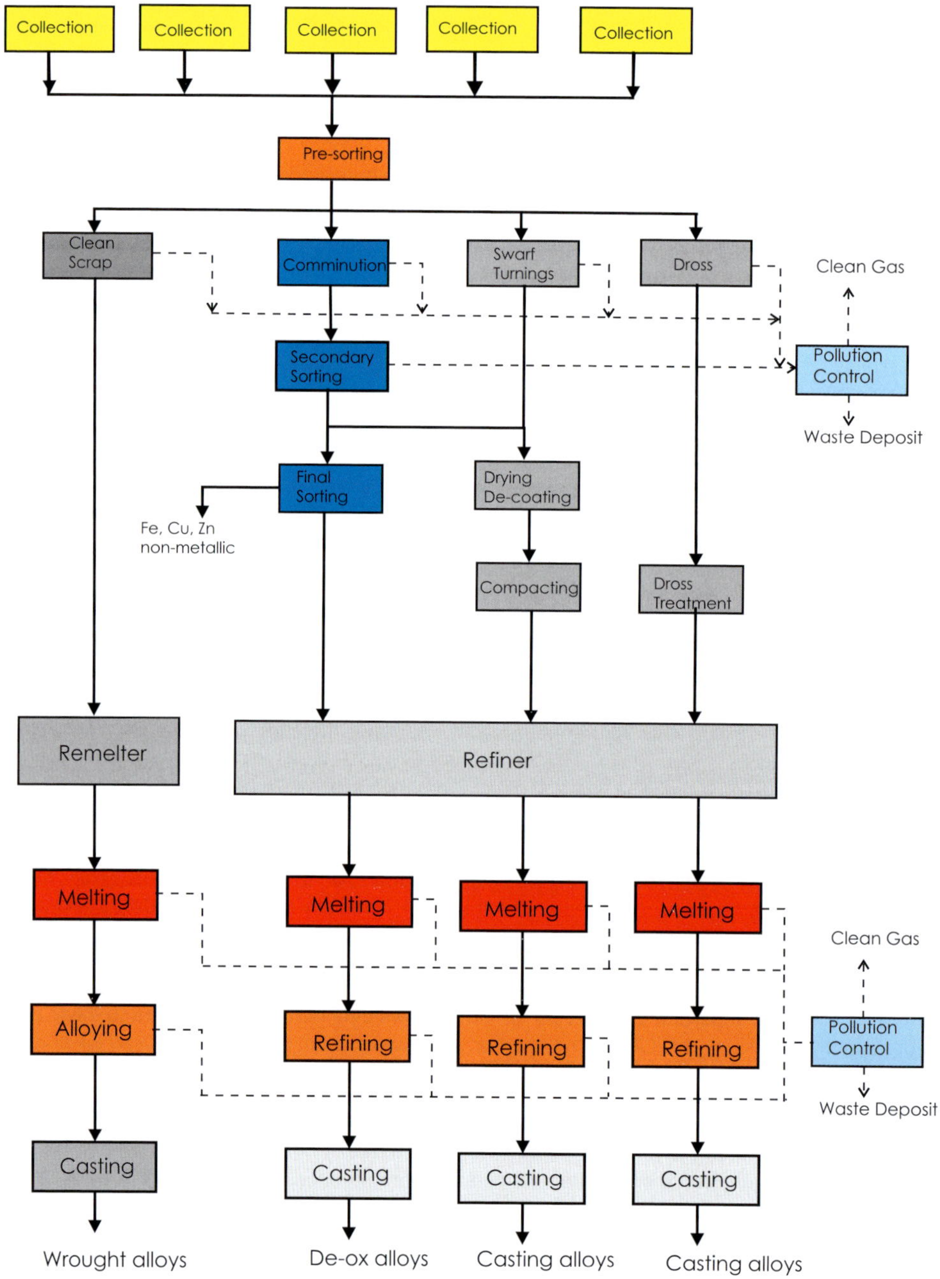

Fig. 5.3: Flow sheet of secondary aluminium recycling

magnets. Since the poles of the temporary magnets are aligned with those of the alternating magnetic field, the net effect is repulsion. The parts are ejected at the end of the separator and enter into a ballistic curve that differs from that of the non-metallic particles.

Depending on the raw material some loose non-aluminium parts are still present. Therefore, the next step is sink float separation. The unit works on the different specific gravity of the metals. The material is charged to a basin containing a heavy media fluid with gravity > 2.7 g/cm^3. This means aluminium floats on the surface while other metals sink to the bottom. Most common units are designed as rotating drum with lifting vanes at the circumference which lift the heavy fraction and discharge it from the drum at the top of the rotation. The aluminium fraction leaves the drum by an overflow dam. The heavy media drained off is recycled to the drum after passing a cleaning stage and a density control unit. Heavy media particles adhering to metal particles are washed off by means of a water spray. The collected liquid is fed into the heavy media cycle. Aluminium is now sent to the allocated storage area ready to be sold to the secondary aluminium smelter. The other non-ferrous fraction can be separated as well and sold to the relevant industry.

5.2.4 De-coating

Painted or material coated otherwise is sent to the de-coating equipment. This process step follows after crushing and iron removal, eddy current separation or sink float separation or by-passes the crushing and sorting line entirely as in the case of turnings and swarf. The collected UBCs usually contain a certain amount of steel cans with aluminium lid. Thus, this is a typical material to go to de-coating after crushing and iron separation. Painted siding will follow the same routing. Sometimes this material is a compound of plastic insulation and aluminium sheets. Therefore, the plastic must be removed in the eddy current separator before the aluminium goes to de-coating.

Thermal de-coating takes place at a temperature of approx. 400 °C. Specifically for magnesium alloys, as in the case of UBC, the temperature is very critical for the formation of aluminium oxide (Al_2O_3). This means metal loss if free oxygen is present in the heat transfer media. Therefore, it is favorable to apply pyrolysis with inert gas or combustion products without free oxygen. This process has some limitations. Depending on the nature of the organic material, three products are obtained during de-coating: gas, tar and char (black carbon). The proportion of the different products depends on heating time and temperature. Generally, at higher de-coating temperature the quantity of char and tar is lower while the gas content increases. Tar and gas are volatilized at the beginning of the process. Since no oxygen is required the pyrolysis process is very favorable. The remaining char adheres to the aluminium and the only means of removal is oxidation with the aid of some mechanical grinding movements of the particles. In order to avoid oxidation of aluminium, i.e. metal loss, very careful oxygen control is essential.

Generally, the organic content of the scrap to be de-coated does not exceed 5 - 10 %. This may not be the case for all materials. Laminated packages may reach a content of up to 50 %. Due to the nature of the product mechanical preparation is not successful. For this type of material a dense phase fluid bed de-coater was developed and is operating successfully. It comprises a perforated drum rotating in a fluid bed with salt or sand as media. Screw-type vanes inside the drum transport the material through the drum. Heat introduced into the fluid bed volatilizes the organics adhering to the aluminium. Controlled oxygen content in combination with the grinding effect of the fluid bed removes the remaining char. The entire gas flow is similar to that of the rotating drum system.

There are other processes, such as dross treatment, that may be applied in a scrap preparation plant. This depends very much on the availability of raw material. This has also a distinct impact on the organization of the overall process. Not every type of scrap has to pass all process steps. Preparation may be complete after every step in the process and the resulting material goes to the allocated stockpile. It may also be required to use an intermediate storage for partly processed scrap in order to maximize the utilization of the individual process units. This requires skilful inspection of the incoming material and careful planning of the production. To assist the management in this task, computer programs must still be developed.

5.3 Thermal processing

The classical meaning of recycling is thermal processing. The three process steps convert scrap to high quality aluminum alloys: melting, refining and casting.

In melting, scrap is charged into furnaces whose design depend very much on the type of available scrap. The size ranges generally from 10 to 30 t bath capacities at a melting rate of 3 to 10 t/h. Energy is introduced by oil or gas burners. For more details kindly refer to part 3 of this book.

The most common type is the reverberatory furnace. It is designed as hearth furnace having a shallow metal bath. The reverberatory furnace is ideally suited for all types of scrap that have sufficient specific surface area for good heat exchange but where the surface is not as large as to have too much contact with the oxygen present in the combustion products causing extensive metal loss. The reverberatory furnace is ideal for melting material like extrusion shapes, ingots or engine housings and gear casings but may not be very efficient for small particles such as UBC, shredded material or swarf. To overcome at least some of the problems, the closed well furnace was developed. This unit allows to process scrap with some contamination and larger specific surface. This furnace represents a variation of the reverb furnace. The ideal choice for small particles, i. e. UBC, shredded scrap and swarf (also for dross) is the rotary drum furnace. This furnace offers proven technology with many metallurgical benefits.

The furnace is designed as rotating drum where the burner is arranged at one side while the flue gases leave the drum through an opening at the other end of the drum. For charging, the burner assembly is swivelled away and the furnace can be charged through this opening or is returned to a duct situated at the door side.

To avoid oxidation, i. e. loss of metal, the batch is covered with salt, usually a mixture of NaCl and KCl. However, due to the necessity of using salts to cover the aluminum when melting small particles some problems arise. First of all, the salt slag containing toxic components, originating from the contamination of the aluminum scrap, has to be deposited. In Europe space for such deposits is getting smaller and smaller and high rates must be paid for the landfills. Recycling of the used slag in central processing plants helps to solve the problem but the remaining cost for the salt is still a major factor of the processing cost of a secondary smelter. In spite of the fact that heat exchange is favorable, the overall heat balance is effected by the problem that the salt and possibly a large quantity of aluminium oxide have to be heated up as well.

To obtain a good product quality, minimize metal losses and reduce environmental problems all melting furnaces must enable good control of the furnace atmosphere. Technologies have been developed for sealing the furnace door and to maintain the pressure within the interior by controlling the exiting flow of flue gas.

After melting, the liquid metal flows to the refining section. This comprises mainly a tiltable reverberatory furnace as well as metal-cleaning and degassing equipment. Although the batch charged to the melting furnace is composed carefully, some adjustments are required by adding some alloying metals or, as done in many cases, a quantity of defined scrap. Aluminum tends very strongly to pick up hydrogen from water vapor unavoidably present in the ambient air and in the combustion products. Since this causes small cavities within the metal, the hydrogen content must be lowered prior to casting. This may be done by introducing some chemicals or, depending on throughput and quality requirements by special equipment, introducing N_2 or Cl_2 into the melt. Oxides formed during de-gassing are skimmed off from the bath surface. These skimmings will be returned to the melting section for recycling. If cost and capacity permit, in-line de-gassing units are installed in the metal transfer line between refining/casting furnace to the casting machine.

As soon as metal temperature in the furnace has reached the required value, the batch is ready for casting. The type of casting equipment again is dictated by the product required. This ranges from very large "sow" ingots having a weight of 500 – 1,000 kg and re-melt ingots of 5 – 20 kg weight to high quality products such as round billets used as extrusion ingots. Lower quality of des-ox pellets and cubes for the steel industry or high quality strips for special purposes are produced in specialized plants. For casting sows large molds are used which may be arranged in a casting circle or a casting line. This product is very simple with regard to the metallurgical requirements and may be cast directly from the melting furnace. Re-melt ingots are used – particularly as casting ingots – in different branches of the industry. This material requires careful composition and treatment prior to casting. The re-melt ingots are produced on chain-type casting machines. A cooling unit is arranged at the end of the chain followed by a stacking device. The casting machine is fed from a casting furnace that is generally a tiltable hearth furnace for good metal flow control. Extrusion billets require accurate melt composition as well. Metal treatment comprising de-gassing and removal of inclusions is essential since the billets are used directly for downstream processing. They are produced on a direct chill vertical or horizontal casting machine with sophisticated mold systems.

6. Material basis for recycling

Christoph Schmitz

Most of the aluminium is returned to the secondary aluminium plants at the end of its service life. The variations of the service life in the individual sectors of aluminium applications cause a difference between the aluminium production and the percentage of aluminium scrap available to the industry. **Fig. 6.1** shows the average service life as experienced in the industry.

Scrap is classified by the German traders on the basis of the different kinds of scrap as well as on the delivery format. The classification of aluminium scrap starts always with A (assumingly for aluminium). A standardization is also being discussed. The format is defined as bales at a maximum size of 100 x 50 x 50 cm, small size (< 10 x 10 x 0.2 cm) ore fines (passing a screen size of 0.84 mm). Foreign matter may consist of non-aluminium metals or plastics that are not permitted expressively in the classification.

For the designation for different aluminium scrap, the metal traders are using some fancy names always starting with A. Some examples are:

- "Abweg" is new wire scrap of pure aluminium free of foreign matter
- "Alter" is scrap of extrusion shapes, free of any foreign matter
- "April" is old scrap of rolled material as used for household appliances, free of Al-Cu and Al-Zn alloys.
- "Apsis" is shredder
- "Assel" is iron-free casting scrap
- "Autor" are mixed shavings

For more detail please refer to literature.

Sources for the aluminium recycling are types of scrap which contain aluminium as metal. Material that contains aluminium in a non-metallic form, such as aluminium oxides, cannot be recycled to aluminium metal. The available scrap is categorized into new scrap and old scrap.

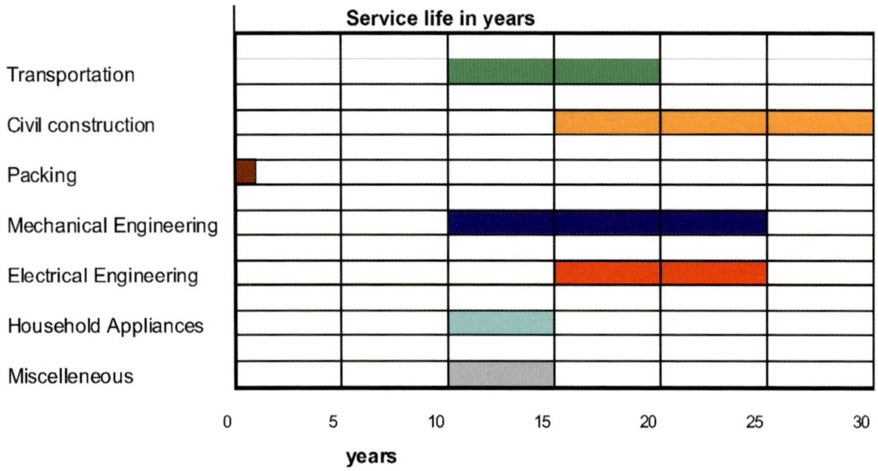

Fig. 6.1: Service life of aluminium alloys

6.1 New scrap

New scrap originates from a production of wrought products such as shapes, sheets and the like or from the production of castings. This scrap is very well defined in regard to the chemical and metallurgical analysis. It can be charged directly to the melting furnace and is then melted with good recovery. It is mostly recycled in the production plant as in-house scrap. Independent remelters are also processing this new scrap to produce remelt ingots or even billets or slabs.

In downstream facilities the wrought aluminium products are further processed to the final product. There will be leftover process scrap that may appear as lumps or pellets or small pieces of thin sheet. Sometimes this material is coated or contaminated by fabrication emulsion or may also have a layer of anodic oxides. If machining is involved, shavings will be generated and also contaminated by cutting oil or emulsion. Generally these materials cannot be processed in the remelting plant and will therefore be supplied to refiners. It could be loose as bulk material or baled.

6.2 Old scrap

Old scrap is aluminium that has been collected after the end of its service life. This ranges from loose collected scrap originating from castings, components of household appliances, vehicle components such as rims, gear boxes, or space frames or building structures such as old window frames or also from used packing material, pie trays and beverage containers. Cable or shredder products from cars are also collected. Old scrap is mostly supplied as loose bulk (**Fig. 6.2**). Very often the scrap is contaminated by other metals, plastics or even oil and dirt. This scrap has to be processed through a mechanical preparation. Collected scrap comprises a wide range of different materials (**Fig. 6.3**). It may contain wrought aluminium of very different alloys as well as castings. There is also a quantity of compound material. To be able to process this material, it must be shredded and then pass the different sorting and classification stages before it can be melted. The target is, besides better handling, the possibility to remove foreign components. But there are still some contaminations. Some may even be caused by the processing in the heavy media classifier which uses a ferro-silica suspension for adjusting the necessary density difference. Some of it may still adhere to the particles which therefore must be washed. If this is not done carefully

Fig. 6.2: Collected scrap

Fig. 6.3: Collected scrap

Fig. 6.4: Baled wrought aluminium scrap

the iron content of the aluminium could increase. In order to have some defined alloy composition the scrap should be sorted prior to shredding. This is a pre-condition for further processing in the various stages of the mechanical preparation. It appears to be a good solution to differentiate at least between wrought alloys and castings.

With the increasing use of aluminium in vehicles the quantity of automobile scrap increases. The wrecked cars are shredded as complete units and the components are separated by air classification to remove plastics, magnetic separation to obtain the steel parts and finally heavy media separation to separate aluminium from other metallic components such as Cu, Pb and Zn.

Although the technology for separating different kinds of scrap has improved remarkably, the result of the mechanical separation may not be a much defined alloy composition so that the shredder scrap requires blending with other scrap or with primary metal.

Wrought aluminium, that is supplied by downstream production plants, is mostly baled (**Fig. 6.4**) when supplied to the melter. Mostly this material can be charged to the melting furnace without problems. Sometimes, however, there is some liquid trapped in the bales. This may originate from storing the material in the open or baling it with all what the scrap may carry. Very often this is the case if the material, for instance, consists of used transformer conductors. There is a potential danger of a severe explosion when charging these bales directly to the furnace or even into liquid metal. Therefore, all bales should be inspected carefully and then handled with caution.

UBC (used beverage container) is mostly processed in a closed loop. But the material also appears on the scrap market as new scrap as generated by the downstream production plant and at an even larger quantity as collected UBC. For the benefit of easy transport the material is usually baled (**Fig. 6.5**). These bales have their shortcomings as well. They may contain a share of steel cans and may contain water or even sand. This is no problem when the UBC scrap passes through a decoating system prior to melting. But in many cases the bales are charged directly to a melting furnace. Thus, the necessary precaution is advisable.

Shavings generated during machining of aluminium are free of any coating and could be regarded as clean scrap. However, the cutting oil or emulsion required during machining is still adhering to the material. It is advisable to remove this contamination by a chip drying process prior to melting. Due to the large specific surface area the individual chips are more or less covered by an oxide skin. The thickness of the skin depends on the chip drying process but also on how long the material has been stored. If the origin of the shavings is not known, the bulk may contain also steel chips. These can be removed efficiently after the drying procedure by magnetic separators. But still the composition of the bulk regarding the different alloys may also not be very clear. Here again careful sampling and inspection is of essence.

Skimmings originating from any liquid aluminium process should theoretically be only aluminium oxides. During the skimming of a liquid metal, bath aluminium metal is also skimmed off and forms the basis for recycling. The oxide portion cannot be recycled. The aluminium metal content may vary from 20 to as much as 80 %. This depends on the skimming procedure and the treatment of the dross thereafter. The OECD defines dross as having a metal content of less than 45 % and skimmings with a metal content of more than 45 %. There is also a difference between white dross and black dross. White dross originates from primary smelters, rolling and extrusion plants and foundries which operate their melting furnaces without using salt. Its color is greyish to white. Black dross comes from all refiners and ore remelters that use salt in their melting process. The black color is generally due to the reaction of aluminium with magnesium that is supported by the presence of sodium chloride. At low aluminium content or at a high content of dust in the dross,

Fig. 6.5: Baled UBC scrap

the material is processed by sieving, crushing and grinding to upgrade the dross. The resulting material has an aluminium content of 70 % or more. The metal entrapped as fines is lost during the process. But this aluminium would go into the dust removed from the furnace with the flue gases anyhow. Since this dust contains a high percentage of aluminium, it can be used as desoxidation media in the steel industry.

Careful sampling and sample melting must precede the processing and blending of scrap in the refining plant. Only this procedure will assure that the client's quality requirements are met. The equipment installed to process the variation of raw material must be selected carefully and correct operation of it is indispensable. Depending on the quantity of a specific material available to one plant site, specialized equipment may be installed and operated economically. But the decision-making for some very specific technology should also consider the future development of the market and the flexibility required to maintain the economical operation in the future.

Part II

Mechanical preparation

1. General

Christoph Schmitz

1.1 Principles of mechanical preparation

The idea of the mechanical preparation is to have the aluminium scrap, as supplied by various sources, either formed into a more handy shape which eases the operation in the recycling plant or to crush it to comparatively small particles to be able to remove components in the material which are either non-metallic or comprise metal elements which are not wanted during the melting process. Compacting and cutting on shears followed, perhaps by bundling, is applied for very well-defined and clean materials. These are generally alloys which can be bent and shaped easily. Other materials, which are generally collected are scraps or castings or compound materials, such as window frames are comminuted and after that classified in various sizes and also separated into different materials.

For efficient operation a shredder plant requires a minimum size that does not always match the size of a metallurgical recycling plant. Only such scrap needs to pass mechanical preparation that consists of compound material and such material that is mixed with other metals such as iron, copper, lead, zinc or similar. In order to justify also downstream classification and separation equipment, a plant size is required to match the throughput of a shredder and to provide sufficient capacity. This kind of plant has to operate with the specific type of raw material continuously to be economical. Shredder throughput may reach 15 t/h, resulting in a daily production of up to 240 t with a two shift operation.

A normal metallurgical recycling plant processes scrap that does not require mechanical preparation besides shredded scrap. Considering these facts, separate plants for mechanical processing and metallurgical processing have to be installed. One shredder plant will, consequently, supply different melting plants with material that matches their product mix. But depending on the type of scrap available, a small shredder may be justified in a secondary smelter. The situation is different for other mechanical preparation equipment such as de-coating or dross treatment. The capacity of this kind of equipment is comparatively small and can be adjusted to the requirements of an individual metallurgical processing plant.

A very different task is the recycling of salt slag, which is a waste product, from melting in the rotary drum furnace. The major components salt, aluminium and aluminium oxide have to be separated in a crushing process followed by leaching. Process plants are most economical for a large size plant and, therefore, they are located in a central area to process the slag provided from different smelters. Smaller units may be justified economically if a central process plant is not in the reach of a secondary smelter.

1.2 Principle plant requirements

The capacity of the shredder must be large enough to be able to process the scrap and also to have a wide feed to accept large pieces of scrap. Major contamination of aluminium scrap is steel which must be separated by a system for small and large steel parts. Stainless steel is not very magnetic and must be separated by strong field magnetic systems. The remaining aluminium is still mixed with plastic and rubbish. This has to be removed first, either by air flow separation, which partly takes place at the shredder waste gas system, or eddy current separation or a combination of both. These materials are discarded as shredder waste. There are still foreign components in the aluminium scrap. These comprise heavy metals such as copper, zinc, lead and the like. To separate these, the shredded iron and waste-free scrap passes a heavy media separation plant. Some traces of iron may be added here again since the heavy medium is a slurry on the basis of ferro-silicon. The final material is a bulk of clean aluminium particles in the fraction 50 – 100 mm and < 10 mm which can be processed very well in rotary drum furnaces. Clean aluminium scrap, provided as shapes, sheet or foil as well as UBC, is generally compacted on lid-type presses to obtain bales of the dimensions of 600 x 600 x 800 mm or similar. Swarf provided as loose bulk can be compacted to pellets of 100 – 200 mm diameter and 50 – 150 mm length. Iron chips and cutting oil are not removed. Since the swarf is provided from various sources the pellets are of doubtful quality.

Dross is another common commodity. Depending on the source it can be of high quality which is usually supplied by primary smelters. It comprises mainly aluminium particles and aluminium oxide (Al_2O_3). The metal content can be as low as 20 % by weight which makes enrichment feasible.

2. Material storage

Christoph Schmitz

Classified scrap is generally delivered to the thermal processing plant and will be received and checked there. This scrap must not obtain too much non-aluminium contents. For easier handling and charging some of this material may be baled or cut.

The situation is different in large plants for mechanical preparations. Scrap provided by different sources is usually a conglomeration of different metals and plastics or other non-metallic materials. The incoming material is usually limited to the two categories steel scrap or non-steel scrap. Large parts, such as steel structures or components of structures, are manually separated units with the aid of heavy lifting equipment. An inspection for radioactive components is carried out by means of suitable equipment when on the scale or during passing the gate. Depending on the kind of scrapyard, clean material may also be supplied. This is classified on the basis of visual inspection and then stored in designated areas. This material is sold directly to melting plants without prior treatment.

The bulk of scrap must be prepared manually, usually by passing through a shredder. To be able to feed the downstream equipment, the material must be in the operating range of equipment for loading the charging system of the shredder. Usually two or three mobile cranes with grabs are located near the feed conveyor to match the large material requirements.

Processed material is collected near the discharge points of the shredder. This will be different for aluminium plus plastic and steel. Some dusty and small lightweight components are discharged through the waste gas extraction system. The different material fraction will go to specific storage areas to be loaded directly on trucks to be supplied to customers or to go to further processing.

Due to the kind of scrap and the volume of waste, the storage area for a mechanical preparation plant cannot be too large. Besides the areas for stockpiling, free movement for the supply trucks as well as for the internal transport must be assured.

Fig. 2.1 shows a typical shredder plant with shredder waste gas treatment and picking belt.

Fig. 2.1: Shredder plant with metal feed (Lindemann)

3. Cutting and baling

Christoph Schmitz

Guillotine-type shears are used for cutting extrusion shapes, pipes and cables. They comprise a feed-bin with lid and a pusher. Cutting is executed by means of a hydraulically-operated knife. Another knife at the bottom of the cutting head supports the shearing effect. All movements are executed by hydraulic cylinders. Material is fed into the feed-bin and the lid of the unit closes. The material is then already pre-pressed into a shape which eases cutting. The hydraulically-operated pusher now moves forward to have the material cut in pre-set length.

Another type of shear is the so-called alligator-shear. Its principle is similar to the sheers for cutting paper and cloths which we all know. The alligator-shear is, of course, of heavy design and is operated by means of a hydraulic cylinder. It can be designed for stationary installation as well as a mobile unit. The size of the material to be cut is somewhat limited and feeding of the machine requires intensive manual work.

For very thin material rotating shears can be used. They are mainly applied for comminution of cables, foils and swarf. Cutting takes place between two rotating shear heads. Another type of shears used in an aluminium recycling plant comprises a rotor with shear knives and a screen through which the cut pieces pass. These machines are used for cables, swarf and also for foils.

The application of different shearing systems in a secondary aluminium recycling plant has some limitations. It does make sense to cut extrusion shapes and pipes as they are supplied mostly as irregular tangled bulks. To have them processed through a cutting shear to get a length good to handle is very convenient for charging into the furnaces.

The other method to get a size easy to handle is the baling press. Its (**Fig. 3.1**) design principle comprises a box with one movable wall which is operated by means of heavy hydraulic cylinder. This box can be closed by a swivel lid which provides pre-compacting already. At the end opposite from the movable wall, the compacting chamber is arranged. It is equipped with a large ram operating from the top. A second pressing ram works horizontally for final compaction of the material. Scrap is charged into the box and will be pre-compacted by closing the lid. The large hydraulic cylinder moves one complete sidewall in direction of the compacting chamber. The now downwards moving ram is the next compacting stage. Compacting is completed by the horizontal ram. Finally, the discharge opening is free and the horizontal ram ejects the bale from the machine.

Fig. 3.1: Baling press

4. Comminution

Christoph Schmitz

4.1 Crusher

Depending on the type of material and the desired grain size distribution, various types of crushing equipment are applied. Most of the equipment is used for processing dross and aluminium slag but also for processing UBC and swarf some crushers are required. This equipment is of fairly small size. It differs very much if large size material, mostly heavily contaminated by foreign material, is to be processed. In such a case large shredder plants are used to obtain the desired results. Due to the large capacities of such shredders, this type of equipment is to be regarded as key equipment for mechanical processing of aluminium scarp.

4.1.1 Impact crusher

The most commonly used crushing equipment in a secondary aluminium recycling plant is the impact crusher (**Fig. 4.1**). It comprises a rotating roll with heavy steel crushing blades made of wear-resistant steel and which are attached to the heavy rotor. These blades are wear parts and can be exchanged easily. Material fed to this rotor is thrown against heavy steel baffle plates which are suspended from a swivel point and supported by buttress rods and, due to the severe impact, they are crushed. They leave the crusher through a bottom opening.

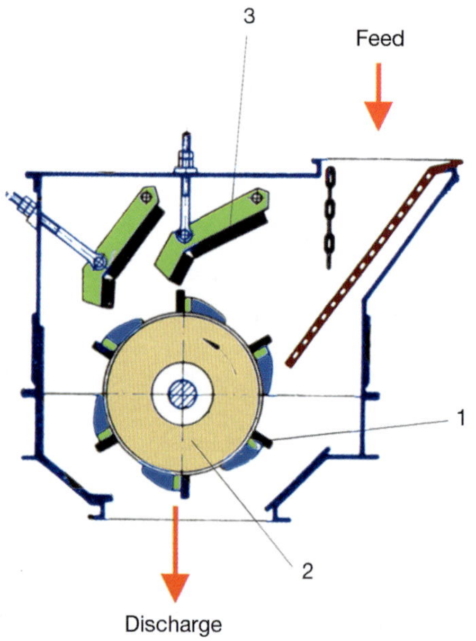

1 - impact,
2 - rotor,
3 - baffle plates

Fig. 4.1: Impact crusher

If the crushing blades are designed as hammers fixed to a rotor and suspended by swivel bolts, the crusher becomes a hammer mill (**Fig. 4.2**). The other difference to the impact crusher is the grid arranged at the crusher housing. The comminution principle is similar to that of the impact crusher. Their major application is for smaller throughput but they produce a more defined grain size distribution than the impact crusher. Having reached a smaller size, the material is discharged

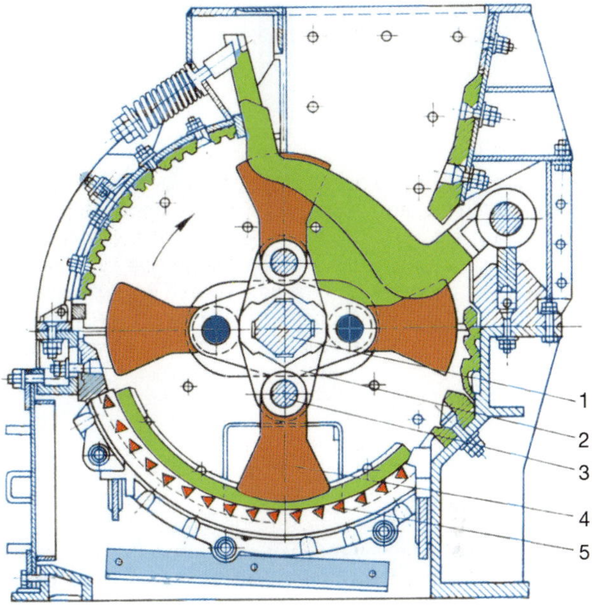

1 - rotor,
2 - hammer boulders,
3 - crushing hammers,
4 - swivel axle, screening baffles

Fig. 4.2: Hammer mill

through the bottom grid or through the grid on top or through both. Smaller hammer mills are mainly used for the processing of dross.

4.1.2 Shredder

Large units with a power of 1,000 kW or more are used for the comminution of aluminium scrap. They are not called hammer mills or impact crushers any more. They are well-known under the classification as a shredder which is key comminution equipment for mechanical preparation

Its concept is basically that of an impact crusher consisting of a rotor and a horizontal axis with a couple of hammers attached to it. The material is broken by the impact of the hammers and by the impact of the broken particles against grid and baffle plates. The principle variants are determined by the size or the type of rotor and its speed.

The shredder has a grid setting for the discharge and the crushed product. Some aluminium alloys and in particular wrought alloys are very tough and cannot be broken easily. Therefore, the top of the crusher is open. The machine with this specific feature is commonly called Zerdirator. **Fig. 4.3** shows the principle design of a shredder. The material supplied is by the feed conveyor and passes via a chute to a pinch roll couple (1). These flatten the scrap and charge it at controlled feed rate into the shredder. Automatic controls switch the pinch rollers "on" and "off" in order to avoid overfilling of the crusher. A stopper bar (2) at the entry point provides first cut-off and start of fragmentizing. The material and hammers pass over the bottom grid (9) and small particles are already discharged. Pieces not fragmented to the required size will hit against the baffle plate edge (3) where also some compacting is done. The baffle plate (4) ensures, together with the hammers, the desired comminution and compaction. The major flow of material passes the top grid (5) and will be discharged. While the crushed aluminium falls through the shredder housing onto the discharge vibrating conveyor (8), the dust will leave the shredder through the open top air discharge (6). The draft is generated by the blower of the filtration plant and by the shredder acting like a

Fig. 4.3: Shredder (Zerdirator) (Source: Lindemann)

blower. It should be noted that the temperature in the shredder may reach 200 – 300 °C due to the work of deformation. Material will leave the shredder through a bottom discharge.

Shredders are available in different sizes (**Table 4.1**) ranging from 370 to 1,030 kW power requirement. The throughput rate depends on the aluminium allowed. Castings require a specific power consumption of 20 – 25 kWh / t · h and wrought approximately 3 times, the value reaching to 50 – 75 kWh / t · h.

4.1.3 Jaw crusher

The jaw crusher (**Fig. 4.4**) is used for comminution of large pieces of scrap material such as cast iron parts (motor-blocks) or large lumps of dross. The jaw crusher comprises one fixed jaw and

Table 4.1: Shredder (Zerdirator) sizes (Source: Lindemann)

ZERDIRATOR Type	Driving power kW	internal width of the housing mm	AL Wrought alloys t/h	AL Castings t/h
ZZ 175 x 160	499 600 750	1600	6-10 8-14 10-16	20-30 25-35 35-45
ZZ 175 x 260	750 920 1030	2600	12-18 14-22 15-25	40-50 45-65 50-70
ZZ 210 x 260	1470 1840 2200	2600	20-36 26-42 32-48	60-100 80-120 100-140

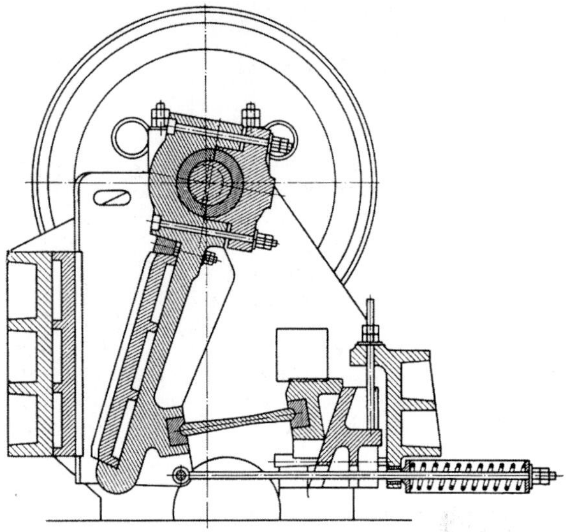

Fig. 4.4: Jaw crusher

one swinging jaw that is supported on an eccentric drive shaft at the upper end and held in position by a sturdy buttress rod at the lower end of the jaw. Due to the eccentric bearing, the movable jaw makes a circling motion which also ensures that the material fed from the top is forced through the crusher. The rod at the bottom is held in position by a spring or hydraulic loaded system that is able to give way if an unbreakable piece of material is forced into the unit. Contrary to the impact crusher, very little fines are generated in the jaw crusher. However, it is limited to processing brittle material. It is the ideal machine to process large cast aluminium parts such as engine housings.

Other crushing equipment, such as gyratory crushers or roll crushers, is not used in the secondary aluminium industry.

4.2 Ball mills

This equipment is applied whenever fines shall be produced. A special case is the ball mill grinding of dross. With this material the aluminium is enveloped by layers of aluminium oxide. In the ball mill the aluminium oxide is crushed and the metallic aluminium is freed. Since the material is very tough it is not destroyed and leaves the mill as solid particles. The ball mill comprises a rotating drum supported by heavy bearings at the entrance of material and at their exit. 30 % of the volume is filled with balls of different sizes.

Due to the rotation of the drum, the balls cascade down onto the material which causes the grinding effect. As material is continuously fed to the ball mill, the processed grinding goods leave the mill via the center opening. Dust and other fine particles can be removed by an air flow through the ball mill. The air flow provided by a large blower passes through the ball mill from the exit to the entrance. The dust-loaded air is then directed to a filtration plant, usually a bag filter. In order to improve the efficiency of the mill, part of the dust and loaded air is recycled to the ball mill before it is taken out of the circuit. This system is very convenient for dross preparation. The drive consists of a speed-controlled motor with gear-box acting on a gear around the furnace drum. Material fed to the ball mill and discharge has to pass through air-locks to avoid leakage of the air flow in the mill. The mill walls are armored with replaceable wear-lining plates.

5. Classification

Christoph Schmitz

During crushing the material will be available in different grain size distributions. To separate these according to the size, classification units have to be applied. There are two main groups of equipment which are commonly used; the vibrating screen and the screening drum.

5.1 Vibrating screens

The vibration screen comprises one or two decks with different mesh sizes. They are designed as closed housing supported on springs. Vibration is generated by unbalanced motors or by an electro-magnetic system. Thus, the vibrating system comprises the oscillating force and the springs. The parameters are selected in such way that overcritical motion is obtained which means the screen operates above the resonance frequency. Since the eccentrical velocity for the screen decks is above the gravity velocity, all particles are lifted and fall back onto the screen deck which keeps the meshes free so the particles of smaller size can then pass the mesh into the lower deck (**Fig. 5.1**). Finally, the fines passing the second deck are collected in the slanted housing of

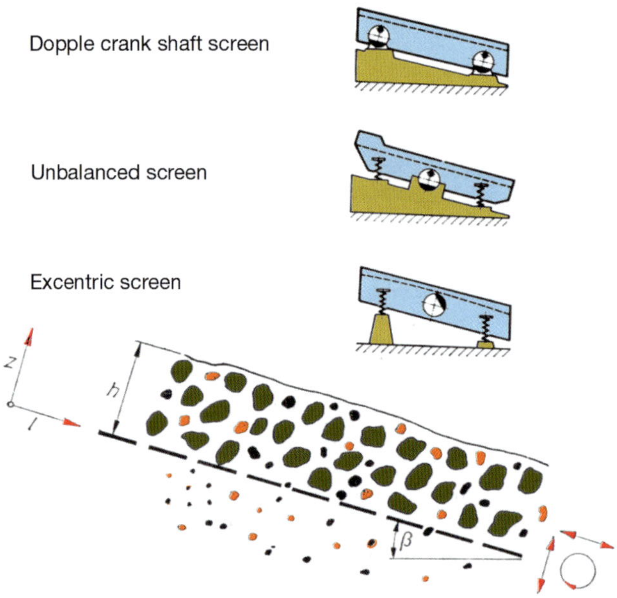

Fig. 5.1: Vibrating screen principle

the vibrating screen and from there they will be discharged, as required. Material not passing the decks leaves the screen decks as well and is also collected and transported away as requested and may be recycled to the crusher. The movement for this charge is supported by the sloped arrangement of the vibrating screen towards the discharge sides (**Fig. 5.2**).

5.2 Screening drums

Another very common means of classification is the screening drum. It comprises a drum with different openings or holes at their shell. Different sections of the drum will have holes of different diameters. Passing through the drum the material can be classified in various fractions. The size

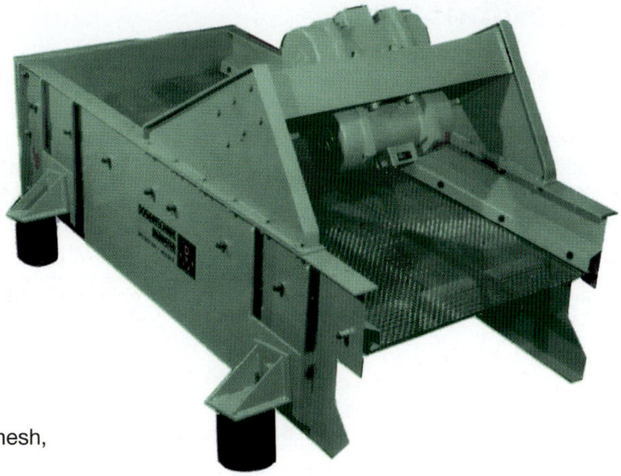

Fig. 5.2: Vibrating screen with screen mesh, unbalanced motor and rubber springs

of the holes increases towards the end or the exit of the material. Thus, in the first section, fine material is separated. The next section separates coarser material, the next section coarser material again until towards the end most coarse material is separated.

At the end of the drum material still not separated leaves the unit as lumps. The screening drum is supported on roller bearings with a supporting ring at the base. The drive is executed by means of a variable speed motor. The different fractions are collected in containers or may be passed onto belt conveyors to be transported for further processing.

6. Sorting

Christoph Schmitz

Screening separates different size fractions of the material processed. There is also the necessity to differentiate between various materials. This is commonly called sorting. The removal of other metals or non-magnetic materials requires different methods. Very light materials, such as paper or light plastic, are already removed by the air flow in the shredder. But heavy metals such as steel, copper, lead, tin, bronze and zinc, which often go with the aluminium as compound materials, are freed from the aluminium but are still present in the mixture of crushed material. To separate these, the gravity and the floating behavior of the materials is used as in the processes of the air flow separator or in the sink-float process or physical characteristics as in eddy current separation or magnetic separation.

6.1 Picking belt conveyors

The simplest sorting equipment is a conveyor operating at slow speed with workers on both sides of the conveyor picking material which was classified before as no-go material. Classification criteria are shape, appearance, color or weight. It can also comprise different metals which can be identified on such a belt. The sorting belt can operate continuously or in stop-and-go mode depending on what is to be removed. For high capacity usually continuous sorting belts are applied. For each sorting position a length of 1.5 to 2 m is allowed on a belt with of maximum 16 m. The velocity of the belt depends on the duty of the sorting and may be between 1.5 and 2 m per second.

On the basis of the chemical composition of the material, detection comes first by laser combined with an automatic sorting system. The different materials are detected by the laser or X-ray unit. A blower nozzle creates a pulse at the right instance after the material has left the belt in the discharge falling curve.

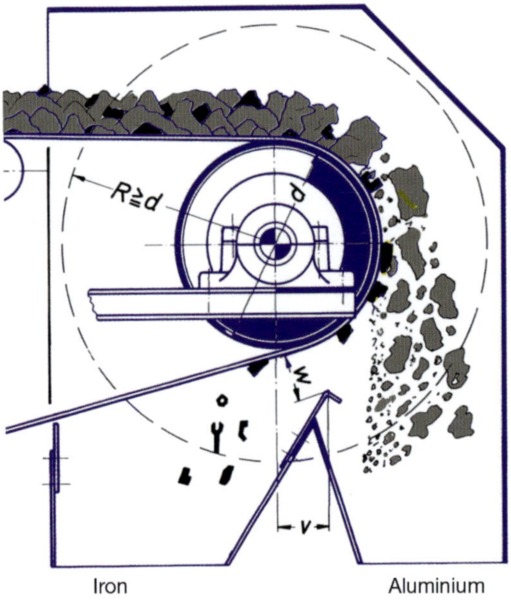

Fig. 6.1: Magnetic separator integrated in the deflection station of a belt conveyor

6.2 Magnetic separators

Iron is removed by drum magnetic separators or over-belt magnetic separators. The magnetic drum is arranged on the upper deflection point of a belt conveyor (**Fig. 6.1**). Thus, the material has to pass a magnetic field which reaches all the way underneath the top of the belt conveyor. Non-magnetic material falls in the normal gravity ballistic discharge curve and can be passed on to a belt conveyor. Magnetic material is carried around the deflection roller and falls off after the magnetic field is not effective any more. Iron pieces are then discharged into containers or to belt conveyors.

A second system is applied for removing larger pieces of iron. This is an over-belt magnet (**Fig. 6.2**). It comprises a large linear magnet in combination with a belt conveyor. This belt of this conveyor is equipped with studs to transport the material on its way passing the magnetic field. The iron pieces are attached to the magnet and then transported by means of the studs to the side where they are discharged into containers or conveyors.

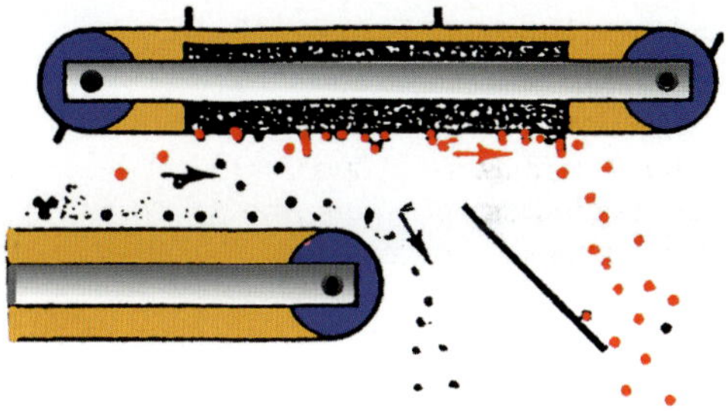

Fig. 6.2: Over-belt magnetic separator

Stainless steel is normally non-magnetic or only magnetic if a very strong field is applied. Therefore, drum separators with an extremely strong magnetic field are arranged like the normal iron separators for removing stainless steel parts. Apart from the strong magnetic fields, the procedure is identical to that of the normal magnetic separators.

6.3 Eddy current separation

The eddy current separator consists of belt conveyor equipment with a multi-pole magnetic rotor fitted to the deflection drum at the exit side of this belt conveyor (**Fig. 6.3**). Due to the rotation of the multi-pole magnetic system, an eddy current is induced in the metal particles which in turn generate a magnetic field which has opposite orientation than the rotor magnetic system.

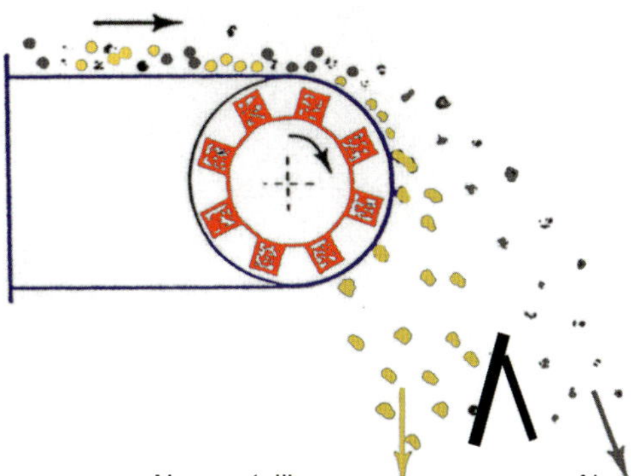

Fig. 6.3: Eddy current separator

As a result, metal pieces will be ejected and thus separated from the non-magnetic material which is not effected and leaves a belt in a normal ballistic curve (**Fig. 6.4**). The ejected metal particles are separated from this material flow by means of a baffle and will be discharged on a separate conveyor. Due to the different physical properties, all metallic particles are ejected and could theoretically be separated from each other. But then the difference in the induced magnetic field is too small to permit defined separation. Since the

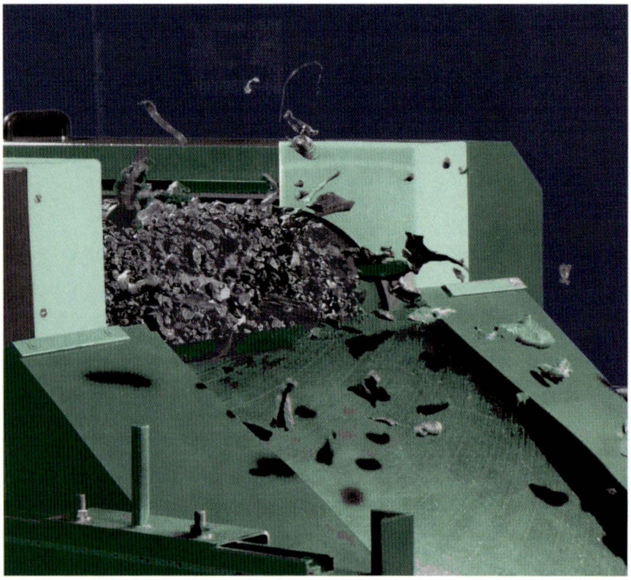

Fig. 6.4: Aluminium separation or by eddy current

eddy current separator is designed as belt conveyor, it can, for instance, be installed as discharge conveyor underneath a screening drum (**Fig. 6.5**). In such a case the material flow is separated into different fractions resulting in smaller material layers on the belt which improve the separation efficienary.

6.4 Air flow separation

The difference in gravity of different materials is used in the air separator. This is particularly effective if the difference is large and also the surfaces offer good resistance to the air flow which is expressed by the factor c. The air velocity must be selected to have sufficient draft for the light particles but let the aluminium particles fall in a counter-flow to the air stream. Every material has a characteristic settling velocity which depends and shape and gravity of the individual grains. Only particles having a settling velocity, that is lower than the air velocity c, are moved by the air

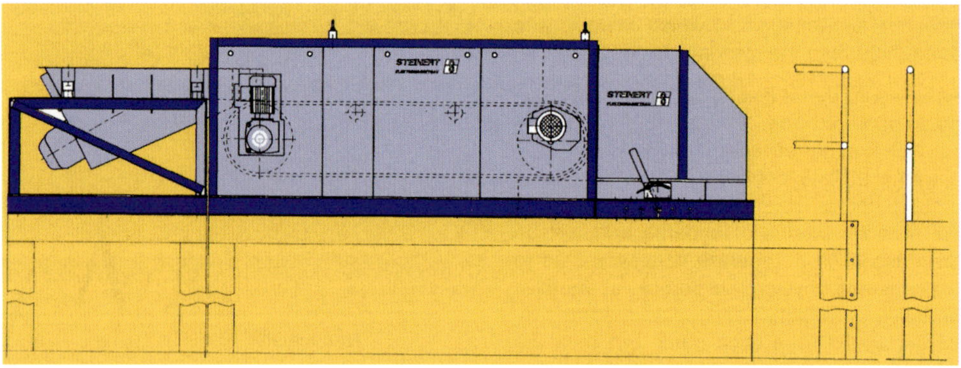

Fig. 6.5: Design of an eddy current separator with feeding vibrating conveyor (Steinert)

stream. This fact is also important for the pneumatic transport of material and for dedusting.

Fig. 6.6 shows the principle design of an air flow separator. The material is fed to the air stream by means of a charging chute (A). The air flowing upwards picks up all particles having a lower settling velocity than the air velocity. These are generally the small paper and plastic materials as well as dust. These particles are thus separated from the particles with higher settling velocity, i.e. the metal pieces. Instead of the vertical reactor, small plastic particles, paper as well as dust can be separated in a horizontally arranged rotating drum. This is standard practise in rotary furnaces or in dryers and de-coaters to remove dust. Due to the rotation of the material in the drum, the inner layers come into contact with the air flow that carries away the fine particles. This is the reason that particularly in shredder plants for steel scrap a de-dusting drum is arranged in-line of the discharge conveyor system. The particles carried away by the air are separated in a cyclone and finally in a bag house as described in part 3, section 6.

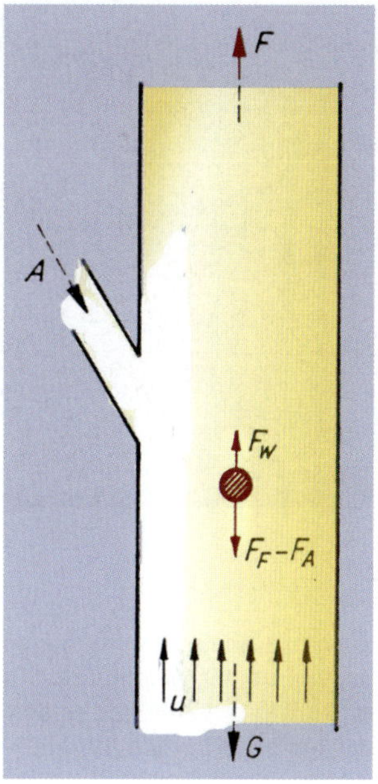

Fig. 6.6: Air flow separation

6.5 Cyclone separator

The principle of the cyclone separator is based on the inertia of the different materials. Heavier particles will flow to the outside wall of the cyclone thus being separated. Details are described in part 3, section 6. Cyclones are efficiently applied for separate coarser particles in a dust-loaded waste gas flow.

6.6 Sink float separation

The separation in the sink float process takes place with the aid of a fluid whose density is between the densities of the components to be separated. The particles having a higher density will sink to the bottom while the particles with lower density will float on the surface (**Fig. 6.7**). This

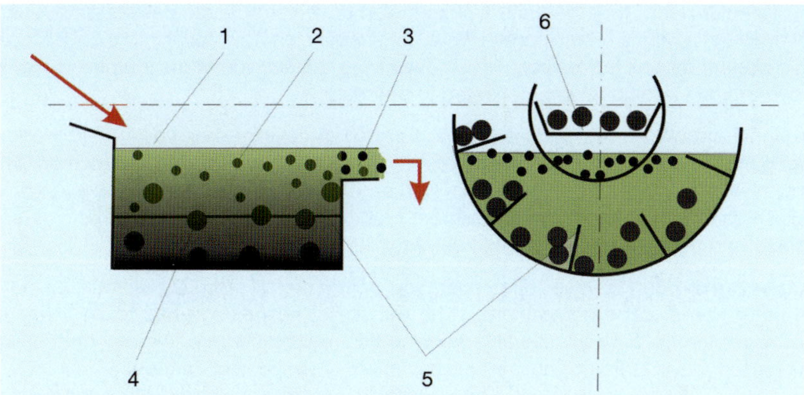

1 - slurry, 2 - aluminium particles, 3 - overflow, 4 - heavy metal particles, 5 - lifting vanes, 6 - heavy metal discharge chute

Fig. 6.7: Principle of heavy media separation

Fig. 6.8: Heavy media plant with sink float drum (Source: Oberländer)

system actually permits the separation of non-metallic components, heavy metals and magnesium alloys from the aluminium. Separation of different aluminum alloys is theoretically possible but too difficult in practice due to the very small difference in density. The difference of densities between the particles to be separated is a ruling factor of the quality of the separation.

Air trapped in the metal, as it may have happened during shredding, will have a negative effect on the quality of separation. As medium a suspension of ferro-silicon is commonly used. Due to its high density of 6.7 kg/dm^3, a maximum suspension concentration of 30 – 35 % solids can be obtained. The other advantage is that it can be removed from the suspension by using a separating magnetic drum (**Fig. 6.8)**.

The most common sink-float separation unit comprises a rotating drum supported on rollers and supporting ring with inlet and outlet opening in the center. The drum is filled with suspension to a level which allows that always a quantity of suspension together with the aluminium floating on top of it leaves the drum through the center. Heavy metals sink to the bottom and from there they are lifted by lengthwise-oriented vanes to be discharged on a chute or on a conveyor above the liquid level. The heavy metal leaves the system via a chute and is collected in containers or on a belt conveyor. The aluminium together with a suspension on which it floats passes onto a separating screen. Aluminium remains on the screen while the suspension now flows into a collecting system. It is very important that all ferro-silicon is removed from the surface of the aluminium because the refiners do not want additional iron collected in the material.

It could happen, however, that ferro-silicon is trapped in cavities created during shredding. The aluminium particles are washed by means of an intensive water-spray on that screen. The water will be mixed with the suspension. A certain portion of the ferro-silicon is removed from the suspension by means of a magnetic drum. Before reaching a mixing tank the density is controlled automatically and ferro-silicon is adjusted as required.

Sink float separation can be used for separating heavy metals from the aluminium but also to separate other non-metallic components. The density of suspension has to be lower than the density of aluminium to allow plastic and wood to float on the surface. Water would be sufficient but some plastic may not float. Thus, a higher density of suspension is required.

As first stage prior to feeding the material to the sink-float drum, non-metallic components may be removed in a settling tank. The slurry density is adjusted to have all non-metallic materials floating on the surface thus being separated. After passing a liquid removal screen

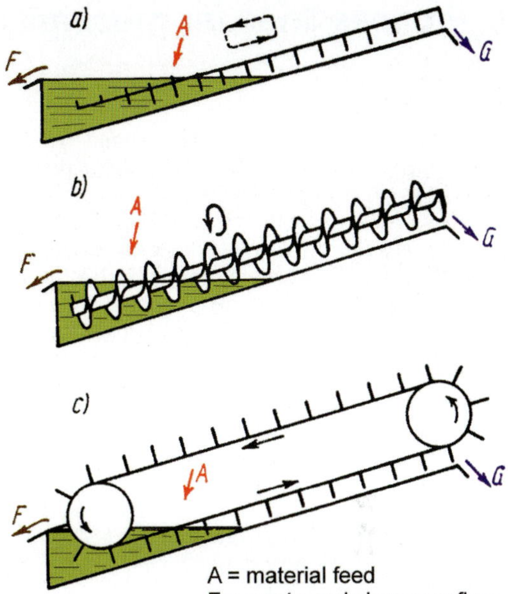

a - catcher rack,
b - screen conveyor,
c - belt conveyor with steel

Fig. 6.9: Sink float basins with different discharging principles

the material goes to deposit together with the shredder waste. Removal of ferro-silica is usually not required since the slurry density is low and ferro-silica is not harmful to the environment. However, if required and economically justified, a magnetic drum could be installed. The metallic material is collected at the bottom of the settling tank. **Fig. 6.9** shows different methods for removing this material. This can be a so-called catcher rack, a screw conveyor or a belt conveyor. The screw conveyor or the catcher are the preferred systems since they are of simple design and easy to maintain. After such rack classifier, the metal can be charged directly to the separating drum by means of a suitable conveyor.

7. Process lines for mechanical preparation

Christoph Schmitz

The individual equipment must be arranged to a complete processing line. Sometimes it is convenient to have different sections of the production separated from each other with some intermediate storage to be able to differentiate between material of different alloy components, shapes and sizes. This depends on the metal market and the requirements of different customers. But it might be necessary to store different shredder products. These could contain a high amount of iron or heavy metal as well as an increased content of plastic or dirt. Therefore, not all of the shredded materials need to undergo all preparation steps. The intermediate storage permits to collect a sufficient quantity of material for the downstream preparation process. It improves the flexibility of the plant which is required for the economically sound processing of different kinds of scrap.

Some material requires mechanical processing to increase the content of aluminium prior to the metallurgical process. This is comparable to processes to enrich minerals which also commonly have to undergo a preparation process to enrich the valuable content of the minerals. Sometimes this process is very complex, as in the case of bauxite, to get the aluminium ore that can be processed to obtain metallic aluminium. The processes of the scrap preparation are not that complex but they are still efficient and generally economically sound. This refers mainly to dross preparation. This material usually has a high share of the raw material mix in a secondary aluminium smelter. Its treatment is often very poor and when the material reaches the melting plant finally its aluminium content has reduced to a low percentage. Therefore, we will look in more detail at the preparation process of this valuable raw material.

Material heavily contaminated by organics sometimes reduces the efficiency of a melting operation. Removal of these contaminations is mostly advisable to arrive at a profitable processing of this material. The same is valid for coated aluminium comprising mainly of UBC which is painted and coated with plastic material, too. If sufficient quantity of this kind of scrap is to be processed, the removal of this coating is advisable. The production line is similar to the process for removal of oil as it is the case for swarf originating from machining.

7.1 Shredder process line

Scrap is reclaimed from the storage area by means of a front loader with grab and placed on a sturdy apron conveyor for feeding the shredder. **Fig. 7.1** shows the principle design of a shredder plant. The shredder crushes the scrap to small pieces of different sizes according to the characteristics of the material. Dust and small non-metallic components will be picked up by the air flow generated by the dedusting blower. These small particles will be separated by means of a cyclone arranged ahead of the de-dusting unit. This may be a dry or wet scrubber. Since the scrap charged to the shredder may be wet or contain a certain amount of oil, a wet scrubber plant is preferred in most cases.

Due to the crushing and compacting energy applied in the shredder, the material is heated up to a temperature that may reach 300 °C. This causes some pyrolysis of organic components. This effect may become critical since the reactive gaseous organics could result in spontaneous ignition and cause an explosion. Therefore, the shredder, but at least the scrubbing system, should be arranged in the open. Noise generated by the shredder could also be a problem and the level certainly exceeds the limits accepted by environmental regulations. A noise abatement system is therefore a standard for shredder plants in the aluminium industry. The material comminuted in the shredder is discharged to a belt conveyor arranged underneath the unit. A magnetic separator is integrated in the deflection drum of the belt conveyor. It removes the iron parts from the bulk of shredded material and passes them to a discharge conveyor. Large pieces of iron are finally removed by an

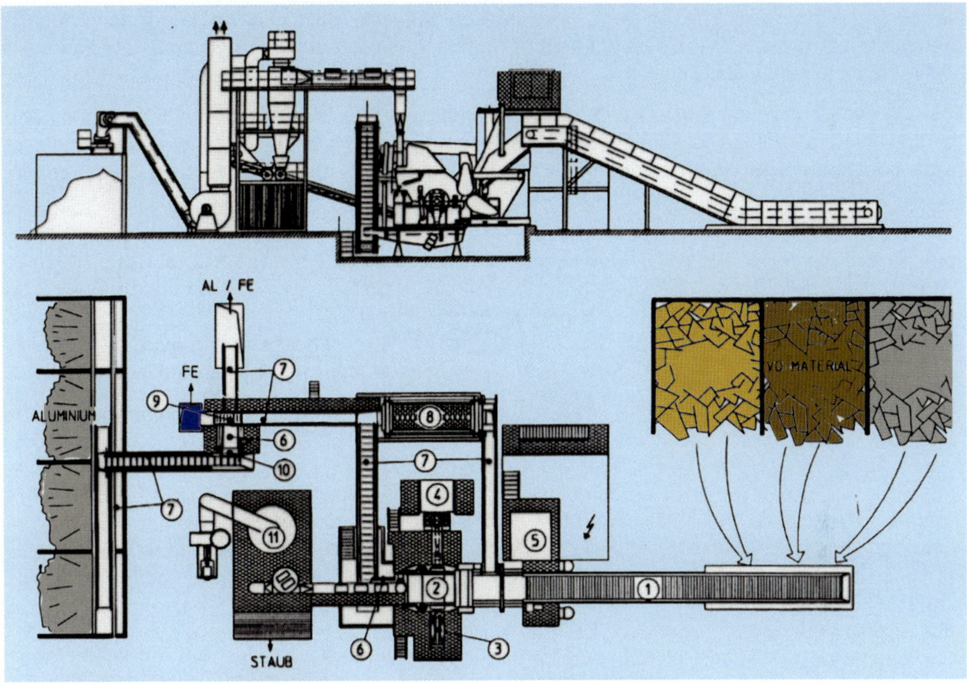

1 - feeding conveyor, 2 - shredder, 3 - drive motor, 4 - hammer dismantling unit, 5 - operator´s cabin, 6 - waste gas pipe, 7 - conveyors, 8 - screening drum, 9 - iron separator, 10 - discharge chute, 11 - waste gas scrubber

Fig. 7.1: Shredder plant layout

overhead belt magnetic separator and will be collected in a container. Stainless steel parts are not very magnetic. They can be removed by means of a strong field magnetic separator as secondary iron removal station. Its effectiveness is improved by the fact that the stainless steel parts are slightly magnetized during shredding. A dust removal drum may be arranged in the conveyor line. The drum is swept by an air flow that picks up dust and small non-metallic components.

The shredded material that is now free of iron can now be classified into different fractions by a screening drum. This is required only if the different fractions are to be sold to specific customers. For direct processing this classification is not required since all fractions can go directly to the rotary drum process.

Sometimes the scrap is heavily contaminated by non-metallic components. The different fractions, as obtained by the screening drum, can be passed over eddy current separators. The result is fairly clean shredder scrap that can be processed without problems.

Additional process steps are required if the metallic contaminations are not only iron but also heavy metals such as copper, lead and zinc. But this is done before the scrap is classified into different fractions. The material first passes a heavy media separation for wood and plastic followed by an eddy current separator or a series of parallel-installed eddy current separators. This will remove almost all plastic and other non-metallic components. The material flow is now directed to a heavy media separation. The result is clean aluminium shredder scrap that can be sent to metallurgical processing.

Not all of the process steps will be realized in every shredder plant. But the importance of such processing will grow with the development of more sophisticated compound materials and more and more increased energy cost.

Fig. 7.2 shows a typical process line for collected scrap. After shredding, iron is separated and the iron-free material is sent to a vibrating screen or a screening drum. Material < 50 mm is sent to the eddy current separator and can now be stored and finally charged to a rotary drum furnace.

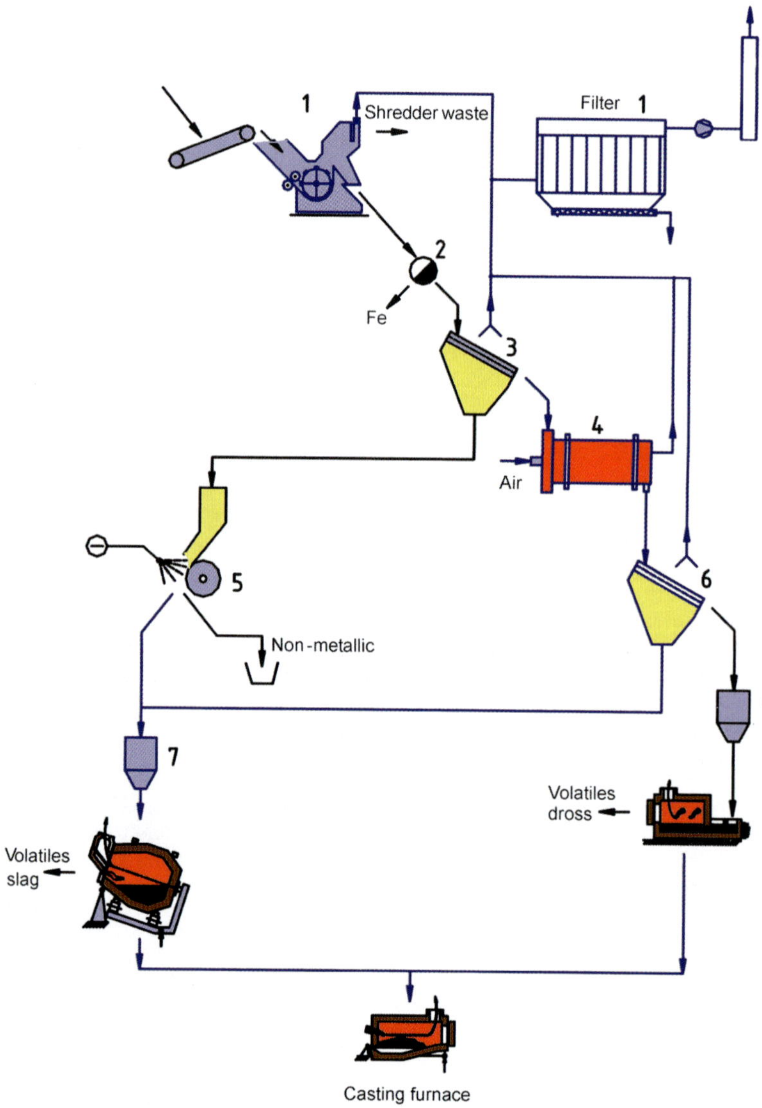

1 - shredder, 2 - magnetic separator, 3 - screen , 4 - separating drum, 5 - eddy-current separator, 6 - screen, 7 - storage bin

Fig. 7.2: Process flow sheet for collected scrap

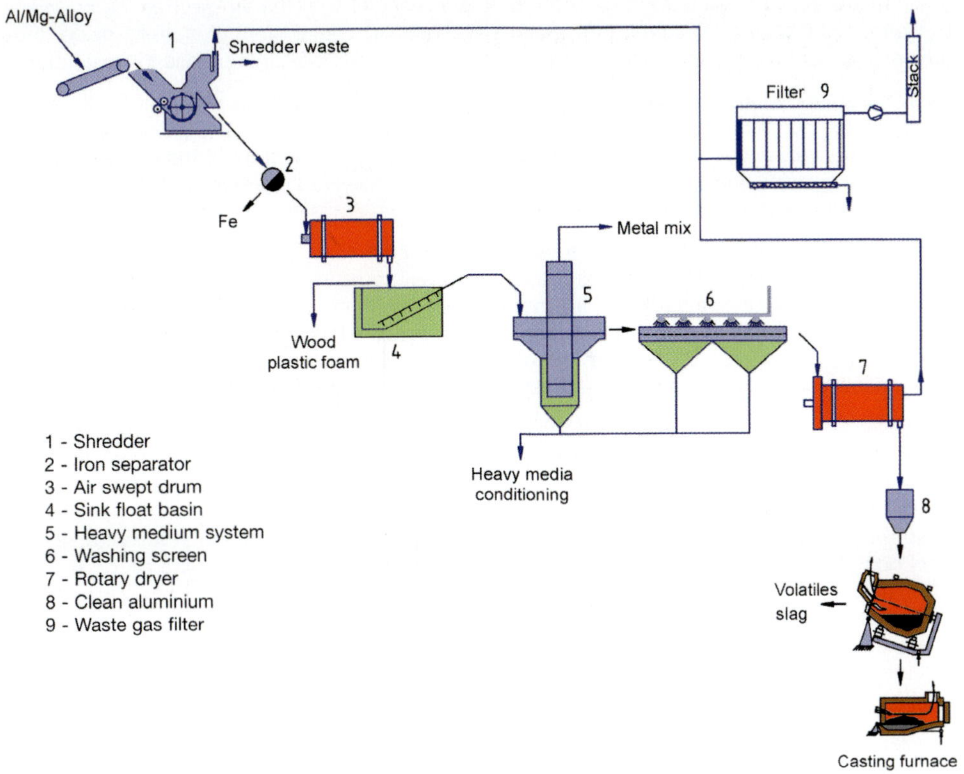

Fig. 7.3: Process flow sheet with sink float separation

The screen overflow is directed to an air-swept separation drum where light material (paper, plastic) and dust is removed. The scrap will pass through a second screen. The mold flow is directed to the eddy-current separator as well while the larger lumps of metal, which consist generally of aluminium, can be charged to a twin-chamber furnace.

A different production line is shown in **Fig. 7.3**. After passing the shredder and the iron separation, the iron-free material flows into an air-swept drum separator to remove the light portion of the non-metallic fraction. Larger pieces of this material (wood, plastic foam) are separated on the sink float basin. The remaining material still contains, additionally to the aluminium, heavy metal such as copper, bronze and the like. These components are separated in the heavy media separation drum. The overflow is sent to a washing screen and finally to a drum dryer. The aluminium is now ready to be processed in the metallurgical plant.

It should be noted that the aluminium is free of non-aluminium components after passing mechanical preparation. However, it is still a mixture of different aluminium alloys. Therefore, sampling to obtain a clear picture of the alloy composition in still indispensible.

7.2 Dross treatment

Dross is a mixture of aluminium and aluminium-oxide. It is generated in reverbaratory furnaces and is removed from the furnace by skimming. With that skimming action a quantity of aluminium is skimmed off as well, representing a final content of metal on the total dross which can reach

70 %. This aluminium will oxidize as soon as it is in contact with the ambient air. Since this is an exothermic process, the metal trapped in the dross will convert into aluminium oxide, thus representing a substantial loss. The operators use different methods to suppress this oxidation.

The simplest procedure is to spread the dross on the operating floor. This means it cools rapidly to ambient temperature which means below the igniting temperature and oxidation stops. This is certainly not a very convenient method since space is required in the front of the furnace and the method will generate some smoke and fumes. The dross collected in dross containers arranged underneath the furnace door could be transported to an area outside and spread out there.

If cooling is very effective, no special treatment is required since the aluminium content in the material is exceeding 60 – 70 %. In most cases, however, operating plants are not very effective in dross cooling. Thus, quite a portion of the metal trapped and the skimming is oxidized and lost for recycling. However, this dross is still a valuable raw material. It has an aluminium content of 30 % in the worst case and 40 – 50 % as an average. If the recycling plant focuses on dross recycling, it is economically feasible to enrich the dross to a content of more than 70 %.

To recover the metal in the rotary drum furnace the entire batch including the salt has to be brought to melting temperature. This also means that the oxide content of the dross must be heated up as well as the metal. Since heat transfer conditions for both components are more or less identical, the heating-up of that material, which means non-metallic material, will result in the subsequent additional input of energy. The specific energy consumption in a well-designed rotary drum furnace is approximately 600 kWh/t of material charge. This refers to aluminium and dross. Thus, if we have 1 t of dross having an aluminium content of 80 %, the energy requirement recalculated for metallic aluminum will be 600 divided by 0.8, equal to approximately 740 kWh. If the aluminium content in the dross is only 40 %, the resulting energy consumption per t of metallic aluminium will be 600 divided by 0.4 equal to 1,500 kWh. This is quite a difference, thus the

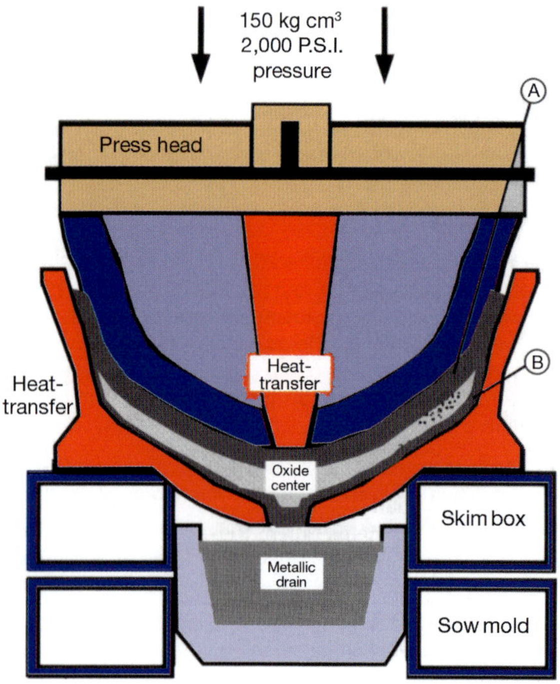

Fig. 7.4: ALTEK dross press

enrichment of the aluminium dross is apparently very economically feasible, particularly for dross with lower aluminium content.

A good and simple solution is to use a hydraulic press (**Fig 7.4**). The dross is placed into a pan having holes on the bottom. A hydraulic stamp now forces the metal out of the dross and the aluminium is collected in the mold underneath this press container. The remaining dross is solidified into a large block with a metal content of less than 40 %. This block can be charged directly into the rotary drum furnace.

Another very efficient system is a dross cooler (**Fig. 7.5**) It requires, however, a certain production capacity. With this technology dross is fed by a vibration conveyor or a chute into a rotating drum. The shell of the drum is cooled by water-spray. The material is transported through the drum by a helical screw and reaches the screening section of the drum. This section is designed as a rotation screening drum. It forms one unit with the cooling section. Generally, three fractions of the dross are obtained which can then be used for different purposes. The system can be connected to a de-dusting system. The dross obtained is very valuable raw material for secondary recycling. Unlike the dross press, all metal remains in the dross but is cooled to stop the exothermic reaction. The resulting dross is very rich with regard to the aluminium content.

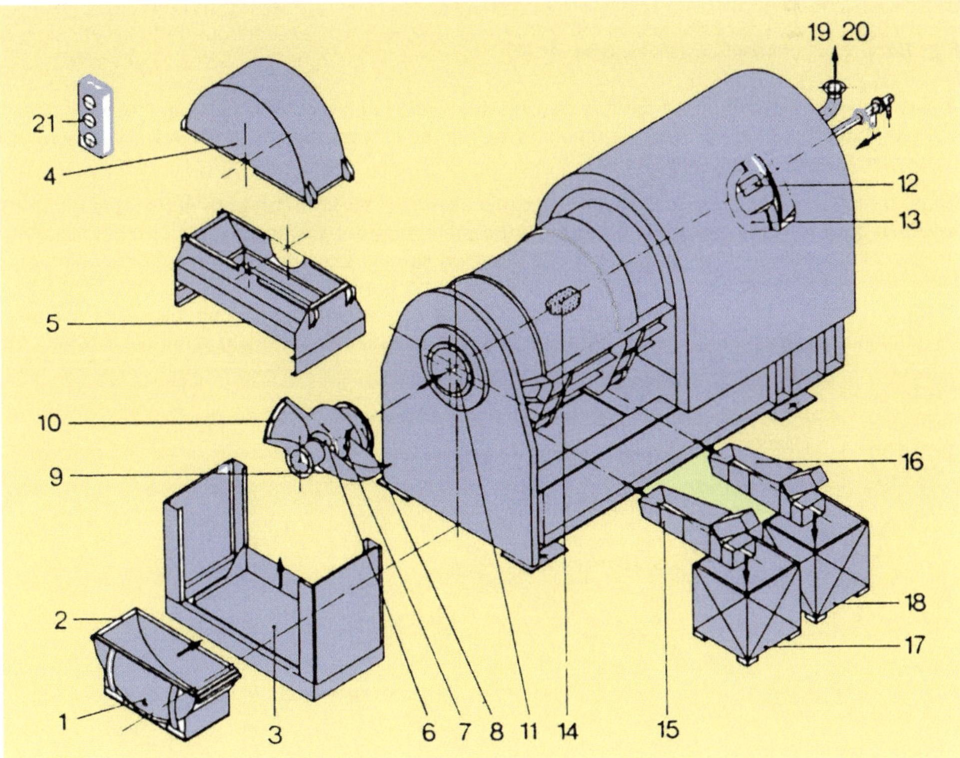

1 + 2 - dross pan, 3, 4, 5 - charging section, 6, 7, 8, 9, 10 - feed screen, 11 - charging opening, 12 - burner, 13 - rotary drum, 14 - screening wall, 15 - discharge conveyor for fines, 1
6 - discharge conveyor for coarse particles, 17 + 18 - dross containers, 19 - air feed, 20 - water feed, 21 - push-button station

Fig. 7.5: Dross cooler

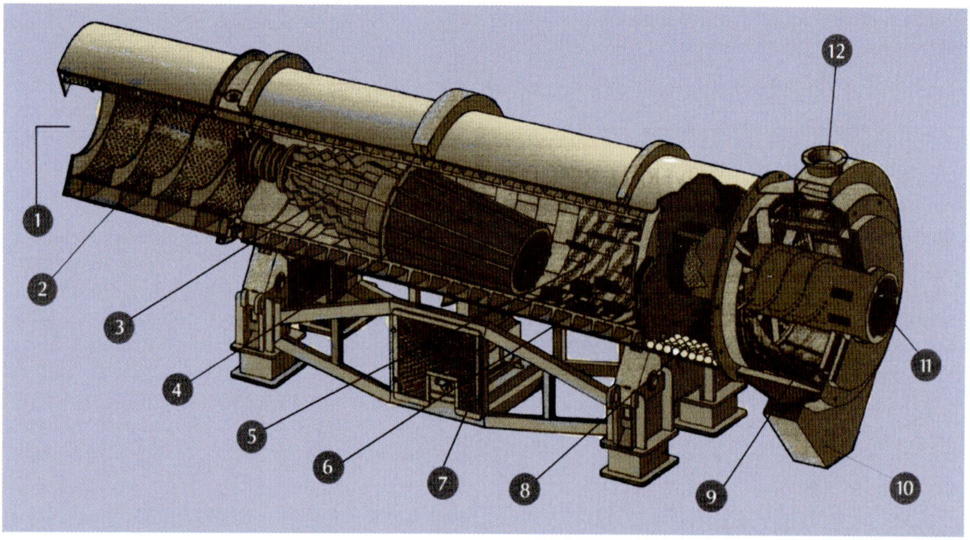

Fig. 7.6: Dross treatment system (Source: ALTEK)

A very interesting system uses a rotary cooling drum with integrated ball mill (**Fig. 7.6**). This unit is able to enrich the dross process additionally but its installation requires sufficient dross quantity to pay for the higher investment cost.

Thus, the method for dross treatment, after it is skimmed off from the bath surface, is either to squeeze out the liquid metal, thus reducing the metal content in the dross, or to cool it rapidly to stop exothermic reaction with some classification already executed in the cooling equipment. But this is done in very few cases. Therefore, the dross supplied to the refining plant has a low metal content. To enrich the dross for more efficient melting, the material should be processed, i.e. mechanically prepared. The principal design of such a preparation plant is shown in (**Fig. 7.7**).

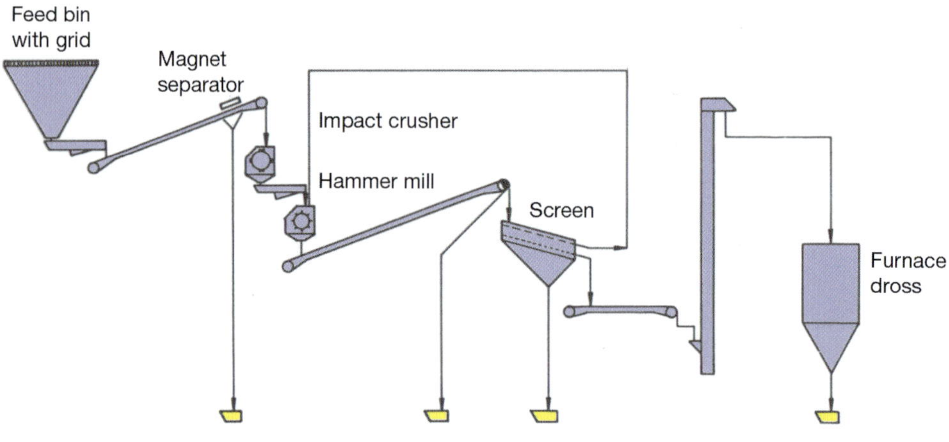

1 - Feed, 2 - Feeding conveyer, 3 - Hammer mill, 4 - Vibrating screen, 5 - Ball mill, 6 - Product bin - 50 mm, 7 - Bucket elevator, 8 - Storage bin + 50 mm

Fig. 7.7: Flow sheet of a dross preparation line

The aluminium-dross material passes through different stages of screening, crushing and grinding. The result will be different fractions of the dross which can be recycled or sold for other purposes, such as de-ox material for the steel industry. Dross is charged to the preparation plant via a grid having a mesh size of 200 mm. Lumps remaining on that grid will be removed and must be crushed manually to obtain smaller sizes. The material passing the grid is collected in a small feed bin and discharged from there by means of a vibration-trough conveyor to the first stage crusher which generally is a hammer mill.

Crushed material is discharged via a vibrating-trough conveyor and sent either by a bucket elevator or inclined belt conveyor to a vibration screen. This screen is equipped with two decks of different mesh size. Aluminium obtained from mechanical preparation is classified according to its size. Pellet size no. 1 is material plus 50 mm, pellet size no. 2 is material minus 50 mm. Material plus 50 mm is separated from the balance of the material flow. It is fed to a belt conveyor equipped with a magnetic drum for separating iron. From there it is fed to the storage for the pellet size no. 1. The under-flow of the screen having a grain size distribution minus 50 mm is sent to a ball mill. This process has the advantage that dross particles are crushed without destroying the aluminium grain.

This process actually frees the aluminium from the crushed oxides and also opens the coarse dross particles to get to the aluminium trapped into that very particle. The mill can be designed as air-swept ball mill which means that an air flow is passed through the mill removing all dust generated. The mill product is enriched dross with a pellet size no. 2. The dust removed cannot be utilized in recycling. It may still contain some metallic component and can be used in the steel industry and in steel cast iron foundries as casting powder. The dust is collected in a bag house. All intermediate conveyors, this means those arranged between screening and crushing, are equipped with magnetic drums to remove all iron parts from the dross. In some cases it may be convenient to have a second screen prior to the ball mill. This is particularly the case if very poor dross is processed. The under-flow of this screen is called the dross-residue and could also be used, as mentioned before, in the steel or foundry industry. The remaining dust separated and discharged from the ball mill has very little metal content and must be deposited.

Dross pellets 1 and 2 are melted in rotary drum furnaces. They usually have an aluminium content of 70 – 80 % or sometimes more and represent a valuable material that can be processed with high efficiency.

7.3 De-coating

UBC returned for recycling are painted and coated with some protective organic material. With the high percentage of these non-aluminium components, the melting efficiency in a furnace will be reduced. It is, therefore, of advantage to have de-coating processing ahead of the melting stage. UBC are usually compacted and then bailed. To get access to the paint and the coating, the material must be crushed to free all surfaces. This results, however, in a material with a very low bulk density. This in turn requires quite complicated handling.

The de-coating equipment should, consequently, be arranged very close to the melting facility. Machining swarf is material that requires similar treatment. It is mostly contaminated with cutting oil or other chemicals that are used during machining of metal. To remove this contamination the swarf must undergo treatment similar to the de-coating process. In both cases organics must be removed by pyrolysis and then burned in an incinerator due to environmental reasons and also to use the energy of the organics for the process whenever possible. Paint adhering to the UBC scrap comprises metallic oxides to obtain the color and an organic binder. The paint particles will either remain on either surface or leave the process as dust. To get a clean surface, a grinding effect must be applied to remove the metallic oxides settled on the surface of the material. A similar effect can be noted when de-coating swarf. This is done by pyrolysis, only char (black

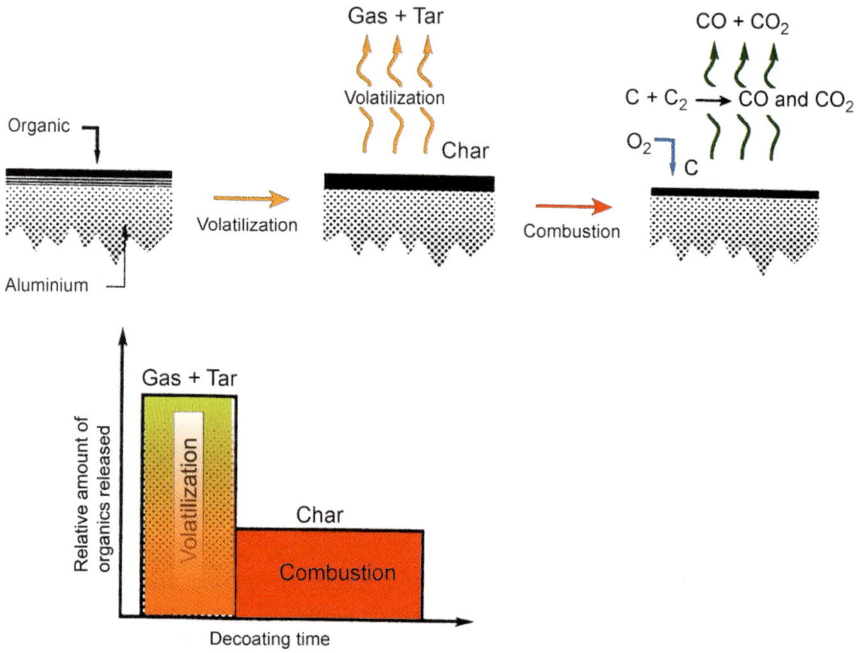

Fig. 7.8: Mechanism of de-coating

carbon) may still adhere to the surface. Most of the organics are removed mainly at the beginning of de-coating by volatilization (**Fig. 7.8**). Following the volatilization step the layer of residual char remains attached to the furnace of the aluminium pieces. This is subsequently removed by oxidation within the de-coater and by the grinding effect, as mentioned above.

In order to initiate oxidation, specific air or oxygen requirements must be met for the required de-coating to occur. However, no matter how the process is controlled, it must be ensured that the temperature of the metal never exceeds 350 to maximum 400 °C. Above this temperature, oxidation of metal will sharply increase thus resulting in additional metal loss. There is quite a variation of scrap material that can be treated in a de-coating process. However, all of it has to be shredded prior to the de-coating to make it treatable in the process and to get access to the entire surface area. (**Table 7.1**) gives an overview about the different scrap materials and their organic content. The question is, however, how some of these materials can be processed in a melting plant. But this is a different subject.

7.3.1 Rotary drum de-coater

The rotary drum de-coater is the most common equipment used in the aluminium industry. It basically comprises a long cylindrical drum rotating at low speed. Material is fed at one end of the drum and discharged at the opposite end. The required energy for de-coating or drying is introduced by means of gas or oil burners at one end of the drum. Fresh air to burn the organic combustibles is usually introduced at the exit end of the drum thus cooling the material that has been passed through the heating and de-coating zone of the furnace. The flue gas leaves at the charging side and is directed to an incinerator and, after cooling, to a scrubbing system. The

Table 7.1: Organic content of different kinds of aluminium scrap

Scrap	Organic content %
Painted siding	4
Mixed low copper clips	5
Extrusions with thermal break inserts	5-10
Mixed alloy old sheet	5
Foam filled extrusion	23
Bottle caps with plastic liners	25
Lithographic sheet with plastic contamination	30
Plastic-filled sandwich panels	55
Yogurt lids and equivalent	20
Heavily contaminated used beverage cans	20
Plastic toothpaste tubes	70
Cable wrap	20
Food pouch	85

principal design shall be explained by an example for a chip-drying plant. As mentioned before, the principal design of a de-coating unit for UBC or other contaminated shredded material is very much the same but may differ in some details.

The plant is designed for drying of swarf for machining with oil as well as high moisture content. Furthermore, de-lacquering (de-coating) of UBC is usually also intended. The swarf is supplied as bulk material, UBC as baled or as bulk material as well. While the swarf can be charged directly to the drying plant, bulk UBC as well as bales have to be shredded prior to the de-lacquering process. The required shredder is integrated in the production line.

The process is based on a technology of indirect heating of the drying drum. This avoids that local superheating occurs which would result in an increased oxidation of the swarf and, consequently, additional metal loss. As mentioned before, it is essential to provide fresh air which is cooling the material after the drying section and, additionally, provides transport air for the generated oil and water vapor which are then directed to the incinerator. The pyrolytic removal of organics has, however, some limitations. Free carbon in the format of char and tar will remain on the surface of the swarf.

Therefore, it is of advantage to start the incineration process already in the drum. This is initiated by a pilot burner at the drum entrance.

For the following description, please refer to **Fig. 7.9**. It is the intention to use the energy of the oil removed to improve the heat efficiency of the process. In case wet swarf with only a little oil content is to be processed, some oil will be added at the feeding vibrating conveyor (1.07) to maintain the smoldering process inside the drum. At the other end, if swarf is to be processed that has an oil content which is too high, water can be added at the feed vibrating conveyor (1.07) in order to slow down the combustion in the drum.

Swarf and crushed UBC will be fed to the plant via the feeding screen (1.03). The material reaches the required charging height via the belt conveyor (1.04) and the bucket elevator (1.05). A surge bin (1.06) compensates for fluctuations in the feed of swarf. Via the discharge conveyor (1.07) and the chute (1.09), the material is charged to the drum (2.01). The double-pendulum flap (1.08) seals the

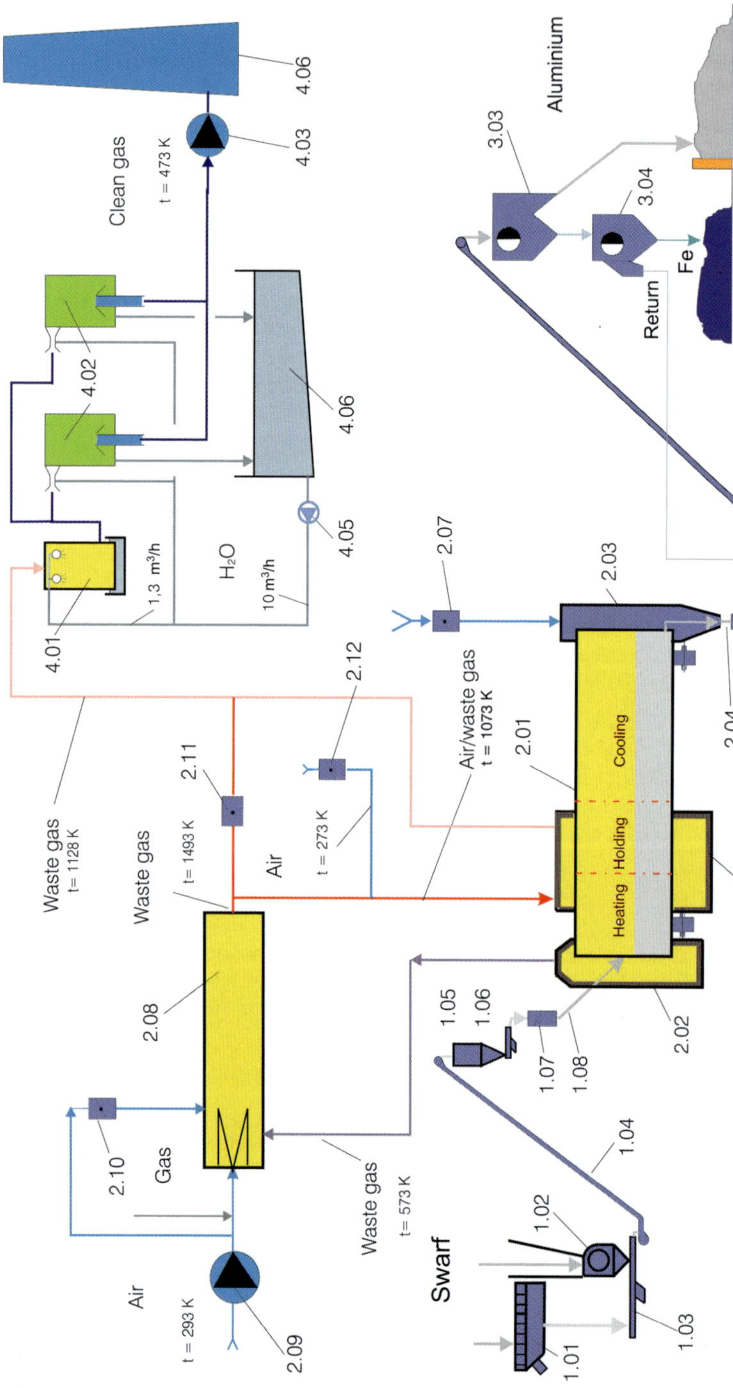

Fig. 7.9: Flow sheet of a rotary drum de-coater

material feed against the atmosphere within the drum. The first section of the drum (2.01) is heated indirectly by the combustion products of the incinerator (2.08). The required heat exchanger (2.06) is designed to improve the heat exchange by creating intensive turbulence. In the first stage, the material is heated up and already reaches a temperature of close to 373 K which causes the water content to evaporate. It is necessary that the material temperature of 573 K is obtained since only then will all oil components evaporate. Evaporation of the light volatiles already starts at the temperature of 240 K, as well as the material travels down the drum it reaches the holding zone, where a complete temperature homogenization is achieved. Carbon, that still may be attached in the format of char, will be removed by the grinding effect of the motion of the individual particles. The resulting cloud of carbon particles will be carried out of a furnace by the already pre-heated fresh air. In order to improve the heat exchange between cooler particles from the inside of the bulk with hotter particles at the outside as well as between the bulk and the drum, the drum inside is equipped with sloped pedals. This also ensures the controlled transport of the bulk along the furnace inside. The remaining length of the drum is used for the cooling of the material. During this phase the temperature of the fresh air introduced will be increased to drying temperature. The volume of the fresh air will be adjusted by the control flap (2.07).

Via the discharge-hood (2.03), the double pendulum flap (2.05) and the vibrating conveyor (3.01) the material is discharged via the chute (3.05) to the bucket elevator (3. 02). By means of the double magnetic separator system (3.03 and 3.04), steel swarf and aluminum swarf will be separated. The different kinds of material can now be stored in boxes or containers. Depending on the elevation of the plant, the aluminium swarf leaving the second separation stage can be lifted by means of the inclined belt conveyor (3.02) or can be collected in a container as well. The Pyrolytic Products are directed via the flue-hood (2.02) to the incinerator (2.08) for post-combustion.

The temperature of the total quantity of the pyrolytic products must be increased to 1,173 K (900 °C) before the combustion of the organic components and to crack furans and dioxins (PCDD / PCDF). For the treatment of PCDD / PCDF, it is actually recommended to hold the temperature lifted to a higher value in order to compensate for heat losses during different process steps and on the route to the quenching chamber.

The energy required for post-combustion will be provided by a separate gas-burner (2.09). The blower for this burner is designed to provide also the combustion air for the oxidation of the organic components in the waste gas. For details of this type of incinerator, please refer to section 3 of this book. After leaving the incinerator, part of the combustion products will be directed via the gate valve (2.11) to the heat exchanger for the drum. The temperature of this partial flow will be adjusted via the gate valve (2.12) by means of adding fresh air. This is required to obtain a reasonable operating temperature for the drum. Since experience has shown that the waste gas leaving the system is heavily contaminated by dust, a cyclone (2.13) is arranged ahead of the exchanger to remove an already substantial amount of dust to avoid too much dust collection in the heat exchanger (2.06). The partial flow will finally be mixed into the main waste gas flow in front of the scrubbing system. For final cooling, prior to entering the waste gas filter, the gas is quenched to a temperature below 473 °C in a quenching chamber with the aid of water spray. Details of such a quench system are also described in section 3 of this book.

7.3.2 Fluid bed de-coater

Crushed UBC, as well as small pieces of aluminum swarf, can be easily transported by air or gas. This is applied in the fluid bed de-coater which is based on a diluted face fluid bed (**Fig. 7.10**). In the rotary drum the heat exchanges within the bulk material and with the drum walls, depending on the internal heat exchange between the individual particles. Even when the bulk is mixed very well, thus exposing its surface to the furnace surroundings, there is somewhat low efficiency of the heat exchange. This is different in a diluted phase fluid bed. The hot combustion products flow around each particle so that there is excellent heat transfer from high velocity gas as long

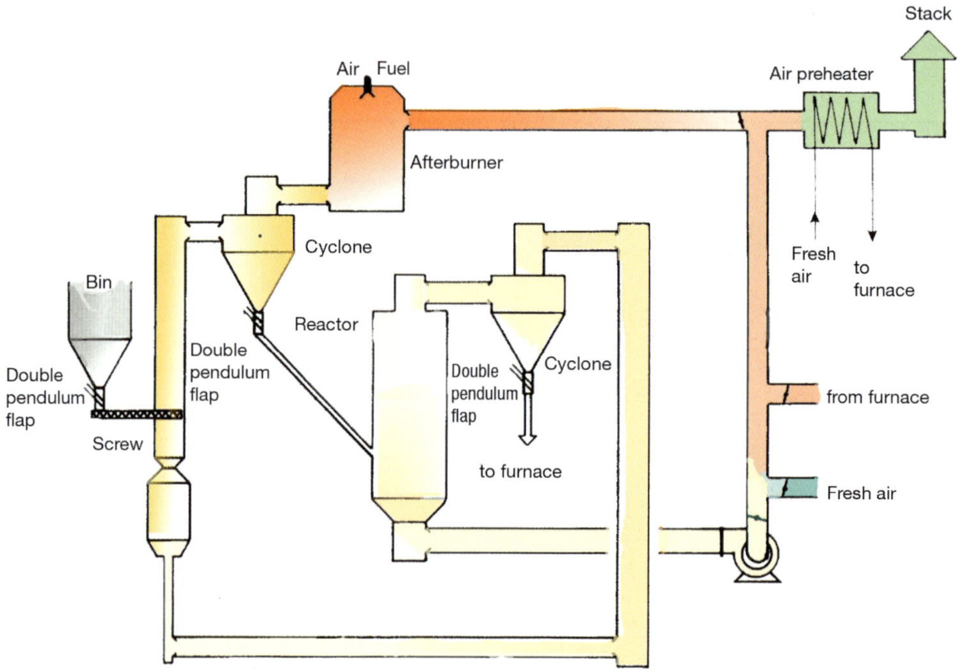

Fig. 7.10: Principle flow sheet of a diluted fluid bed dryer

as the velocity difference is high enough and solid material with the good heat conduction within that solid material.

The fluid bed de-coater has as only moving part: the air blower and maybe some motors for gates and flaps. It therefore requires very low maintenance and combines low space requirement with high capacity. Similar to the system of the rotary drum de-coater, material is supplied via a surge-bin and double pendulum-flap. It is now transported into the suspension dryer for pre-heating. This dryer has a comparatively low cross-section in order to obtain a high gas velocity. Below the feed point the diameter of the shaft is reduced. The resulting very high gas velocity assures that no swarf can fall below this point during charging. Due to the contact of hot gasses and material, the particles are pre-heated. Gas and solid material are separated in a cyclone. The discharged material particles are directed via a chute in the suspension reactor. This unit is a steel vessel which receives hot combustion products from the bottom.

The material, getting into that reactor by means of the chute, is picked up immediately and will be carried by the hot gasses to the exit of the vessel. Due to the intensive turbulence in vertical direction, an intensive contact between hot gases and material particles is assured. The particle-gas-mix leaves the reactor to reach a second cyclone. The material, which is now completely de-coated and dry and free of any foreign matter, is discharged from the cyclone by means of a double pendulum-flap. Since it has a temperature of approximately 350 °C it can be charged to a twin-chamber furnace with a rotating pump immediately to make use of the heat of that material. The gas leaving the cyclone is directed to the very first stage of the feed dryer with the venturi system. The gas leaving the first cyclone reaches the incinerator where the organic components are burned, as described before. Parts of the hot gases are recycled to the top of the reactor. Their flow can be controlled by the hot air slide gate. The fractions of the hot gases are now released to the open and pass the first recuperator to heat up combustion air either for the incinerator or for the melting furnace. To improve heat efficiency, hot combustion products leaving the melting furnace can also be used as hot gas supply to the system. Fresh air has to be added to provide some cooling since the combustion products, leaving the furnace as well as leaving the incinerators, are too hot for the treatment of the small metal particles. At the same time, the fresh air provides the oxygen or at least a part of the oxygen required for the combustion of the organic contamination adhering to the charged metal.

Conventional thermal de-coaters for aluminium have been designed to operate with scrap having a level of organic material lower than about 10 % per weight. The main practical application is the recycling of used beverage cans, where the level of organic material does not exceed 8 %. Fluid bed technology could assist to have material de-coated to free it from other plastic compound material up to an average content of 50 %. This is possible by a fluidized bed operation (**Fig. 7.11**).

This is made up of small particles (the fluidization medium) maintained in a suspension in an upward-directed gas flow; when fluidized it assumes the properties of a boiling liquid. When immersed in a fluidized bed, aluminium having a greater density than the fluidized bed sinks to the bottom. The bed is heated with a freely determined temperature. Organic components attached to the scrap decompose on contact with both the fluidization medium and the stream of air which is being used to fluidize the bed. The temperature is chosen to ensure complete decomposition of organic components while avoiding melting of any of the various alloys being processed. The purpose of the medium is to provide both efficient transfer of heat to the scrap to promote rapid de-coating and to dissipate the heat released from organic components when they combust.

For a practical application the material is supplied from a surge bin via double pendulum-flaps to a vibrating conveyor which charges the scrap into a drum. The shell of this drum is perforated with an internal screw (de-coater drum). The material passes though the de-coating drum which operates in a fluidized bed generated by the fluidizing media via nozzles in the bottom of a large housing. The upper part of the housing is designed as thermal oxidation zone. At the exit of the bed, the diameter of the drum is reduced and acts for primary separation of the media from the scrap. After discharge from the de-coater through the double pendulum-flap, a secondary separation of media from the scrap is completed with a multi-deck vibrating screen. Heat and oxygen for the removal of moisture and organics from the scrap are provided by the injection of compressed air and natural gas into the bed through the fluidizing air tubes. Additional heat is provided by the oxidation of the organics in the bed and heat transferred from the free board area above the drum (thermal oxidation zone). To reduce hydrocarbon and CO-emissions and to ensure ignition of combustible gases, the oxidation zone utilizes a series of pilot and high velocity burners. Due to the presence of acid gases and particulate emission in the waste gas stream, a gas treatment system is provided. The exiting gas temperature is reduced by the injection of water in a quenching chamber. A dry neutralizing agent is injected as additive to the filter air. The final stage is a bag house.

The unit is operated very efficiently with a mixture of different types of scrap, i. e. a certain portion of scrap with a low organic content and scrap with a high organic content. It is noticed that the de-coating efficiency is much better when using this mix.

Apparently this type of equipment is very useful with regard to larger particle size but may not be very efficient if very small aluminium pieces, such as shredded UBC and swarf, have to be processed.

7.4 Slag processing

When melting aluminium alloys during the refining process of the secondary aluminium industry, slag is generated having a high content of salt as a mixture of NaCl and KCl. Since salt is soluble in water this slag is highly hazardous to the environment on a landfill and this material may contami-

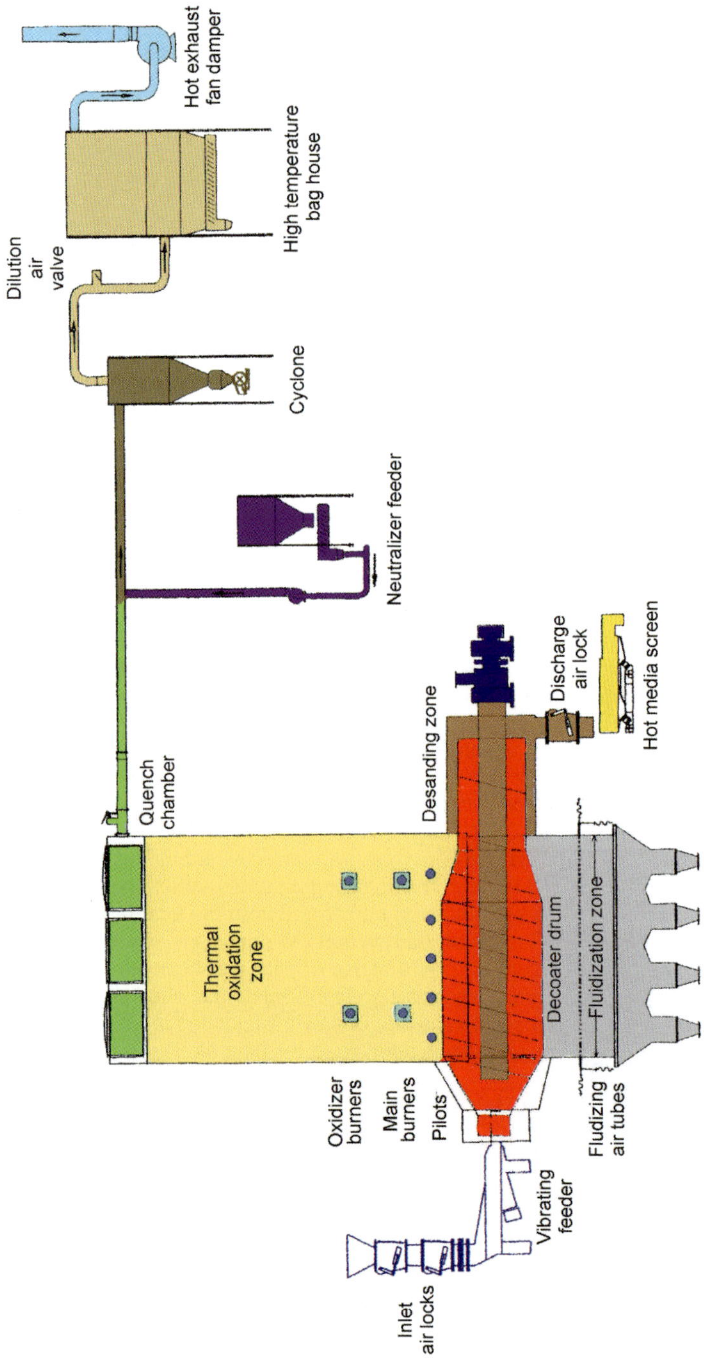

Fig. 7.11: Fluidized bed de-cooler (Source: Alcan)

nate the groundwater. Generally, authorities do not permit dumping on open stockpiles. Consequently, the material needs to be processed with the target to recycle the salt for use in the melting process. Slag also contains a certain amount of metallic aluminium that can also be recovered and finally sold. The remaining residue mainly consists of aluminium oxide that can be deposited and traded to the cement industry or to the refractory industry.

Generally, the process is performed in large plants which are located centrally for easy access during melting operations. However, transport becomes more and more a remarkable cost factor due to increasing fuel cost. Also, in countries having a wide spread distribution of operating melting plants, supply of slag to a central processing unit is not very attractive, if not inacceptable. In order to serve the individual melting plants it is obviously of advantage to install smaller slag preparation units customized for the production rate of the aluminium production. However, some constrictions due to size of the preparation plant have to be accepted. This must not refer only to the remaining salt content of the residue, that can be tolerated, since small plants as well have to fulfil the conditions of the environmental regulations. It also refers to the values for water drainage and clean air. Consequently, the water circuit has to be designed for release to the environment. All water used in the process for leaching has to be recycled in a closed loop within the process. Only the comparatively little quantity lost to the moisture content of the product and evaporation has to be replaced by fresh water. Dust generated during the mechanical comminution and screening of the incoming slag will first pass a filter plant and finally be recycled to the process. The leaching process produces some gases, mainly hydrogen and methane. These will be directed to a post-combustion. The heat generated will comfortably be used to improve the energy balance of the process. **Fig. 7.12** shows a block diagram of the process with indication of the different material flows based on a slag throughput of 4 t/h.

7.4.1 Characteristics of slag

In general, the composition of the various slags is quite similar and follows a defined base pattern. For details refer to Section 1.3 "Melting Additives". Depending on the type of scrap processed, substantial deviation from this pattern may occur, however. Mostly the components are NaCl, KCl and Al_2O_3. Other additions are originating from components of the scrap processed, such as painting pigments, composite materials, coatings or just dirt. For improving the metal recovery during the melting process NaF, KF or CaF_2 are used in the order of 5 – 10 % which also do have to a certain extent some impact on the characteristic of the salt slag. Additives to the salt applied are generally natural products which may carry some contaminations mostly as sulfides. For details of the slag contents refer to Section 1.3.1.6, Fig.1.1.

The portions of salt and oxides in particular depend very much on the material to be processed since scrap varies with regard to metal content, contaminations and salt mixture applied. Important factors for the equipment selection are the aggressive salt and the very abrasive aluminium oxide. Depending on the melting process, the slag will contain 5 – 8 % of aluminium metal.

Salt and aluminium are recycled. Apart from the economical features the effect on the environment is to be considered.

All components of the slag react already during melting but to a higher extent during slag preparation

For the plant design it is important to consider that the slag reacts very aggressively due to the high salt content while the content of aluminium oxide, particularly the α - Al_2O_3 (corundum), is very abrasive. However, most of the organic components of the scrap are removed within the melting process. During the slag preparation very small particles of aluminium react with the solution resulting in a certain quantity of H_2, CH_4 and NH_3 which will appear in the waste gas.

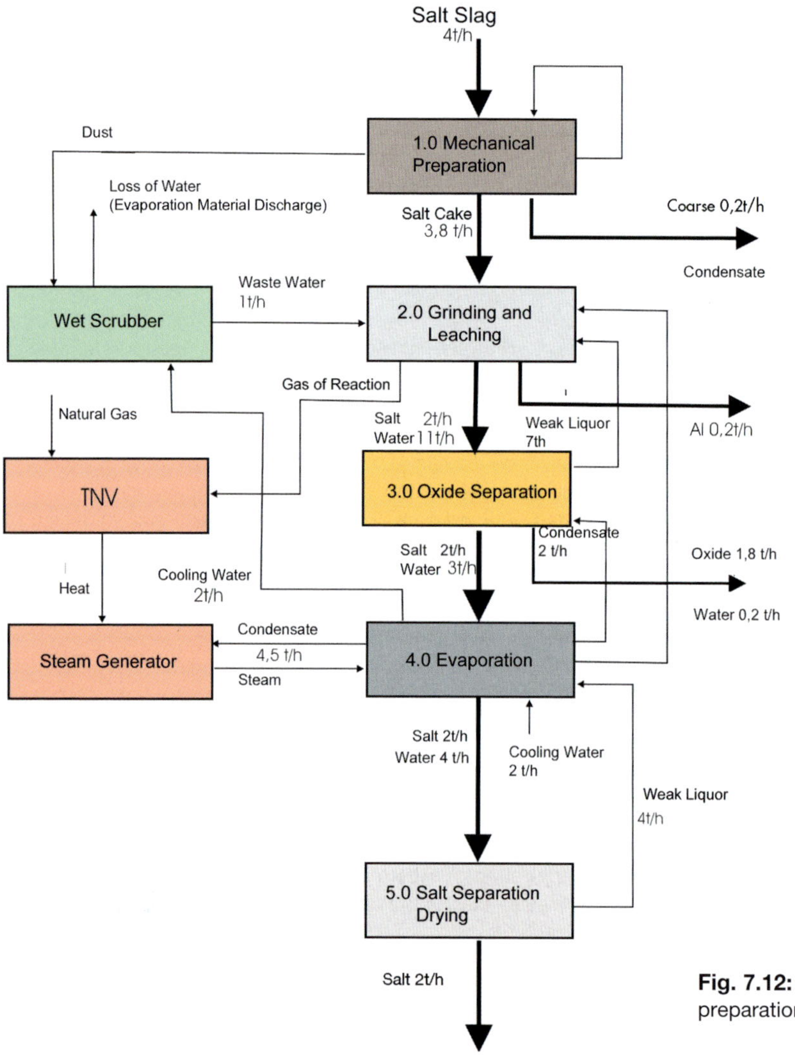

Fig. 7.12: Flow sheet slag preparation

7.4.2 Plant capacity

In Western Europe, slag recycling plants are operating in central areas. Aluminium plants deliver their slag to those plants and get salt in return. However, the client has to pay for the disposing of the scrap and finally for the salt. The aluminium recovered is sold by the recycling plant. Considering these cost items and further considering the steadily increasing transport cost due to high gasoline prices, the system becomes rapidly inefficient for the aluminium melting plant. Furthermore, if long distances are involved the logistics are difficult.

Therefore, it is obvious that a customized small capacity plant, meeting the requirements of a melting plant with the data listed in **Table 7.2**, offers advantages.

The capacity of a slag recycling plant adjusted to such a recycling plant should have an annual capacity of 25,000 t. This results in a throughput rate of 4 t/h based on the data in **Table 7.3**.

Table 7.2: Typical throughput data of a secondary aluminium plant

Slag preparation plant	
Annual throughput	23,000 t/a
Working days per annum	300 d/a
Availability	90 %
Working hours per day	24 h/d
Throughput per hour	3.85 chosen 4.00 t/h
Quantity of salt	2 t/h
Quantity of aluminium	0.2 t/h
Quantity of oxide	1.8 t/h

Table 7.3: Throughput data of a slag preparation plant based on an annual capacity of 23,000 t. p. a. of salt slag

Recovery (typical scrap mix)	Feed	(%)	Burn Off (%)	Aluminium (%) in slag	Total %	(t/a)
Chips and turnings	12,000 t/a	98	5	5	8.6	1,029
Shredder	12,000 t/a	95	3	6	9.7	1,159
Ingot/bales	8,000 t/a	96	2	5	7.2	577
Dross	12,000 t/a	56	2	8	51.9	6,230
total charge total aluminium	44,000 t/a 35,005 t/a					8,995
Salt consumption (typical scrap mix)	**Feed**	**Salt (kg/t)**	**Salt (t/a)**	**Slag (t/a)**		
Chips and turnings	12,000 t/a	180	2,160	3,789		
Shredder	12,000 t/a	150	1,800	3,319		
Ingot/bales	8,000 t/a	150	1,200	1,937		
Dross	12,000 t/a	75	900	7,370		
	44,000 t/a		**6,060**	**16,415**		

7.4.3 Environmental aspects

The most important reason to install a preparation plant is to obtain an environmentally acceptable refuse deposit. If stored in the open, the salt content of the slag will be washed out by rain and penetrate into the groundwater. Removing and recycling the salt content of the slag is therefore the major and most important process step. Aluminium contained in the slag represents no environmental hazard but may disturb the salt recovery.

Remaining waste are mainly environmentally neutral metal oxides. It may even be sold to the refractory industry. Water required is recycled to the process. Dust generated during the mechanical part of the preparation is collected at the various equipment and separated from the process gases in a filter unit to obtain clean air with a dust content of < 5mg/m^3 according to the environmental regulations. The dust is recycled to the process.

Some gaseous emissions result from the leaching process. They are directed to a combustion chamber. The hot combustion products are used in the process to the benefit of the total energy balance. The final waste gases are collected and directed to the filter plant.

No other waste and emissions need to be considered.

7.4.3 Process

The slag preparation comprises three steps with the addition of the waste gas cleaning.

During the first step the slag will undergo a mechanical preparation to obtain the required grain size distribution for leaching of salt. The material will be comminuted and classified in order to remove the metallic aluminium.

Salt leaching and separation of oxide takes place during the next process step. This is the major process since it is the actual reason for the entire preparation. In order to avoid precipitation of salt during oxide separation, the concentration of salt must be below the saturation point at the actual temperature. The situation is different during the precipitation of salt. Therefore a concentration far above the saturation will be obtained. **Fig 7.13** shows the concentration of salt and salt mixtures at different temperatures.

During the final process step salt and oxides will be separated by evaporation to obtain the largest possible salt quantity.

For the protection of the environment the waste gas cleaning part of the process is very important. As an indication **Fig. 7.17** gives major data of the process.

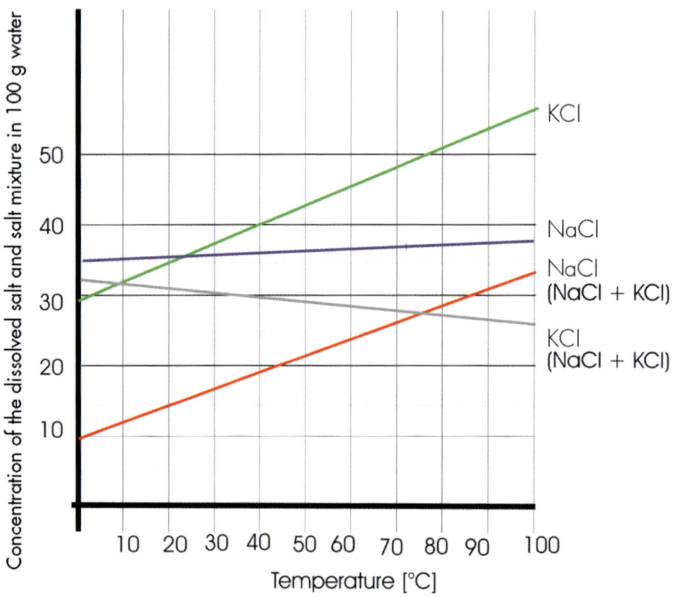

Fig. 7.13: Solution diagram of salt and salt mixtures

7.4.4 Mechanical preparation (Fig. 7.14)

Slag will be charged to the plant by means of a front end loader onto the charging screen having a mesh size of 150 mm. Very coarse pieces, that cannot pass the screen, will be crushed manually by using suitable tools. The material will be sent by unbalanced drive conveyor 1.02 onto belt conveyor 1.03 that is equipped with a magnetic separator to remove large iron lumps and charged to primary crusher 1.04. Due to its characteristics, the crusher 1. 04 produces a fairly high content of fines. Due to its high ductility, aluminium will not be comminuted. The material now reaches via unbalanced drive conveyor 1.05, bucket elevator 1.06 and iron separator 1.07 the vibrating screen 1.08. Here it will be separated into the fractions 5 – 20 mm, 10 mm – 20 mm and < 10 mm. The screen oversize > 50 mm is recycled to the crusher 1.04. The fraction 10 – 20 mm comprises approximately 70 % of aluminium that can be charged to a melting furnace as coarse slag to a melting furnace.

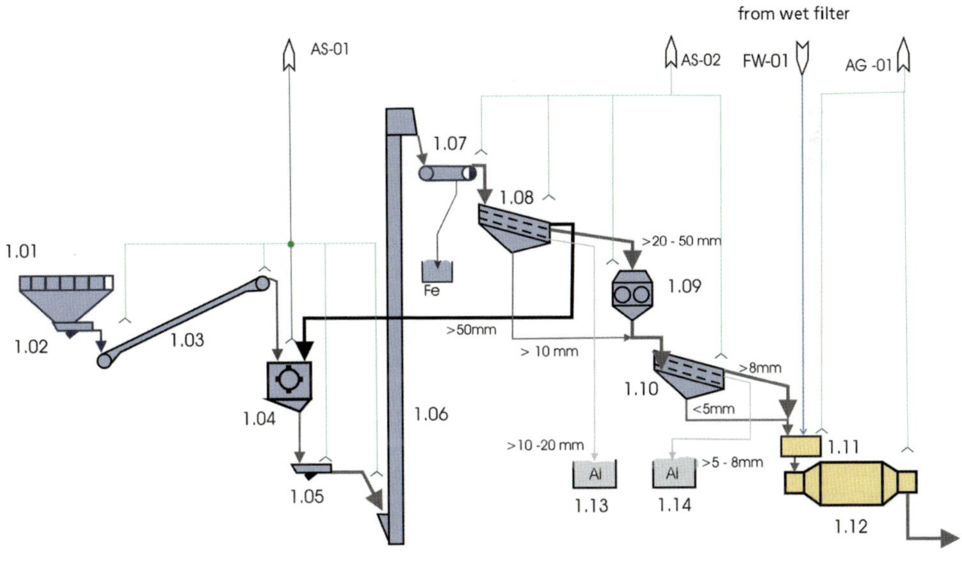

Fig. 7.14: Mechanical salt slag processing

A major portion of the aluminium is < 8mm. For additional classification this portion is charged to the roll mill 1.09 followed by vibrating screen 1.10 for separating the fraction + 8 mm. The underflow of screen 1.10 will be added to underflow of screen 1.08 and charged together with the fraction > 8 mm to the ball mill 1.12. This ball mill operates as wet mill with rubber lining. Before the material reaches the ball mill it passes through a mixing bin 1.11 where dust containing slurry as well as a certain portion of circuit water is added. The product of the ball mill comprises material < 1 mm. Due to its high ductility, aluminium still contained in the mill feed is not ground and appears in the mill product as a fraction > 1mm.

7.4.5 Leaching and oxide separation (Fig 7.15)

From the ball mill 1.11 the material now reaches the screening drum 2.03 via a chute where the small aluminium particles not ground in the ball mill are separated. Screening is supported by wash water from the lamellar separator 2.08. Non-aluminium material passes the screen and is

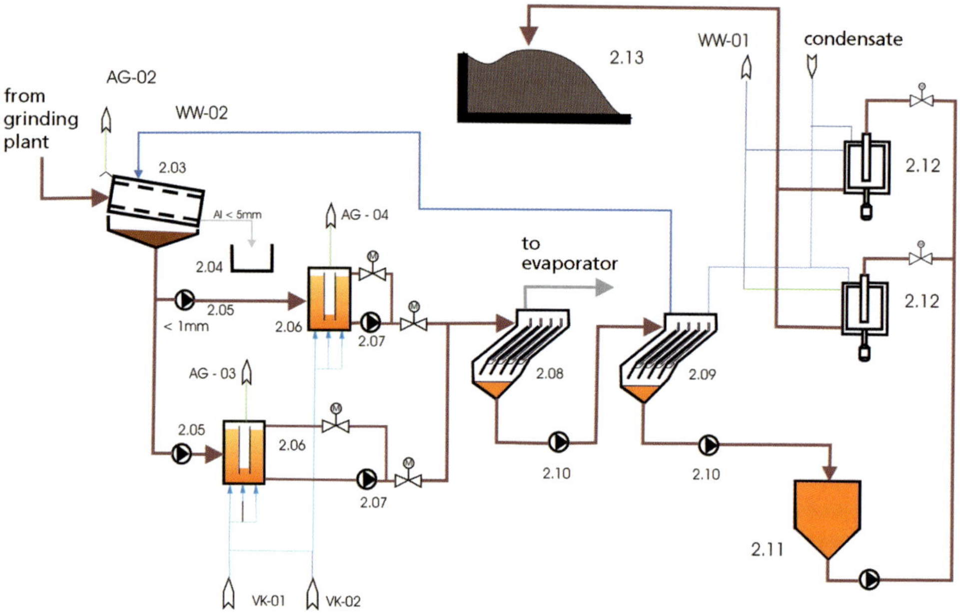

Fig. 7.15: Leaching and oxide separation

collected in a bin arranged underneath the drum. The suspension is pumped either by the pulp pump 2.04 or 2.05 to one of the mixing reactors 2.06 operating in tandem cycles. In these reactors the required pulp density needed for the downstream process is adjusted by feeding warm condensate from the evaporation stage. The concentration of salt is adjusted to remain below the maximum solubility of the salt/water solution in order to avoid crystallization of salt at the present process step (refer to Fig. 7.13). For the design of the mixing reactors a low wear system should be applied. This may be achieved by a special feeding system of condensate to assure a thorough mixing of pulp and liquid. Dissolving of salt requires some time. Therefore, the two mixing reactors 2.6 operate in a tandem cycle. The pulp obtained consists of the salt solution and the solids in the feed material which means mainly of metal oxides and here again mostly of aluminium oxide (Al_2O_3). This pulp now reaches the first lamellar separator 2.08. The overflow comprising the salt solution as pregnant liquor is pumped to the evaporator section. The underflow still contains some salt adhering to the solids and, therefore, needs additional washing. This is performed in the second lamellar separator 2.09. The resulting overflow is recycled to the screening drum 2.03.

The lamellar separators may be replaced by any efficient means of separators such as vacuum filters or centrifuges. This is mainly a question of energy balance and investment cost. The thickened solids are now pumped to the surge bin 2.11 and from there to the centrifuges 2.12. The liquid phase will now be removed down to a liquid concentration of 4 – 6 %. This permits safe depositing. The two vertical centrifuges operate batch-wise which requires the arrangement of the surge bin 2.11.

7.4.6 Evaporation (Fig. 7.16)

Leaving the oxide separation section the pregnant liquor is directed to the two stage evaporation. The first stage 3.01 reduces the salt concentration to a nearly saturated solution which reaches supersaturation in the second evaporation stage 3.02. Generally, the first evaporation stage is

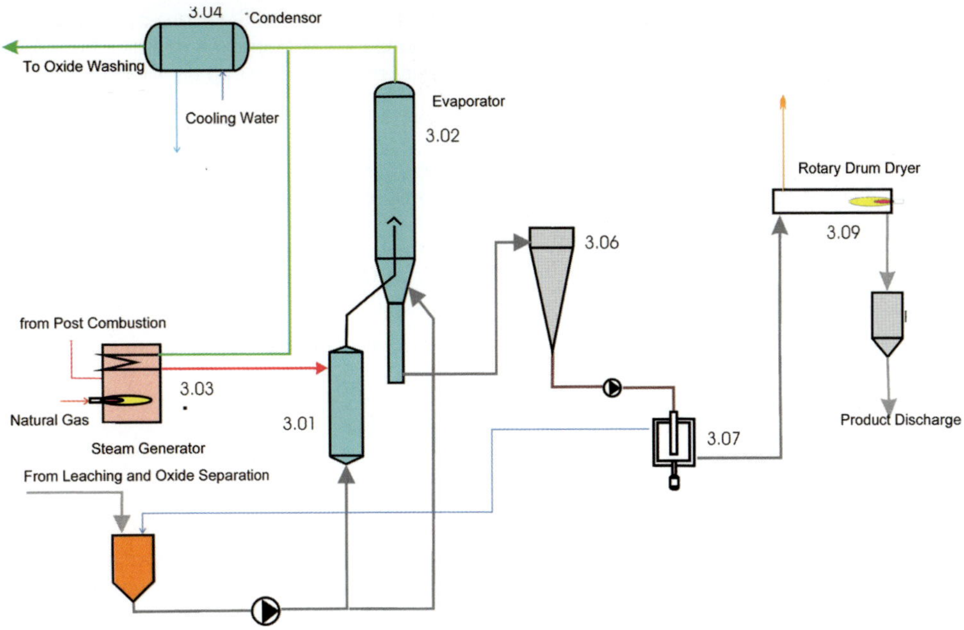

Fig. 7.16: Evaporation of salt solution

designed as plate heat exchanger that is fed with hot steam from the steam generator 3.03. The preheated solution flows to the second stage of the evaporator.

This is heated by the remaining quantity of the steam generated in 3.03. The separated concentrate is collected in the lower part of the stage 2 evaporators and fed to the hydro-cyclone 3.06 and

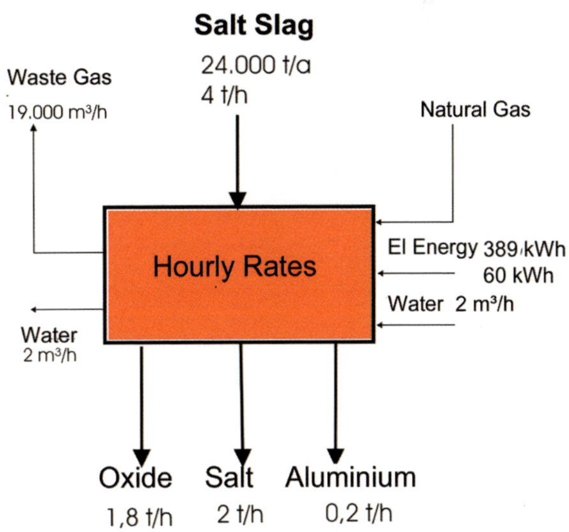

Fig. 7.17: Major plant data

from there to the continuously operating centrifuge 3.07. The overflow of the hydro-cyclone 3.06 is returned to the evaporator stage 2. Salt leaving the cyclone 3.07 still contains approximately 4 % water. If lower moisture content is required, a rotary drum dryer 3.09 may be installed.

7.4.7 Waste gas cleaning

Two major emissions appear during the various process steps. The mechanical preparation part generates dust during crushing and screening of the dry slag. The dust is collected by special dust hoods at the relevant pick-up points and directed to the scrubber. Since water is available within the process, wet scrubbing appears to be a favorable solution for the dust removal. Separated dust is returned to the process as slurry whereby the wash water is used to dissolve the salt. Except clean air no emission occurs.

During the leaching process some gases are generated particularly due to aluminium fines acting as catalyzer. The main gaseous fraction comprise H_2 (20 – 25 m^3), CH_4 (12 – 15 m^3) and NH_3 (5 – 10 m^3). These gases are collected at the various pick-up points and directed to a post combustion chamber. The energy generated there is used to improve the energy efficiency of the steam generator. NH_4 does not contribute to the heat balance but reduces NO_x to a certain extent.

7.5 Mechanical processing

The target of the mechanical preparation is to obtain clean metal with low content of contamination that can be processed in melting furnaces. Some types of scrap, particularly wrought alloys such as window frames, can be processed to arrive at a final product without thermal processing. This is limited to de-ox aluminium.

In this production line the scrap is crushed and classified to remove the plastic compound material. This classification undergoes various steps using screening, eddy current separation and magnetic iron removal. Crushing takes place at various steps to obtain the required grain size distribution.

For a high content of non-metallic components, a specially developed rotor mill is applied (DARMETALL). This system allows processing of a wide range of scrap which is normally difficult to process such as bottle caps, foils, sandwich panels and door sealing of cars. The technology permits processing of materials with a non-metallic content that may reach 90% without using chemicals. The technology may be used as basis for developing a process of recovering aluminium from household waste.

Mechanical processing may also comprise the dismantling the certain types of scrap prior to further treatment. In some cases the clean parts may be used without melting. This is the case with automotive engine parts or gear boxes that can be revamped for use providing low cost spare units.

Part III
Metallurgical processing

1. Melting process

Christoph Schmitz

1.1 Principle considerations

Aluminium returned for recycling can be converted into valuable alloys after its first period of service life. It is not possible to obtain pure metal. This can only be provided by the primary aluminium smelters. But valuable alloys can be produced in the secondary smelter. In different process steps the aluminium is separated to the most possible extent from the accompanying materials and finally the remaining scrap is melted to get access to the metal for further refining and processing.

In the good old days aluminium recycling was simple. Good scrap was collected – and there was not too much in the early days of aluminium – and remelted. The resulting alloy was more or less accepted. The final alloy was mixed in the aluminium foundry. Also energy consumption was not a challenge. With increasing use for different applications, recycling became more sophisticated. Clients now want alloys which are composed to their very specific requirements so that only melting is necessary in their casting plant. Sometimes they want even liquid aluminium delivered in special crucibles suitable for road transport over a remarkable distance.

To be able to produce the required alloy, the scrap aluminium must be melted. This allows separating metal from contaminations and also from some alloying elements. The necessary elements can be added thus arriving at the target analysis.

Since aluminium has a high affinity to oxygen it reacts spontaneously with the ambient air to form aluminium oxide. In the solid state this is no problem since the oxide forms a very tight skin that stops any further oxidation. The skin acts as corrosion protection. In the liquid stage the oxide skin is also formed but it is continuously disturbed by thermal forces with the result of ongoing oxidation. It becomes particularly intensified when melting metal with large specific surface area. The reaction must be limited during melting since the oxides are lost during the recycling process and thus they are definitely a costly loss in the process. Processes have been developed to suppress oxidation during melting. They either use furnaces, whereby the scrap is submerged into a liquid aluminium bath, or melting takes place in a furnace sealed against the ambient air. Most common is the process of melting under a protective layer of flux in the rotary drum furnace, particularly when scrap with a high content of contaminations and aluminium oxide has to undergo processing. The flux used does not only protect the metal against oxidation but acts also as aid to strip the oxide skin off the metal and to clean the melt. Beside the application of flux during melting, additives are used to clean the metal or to add small quantities of alloying elements.

There are other inevitable metal losses in the production line. Some are caused by handling of liquid metal or no-go production; others are generated during metal refining in the casting furnace or in cleaning equipment such as inline de-gassing or in-line filtration. Also starting and stopping of casting equipment causes a certain amount of metal loss. Most of the metal lost in the production line can be recovered and internally recycled.

The specifications for metal quality, set by the clients of the secondary aluminium industry, are very stringent. Thus, all of the material handling from scrap storage to expediting the finished product requires careful attention and permanent monitoring. It is rather simple to produce the required alloy from primary metal just by adding the necessary alloying elements and the high quality ingot is finished. Sometimes this procedure is followed up by primary smelters. For secondary aluminium smelters such a procedure is unthinkable. The available scrap is not pure aluminium and clean without contaminations. Fortunately, a limited range of required alloys is in use and they end up finally in the secondary recycling plant.

The production of secondary aluminium alloys will start with the use of scrap whose analysis is close to the one for the current order of a client. Selecting the proper material from the storage area of the plant should be efficient and correct. Therefore, it is irrevocable that the location of the individual material, its quantity and its quality are precisely known and registered. Monitoring of the different materials starts at the plant gate where the trucks with the scrap arrive. Thorough inspection and registration of the material by experienced personnel is a prerequissite for a correct definition of the scrap. This is certainly easy if the scrap supplied can be defined as castings or bales of UBC or similar. It becomes more difficult with collected scrap. To get a clear picture all of the material supplied with the incoming truck load should be molten and analyzed. This is possible and justified only if the batch can be processed immediately. Thorough inspection and selected samples must suffice. This is then molten in a sampling furnace and the result is the basis for further handling.

Based on the data obtained, the furnace charge is calculated following established procedures and batches are prepared.

In most cases the material charged does not fully meet the requirements for the new alloy. Therefore, additional alloying elements must be added. These could originate from other scrap or must be added as pure element or master alloy. Some alloying can be done in the melting furnace and some final adjustments have to be made in the casting furnace. So these furnaces should be designed to permit this. Please refer to section 2 for details on furnaces. Special additives are used in the casting furnace to protect the metal from oxidation, remove impurities or obtain a fine grain size distribution. The additives are provided by specialized companies as ready mix for the specific application. Also treatments by gases, such as argon, chlorine or nitrogen or mixtures of these gases, are performed in the casting furnace or in-line in the transfer launders to the casting equipment.

1.2 Oxidation during melting

Due to its position in the electro-chemical series, aluminium is a base metal (-0.5 V, magnesium as important alloying metal has an even lower potential of -1.87 V). Therefore, like most of the other metals, aluminium appears in nature as very stable oxide. Chemically, this is regarded as having the most stable state at the lowest possible energy level. During the electrolytic process the metal is forced to separate from the oxygen by lifting the energy potential. Upon contact with oxygen, aluminium tends to return to the lower energy level as aluminium oxide. Due to its high affinity to oxygen aluminium reacts immediately.

The reaction follows the equation

$$2Al + \frac{3}{2} O_2 \rightarrow Al_2O_3 + \Delta H \qquad (1.1)$$

The resulting mass balance is

$$2 \times 27 + \frac{3}{2} 32 = 104 \qquad (1.2)$$

The positive value for the energy ΔH indicates that the oxidation of aluminium is an exothermic reaction. This is logical since the aluminium has now settled on a lower energy level. The affinity of aluminium to oxygen causes a rapid oxidation of the metal at a velocity that is expressed by an empirically developed Arrhenius equation

$$\dot{m} = k_s \cdot \exp\left(-\frac{E}{R \cdot T}\right) \qquad (1.3)$$

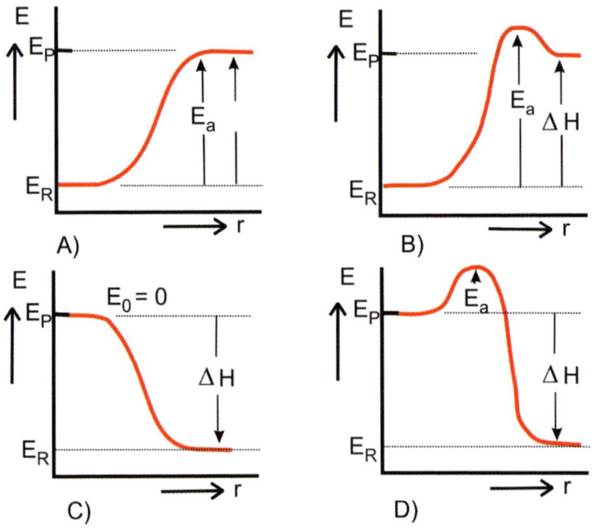

Fig. 1.1: Energy of activation
A) + B) endothermic processes
C) + D) exothermic processes

with $\dot{m}$. = velocity of oxidation, k_s = velocity factor, E = activation energy of reaction, R = gas constant, T = absolute temperature.

E is the energy required to start a reaction. For an exothermic reaction no additional energy of activation is required to initiate a reaction (**Fig. 1.1**). The reaction velocity is now determined by the velocity factor ks only, which may also depend on temperature.

Introducing the specific surface area

$$a_{spec} = \frac{m}{A} \tag{1.4}$$

with m = total mass A = surface area.

With the introduction of the ratio between the oxidation velocity m· and the specific surface area a_{spec} in relation to the total mass m the equation now reads

$$\frac{\dot{m}}{m} = k_s \cdot a_{spec} \cdot \exp\left(-\frac{E}{R \cdot T}\right) \tag{1.5}$$

The quantity of oxides produced can be calculated by multiplying equation 1.5 by time t. The specific density of aluminium oxide is twice as high as that of the aluminium (104/54). Thus, the generation of oxides can be determined by the increase in weight. **Fig. 1.2** gives the increase in weight obtained during tests. The graph shows a steady increase up to a temperature between 400 and 500 °C. Then a discontinuity in oxidation occurs, expressed as sudden and sharp increase in skin thickness. The reason is most probably the change from an amorphous oxide structure to a crystal structure. As consequence, for practical purposes small aluminium particles should not be heated to a temperature exceeding 400 °C during drying or de-coating in order to avoid excessive oxidation.

There is a positive effect in the solid state of the metal since the oxide skin comprising γ-Al_2O_3 forms a thin sealing layer whose thickness is in the range of some µm only. It seals the surface very effectively thus stopping any further oxidation. The thickness of the skin grows first very rapidly but with time the curve flattens soon and further oxidation is terminated. An interesting phenomenon related to this effect is noted when storing swarf. This kind of scrap is generated during machining of aluminium and comes to the recycling plant as turnings, borings or other chips. They come off machining with clean surface and, consequently, will start corroding immediately. Since they are stored as stockpile only the outer layer should oxidize and thus protect the swarf inside the pile. But it can be noted that oxidation continues with time. The reason is that there are gaps and empty space in the pile of material into which air can slowly but surely penetrate. Most of its individual particles are very thin and the oxide layer, although very thin, represents a remarkable share of the total weight so there is substantial metal loss caused by just doing nothing. The consequence should be that swarf is processed without delay.

Besides surface protection there are other desirable effects. By controlled oxidation in an aqueous solution a colored surface of the aluminium can be obtained. This so-called anodic oxidation is used in many decorative applications.

As shown, the specific surface area is a critical value that increases with smaller particle size. Thus, the specific surface area of a cube with a side length of 10 cm equals 600 cm^2 (6 x 10 x 10 = 600 cm^2). That is only 1/10 of the surface of 1,000 cubes with a side length of 1 cm at the identical total mass (6 x 1 x 1 x 1000 = 6,000 cm^2). Applying equation 1.5 to this example shows that the ratio oxidation velocity to total mass increases by a factor 10. Aluminium powder, for instance, has a very large specific surface area. The reaction velocity could arrive at a very critical value and result in an explosion at a very high temperature since the reaction energy is freed on spot. Although not wanted in general, the effect is technically used in the process of alumino thermic welding whereby the reaction not only provides the energy for melting the iron but provides sufficient energy to heat up the steel parts to be connected.

Fortunately, we do not experience an explosion due to oxidation even when melting small particles although explosions may occur sometimes in the melting plant for other reasons. Oxidation in general is not really wanted since it means losses of metal and the reaction product is lost to recycling. Thus, plant operators are keen to use processes and technologies which reduce the unavoidable oxidation as much as reasonably possible.

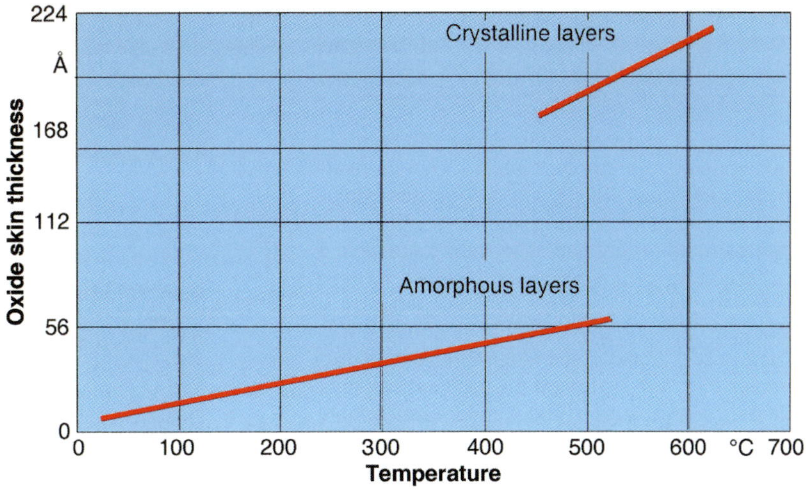

Fig. 1.2: Growth of oxide skin

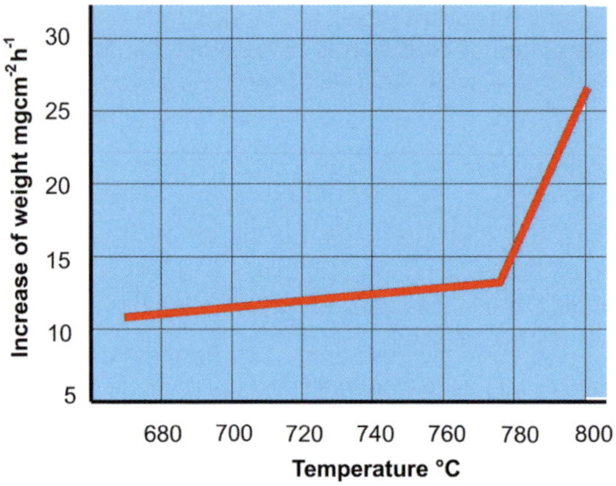

Fig. 1.3: Increase of oxidation with temperature

Apart from the de-coating and drying processes, the oxidation takes place when the aluminium is in the liquid state. During melting the protective oxide skin is destroyed and the oxidation starts again at a higher temperature level. At an undisturbed surface a stable layer of oxides is established that grows very slowly in time.

The oxidation velocity factor ks, as used in equation 1.5, does not remain constant with temperature variations. The increase in weight, i. e. growth of the oxide skin, of a liquid metal bath follows a steady line until a temperature between 760 °C and 780 °C. Then an abrupt increase of oxidation velocity is noted (**Fig. 1.3**). The reason may be a selective oxidation of alloying elements.

Tests have shown that almost all of the alloying elements increase the oxidation velocity:

- Se, Ca, Na, Mg at 700 °C
- Cu, Fe, Si, Mn, Zn, Ca,
- Na, Se, Mg at 800 °C

Almost all of the alloying components of the aluminium increase oxidation. This grows rapidly if the temperature of the melt reaches 800 °C. Thus, if there are no other reasons for a high metal temperature, which could be long distance metal transport, the metal should be heated to a temperature just permitting efficient casting. In most cases this could be a temperature around 730 °C or 750 °C.

Magnesium is always critical. This metal has an even higher affinity to oxygen than aluminium. When melting magnesium alloys a selective oxidation of this metal takes place with the result that most of the magnesium originally contained in the alloy ends up in the dross.

The dependency on surface area and time will be unfavorably influenced by yet another factor. During a melting process, heat must flow from the furnace atmosphere to the interior of a block via the surface until melting temperature is reached. The heat transfer is a function of the temperature difference between furnace atmosphere and metal, the heat convection coefficient λ, the transfer surface A and the heat transfer coefficient α. Looking at a bulk of material the heat transfer becomes more complex (**Fig. 1.4**). Although the total surface area is quite large the exposure to the furnace atmosphere is very limited and so is the portion of heat transmitted from the furnace. Furthermore, the heat flow inside the bulk is poor due to the numerous particle boarders and the free space between the individual particles. At

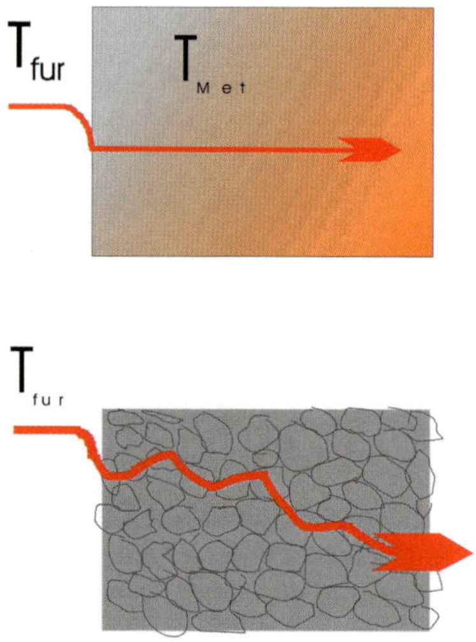

Fig. 1.4: Heat transfer in the furnace charge

the outer limits of the bulk the metal becomes liquid and flows over cold area where it solidifies again. Thus, the metal passes through multiple stages of melting and solidifies with the result of excessive oxidation. These conditions must be considered when selecting the melting equipment for such material.

The melt must be heated to the required casting temperature. Thus, the protective surface is destroyed due to the bath movement by convection and the unavoidable action of the heating system. The surface skin breaks up and fresh metal comes in contact with the furnace atmosphere. With increasing temperature the crystal structure of the aluminium oxide changes. Due to burner flame input at spots, the temperature may exceed even 1,000 °C and in this region α-Al_2O_3 changes to γ-Al_2O_3 that goes in hand with a change in volume. This again creates cracks in the oxide layer in which fresh aluminium is able to penetrate and react (**Fig. 1.5**). Before casting, the bath surface in the furnace will be skimmed. To be sure that the result will be a clean surface, some liquid metal is necessarily skimmed together with the aluminium oxide and the metal trapped herein. Having access to the oxygen-rich air outside the furnace, the reaction of the liquid aluminium in the skim is accelerated and being an exothermic process oxidation continues at quite a substantial rate. To stop this process the skim must be cooled immediately to a temperature below the melting point of aluminium to cause the metal to solidify and to form a stable protective skin. The simplest method to cool the dross is to spread it out and offer sufficient surface area to the ambient air. The metal in the dross can be recovered before it oxidizes by putting the hot dross into a press that squeezes the aluminium out. The result is a block with high aluminium content that can be processed in a melting furnace. Liquid metal coming out of the process is collected in a small liquid metal pan. Another technology uses a rotary drum in which the hot dross is exposed to a counter current air flow to cool. At the end of the drum, the dross is classified in various fractions that can be processed further. Details of the systems are described in part 2 of this book. The hot dross skimmed from the bath surface can also be charged directly into a melting furnace, preferably a rotary drum furnace and gets processed with other scrap. One other technology sup-

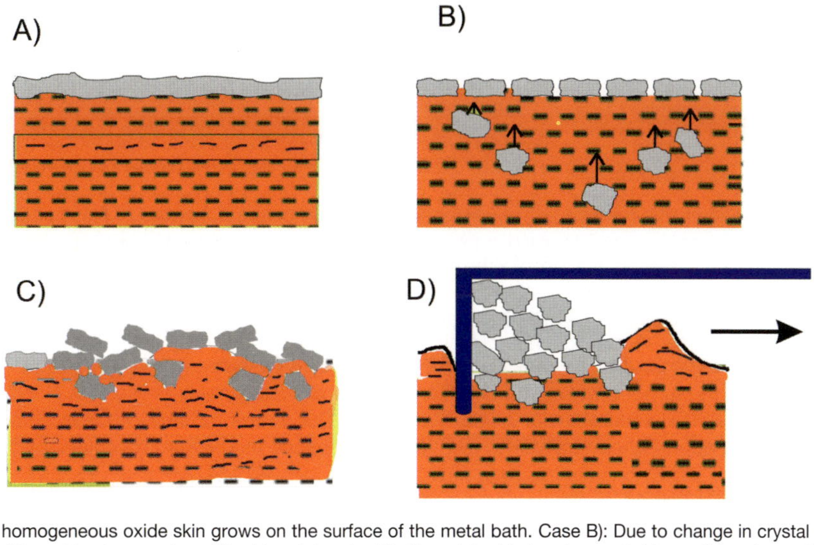

Case A): homogeneous oxide skin grows on the surface of the metal bath. Case B): Due to change in crystal structure and the different heat expansion of liquid metal and oxide the oxide skin breaks up. Oxide particles charged with the scrap float to the surface. Case C): The breaking of the oxide skin is additionally supported by bath movement. Case D): During skimming metal is trapped in the oxides being removed.

Fig. 1.5: Formation of oxide skin on a liquid metal bath presses the contact of the skim with oxygen. The dross pan filled with fresh skimming is placed in a chamber into which nitrogen is introduced until the dross is cooled to the required temperature. The resulting large blocks are, however, somewhat difficult to handle in a melting furnace.

1.3 Melting additives

1.3.1 Usage of salt in the rotary drum furnace

Besides energy consumption the loss of metal by the process is a factor that has substantial influence on the efficiency of the process. Due to its high affinity to oxygen, aluminum tends to react, thus forming aluminium oxide. This is very stable and, consequently, the metal cannot be recovered in the secondary process and the relevant aluminium is lost. For large pieces of scrap, such as sow-ingots or normal ingots, castings, crop ends, no go billets and slabs or even cut and bundled extrusion shapes or even bales of shapes or, with limitations, compacted swarf, this is no problem since the percentage of oxides compared with the total weight of the piece is very small. Also heat transfer into the block is very good as explained before. Therefore, this material can be melted efficiently in the reverberatory furnace with very low melting losses. Having small particles, i. e. material with high specific surface area, the situation is different. The portion of oxides compared with the total mass is already high before the material even touches the melting plant. As discussed before, the rate of oxidation is also very high. A substantial quantity of metal could be lost during metallurgical processing. The oxidation velocity of aluminium increases due to alloying metals since almost all of the alloying elements decrease the metal yield. The effect on aluminium oxidation of most of the elements is lower at a temperature not higher than 700 °C. Thus, the melting process should make use of this and should be run at lowest possible temperature. The tiltable rotary drum furnace permits this procedure since removal of used salt, i. e. slag, is very easy. The reason for this is explained in section 2.

To reduce oxidation the surface of the aluminium must be protected which means access to oxygen must be suppressed or the material must be covered by non-reactive flux. There are various technologies available that take care of these requirements. Please refer to section 2.1. The most flexible furnace technology is the rotary drum furnace since all of the material with high specific surface area can be processed even with a large quantity of metallic and non-metallic contaminations. It is the only technology that is able to process dross or skimmings besides the other material.

1.3.1.1 Application of salt

One characteristic of this furnace is the use of salt. This comprises a mixture of NaCl and KCl with addition of other chemicals. This mixture fulfils the requirements such as thermal stability, high resistance to aluminium, no attacking of the refractory lining, optimal interfacial tension between aluminium and flux as well as oxide and flux, sufficient difference in density between aluminium and flux and, last but not least, good solubility or suspendability for contaminations.

Besides its duty as cover flux, the salt has other applications as well and is also used as cleaning flux (for cleaning the furnace walls), refining flux (removal of metals such as Mg, Ca, Li, Na, K) and de-drossing fluxes (for separation of oxide and metal). During the furnace process components of the salt may react exothermically with aluminium and cause a so-called “burning” of the salt.

Main duties of the salt flux are to

- cover aluminium to prevent further oxidation
- strip away the oxide layer from molten metal
- promote coalescence of metallic droplets
- dissolve or suspend other materials attached to the metal.

Numerous studies have been made to determine the most effective mixture of salt with various additions to the basic NaCl/KCl mixture. There are more than 200 patented mixtures and certainly some more kept secret by specific companies. To obtain at least a general understanding of the mechanism between the components of a batch of scrap, certain influences had to be eliminated for the tests. Mostly clean alloys or even pure metals have been used with a very high content of different fluxes to get a tendency of the influence of the different additives. The problem is that there are too many components involved in a production-type melting process and that flux ratios have to be kept at an economical level. Practical experience with production conditions has resulted in some basic mixtures which are applied in the specific production plant with all pros and cons.

But to understand the basic effects of flux and its additions, the principles shall be discussed briefly.

1.3.1.2 Protection of metal

Although salt is used for various duties the most important task is to cover the aluminium pieces and hence to protect them against the furnace atmosphere in order to minimize oxidation. This requires a minimum of salt. In general, the salt quantity is defined by applying the so-called salt factor.

$$m_{salt} = m_{scrap}\left(1 - \frac{\eta_{rec}}{100}\right) \cdot f_{salt} \qquad (1.6)$$

m_{salt} means the quantity of salt required, m_{scrap} the quantity of scrap to be charged, η_{rec} the expected metal recovery as per test charge and f_{salt} the salt factor. Thus, the salt factor indicates how much salt per unit of weight of contamination included in the scrap is required. Usually a salt factor of 1.0 to 1.5 is mentioned for the fixed axis rotary drum furnace while a salt factor of 0.5 to 1.2 is regarded as typical for the tiltable rotary drum furnace.

However, there is a problem!

For materials with high oxide content a salt factor of 0.5 to 1.5 may be quite meaningful. With scrap having a smaller quantity of oxides, as an example, applying a salt factor of 0.8 at an envisaged metal recovery η_{rec} of 90 % or even 95 %, the calculated salt quantity would be 45 – 80 kg per ton of scrap. Even with the reduced salt requirement of the tiltable rotary drum furnace, this quantity is insufficient to cover the melt. On the other hand, with high oxide content, sufficient covering material is present for coverage but still there is some salt required to separate the different phases of the melt.

It appears to be advisable to extend the equation by a factor considering a corrective element with high metal content as well as differences in the reactive surface of the crap charged. However, looking at the multitude of factors to be applied using this equation may be somewhat unhandy. On the other hand, it seems to be much simpler to use the specific salt value q_{salt} obtained by practical experience during operation. This permits to arrive at a reliable calculation with minimal efforts. **Fig. 1.6** shows different salt factors compared with the salt requirement as obtained empirically. With oxide content in the range of 75 to 90 % the salt quantity by applying the salt factor and the equation comes close to the values found by experience. Particularly in the lower ranges of the oxide content, it appears to be more reasonable to calculate with a specific salt content per ton of scrap charged. As already mentioned, the specific content depends also very much on the type of alloy.

Typically, material with a very high content of oxides is dross. This is material skimmed off the surface of liquid metal bath in a hearth furnace. There are different fractions appearing as certain grain size distributions. There is a fines fraction (dust) medium grain size and there is also coarse

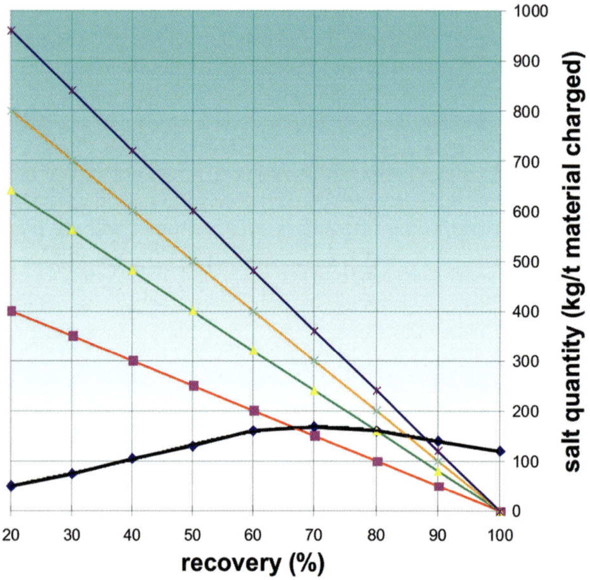

Fig. 1.6: Salt factor for a tiltable rotary drum furnace

material. The dross comprises mainly aluminium oxide that does not react with liquid metal or the refractory lining. It is a very good medium to cover the metal thus protecting it from contact with oxygen. Prior to skimming a reverberatory furnace, some salt is distributed over the surface of the metal bath. This helps to reduce the formation of new aluminium oxide in the hearth furnace. This also results in a more or less even distribution of the dross on the surface which in turn assists the skimming in a way that not too much of the aluminium is skimmed off as well. Coming back to the rotary drum furnace, dross is a good covering agent so that only little additional salt needs to be added to provide protection. In spite of that, there are other useful effects of salt that may offer some benefits if more salt is added. We will discuss that a little bit later. Oxidation must also be stopped on the large lumps of dross protruding out of the batch. The covering material should act as kind of "fire extinguishing sand" and due to the movement of the batch, caused by the movement of the drum, any oxidation is stopped from the beginning. To make best possible use of the aluminium oxide as covering material, dross should be charged to the furnace first so that the coarser material is comminuted at high speed of the furnace drum in a grinding action. The finer grain size obtained is very well suited now to cover and surround the metal particles in the dross. This takes place during the solid state of the metal that has not yet reached melting temperature.

Prior to melting, salt must be added. The mixture of salt has a lower melting temperature than metal. Thus, when the temperature of the batch approaches the melting temperature of salt, a part of the batch may become semi-pasty and stick to the wall of the drum. As shown in **Fig. 1.7**, the lowest temperature of the salt mix is lowest at an equimolar ratio. Pure NaCl has a melting temperature of 805 °C and that is far above the temperature of the liquid metal the operators are aiming for. Thus, the first charge of salt should be NaCl only with the KCl quantity and the other additives charged just before the metal starts melting. Thus, even if only dross is processed, the sequence of the furnace cycle should have the attention of the plant operators. For obtaining an efficient melting process with the maximum possible metal recovery, it appears not to be sufficient just to add a certain quantity of salt and accept the result.

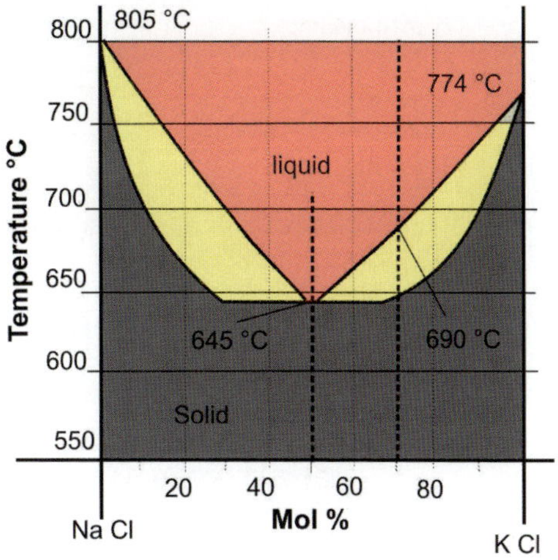

Fig. 1.7: Phase diagram NaCl – KCl

Scrap with less oxide content does not provide sufficient covering medium by itself. Here again, only NaCl should be charged at the beginning. Particularly if the scrap has a high content of organics, it should be removed before the melting step. This material tends to start "burning" which in fact means that the oxidation is initiated and will become self-propelling. If this happens, salt, i. e. NaCl, must be added immediately to stop this process. The scrap of this kind has only little oxide content. Thus, the other functions of the salt mixture gain more importance. By controlling the process, the content of organics of the scrap may be used to improve the energy balance. This is described in more detail in Section 7.0 "Furnace Control".

1.3.1.3 Stripping of oxides and coagulation

Due to its high affinity to oxygen, each aluminium particle is surrounded by a thin oxide skin that must be destroyed after melting and the individual particles must be separated. This is trivial knowledge and it is common knowledge that metal and contaminations separated after melting and metal collects at the bottom of the batch. But why does it happen? Just gravity is certainly not sufficient to explain the observable fact. The intensive mechanical movement of the rotary drum furnace is a factor and particularly the tiltable rotary drum furnace supports separation. Also movement by metal circulation generated by metal pumps or in the induction furnace helps.

And it is to be achieved by the flux.

First explanation is that metal obviously does not wet the oxide. Everybody knows the phenomena: When a liquid comes in contact with a solid surface it may be distributed evenly over the surface and the solid material gets wet. Sometimes the liquid forms individual little drops appearing as small spheres. These pearl off the solid surface which remains dry. Liquid and solids are separated. We know, to be able to do so, the liquid must form a droplet, i. e. it must coagulate. The subject becomes more complex if a third phase is present. In the case of metal melting, the three phases are metal, flux and oxides.

Border skin phenomena play the decisive part in the action of the flux and these phenomena do have distinct influence on the reaction between the three phases metal, aluminium oxide and flux.

In the first stage of melting metal, the oxide skin must be removed. During the melting process the original shape of the metal particle is maintained since the oxide skin prevents it from defor-

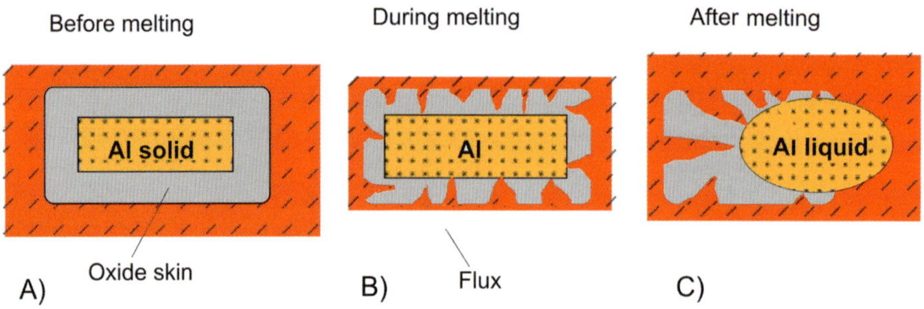

There are three phases for the removal of oxide skin from the metal. Phase A): The metal particle is enveloped by a solid oxide skin. Phase B): As the material is heated, its higher expansion rate causes cracks developing in the oxide the oxide skin. Phase C): The now liquid metal forms a spherical droplet and is separated by mechanical forces created by spinning of the droplet and mechanical movement of the batch.

Fig. 1.8: Stripping of oxide skin

mation (**Fig. 1.8**). With increasing temperature, a piece of aluminium expands at a faster rate than the oxide skin causing it to crack and allowing the metal to react with the flux. Finally, as soon as the metal is molten, a droplet having a spherical shape is developed. However, this process step does not yet remove the oxide skin. The components of the flux act as surface-active elements and will be adsorbed onto the metal surface. Since the distribution of these elements is not uniform, along the contact surface a concentration gradient occurs and the droplet will want to spin in an interfacial movement. This creates a stripping force that removes the oxide skin, similar to the agitation of the metal bath. If additives are charged to the flux, the interfacial tension could decrease locally and thus the spinning of the metal drop could be accelerated (**Fig. 1.9**). For example, in the equimolar NaCl-KCl salt system, the interfacial tension between salt and metal is 710 Nm^{-1}. When NaF is added to this system, this value decreases to 450 Nm-1. The droplet will spin towards the area where the interfacial tension between flux and aluminium is 710 Nm-1 and will lower the overall energy of the entire system.

Wetting occurs if the interfacial tension σ, expressed in Nm^{-1}, is high which actually means that the energy difference between the two partners is high. Since nature always tries to settle on the lowest possible energy level, the liquid is soaked into the surface, i. e wetting it. Due to internal forces directed towards the center, a liquid tries to form a stable sphere becoming a ball if there is no disturbance from the outside.

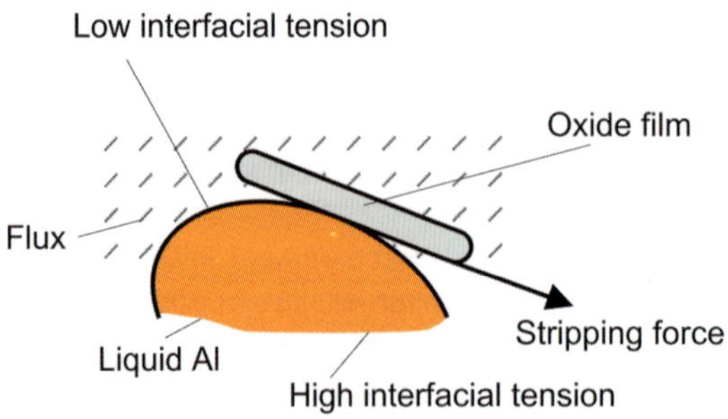

Fig. 1.9: Spinning of liquid metal particle

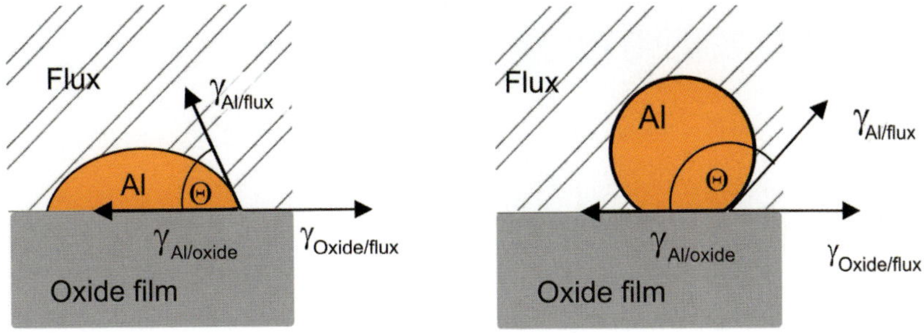

Fig. 1.10: Wetting of a surface

Once a liquid metal drop rests on a solid surface (i. e. oxide film) in flux, three tension forces will act at a common point as shown in **Fig. 1.10**. By equating the horizontal components and maintaining equilibrium, the following equation is obtained

$$\cos\theta = \frac{\sigma_{oxide/flux}}{\sigma_{Al/flux}} \tag{1.7}$$

The contact angle Θ can be determined by different methods. One method is to have molten flux in a quartz crucible that has an Al_2O_3 single crystal at the bottom. When the required temperature of the salt is obtained, a sample of 1 gr or 2 gr is dropped into the crucible. After a holding period (i. e. 30 minutes), a photograph is taken and the angle can be measured.

The contact angle Θ between the three-phase contact line and the solid interface is a measure for wettability (Fig. 1.10). It is the angle between the tangent at the three phase contact line and the solid interface. The contact angle between metal and oxide in various salts with and without a fluoride content ranges between 132° and 140°. Assuming a thin oxide layer was stripped away from a metal droplet leaving the droplet to have area S in contact with oxide, the change in free energy ΔG is given as

$$\frac{\Delta G}{S} = \left(\sigma_{oxide/flux} + \sigma_{Al/flux} - \sigma_{Al/Oxid}\right) \tag{1.8}$$

If the angle is high, the wettability is poor. This is desired for the system flux/aluminium while the wettability shall be good (high angle) for the system flux/oxide. Thus, when the oxide skin cracks, the flux penetrates into the fissures and can now remove the oxide.

If the interfacial tension between oxide and metal increases, the change in energy becomes more negative and the oxide removal tends to be more spontaneous. Actually, the above equation 1.8 shows that the removal of the oxide film is not spontaneous and has to be supported by mechanical agitation. Larger contact angles between metal and oxide film and lower interfacial tension will be needed to promote oxide film removal and metal coalescence.

Additives to the NaCl-KCl mixture have an influence on the coalescence. Particularly surface active fluorides reduce the interfacial tension, thus improving the metal recovery. Their effects can be classified as follows:

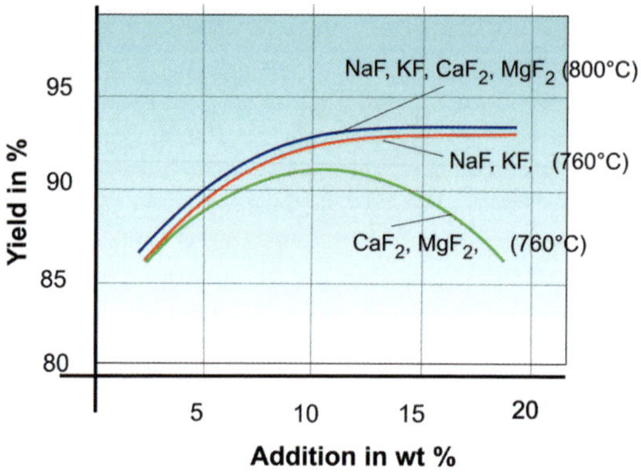

Fig. 1.11: Influence of different additives on metal recovery

excellent	Na_3AlF_6 < LiF < NaF < Kf
good	CaF_2 < MgF_2
moderate	AlF_3 MgF_2
poor	LiCl < $CaCl_2$ < MgC_{l2}

Tests have shown that there is a positive influence of NaF, KF, CaF_2 and MgF_2 shows similar effects at a temperature of 800 °C. With increasing content up to 10 % an increase in recovery is noted. At a content exceeding 10 %, recovery remains constant. At 760 °C, MgF_2 and CaF_2 show a maximum content of 10 % (**Fig. 1.11**).

Cryolite seems to have good dissolving behavior for aluminium oxide. This is quite logical since it is used to dissolve aluminium oxide for the purpose of electrolytic smelting of aluminium.

The addition of fluorides requires careful considerations since components of the alloy may react with the salt. Particularly Mg, as component of many alloys, is regarded as oxide former resulting in more metal loss and in poor coalescence of metal droplets in molten flux. The presence of $MgCl_2$ in NaCl, KCl and NaF/KF salts reduce the coalescence which may be caused by the formation of MgF_2. following the reaction

$$Mg + 2NaF = MgF_2 + 2Na \qquad (1.9)$$

Al and Mg go from molten metal to the slag as halides and contribute to the metal loss. Mg dissolves into the flux from the Al-Mg alloys.

MgF_2 is an effective addition to the salt for improving the coalescence of metal but should be avoided when processing alloys with high Mg content such as UBC. In this case, CaF_2 or cryolite (Na_3AlF_6) should be used.

A dissolving effect of fluorides is assumed, too, based on the known solubility of aluminium oxide in molten cryolite. A positive effect of addition of fluorides could be noted when melting UBC and dross. The high magnesium content of UBC alloys does have a negative influence because of a very stable oxide layer that is difficult to remove.

Coalescence starts already as soon as the metal particles are partly melted. This is supported by the movement of the bath in the rotary drum furnace. In addition to the rotating movement, an axial movement is applied to the batch by the tilted axis in the tiltable rotary drum furnace.

1.3.1.4 Dissolving and suspension of other materials

Contaminations carried into the batch charged to a furnace and impurities of salt and additives do have an influence on the metal treatment. This means purity of metal as well as negative effects on the metal recovery. The variety of contaminations in scrap is numerous. This does not refer to metal as part of the different alloys but as metallic or non-metallic materials that have been collected with the scrap or which are attached to the aluminium parts during the service life of the aluminium. Some additions may even have been added purposely to increase the weight of the material supplied in order to receive more money. Sometimes this is water or sand or steel pieces hidden in the bales. But mainly the contaminations are due to the collection of material from various scrap yards. Collected beverage containers and pie trays may carry some leftovers from the previous beverage or food or cooking oil. More organics are added as binders for paint, plastic coatings or plastic components, i. e. plastic heat bridges in window shapes. The organics will burn during the melting process but there will be residues, mostly metal oxides originating from paint. These will be separated from the aluminium and go to the slag. There are also reactions between aluminium and contaminations. Water introduced by wet material will react with aluminium to form aluminium oxide and hydrogen.

$$2Al + H_2O = Al_2O_3 + 3H_2 \tag{1.10}$$

This reaction is very dangerous and has caused severe explosions. Operators should take care when charging wet material to avoid such spontaneous reaction which is accelerated by the sudden evaporation of water causing an explosion like eruption of liquid metal. The better way is to avoid wet material in general. A similar effect will occur when feeding a furnace with scrap containing oil. The oil will start burning immediately and may cause a severe blow. The reaction of organics with metal results in carbides such as Al_4C_3, SiC, TiC and free C. Although the quantity of the individual reaction products is very limited they all can go to the liquid metal and become trapped as impurities. In saltless processes they have to be removed by other methods such as filtration or bath treatment with reactive chemicals. In the rotary drum furnaces these impurities are suspended by the flux or they are dissolved in it. In any case they contribute to the metal loss.

Nitrogen originating from the combustion air reacts with aluminium and its alloying elements, too, thus forming nitrides. Especially alloys with high content of magnesium tend to react with the nitrogen to form nitrates. To a lower extent aluminium reacts also to form AlN which finally ends up in the slag. Some alloys may still contain some nitrogen originating from the previous degassing process. Nitrates react also with components of the flux and may cause “burning” of salt.

We are already aware of the influence of magnesium and its reaction with fluorides, except CaF_2 and LiF thus forming MgF_2. $MgCl_2$, however, reacts with CaF_2 and other fluorides to form MgF_2 which does change the characteristics of the flux by increasing melting point and viscosity. As mentioned before, Mg also forms a very stable oxide skin that is difficult to remove. There is a dramatic loss of magnesium in the melt due to the selected reaction of this metal whose affinity to oxygen is even higher than that of aluminium. This effect, which is called “demagging” by the metallurgists can in certain cases be regarded as positive since most of the casting alloys require a reduced Mg content.

Salts and additives are obtained from natural resources or as slag recycling product. They contain a small but effective quantity of impurities, mainly sulfate.

Sulfates reduce the metal recovery and with growing sulfate content metal loss grows as well. They react with aluminium (as well as with magnesium) with the result of a “skin” formation comprising layers of oxides and sulfides over the molten metal. With increasing sulfate content metal loss increases as well. This skin is not very strong and can be removed by mechanical agitation.

Skin is formed due to the following reaction

$$^{32}/_3\ Al + 4CaSO_4 = CaS + {}^{13}/_3Al_2O_3 + Al_2S_3 + 3CaO \qquad (1.11)$$

The effect of sulfates on metal recovery is quite small and is in the 0.01 % range. In operating plants this is of no significant influence to the total recovery.

The obviously unavoidable fluorides and in particular calcium fluoride contain comparatively large quantities of contaminations such as CaO, Ca SO_4, and $CaCO_3$. In Germany (this should be similar in other countries), fluorspar (CaF_2) is traded in different qualities. Industrial spar (>76 % CaF_2), ceramic spar (>95 % CaF_2, max. 3 % SiO_2, 1.5 % $CaCO_3$, 0.12 % Fe_2O_3), acid spar (>97 % CaF_2, max. 1 % SiO_2) and optical spar (100 % CaF_2). For cost reasons, aluminium melters use industrial spar. Sometimes the content of CaF_2 is even below specification for the industrial fluorspar and the contents of $CaCO_3$ and SiO_2 are higher than should be. Both could reduce or even compensate the positive effect of CaF_2. The stoichiometric calculations based on tested fluorspar as supplied to a production plant (65 % CaF_2, 11 % $CaCO_3$ and 6.5 % CaO) resulted in a calculated loss of recovery of only 0.1 %. The difference actually measured was 0.5 %. With these contaminations prevailing in the fluorspar, addition of more CaF_2 does not have effect on the coalescence behavior of the flux anymore, since the foreign components in the fluor spar mix, this is $CaCO_3$ in particular, eliminate the positive effect of CaF_2. In other words, the recovery of metal will become independent from the fluorspar addition. This fact should also cause the plant operators to have a close look not at the scrap only but also at the additives supplied. Sand (SiO_2) and carbonates (i. e lime) may be present in the scrap to some extent.

1.3.1.5 Characteristics of salt

In order to get in contact with metal, salt must be liquid. Preferably this state must be reached before the metal melts. That is certainly one reason for the mixture NaCl/KCl of the salt. The phase diagram of NaCl-KCl (Fig. 1.7) shows the melting point of different Mol ratios in a salt mixture. The melting point of 100 % NaCl is 805 °C. Pure KCl has a melting point of 774 °C. At an equimolar ratio, that means the molar ratio is 50/50 %, the mixture has an eutecticum at 645 °C which means the lowest melting point of the mixture is at this temperature. This is below the melting point of pure aluminium (658 °C). This may be the reason that the equimolar ratio of both salt components is preferred by some operators, mainly in the USA. European smelters use a salt mixture with a ratio of 70 % NaCl and 30 % KCl having a melting point of approximately 690 °C. This mixture may have been obtained by practical experience providing good covering of the metal at acceptable coagulation. Addition of Ca F_2 results in an increase of the melting temperature by 20 – 30 °C (The melting temperatures by the system NaCl-KCl-CaF_2 may be taken from the phase diagram (**Fig. 1.12**). At lower metal temperature this may have a positive influence on the salt as cover

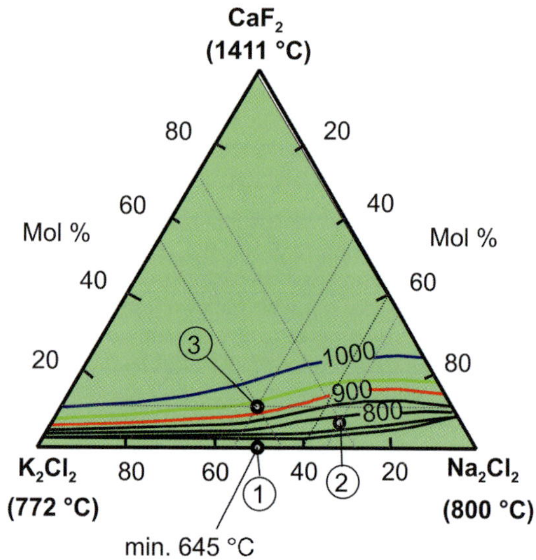

Fig. 1.12: Ternary phase diagram NaCl – KCl – CaF_2
1 - equimolar ratio NaCl-KCl, 2 - European ratio 70/30,
3 - Addition of 5 % CaF_2 to European ratio

flux since it now acts as "extinguishing sand" but reduces the effect on coagulation. In plant operation this is quite advantageous during the phase of de-coating and pre-heating. Before metal starts melting, the temperature of the batch should be increased rapidly to have the salt in the liquid stage to make use of good stripping behavior and coagulation.

As soon as the metal is melted, gravity causes the coagulated metal droplets to precipitate and collect at the bottom of the batch. The velocity of this process depends on the difference of specific density of metal and flux.

Density of liquid aluminium at a temperature of 800 °C is 2.3 g/cm3. The density of both liquid NaCl and liquid KCl change with temperature. With increasing temperature the density decreases. Since a linear relationship between temperature and density has been observed, an empirically found equation to calculate the density can be applied using density $\rho_{(\tau)}$ and temperature τ:

$$\text{NaCl:}\quad \rho_{(\vartheta)} = 1.991 - 0.543 \cdot 10^{-3} \cdot \vartheta \tag{1.12}$$

and

$$\text{KCl:}\quad \rho_{(\vartheta)} = 1.977 - 0.583 \cdot 10^{-3} \cdot \vartheta \tag{1.13}$$

The density of the mixture NaCl-KCl is slightly lower. With the addition of fluorides the density increases depending on the percentage of additives. With an addition of 8 % the density increase varies between 1.58 g/cm^3 (CaF_2) and 1.54 g/cm^3 (KF and AlF_3). The density difference between flux and liquid aluminium (2.3 g/cm^3) at 800 °C is in the range of 0.7. This is very convenient and sufficient for the separation of the two phases.

As more and more oxide is stripped away from the metal, the concentration of oxides suspended in the flux increases. This has one more unwanted effect on the separation of flux and liquid aluminium. With increasing oxide concentration the viscosity of the flux changes.

The viscosity has a decisive influence on the precipitation velocity. High viscosity of the flux slows down the separation of the phases. The viscosity is expressed as dynamic viscosity h expressed as mm^2 · s-1 or as kinematic viscosity $\nu = \eta/\rho$ expressed as Pa · s. The viscosity decreases with higher temperatures (**Fig. 1.13**). Mixing the salt results in decreasing the viscosity, depending on the ratio of NaCl and KCl. With the exemption of cryolite, addition of fluorides increases the viscosity. With NaF as well as KF the viscosity goes up until a peak at a content of 5 % and then decreases. The content of solids suspended in the flux controls the viscosity as well. With growing

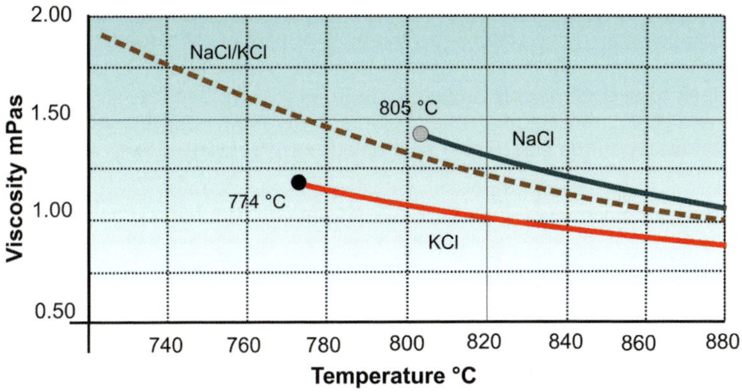

Fig. 1.13: Viscosity of salt and salt mixtures

portion of solids the viscosity increases. Furthermore, the particle size and particle morphology (i. e shape and structure) influence the viscosity remarkably.

A very significant factor is the portion of aluminium oxides in the flux. The dissolved oxide skin forms a suspension whose viscosity rapidly increases sharply even with small oxide content. Studies of the system NaCl-KCl have shown that the addition of only 2 % of γ-Al_2O_3 (smelter grade alumina) to an equimolar flux caused the dynamic viscosity to increase from 1.3 mPa · s to 140 mPa · s. This means the viscosity jumped by a factor 100. Other solid particles, such as magnesium oxide or silicon oxide, have a similar effect on viscosity. In practise, this means that during the melting process the viscosity of the flux increases rapidly and the precipitation of metal droplets becomes more difficult. In this stage increase of temperature has no effect any longer. Only additional charges of salt reducing the oxide concentration show a positive effect. This is also the only means of discharging salt from the standard rotary drum furnace. The situation gets worse when processing dross. The only solution seems to be charging more and more salt with increasing oxide content. This does not really result in a better metal recovery but helps to discharge the slag from the furnace. The state-of-the-art rotary drum furnace permits the discharge of even more or less "dry" salt very easy. Even dross with a metal content as low as 20 % can be processed efficiently if cost for the raw material and – of course – energy permit.

The aluminium droplets suspend in the flux and will be precipitated by gravity thus collecting at the bottom of the furnace charge. This would be very convenient if the aluminium is dissolved in a clean and well defined liquid flux. Unfortunately, the real conditions in a furnace are different. It is not possible to have an excess of salt for economical and also for ecological reasons. Salt cost adds to the production cost and so does the disposal of salt slag removed from the furnace at the end of the cycle. Thus, the operators have to find a compromise which ensures that the inevitable metal loss is kept within acceptable limits at bearable cost. When the metal is molten finally it must pass many obstacles on its way to the bottom of the furnace. There is first the slag comprising a pile of solid aluminium oxide particles, molten and semi-molten salt with probably high viscosity and reaction products originating from contaminations of the scrap, salt, additives and other not defined material charged with the scrap. Thus, a certain percentage, that may range from 4 to 10 %, remains in the slag.

1.3.1.6 Characteristics of slag

After the melting process the remaining slag contains all impurities, contaminations and reaction products. It certainly varies with the type of scrap processed and the salt mixture. **Table 1.1** gives some data on the composition of slag. They indicate that the organics are burned off completely and the remaining mixture comprises the different combustion products, such as Al_2O_3, SiO_2 as well as $SiO_2 \cdot Al_2O_3$, and MgO Al_2O_3. Without knowing the exact details of the material charged, no determination is possible as to the original content of oxide and relationship to the newly formed metal oxides. Apparently both materials did have a fairly small amount of original oxides since their total content only ranges from 20 – 25 % in the slag. The Type A is regarded to be typical for salt slag recycled by special plants in Germany. Also of interest is the content of 4 – 10 % of aluminium in the slag. This metal could not precipitate during and after the melting process.

There is also a certain quantity of material that is discharged with the flue gases. The amount can be as high as 6 % of the quantity charged if, for instance, poor dross with a high content of fines and a low metal content is processed. The content of solids in the flue gas contains almost all of the solids present in the slag including tiny droplets of aluminium commonly called "spray aluminium". For some more detailed calculations of the slag content please refer to section 1.4 below.

In many countries the salt slag has to go to deposit sites since there is no further use. However, in developed countries salt slag preparation plants are in operation at large scale processing plants. Secondary smelters supply their slag to these plants and get back a salt mixture for flux. In the

Table 1.1: Slag composition

	wt % type A	wt % type B
NaCl	40 - 50	48.5
KCl	18 - 20	17.1
Other chlorides	<10	0.2
Al_2O_3, SiO_2, $SiO_2Al_2O_3$, $MgOAl_2O_3$	20 - 25	
CaF_2	1.5	0.2
Al	4 - 10	
PCDD/F	20 ng ITE/kg	
Water solubility	40 - 60	
C		1.6
Si		1.0
SiO_2		6.1
a-Al_2O_3		8.1
Spinel, Mg, Fe, Zn, Mn-Al $FeFe_2O_4$		9.2
$Al(OH)_3$, AlOOH		2.1
AlN		0.1
Silicates		1.5

preparation plants the salt mixture is separated from the other components of the slag. A certain percentage of the aluminium is also recovered and sold as granulate.

The residual matter comprises just above 40 % of aluminium oxide. Various applications have been investigated. The use in the primary aluminium industry is not favored because of the high content of other contaminations. In the contrary, the cement industry is a good customer for the material.

1.3.1.7 Summary and recommendations

Numerous tests have been conducted to determine the ideal salt mixture. It can only be repeated what was said before. Each company, every process and any material have an optimal mix for the flux. Focus may be placed on the duty of the salt acting as coverage to protect the metal to be molten or the removal of unwanted alloying elements, such as magnesium. There is some general recognition that can be used as starting method.

Dross with a high content of aluminium oxide does not require much salt as covering agent to protect the metal. Some salts with additives of aluminium fluoride, preferably cryolite, are very useful to crack dross grains trapping aluminium. As purity of scrap increases, coagulation as well as protecting requirements increase. Thus, the quantity of salt charged becomes more important.

Some examples resulting from tests may illustrate this.

An addition of cryolite, when processing dross, is significant. The content should be as high as economically acceptable. It should be 5 % at a minimum. Addition of CaF_2 up to as much as 10 % does not increase the metal recovery.

Granulate, that is processed dross with a metal content of 60 to 70 %, shows very good results up to an addition of 8 % CaF_2. Further increase of this content, even up to 10 %, does not improve the results. Addition of cryolite does not show any effect.

Melting swarf is very efficient with a NaCl-KCl mixture. No fluorides are necessary and their addition does not improve metal recovery. The reason might be that this material was very pure and dried.

Materials, such as castings and large pieces of scrap, do not require any salt. But there are advantages if small quantities are added to dissolve and strip the oxide skin.

As general rule we should define some basic conditions:

- Scrap should be as clean as possible.
- The content of carbonates and silicon oxide should be limited as far as possible.
- Charge only dry scrap. This does not only improve the recovery but also prevents dangerous situations.
- Charge as quick as possible.

And there is one more point to be considered. Salt is a costly item of the production. Thus, many investigations have been undertaken to find processes that do not require any flux. Some of them will be discussed in one of the following chapters. Particularly when discussing the rotary drum furnace technology a frequent statement is made that a new process would not require any flux. This might even be true to a certain extent, if processing dross. Since it has a high content of non-reactive components, particularly aluminium oxide, the contaminations provide a fairly good coverage for the metal. There is also a sometimes high content of salt originating in the previous furnace process. Thus, in some cases the addition of salt can be low, if only the function to protect the metal is considered. But, as we have stated before, there are other functions of the flux, which can by no means be fulfilled by aluminium oxide. The result will be reflected by the lower recovery. Thus, when planning to use a certain type of dross the production manager should carefully compare the savings of salt and the related cost for deposit of slag with the loss of metal to be expected.

For all other materials with large specific surface area, the use of salt in the rotary drum furnace is imperative if an acceptable level of metal recovery is to be obtained.

1.3.2 Usage of chemicals in the hearth furnace

The specifics during melting and transport of liquid metal, such as generation of dross, forming of non-metallic inclusions as well as scaling at furnace walls, depend very much on the furnace design. Furthermore, burner arrangement and burner adjustment as well as door design and door sealing also have an effect on these parameters. But also the impact of operation of the furnace should not be underestimated. To reduce cost involved and negative effect on the quality of the metal, chemicals are used not only in the rotary drum furnace but also in casting furnaces and transport crucibles. Additionally, the furnace walls must be kept clean to avoid contamination of the metal by traces from the previous alloy.

This is not the only purpose of salt treatment in a hearth or crucible furnace or even in transport crucibles. Especially during in-house recycling, aluminium with some contaminations is charged to the hearth furnace. They comprise mainly oxides which are distributed as insoluble inclusions in the melt. These mechanical inclusions do have a negative effect on the mechanical characteristics and the corrosion resistance of the final casting. But also other unwanted contaminations should preferably be removed. Besides oxidation, liquid aluminium reacts with water from humid ambient air or from combustion products to form free hydrogen that is absorbed from the liquid aluminium. This hydrogen can be removed by cleaning salts or by gas treatment.

Depending on the type of alloy or on the solidification sequence, coarse grains could be formed in the aluminium that have a negative effect on the mechanical strength of the final product. To avoid the coarse crystal structure from forming seeds, chemicals are injected into the melt.

The oxide layer that floats on top of the aluminium melt needs to be removed. To aid de-drossing some salt is very efficient.

Another application is the cleaning of furnace walls. Aluminium tends to form α-alumina layers on the walls of the furnace which need to be removed from time to time. This is necessary to maintain proper functioning of the furnace, prevent decrease of furnace volume and to avoid contamination of the new batch by the previous alloys that may contain unwanted elements.

After this short summary we will look at some more details.

1.3.2.1 Covering of melt

We have noticed that any time an unprotected surface of aluminium is exposed to air, oxidation will start. The velocity of this reaction increases with higher temperature. Due to burner input the melt will heat up and create a movement of metal that disturbs the stable oxide layer. Actions of the operator, for instance stirring of the bath to homogenize it or tapping, metal transfer as well as charging of solid and liquid metal, disturb the surface as well with the result of additional oxide generation. There is yet another effect once the bath surface is disturbed. The severe movement of the bath surface not only results in cracking of the oxide layer but also traps a large quantity of liquid metal. This creates a "wet" dross. The aluminium content of this wet dross may reach 60 – 80 %. This dross has a metallic and light appearance. Charged into a dross pan and having the exothermic reaction stopped, this dross looks like a solid block of aluminium.

Not only in a rotary drum furnace but also in every furnace containing liquid metal, the operators have to take measures to keep oxidation as low as possible. They do this by covering the surface with salt. The salt mixture is similar to that for the salt flux in the rotary drum furnace. It is a mixture of NaCl and KCl with addition of fluorides. We have seen that alloys with a magnesium content exceeding 3 or 4 % tend to react easily with the nitrogen of the combustion products to form nitrides. Besides the chloride content, the covering flux for these alloys contains magnesium chloride that suppresses the burn-out of magnesium. We have discussed the formation of the oxide layer on the surface of a liquid metal bath in section 1.2. (Fig. 1.5). Covering salt may also assist de-drossing. Due to its content of fluorides, it strips the oxide layer from the trapped aluminium particle and effects coalescence. The result is that the aluminium content of the dross decreases and it shows a more blackish color. Now the metal content of the remaining dross could be as low as 40 %.

1.3.2.2 Refining

Oxides floating on top may submerge in the bath thus forming unwanted inclusions. This is also the case when turbulences in the launder disturb the protective oxide skin or surface oxide is drawn into the liquid metal during pouring. To remove these inclusions from the metal bath, salt is used again. This refining salt comprises a mixture also of chlorides of calcium and potassium with sodium and potassium fluoride additives. The cleaning effect is mainly due to a flotation effect. Removing the inclusions shortly prior to casting improves the castability of the alloy. To be effective, the required salt has to be mixed with the liquid metal by stirring.

A common process uses gases to be added to the melt. The gas is brought into the melt via special tubes or lances. Also special porous bricks may be applied for gas injection. It is essential that the gases develop many small bubbles distributed as much as possible to cerate a flotation effect as well. A certain amount of reactive Cl_2 may be used at a stoichiometric ratio. Operators are not fond of chlorine due to its poisonous character. Thus, the application of this gas is very limited today. Reactive salts or gases or combinations of both are applied for removing alloying components such as magnesium or zinc unwanted in a specific alloy.

Refining of the melt has yet another purpose. Aluminium not only reacts with the oxygen of the ambient air or the furnace atmosphere but also with water which appears as humidity of air. The reaction products are aluminium oxide and hydrogen (equation 1.10).

Liquid aluminium will absorb the hydrogen as a much undesired contamination since it results in the generation of pores and fissures in the solidified aluminium product. Treatment for refining of the melt also removes hydrogen. To avoid reaction with air again, the degassing should be done shortly ahead of casting.

For more details on metal treatment please refer also to section 5.

1.3.2.3 Grain refining

Certain alloys and their particular solidification characteristics tend to develop very coarse crystals appearing as large grains. As we have mentioned before, this coarse structure may have a negative effect on the mechanical characteristics of the cast product. This applies particularly for wrought alloys where coarse grain may cause disturbance in the production of printing foil or beverage containers or may cause color differences during anodic oxidation. Therefore, grain fining salts introduced into the molten metal create many solidification seeds and lead to a very fine grain size. The interaction of the seeds, mainly consisting of aluminium boride, titanium boride, titanium carbide and zinc carbide, is very complex. Grain refining of eutectic silicon alloys is obtained by addition of phosphor-containing tablets which do also have a de-gassing effect.

Hyper and hypo-eutectic AlSi alloys require a special procedure. Very small quantities of sodium – in the range of some hundredth percent – cause the development of very fine grains by super cooling. The thus treated alloy is characterized by very good mechanical strength. Sodium is introduced into the melt by salt or tablet. But also metallic sodium can be charged by means of a plunger. Alloys for gravity or low pressure die-casting are refined with strontium. This has, compared with the refining with sodium, a long-term effect that can be noticed even after repeated remelting. Thus, the secondary melters are able to supply ready-made and Sr-refined alloys for the use in foundries. A common application of these alloys is the use for car wheels which are usually made of near-eutectic Si alloys.

Since advantages always go in hand with disadvantages, some problems may have to be expected if the recycled material is used for other applications.

Refining with strontium requires that no NaCl or free chlorine is present in the melt. These should be neutralized prior to introducing Sr by adding small quantities of metallic Na.

As an example, the following procedure for refining in a reverberatory furnace with a bath capacity of 20 t could be applied.

One kg of metallic sodium is charged to the metal by means of a plunger. This is followed by a short flushing with nitrogen or argon. Thereafter, strontium is charged by means of tablets followed by flushing again for 30 minutes. The metal temperature should be set at 740 °C and maintained for the casting to follow.

For the refining with Na and Sr chemicals in the format of tablets are available on the market.

For further details please refer to section 8: "Quality assurance"

1.3.2.4 Wall cleaning

The sidewall area just above and below the bath level is very critical for scaling. Oxide builds up a sturdy belt comprising a-alumina and trapped metal. For ease of operation and also to avoid contamination, this layer must be removed from time to time. The treatment with salt is followed by the mechanical removal of the scaling while the wall is still hot.

Some reactions of the aluminium with additives are exothermic and generate local heat. Thus, for example

$$2Al + KNO_3 = Al_2O_3 + 1/2N_2 + K \qquad \Delta H° = -1230 \text{ kJ/Mol} \tag{1.14}$$

or

$$6Na_2SiF_6 + Al_2O_3 = 4Na_3AlF_6 + 3SiO_2 + 3SiF_4 \tag{1.15}$$

This exothermic reaction may improve separation of aluminium and oxide locally. Controlled addition of such additives may result in a flux that is very useful for cleaning the furnace walls in the area around the bath level.

The cleaning agent comprises fluorides which penetrate into the scaling and help to dissolve some components. A certain quantity of oxidizing agents ensures generation of local heat so that combined effect of the fluorides and the melting of some of the metal trapped soften the scaling, thus helping the removal. Care must be taken to keep the content of oxidizing agents low since an excessive oxidation of aluminium is not desired. Thus, the components of the cleaning salt have to be composed with care.

It will be indispensable to scrape off the scaling as long as the oxide is hot. Once cooled down it becomes very tough and is extremely difficult to remove.

1.3.2.5 Treatment chemicals

The treatment salts are available as powder, granulate and tablets. They are distributed over the surface of the bath. Powder creates a dense cloud in the furnace which may partly leave the furnace through the charging door or is released into the open in the case of the crucible furnace. This is a hazard for the operators. For metal-cleaning the salt must be brought into the liquid metal bath. This takes some time so that the plant personnel is exposed to the salt cloud for an extended period of time. Fluor-containing salts are regarded as harmful to the environment. The industry tried to develop a salt mixture with the target to avoid fluorides altogether. This was not very successful because these salts are much less effective than those with fluorides. One way out is the development of granulates. They do not generate a cloud of salt powder so that working next to the furnace is much more convenient. Charging of the salt is also much easier and the efficiency is better since granulate has immediate contact with the bath due to its gravity. Finally, consumption of salt is lower since the losses through door and stack are much lower, too.

Wherever possible, gas fluxing in the furnace is the preferred technology today. The equipment is easy to operate and not harmful to the environment and to the plant operators.

1.4 Summary of process losses

1.4.1 Melt losses

When discussing furnaces many arguments are focussed on melt losses of the specific technology. There are frequent misunderstandings as to what the expression "melting losses" actually means and agreements are sometimes difficult to obtain. Therefore, before we go into more details of oxidation and losses occurring during the recycling processes, we should define the term "melt loss" somewhat more precisely.

It is certainly common understanding that scrap supplied to the plant is a mixture of aluminium alloys with a variation of different contaminations. Furthermore, it is quite obvious that the weight of metal finally ending up as sellable production is lower and sometimes even very much lower than the weight of metal charged into the system. The total yield could therefore be defined as

$$\eta_{tot} = \frac{m_{prod}}{m_{scrap}} \tag{1.16}$$

or as percentage:

$$\eta_{tot} = \frac{m_{prod}}{m_{scrap}} \cdot 100 \qquad (1.17)$$

In this equation m_{prod} is the quantity of sellable product, m_{scrap} is the quantity of scrap charged and $\eta_{\tau ot}$ the total yield. One should use another Greek letter for the recovery but h is very common for any kind of efficiency. Thus, we do not like to create confusion. For the sake of clarification, the designation "recovery" shall be used to describe the metal content of the scrap when supplied to the plant. For the metal obtained during melting and casting we will use the expression "yield".

The metal content of the scrap supplied to the plant is determined as soon as the scrap arrives at the plant gate or, where this is not possible, shortly thereafter. The loss of weight is defined before the material will be charged to the melting furnace and is considered in the batch calculation. In the first step of metallurgical processing the aluminium will be freed from the contaminations. With the weight of scrap being mscrap and the weight of recovered aluminium m_{Al} we obtain for the recovery η_{rec}

$$\eta_{rec} = \frac{m_{Al}}{m_{scrap}} \qquad (1.18)$$

Some scrap is heavily contaminated by aluminium oxide. In the case of dross, the oxide content ranges from 20 to 80 % by weight. But also "clean" scrap carries a load of oxides. This is due to the formation of the skin as mentioned already. For large pieces of aluminium the related content of oxide is fairly small and does not exceed 2 % by weight. Swarf or turnings are very clean in the beginning with an oxide content of 2 or 3 % by weight. During a lengthy period of storage time, oxidation extends to lower layers of the material pile. Since the ratio of oxides to metal is high for material with large specific surface area, the oxide portion increases and may reach 6 or 7 % by weight. There are not only aluminium oxides present as contaminations. Painted material contains other metal oxides as pigments for the different colors.

But it is not only oxides that contribute to the weight of scrap. Other contaminations comprise organics, such as oil, plastic coatings or plastic compound material. These will burn-off during melting. Metals used as compound material will also contribute to the burden of non-aluminium parts in the scrap. But also tramp metal, wood, sand, dirt, plastic and the like are sometimes collected together with the scrap.

During melting, some losses to the valuable metal are inevitable. A portion of the aluminium oxidizes to Al_2O_3 and also some alloying metals, especially magnesium, will burn off. Further losses occur by the escaping of flue gases. Some aluminium is contained in the dust and some very fine aluminium particles which are formed during melting. These will escape together with other dust and can finally end up in the filter dust. These losses are difficult to determine in a production plant. Thus, they are summarized as burn-off that is related to the metal recovered m_{Al} and expressed as burn-off factor η_{burn}:

$$\beta_{burn} = \frac{m_{burn}}{m_{Al}} \cdot 100 \qquad \text{expressed in \%} \qquad (1.19)$$

From that the melting efficiency η_{melt} can be calculated:

$$\eta_{melt} = 100 - \beta_{burn} \qquad \text{expressed in \%} \qquad (1.20)$$

Now, this now does not consider all the losses of metal occurring during melting. Depending on the melting technology, some metal is lost to skimmings or even to slag. Usually dross is fed into a closed loop plant recycling. Thus, the metal trapped in the dross is recovered. That is different to slag which normally leaves the plant and is, consequently, lost for production. Since the quantity of metal recycled internally will reach equilibrium, it is not to be taken in consideration for the total production (**Fig. 1.14**). Looking at a single piece of equipment, i. e. a melting furnace or a casting furnace, the point of view could be quite different. The loss occurring in a furnace for instance depends on a multitude of factors related to design and operation (and maintenance).

The metal trapped in dross is expressed in equation 1.19. But how do we get to the quantity of oxide? In case of having just dross generated by a clean bath, as it is the case in a casting furnace, the weight of dross is almost exactly twice the weight of aluminium that has reacted with oxygen. For each kg of aluminium lost two kg of aluminium oxide are produced:

$$m_{dross} = 2 \cdot m_{Al} \cdot \beta_{burn} \tag{1.21}$$

Calculated with the efficiency factor we will arrive at

$$m_{dross} = 2 \cdot \left(1 - \frac{\eta_{melt}}{100}\right) m_{Al} \tag{1.22}$$

Example: A 20 t reverberatory furnace will have a burn-off of $\eta_{burn} = 2$ %. According to equation 1.21 the melting efficiency will be

$$\eta_{melt} = 100 - 2 = 98 \text{ \%}$$

From that the quantity of dross is calculated to be

$$m_{dross} = 2 \cdot 20{,}000 \cdot 2/100 = 800 \text{ kg}$$

Calculated with equation 1.21 we obtain, of course, the same result

$$m_{dross} = 2 \cdot \left(1 - \frac{98}{100}\right) \cdot 20{,}000 = 800 \text{ kg}$$

The situation is a bit more complex in a rotary drum furnace. The oxides generated can be calculated using the above equations 1.20 or 1.21. We recognize, however, that this procedure is not quite correct since we do have other metal oxides, reaction products and metal loss due to dust removal. Considering the circumstances of a production plant and also considering how accurate weights in the furnace area are available, the small deviation from the scientific summary of losses can be excused. The dross is part of the oxides, the non-reactive contaminations of the scrap and the salt charged. To get to the total amount of slag we have to add up the various components.

This should be shown taking an example.

A 20 t rotary drum furnace is to process dross with a metal content of 60 % which means the recovery η_{rec} equals 60 %. The non-aluminium content comprises aluminium oxide and some steel parts. The melt loss shall be 3 %. For efficient melting 75 kg of salt per ton dross charged shall be used for the process. Tests have given the result that the slag contains 6 % of aluminium.

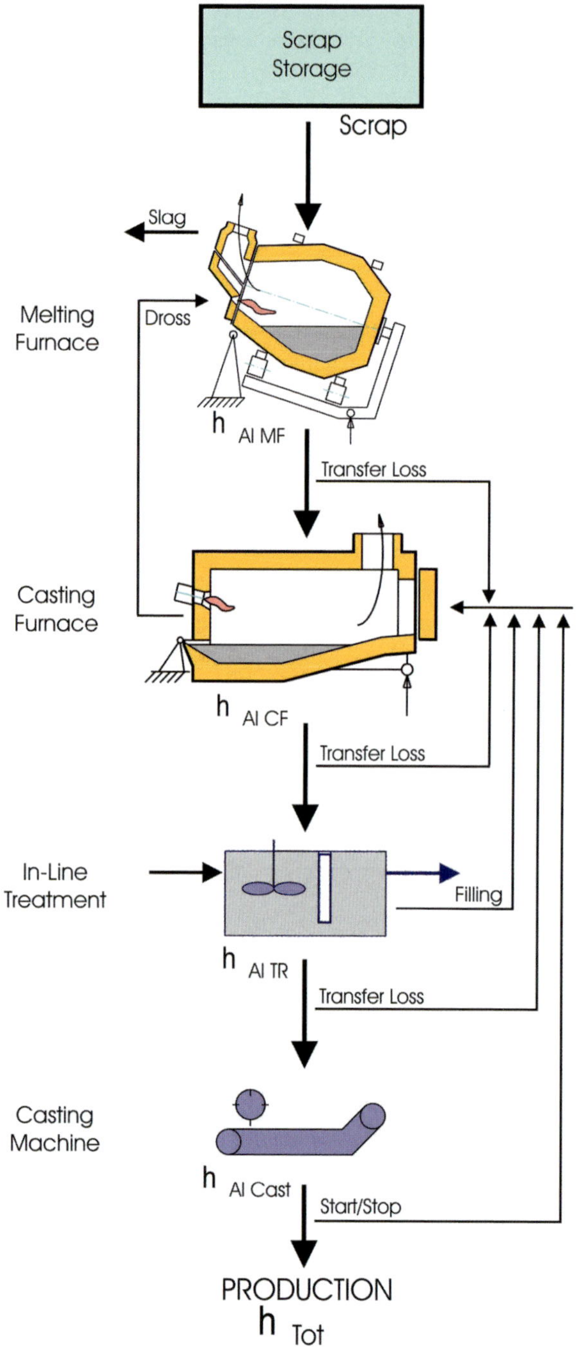

Fig. 1.14: Metal flow in a melting plant

Thus, we first calculate the quantity of metal:

$$m_{Al} = m_{scrap} \cdot \eta_{rec} = 20{,}000 \cdot 60/100 = 12{,}000 \text{ kg}$$

The non-reactive portion of the scrap is also easy to obtain. It is the difference of scrap charged and the metal = 20,000 – 12,000 = 8,000 kg.

Now we should have a closer look at the losses. Melting loss is correlated to the actual metal processed:

$m_{burn} = m_{Al} \cdot \beta_{burn} = 12{,}000 \cdot 3/100 = 360$ kg and from this we obtain

the remaining quantity of metal $m_{Al}m = mAl - mburn = 12{,}000 - 360 = 11{,}640$ kg

The quantity of dross will be

$$m_{dross} = 2 \cdot 360 = 720 \text{ kg}$$

Together with the dross portion of the scrap we have a total quantity of 8,000 kg + 720 kg = 8,720 kg.

To get the total quantity of slag we need to add the amount of salt.

$$m_{salt} = 20{,}000 \cdot 75 = 1{,}500 \text{ kg.}$$

The weight of slag is summarized by adding the salt quantity to the amount of oxide:

$$m_{slag} = 8{,}720 + 1{,}500 = 10{,}220 \text{ kg.}$$

The content of aluminium in the slag was defined as 6 % of weight of the slag quantity. Thus, the metal lost by the slag will be

$m_{Alslag} = 10{,}220 \cdot 6/100 = 613$ kg. From this we calculate the slag efficiency

$$\eta_{slag} = \frac{m_{Alslag}}{m_{Alm}} = \frac{613}{11640} = 0.053$$

The remaining amount of metal m_{Alm} will be

$$12{,}000 - 360 - 613 = 11{,}027 \text{ kg.}$$

This is the quantity of aluminium obtained from the scrap quantity of 20,000 kg. Thus, the total loss of weight adds up to 20,000 kg – 11,027 kg = 8,937 kg.

Now we are able to calculate the total yield in the melting furnace by comparing the scrap charged with the furnace output $m_{Alm}r$:

$$\eta_{tot} = \frac{m_{Almr}}{m_{scrap}} = \frac{11027}{20000} = 0.55$$

Considering all losses as well as the contaminations of the material charged to the melting furnace, 55 % of weight is the output as sellable product. If we compare the metal charged to the melting furnace without contaminations, the percentage of metal lost is lower and the efficiency will be better since the contaminations do not have to be taken in consideration:

$$\eta_m = \frac{m_{Alm}}{m_{rec}} = \frac{11027}{12000} = 0.91$$

Considering all losses of metal in the furnace, 9 % is lost as melting loss and metal discharged with slag.

The example illustrates very well the different losses in a melting furnace. The calculation may now continue for the casting furnace or casting furnaces to arrive at the final production output. There are losses during the processes which do not influence the output. This is seen by metal that is recycled internally (Fig. 1.14). During metal transfer some aluminium will solidify at the walls of launders or transfer and in casting tundishes. Also metal filling of in-line refining equipment, like metal filters or in-line degassing units, are lost to the actual production stage. Also upon starting and stopping a casting process some metal cannot be used. All of this material is, however, not an actual loss for the plant since it is recycled internally to relevant production stages and remains in the production cycle in a closed loop. It would only be a loss if taken out of the production entirely as it is the case with slag. Dross skimmed off from the casting furnaces (and the launders) is also recycled internally. The aluminium skimmed off remains in the process while the oxide content cannot be recovered and will be removed from the process as a definite loss. Thus, operators and plant designers must try to avoid generation of dross as much as possible. One source of dross, whose effect is underestimated very often, is the transfer of metal. Having the aluminium discharged from a furnace into a launder or from a launder into a furnace or casting machine as cascade will result in disturbing the protective oxide layer, thus increasing metal loss and undesired inclusions in the metal.

All aluminium in internal closed loop recycling does not have an effect on the actual production rate. Equilibrium is established for this metal to be returned to the relevant production stage. Dross will go to the melting furnace while metal spillings and start and stop metal can be charged directly to the casting furnace to be part of the new batch. Although the total production rate is not affected by the material internally recycled, the plant equipment must be designed for a higher production rate which can reach substantial quantities. This may not be much if the casing plant only is considered. Downstream facilities may result in large no-go production or large quantities of material required for start and stop. In these cases the recycling rate may reach 10 to 15 % of the entire production.

A step by step evaluation of metal losses and efficiency is helpful to go through the casthouse process steps. **Table 1.2** uses the above example to summarize the metal flow.

1.5 Energy consumption

Input of energy is imperative to convert solid metal into the liquid state, if required. As outlined in section 3.1.4, the theoretical increase of enthalpy is 316 kWh/t. This is the energy required to heat up solid metal to melting temperature, the melting enthalpy and to increase the liquid metal temperature to 720 °C. Since energy losses have to be taken into account, the actual energy required is higher. This is expressed by the furnace efficiency η_W. For some details of furnace efficiency please refer also to section 3. It is standard practise in the industry to define specific energy consumption for a process since calculating with theoretical energy and furnace efficiency is somewhat unhandy. This factor includes all losses of the process and all energy recycled

Table 1.2: Metal quantity process during different production stages

1	Charge	Melting furnace dross	Casting furnace ingots	alloying metal
2	tested recovery (%)	60	98	99
3	quantity (tons)	20	4	4
4	melt loss (%)	3	2	2
5	metal in slag (%)	6	2	0
6	salt quantity (kg/t)	75	0	0

7		batch (kg)	non- metallic (kg)	metal content (kg)	metal loss (kg)	remaining metal (kg)	efficiency
8	**batch charged**	20000	8000	12000	0	**20000**	
9	metal recovered					12000	0.60
10	melt loss		720		-360	11640	0.97
11	salt	1500	1500				
12	slag (10+11+12)		**10220**		-613		
13	**metal yield**					**11027**	**0.55**
14							
15	**batch charged**	8000	120	7880		**19027**	
16	melt loss		761		-381	18646	
17	metal trapped	0	0		-373	18273	0.96
18	dross		753				
19	**metal yield**					**18273**	**0.53**
20							

to improve the efficiency. A good hearth furnace will have a specific energy consumption of approximately 1,000 kWh/t if used as casting furnace only. In this case the furnace must hold the temperature at a certain level and only little quantities of material have to be melted. Introducing fancy systems for energy recycling with sometimes substantial investment and high maintenance cost are not justified economically. The situation is very different if a hearth furnace is used as melting furnace. Now the installation of energy recovery systems pays off and the specific energy consumption can be reduced to approximately 600 kWh/t. The technology of the tiltable rotary drum furnace requires approximately 600 kWh/t with gas-air burners and approximately 400 kWh/t if oxy-fuel burners are used.

Compensation of the furnace losses is only one aspect when defining the energy requirements of the melting process. Since most of the scrap we have to process is accompanied by contaminations, additional energy is necessary to increase the temperature of these components to melting temperature – and above – of the aluminium. If the contaminations are organics, then their heating value may contribute to the energy provided by the heating elements, i. e. the burners.

When evaluating the energy requirements of a process all these factors need careful consideration. The specific energy consumption, as typical for a process, is applied for the complete batch charged into the furnace including all contaminations. Thus, with increasing oxide content, the amount of energy referring to the metal quantity increases as well. If salt is used – as is the case if scrap with high oxide content is to be processed – the energy to melt the salt has to be added to the total quantity.

Example:

15 tons of dross with a recovery of 70 % are to be processed in a tiltable rotary drum furnace. The quantity of salt for this type of material is defined to be 75 kg/t of material charged. The specific energy consumption of the tiltable rotary drum furnace is 600 kWh/t. The total melt losses, including slag loss will be 8 %.

Metal available will be

$$m_{Al} = 15{,}000 \cdot 0.70 \cdot (1 - 0.08) - 9{,}669 \text{ kg}$$

The energy consumption is calculated to be

$$P = 15 \cdot \left(1 + \frac{75}{1000}\right) \cdot 600 = 9675 \text{ kWh}$$

From this we get to the specific energy consumption based on the metal produced of

$$p_{specAl} = \frac{P}{m_{Al}} = \frac{9675}{9660} = 1002 \text{ kWh/t}$$

If no salt is used (however, this might be possible to obtain the same recovery)

$$m_{Al} = 15{,}000 \cdot 0.70 \cdot (1 - 0.08) = 9{,}660 \text{ kg}$$

For the total power consumption we will now have

$$P = 15 \cdot 600 = 9{,}000 \text{ kWh}$$

Batch Calculation

Prepared by

Date

Batch No.

Cast Date

Cast No.

Furnace

Quality 226

Remarks

Material analysis	Batch kg	%	Recovery kg	Si %	kg	Fe %	kg	Cu %	kg	Mn %	kg	Mg %	kg	Zn %	kg	Ti %	kg	Pb %	kg	Sn %	kg	Ni %	kg	%	kg	Al %	kg
Analysis required				**8,0-11,0**	**0,00**	**1,00**	**0,00**	**2,0-3,5**	**0,00**	**0,1-0,5**	**0,00**	**0,1-0,5**	**0,00**	**1,20**	**0,00**	**0,15**	**0,00**	**0,20**	**0,00**	**0,10**	**0,00**		**0,00**		**0,00**	**97,35**	
	0	0,00	0		0,00		0,00		0,00		0,00		0,00		0,00		0,00		0,00		0,00		0,00		0,00	**0,00**	**0,00**
Dross 1	5500	26,40	1452	12,40	180,05	0,66	9,58	0,64	9,29	0,22	3,14	0,03	0,44	0,16	2,28	0,09	1,31	0,01	0,15	0,00	0,00	0,21	3,05		0,00	**85,38**	**1239,67**
Dross 2	2000	17,20	344	7,96	27,38	0,83	2,86	2,12	7,29	0,23	0,80	0,04	0,14	1,08	3,73	0,04	0,12	0,03	0,10	0,00	0,00	0,09	0,30		0,00	**87,49**	**300,98**
Dross 3	460	29,00	133,4	0,50	0,67	0,60	0,80	0,03	0,04	0,01	0,01	0,13	0,17	0,02	0,03	0,03	0,04	0,06	0,08	0,00	0,00	0,05	0,07		0,00	**98,52**	**131,43**
			0		0,00		0,00		0,00		0,00		0,00		0,00		0,00		0,00		0,00		0,00		0,00	**0,00**	**0,00**
Cu	40	100,00	40		0,00		0,00	100,00	40,00		0,00		0,00		0,00		0,00		0,00		0,00		0,00		0,00	**0,00**	**0,00**
			0		0,00		0,00		0,00		0,00		0,00		0,00		0,00		0,00		0,00		0,00		0,00	**0,00**	**0,00**
			0		0,00		0,00		0,00		0,00		0,00		0,00		0,00		0,00		0,00		0,00		0,00	**0,00**	**0,00**
	0		0		0,00		0,00		0,00		0,00		0,00		0,00		0,00		0,00		0,00		0,00		0,00	**0,00**	**0,00**
Weight of Metal	**8000**	**24,62**	**1969,4**	**10,57**	**208,10**	**0,67**	**13,24**	**2,88**	**56,62**	**0,20**	**3,95**	**0,04**	**0,74**	**0,31**	**6,04**	**0,07**	**1,46**	**0,02**	**0,33**	**0,00**	**0,00**	**0,17**	**3,42**		**0,00**	**84,90**	**1672,08**
NaCl	405	100	405																								
KCl	165	100	165																								
CaF2	30	100	30																								
			0																								
			0																								
total	**600**																										
Adjustments																											
			0		0		0		0		0		0		0		0		0		0		0		0	**0,00**	**0**
			0		0		0		0		0		0		0		0		0		0		0		0	**0,00**	**0**
			0		0		0		0		0		0		0		0		0		0		0		0	**0,00**	**0**
Grand total	**8600**	**0,00000**	**0**	**0,00000**	**0**		**0**		**0**		**0**		**0**		**0**		**0**		**0**		**0**		**0**		**0**		**0**
No-go material	1	0,00000	-600																								

Fig. 1.15: Combustion of Organics in the Rotary Drum Furnace. The spontaneous combustion may be delayed by controlling the operating parameters of the furnace

resulting in a specific power requirement of

$$p_{specAl} = \frac{P}{m_{Al}} = \frac{9000}{9660} = 931.7 \text{ kWh/t}$$

We notice that there is a power-saving of roughly 7 % if we would be able to melt without salt. So the salt quantity should be as low as possible in order to save energy and this apart from the cost for salt and deposit.

Sometimes the scrap available contains a high amount of organic components. Usually these organics comprise - besides oil – PVC, PVF and polyethylene. Paints adhering to the metal need not be considered since the organic binders have volatilized during drying of the paint. Usually the organics are just pyrolyzed during heating up of the batch with the result that the flue gases have to pass an after-burning system to remove unburned carbon. In most cases this process step is not energy-efficient. The technology of the rotary drum furnace offers the advantage to use the latent energy of the organics to improve the energy efficiency of the furnace. This requires introduction of additional air to the furnace in order to provide the oxygen required for the reaction. This takes place immediately after charging the scrap which means that spontaneous combustion will occur (**Fig. 1.16**) and the heat generated cannot be transferred to the batch totally due to the given heat transfer mechanism (see also Fig. 1.4). By setting furnace parameters, such as air quantity, furnace pressure, energy input by the main burner and rotating speed, the pyrolysis of the organics can be delayed to a certain extend. But although most of the energy is lost, the total energy balance is improved. **Table 1.3** gives an example of such procedure.

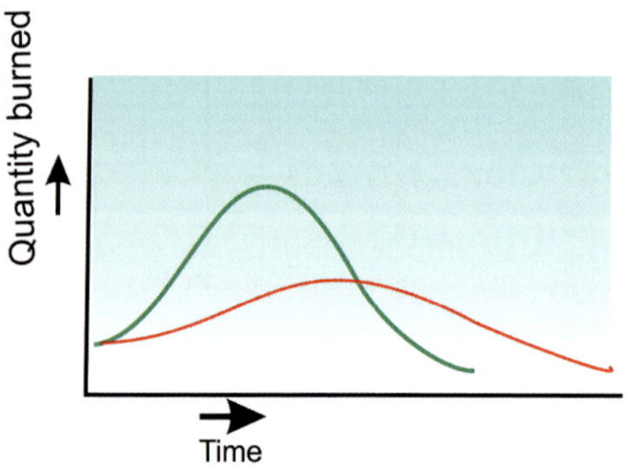

Fig. 1.16: Time-dependent combustion of organics in a rotary drum furnace

The example now shows clearly: When comparing power consumption of different melting technologies it will be indispensable to look closely into the basis of the relevant statements or expressions. It is of very limited use to have a consumption figure but not knowing exactly how it is obtained and what is the basis for such evaluation. Furthermore, it is also not possible to have a general figure based on the metal quantity processed since the components of the scrap melted play a very important role. It is also quite clear that there is a limit to the lowest recovery, i. e. the lowest metal quantity contained in the scrap. It will be unavoidable to evaluate the economics using factors such as cost for scrap, energy, additives and deposit. In an existing plant meaningful data can only be obtained on the basis of long-term operation data. Necessary calculations can be made, however, if reliable data are available. This refers to the overall economics. If considering data for the entire batch to be charged to the furnace, reliable energy data can be calculated. These together with empirical data can provide a trustworthy picture about a specific technology.

Table 1.3: Example of organic combustion, 10 t of scrap, organics 10 %

Annual throughput	23,000	t/a
Working days per annum	300	d/a
Availability	90	%
Working hours per day	24	h/d
Throughput per hour	3.85 chosen 4.00	t/h
Quantity of salt	2	t/h
Quantity of aluminium	0.2	t/h
Quantity of oxide	1.8	t/h

1.6 Alloying

One of the keys to a good product quality – if not the most important one – is to finally obtain the alloy as required by the customer. The contents of the charge have to be composed carefully to reach that target. The simplest procedure is practised in a remelting plant. Mostly the alloys to be produced are available as recycled material as purchased on the scrap market or recycled in a closed loop as in-house scrap.

For the process in a refining plant, scrap having an analysis as close as possible will be selected to be charged. On this basis, other types of scrap containing a large quantity of desired alloying metals are blended to that specific scrap. If this does not suffice to obtain the target analysis, pure alloying elements, such as silicon or clean copper scrap, are charged at defined quantities. Sometimes it is necessary to add primary aluminium as well. When selecting the scrap also economical factors have to be considered since the plant has to achieve the product at cost as low as possible. Pure alloying elements are expensive and their addition should be reduced to the absolute minimum. Also tolerable additions of metals, such as Zinc for instance, should be provided by the scrap.

Knowing the composition of the different scrap qualities, a batch calculation will be prepared. It is based on the quantity of one complete furnace batch. Mostly more than one charging is necessary to fill the entire batch into the furnace. The complete quantity of material must then be prepared for the individual chargings whereby also the specific requirements of the melting process are considered. It could, for instance, be necessary to charge big lumps of scrap with the first loading or to add all of the salt at this stage.

In principle, the procedure for mixing the proper batch composition is simple.

First step is to choose as basic material a scrap with an analysis as close as possible to the alloy to be produced. Now alloying elements need to be added.

For example, according to the available data, the copper content of a certain scrap may be 1.8 %. But the target to obtain, for instance, a 226-alloy would be 3 – 3.5 %. Consequently, 1.2 – 1.7 % of copper have to be added. To have some flexibility 1.5 % may be chosen. The entire calculation has, of course, to be based on the required batch size, i. e. the furnace capacity.

For this example the calculation is made for a furnace with a bath capacity of 10 tons.

We first calculate the weight of the 1.8 % copper in the selected scrap:

$$1.8/100 \cdot 10{,}000 = 180 \text{ kg.}$$

To arrive at the total copper content of 3.3 %, a quantity of 1.5 % is added:

$$1{,}800/1.8 \cdot 1.5 = 150 \text{ kg.}$$

This calculation is, however, not quite correct since we exceed the capacity of the furnace. We either accept the overloading of the furnace (what most production people do anyhow) or we reduce the scrap weight by 150 kg to 9,850 kg. The final weight of the copper content should be

$$3.3/100 \cdot 10{,}000 = 330 \text{ kg}$$

Original copper content is

$1.8/100 \cdot 9{,}850 = 177$ kg and to that we add $330 - 177 = 153$ kg of Cu to arrive at the complete furnace batch of 10,000 kg.

We finally end up with a copper content of

$(177 + 153) \cdot 100/10{,}000 = 3.30$ % which is very well within the target range.

If we proceed as outlined first by overloading the furnace we obtain a copper content of

$(180 + 150) \cdot 100/10{,}150 = 3.25$ % which is also close enough.

In case other scrap is used to adjust the analysis, the calculation is a bit more complex. Since the content of a specific element in scrap alloy is far below 100 %, the quantity of material added is much higher than using pure alloying metal. Other elements are charged with the scrap as well and will have a distinct influence on the total analysis. Therefore, the charge of the other alloying elements must be considered, too.

The element with the highest content will rule the calculation. With smaller additions of scrap from the scrapyard catalogue or even pure elements, the plant personnel will approach the target step by step. The components to be blended will be calculated on the basis of a trial and error procedure. The person responsible to do such preparation work requires thorough experience and sound knowledge. A computer program or a simple calculation using spreadsheet software is mostly used. Fig. 1.15 shows such a batch calculation. It first lists the different alloying elements of the target alloy and their percentage. Scrap intended to be used is filled in together with the tested analysis of this material. Based on experience, a quantity of the main scrap is estimated and filled into the relevant column of the spreadsheet. Also the tested recovery is filled in to obtain the actual metal quantity. The individual weights of the various alloys are automatically calculated by the program. The next quantity of material is also filled in using the same procedure. All values are totalled and the new analysis is calculated. The total charge is summarized at the bottom for easy comparison with the furnace capacity given in a bottom line. Weights and materials are adjusted to arrive at the desired data and keeping in mind the furnace capacity as well. If calculated for a rotary drum furnace, the required quantities of salt are also given in the sheet. Finally, the metal treatment, such as refining and flushing, is defined to give clear instructions to the operator. The batch calculation sheet follows the metal through all process steps. In most cases it will be necessary to adjust the alloy finally in the casting furnace. Calculation will be made following the procedures as described above. The sheet will finally be filed for possible reference at a later stage, i. e. in case of a customer complaint.

The furnace batch is composed of scrap, pure alloying elements and master alloys.

We have noticed already that the operators try to start with scrap that has an analysis close to the target alloy. When selecting scrap, the very specific characteristics of this material need due consideration. This is in general not very difficult. But some types of scrap do have some particular features. We will discuss this using three examples.

- Swarf comprising turnings, borings and the like are generally a low-cost material and, therefore, they will be used as typical for material with high specific surface area. Very often the material is collected from different sources. Thus, its use requires some care. Sometimes the materials will contain too much zinc or iron and also, particularly, if not passed through the swarf drying plant, a high percentage of iron. Thus, for instance, their use for alloy 226 is very limited. They have to be very clean, if not processed in a rotary drum furnace. Other material with large specific surface area swarf should be blended always with other material, if good quality metal is required. Exemptions are, however, swarf supplied from very well-known sources and defined alloy composition.
- Dross is a very particular raw material. It can only be processed in a rotary drum furnace. Dross may be of very different qualities regarding analysis, contaminations and metal content. Primary smelter dross has a white to metallic image and is very clean and its metal content is high. It can be used for any alloy and could even replace primary ingots if the plant is able to process it. Other dross is collected from different sources. It may comprise various alloys and generally has a high content of salt. Due to the salt it looks black. Used in the production of high quality alloys, it must be blended with other elements.
- Used beverage containers (UBC) are a valuable material. They have a high magnesium content that is mostly burned off during melting. The magnesium content is also high. One disadvantage is the substantial amount of paint and plastic. The body became continuously thinner as the quality of the aluminium alloys were improving. This saves energy required for transport but the portion of paint and plastics increased. For efficient processing in a melting plant the UBC material should be de-coated prior to melting. This is important if the material is used to produce body stock again. In refining plants UBC scrap is added to produce different alloys. Therefore, it will not be de-coated and processed as supplied.

There are different methods to add the alloying elements to the batch to arrive at the required analysis. The most important alloying elements are silicon, iron, copper, magnesium and manganese.

- If casting alloys are to be produced, silicon is generally the element representing the largest percentage in the alloy (obviously with the exemption of aluminium). But still it may be necessary to add some pure silicon or some silicon master alloy if no adequate scrap is available. Ferro-silica is used as master alloy if a certain percentage of iron is wanted or tolerated. Usually the ratio aluminium to iron is 75 / 25. Using this master alloy the contents of Si and Fe must permit addition of iron in this ratio. Producing, for instance, the alloy 223 with 3 % Si and 1.1% Fe, there is now danger of spoiling the charge with too much iron. The alloy 260 has 12 % Si and 0.7 % Fe. Thus, assuming that there is no iron content in the scrap selection only 2.1 % of silicon can be added as ferro-silica. The balance of approximately 10 % Si must either come from the scrap charged – which is most likely the case – or charged as pure silicon. There is also ferro-silica with much lower Fe content available that could be used. There is also the possibility to add silicon as master alloy with aluminium. Ferro-silica is also used when the iron content of the alloy needs to be adjusted and silicon is also part of the analysis.
- Magnesium is easy to be charged as master alloy with aluminium since metallic magnesium oxidizes very fast, i. e. burns off. It can be charged as pure metal if the necessary precautions are taken.
- Manganese is added as pellets or round piglets. The material can also be charged as thin plates.
- Copper is mostly charged as cable scrap, sheet material or as "candy", which is a mixture of wire and sheet scrap.

Due to their different melting points and also different density there are various procedures for adding these elements to the batch. These elements are usually not melting in aluminium. Thus, they dissolve over a longer or shorter period of time. Obviously this process accelerates at higher temperatures of the aluminium. This has a distinct influence on the method of charging them to the furnace.

There are two stages where the necessary alloying elements are added to the aluminium scrap selected for production.

The first stage is the melting furnace. All of the scrap will be charged together with copper and iron. These metals go into the furnace with the first batch to ensure maximum residual time for dissolving. Zinc is preferably charged as part of scrap. Particularly in a rotary drum furnace the alloying components are intimately mixed with the furnace charge thus getting evenly distributed in the aluminium.

Addition of silicon has to be done in the casting furnace. It is good practise to charge 80 – 90 % of the calculated quantity of silicon into the empty but hot furnace. Then the liquid metal is transferred from the melting furnace. The silicon is now completely covered by liquid metal whereby the temperature should not be less than 740 °C to allow the silicon to dissolve in a reasonable period of time. If the residual time is too extended the melt losses in the furnace become very high. After thorough stirring a sample will be taken and the remaining quantity of silicon is added. Thorough stirring is required again to cover the silicon with metal thus avoiding spontaneous oxidation. If the lumpy silicon is humid it should be dried on the pre-heating bridge of the hearth furnace prior to charging in order to avoid an explosion. Drying is evidently not necessary if charged into the empty furnace.

Magnesium has a very high affinity to oxygen. Therefore, it can only be charged into the casting furnace. If pure metal is charged, measures have to be taken to avoid direct contact when placing the metal into the furnace and when submerging it rapidly into the aluminium bath. For this purpose operators use a plunger which they put into the furnace swiftly followed by agitating the bath.

A good stirring system will definitely be of assistance, not only when charging magnesium but in all cases where thorough stirring is required. It is very convenient to add magnesium in the format of a master alloy with aluminium. Alloys with sufficiently high magnesium content are also a means of adding this metal. In former times engine blocks of the good old Volkswagen Beatle did have very high magnesium content. But this resource is not available anymore. The trend in the automotive industry to use more magnesium alloy for car wheels or for interior parts like dashboards or seat supports in order to reduce weight further may open a new source.

What are the requests to the designer for the furnaces to allow easy and effective alloying?

Sufficient rotating speed of the rotary drum furnace, large charging doors, an amply sized preheating and drying bridge next to the charging door, good temperature distribution within the liquid metal by shallow bath or by efficient agitation and, last but not least, sufficient burner capacity to provide fast heating of the metal.

2. Overview of melting technologies

Christoph Schmitz

The key equipment for the metallurgical processing of aluminium scrap is furnaces. Contaminations and foreign matter can be removed as soon as the metal reaches the liquid stage. It is possible now to add any alloying element and mix it with the original batch. To obtain the desired alloy within defined tolerances of the metallurgical analysis, different types of scrap are blended with the alloying components. These targets cannot be obtained with a kind of "universal furnace technology". Different types of scrap present particular problems to the melting process. Block material, for instance, requires far less protection against oxidation of the metal than small particles do and scrap with a high content of organics needs to be processed differently from material such as dross that has an extensive oxide content right from the beginning. To achieve the most effective technology, particularly in view of economy and ecology, very different processes have been developed, tested and finally operated during the past years. Thus, equipment selection for a plant must require careful investigation, taking into account material available for the process, product mix, availability of resources, main energy and type of fuel (oil or gas), infrastructure of the plant ambience etc. The envisaged production capacity is also important when technology is selected.

These all are ruling factors to define the melting technology for a particular plant.

During the first process step the furnace must permit efficient melting of the metal as well as separation of unwanted components attached to the aluminium that have not been removed during a mechanical preparation of the scrap. Steadily increasing cost for energy, together with the more and more stringent regulations for environmental protection, have been the driving force behind the development of today's efficient melting technologies. Because of the also rising cost for raw material, it is essential for an economically successful process not to lose too much of the valuable material during the furnace process.

The furnaces used in the first production step are the "melting furnaces".

In the second stage of metallurgical processing, the refining stage, the final characteristics of the alloy will be set. The necessary corrective alloying elements are charged and mixed with the liquid metal in the furnace. Therefore, the furnace for this production stage is called "mixing furnace" or just "mixer". The furnace must also be able to rise the temperature of the metal, let it cool down to a defined temperature or hold the batch at a certain temperature for metal refining. Now the unit becomes a "holding furnace".

There are other characteristics to be considered for the second production stage. It is a standard practise in recycling plants to add no-go material produced before, large pieces of scrap or purchased ingots to the furnace in the refining stage. Since this material needs to be melted, the furnace requires a certain melting capacity. One quality factor is the amount of gaseous or mechanical contaminations in the melt. Therefore, the furnace should permit treatment with chemicals or gas to reduce the gas content, i. e. hydrogen and metallic inclusions such as sodium or potassium and the like, or as may be required in some cases, magnesium. Although there are very efficient inline treatment systems available, treatment of the melt in the furnace offers advantages, if many alloy changes are required. In such case the inline unit must be emptied completely or a standby unit must be kept ready for operation and preheated. Since the furnace is used for metal treatment and to "convert", the metal composition of the batch in the furnace becomes a "converter".

The next production step is casting. Here again different characteristics of the furnace are required.

Generation of dross by oxidation must be minimal during the casting period and a precisely controlled metal flow is essential. Generally, mixing/holding furnaces are identical to the casting furnace and also – with limitations – melting and casting furnace as well. From the "mixing/holding" furnace metal is transferred at a steady flow to the casting equipment. Now we have a new name for our furnace and we call it "casting furnace".

Thus, identical furnaces may have different names according to their application. But there is no reference to the furnace design, may it be a rectangular or round stationary reverberatory furnace or a rectangular, drum-type or oval-shaped tiltable reverberatory furnace. The industry uses various types of furnaces; some of them are designed for very particular applications. That could be depending on the product mix or on a very particular type of scrap to be processed.

2.1 Classification of furnaces according to design characteristics and application

2.1.1 Basic requirements

With the increase of plant capacities and the modern furnace technologies, the bath capacity, i. e the quantity of liquid metal a furnace is able to accommodate, is growing, too. Small furnaces having a capacity of only 5 or 10 tons are used for very particular applications. This refers to melting furnaces and casting furnaces alike. The smaller furnace size leads to many service operations for a given metal throughput and this time is lost for the production. Today the most suitable furnace may have a capacity of more then 25 tons. This bath capacity is very convenient for operation and efficiency. Since many customers ask for a minimum charge of 20 tons produced from the same batch, marketing may also be a factor for selecting larger furnace capacity. Another factor to be considered is ease of operation. Dross generated during the furnace process, be it melting or holding or casting, tends to collect at furnace walls or settles at the furnace bottom, sometimes even on the roof. Even using the most optimal furnace design, this occurrence cannot be avoided entirely. The growth of the oxide layer decreases the furnace capacity. Therefore, the dross layer must be removed from time to time. This is even more important when the alloy changes. Consequently, the design must facilitate the operation to remove the dross.

The type of energy input is a ruling factor for the furnace design as well. There are different technologies available.

Fuel-heated furnaces

The energy is provided by fossil fuel as natural gas or heavy fuel oil, sometimes called bunker C oil or masut or as light fuel oil which is identical to diesel fuel. Gas generated by coke or coal is used in other industries but not in the aluminium industry. The application depends on availability. Although gas is very convenient to use in furnaces, it is only available where a connection to a gas pipeline exists. If not, oil must be used. This could be light fuel oil, which is mostly expensive or heavy fuel oil that is less expensive but difficult to handle. At ambient temperature the viscosity of this type of oil is very high. To get a good combustion, the fuel oil must be brought to a temperature of at least 80 °C. For starting and stopping the firing equipment, light fuel oil is used since heavy oil cooling down in the system may clog pipes and nozzles and a restart is very difficult to achieve.

The energy of the fuel is used to emit heat by rising its temperature to a value above the igniting temperature in a burner that ensures a delayed contact between oxygen and fuel. Immediate contact of the partners participating in the reaction would result in an explosion.

Fuel-fired furnaces are designed as the variations of the traditional hearth furnace, crucible furnaces or rotary drum furnaces. They are most commonly used in the aluminium industry with very good success with regard to metal recovery and energy consumption. Due to the direct heat emission, the efficiency of the firing systems is very good and the transfer of heat to the product is excellent. Cost of energy is also much lower than cost for electrical energy.

Electrically-heated furnaces

These furnaces are designed as induction furnaces or resistor-heated furnaces. Electrical heating seems to be the most convenient means of heating if this energy is available at low cost. This is obviously a precondition for the operation of a primary aluminium smelter in the specific area. But the conditions may be quite different at a location for an aluminium recycling plant. The absence of flue gases during furnace operation ensures that metal loss due to oxidation is low. There is also no contamination by flue gas in the vicinity of the electrically-heated furnace.

Crucible induction furnaces can be charged with clean scrap, ingots or even liquid aluminium. Unfortunately, they have a design size limitation at 8 – 10 tons that is too small for the requirements of a modern secondary smelter. This capacity requires a large furnace diameter and is operated from above, resulting in a potential danger to the operators working in the uncomfortable environment. Generally, induction furnaces are characterized by substantial investment cost and high cost for maintenance and labor. Therefore, in a secondary smelter the application for this type of furnace is limited to melting swarf generated by machining.

Channel induction furnaces are generally designed as round top-loaded design with one or more inductors attached to the furnace bottom. Due to the circulation of the metal generated by the induction process, the temperature distribution and homogeneity of the aluminium are excellent. Since there is also no contact with flue gas metal, loss due to oxidation is very low. In some instances channel induction furnaces may be designed as rectangular box. Thus, the danger of operating from the top does not exist. Channel induction furnaces of more than 40 tons bath capacity have been installed and operated successfully. Since this type of furnace has to be operated with a liquid metal heel, the application as melting furnace is very limited and so is the application as holding/mixing furnace due to the somewhat difficult alloy change. However, if economic considerations permit, it may be an option.

Resistor-heated furnaces are designed as stationary or tiltable reverberatory furnace. The resistors with protection tubes are inserted and installed in the ceiling with heat transfer to the metal by radiation only. The service temperature for the refractory material suitable in an aluminium furnace for roof and walls can hardly exceed 1,200 °C which is the limiting factor for this type of furnace. The energy transferred by roof radiation does not permit efficient melting of solid material. The application of the furnace is limited to casting and holding. Due to the absence of flue gases, metal loss is very low but higher cost for investment, maintenance and operation have to be evaluated from case to case. Generally, this type of furnace is not used in the secondary industry.

Summary

The flue gases leaving the furnace are generated at the location of the furnace and have to be treated on site. This is different with electrically-heated furnaces. Here the flue gases are produced at the place of the power generation and treatment has to take place there. This must be considered in the environmental balance when comparing the two systems. In fact, this balance appears to be more favorable for the directly fuel-heated systems even if the energy consumption of electrically-heated furnaces appears to be lower since no waste gas losses must be considered. But the contrary can be observed in production plants. All electrical furnaces have in common that the energy cost is higher than for furnaces using oil or natural gas. The absence of stack losses in energy is countered by energy loss for cooling the inductors – which may reach 35 %, a value comparable to the stack loss of a modern fuel-heated furnace.

One other aspect should be considered as well. Mostly, electrical energy is generated by thermal power stations using fossil fuels or nuclear power. It is obvious that there is no flue gas emission in the vicinity of an electrically-heated furnace. These emissions occur at the power plant in the same quantity. Thus, there is no advantage with regard to ecology.

Looking at the overall ecological balance, the unavoidable losses are generated at a different place in the power plant. Also the total efficiency of the electrical power generation is to be taken into consideration. To generate electrical power, heat must be generated first and is then transferred by means of turbines and generators to electrical power. The overall efficiency of these thermal processes is in the range of 33 %. This is reflected by the price of the electrical energy. In the furnace this is transferred into heat again. Hence, the total efficiency of the energy chain is definitely lower than the one obtained in direct heating.

The situation is much better if hydropower is available. There will be no calculated heat efficiency to be considered and waste gas emissions do not exist. In such cases the use of electrically-heated furnaces may be a good option for certain types of scrap and specific process conditions.

Considering these facts, gas or oil-heated furnaces appear to be the most heat-efficient process for the aluminium metallurgy with a wide range of applications. But there are certainly cases which favor the use of electrically-heated furnaces, particularly induction furnaces.

2.1.2 Classification of furnaces

In the following we will define furnaces according their design principle. **Fig. 2.1** gives an overview of furnaces used in the aluminium industry. The denomination for the different production

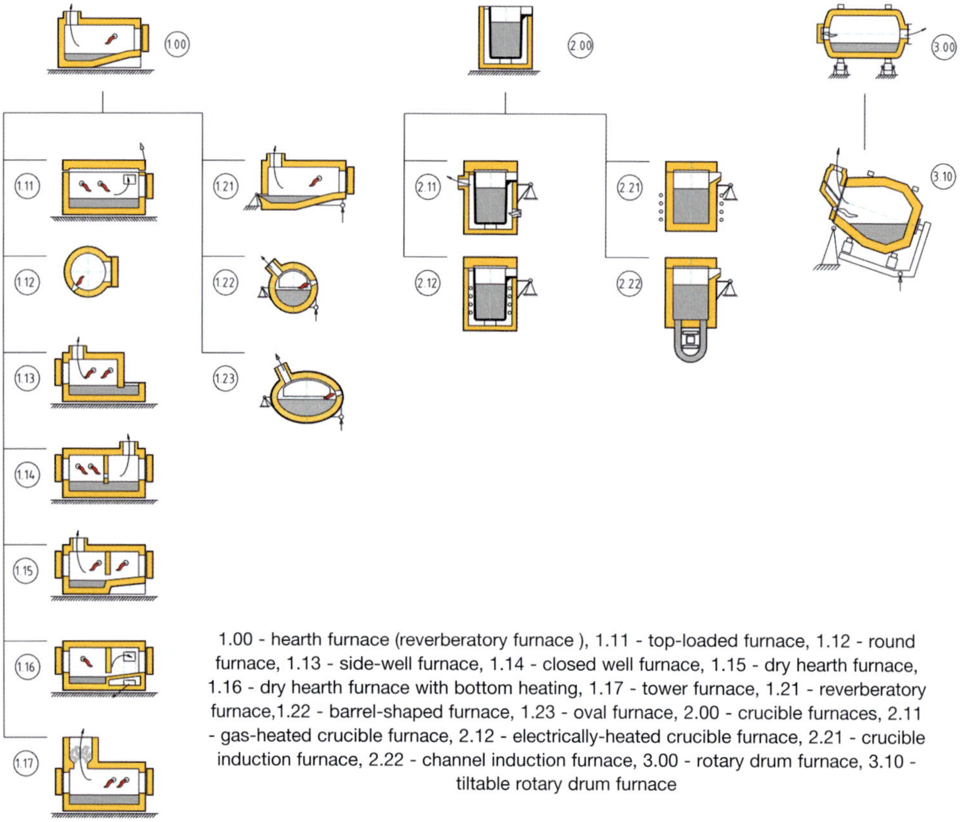

Fig. 2.1: Family tree of furnaces used in the aluminium industry

Table 2.1: Application of furnaces in the aluminium industry

No.	Type of furnace	Remelter	Refiner	Foundry	Primary
1.00	Reverberatory furnace, stationary	4	3	3	0
1.11	Top-loading furnace	2	1	0	0
1.12	Round furnace	1	0	0	0
1.13	Open side-well furnace	0	0	2	0
1.14	Twin chamber furnace	3	1	1	0
1.15	Dry-hearth furnace	0	3	0	0
1.16	Quick-melting furnace	0	0	3	0
1.17	Tower furnace	0	0	3	0
1.21	Reverberatory furnace,tiltable	4	3	3	4
1.22	Barrel-type furnace	2	3	0	0
1.33	Oval furnace	2	3	0	0
2.00	Crucible furnace	0	0	3	0
2.21	Crucible induction furnace	2	1	3	1
2.21	Channel induction furnace	2	0	1	2
3.00	Rotary drum furnace	0	4	0	0
denomination: 0 - not used, 1- used in special cases, 2 - frequently used, 3 - standard technology, 4 - key technology					

steps in view of the application is discussed in one of the following sections with more details in the description of the different furnace designs. The most common furnace in the secondary aluminium industry is the hearth furnace and to some extent all furnaces developed from the basic concept for special applications. Secondary smelters operate rotary drum furnaces as melting furnace, particularly for processing scrap with large specific surface area and mostly heavily contaminated.

The family tree of furnace design will provide an overview of furnaces used in the secondary aluminium industry. Some of the technologies are limited to very particular applications. There are different applications throughout the industry depending very much on the kind of production. According to common understanding, the aluminium recycling industry comprises remelters and refiners. Here large capacity of the plants is essential and very specific material needs to be processed. For remelters, mainly hearth furnaces of different and sometimes special designs are used. Rotary drum furnaces are generally limited to refiners. Downstream facilities, such as rolling mills or extrusion plants, use hearth furnaces as well to process their inhouse scrap and fresh metal. The capacities of such plants are also high. In general, foundries have a fairly small production rate and process metal. The foundries purchase ready-made alloys from the refiners. They use furnaces which best suit their requirements.

Table 2.1 gives an overview of the application of the different furnaces in the aluminium industry. The shadowed columns refer to the secondary industry. It is the intention of this book to focus on recycling of aluminium. The book focusses on the technologies, processes and practises applied in this industry. Although the other furnaces are discussed briefly, a more intensive look is directed to the relevant furnaces, i. e. hearth furnaces in their different forms of design and technology, rotary drum furnaces and to less extent on induction furnaces. Other types of furnaces will certainly appear in secondary recycling plants because the plant owner may have purchased them for a certain line of production or reasons which are certainly not known to people outside their line of business. It may not be general practise to have such equipment in a secondary aluminium plant but there is certainly no rule without exemptions.

More details of typical design features or applications are covered later on.

For the classification of the different furnaces please refer to Fig. 2.1 which tries to define genesis and evolution of the different furnaces. The numbers used in the description below refer to the item numbers in Fig. 2.1.

Hearth furnace (1.00)

The hearth furnace is certainly the "classical" furnace in metallurgy and has been used in early times by our forefathers for processing their bronze or, in later periods, their iron. From this design many types of furnaces have been developed over the centuries to serve the individual applications. The more special designs evolved, the narrower became the field of use. A parallel development took place when operators realized that some materials, mainly scrap with high oxide content as well as scrap with large specific surface area, could not be efficiently processed in the hearth furnace in the traditional way. To overcome the problem they covered the surface with a protective flux. But the heat exchange was very poor. Thus, the batch required movement to allow the heat to reach all of the material. Moving the content of the furnace with muscular power was very exhausting. Thus, the idea came up to move the furnace instead and the rotary drum furnace was born.

As already mentioned, different furnace designs based on the traditional hearth furnace are getting more and more specialized.

The reverberatory furnace, (1.00) may be regarded as the basic concept. Its use as casting furnace results in the design of the tiltable furnace (1.21). Tilting at controlled rate is required for a steady metal flow during casting.

In case the rectangular shape of the furnace appears to be more costly, a deep bath is more advantageous. Particularly if the metal temperature is to be maintained only, a furnace with cylindrical cross-section (1.22) is functional to fulfil the required duties. The design combines low wall losses with good conditions for metal refining. Because of the furnace shape, the size of the skimming door is very small and obstructs operation. To reduce the impact of the unfavorable door design, the furnace cross-section is flattened and becomes an oval shape (1.23). Both furnaces are always designed as tilting furnaces.

The casting/holding furnace design does not permit many variations. This is very different for melting furnaces. The standard hearth furnace is loaded through a door in one of the sidewalls. Particularly in rolling and extrusion plants, inhouse scrap is collected in tangled piles. To charge this with side doors is somewhat difficult. Thus, top loading furnaces (1.11) are quite useful. If the piles are very irregular, the design is a circular-shaped top loading furnace (1.12) with one or more small skimming doors at circumference. More and more stringent environmental regulations today and propelling energy cost limit the application of top loading furnace to very specialized operating plants and push the operations towards more sophisticated scrap handling.

Energy cost is the leading consideration for the design of the tower furnace (1.17). It uses the comparatively high quantity of heat of the flue gas leaving the furnace through the stack. Material charged to the furnace is placed into the flue gas stream in a so-called tower prior to being charged to the furnace. Since capacity and melting rate of furnaces of this design are comparatively low, the application is generally limited to diecasting plants. A similar design is used in the dry hearth furnace (1.15) that is used when aluminium must be separated from other metals having a higher melting temperature than aluminium. This is generally brass, copper or iron. The combustion products melt the aluminium first and the liquid metal now flows over the sloped ramp to the holding section of the unit. The furnace operators are now able to remove the tramp metal from the ramp through a large door at the upper end of the dry hearth. Variations of this technology incorporate different flue routes of the flow gas and various arrangements of the heating systems by different types of burners.

The furnace shown as item (1.16) may be regarded as a combination of the tower furnace and dry hearth furnace. It uses the energy of the flue gas to heat up the batch of material placed on the ramp. The hot flue gases are directed to flow first to the metal batch on the ramp and are then sent through a chamber underneath that ramp. Here again, the design limits the application of this furnace more or less to diecasting plants.

The standard hearth furnace is very suitable when melting large pieces of metal such as sows, ingots, compacted bales, bundled extrusion shapes and the like. Melting of material with large specific surface area will invariably result in excessive metal loss due to oxidation. If charged into liquid metal, contact with flue gas and the ambient air is limited. Consequently, the hearth furnace was equipped with an exterior side well (1.13). This well permits very easy charging into the aluminium bath. But even equipped with a lid, the energy loss of the side well is considerably high and the melting rate is pretty low.

Thus, the next step of development incorporates the side well in the furnace interior by simply separating the original furnace room as heating chamber from the charging well. The furnace now becomes a closed well furnace (1.14) or, as it is called as well, a twin chamber furnace.

Both chambers are "communicating" via the liquid metal bath. Scrap is loaded via the side door in the wall of the charging chamber and melts due to contact with the liquid metal. The metal will be tapped from the bottom of heating chamber. Due to the typical features of the design, the furnace always operates with a heel of liquid metal. If the alloy of the feed material does not change and the required metal analysis remains constant, the furnace could be operated continuously. Modern design incorporates sophisticated flue gas flow and effective burner systems which handle the individual conditions of the production site. A liquid metal pump creates a forced circulation of liquid metal for improved heat transfer to the solid scrap. Different pumping systems may be used. They are arranged as part of the charging well or a separate pump chamber attached to the furnace or, in case of an electro-magnetic stirrer, they can be arranged at the side or underneath the furnace without contact to the liquid metal. The arrangement of the pumping system or the method of introducing heat to the furnace does not change the general technology.

Crucible furnaces (2.00)

These furnaces represent a development originating from the hearth furnace. They are round furnaces that are loaded from the top. The lid provides an excellent sealing of the liquid metal against the atmosphere. A typical design feature is the indirect heating of the batch (2.21). A crucible to receive the metal is arranged inside a refractory-lined steel housing. A burner in the furnace housing heats the crucible from its outside and the flue gases leave the furnace without direct contact to the metal. The crucible is the critical item in this design. It must be made of heat-conducting material that also possesses sufficient mechanical strength to carry the load of the metal inside. It is not possible to design a supporting steel structure because of the burner impact outside of the crucible and the metal contact inside. Silicon carbide is a very good material to be chosen for the crucible. But size is limited due to the mechanical strength and for cost reasons as well. The crucible furnace may also be heated by electrical resistors (2.12).

Induction furnaces are also crucible furnaces. The crucible induction furnace (2.21) is designed as a refractory-lined crucible surrounded by an induction coil. A steel frame supports the entire unit. Alternating electric current flowing through the coil induces eddy current in the aluminium batch that is, according to Joule's law, heated up. Thus, the heat is generated directly in the batch without excess temperature. The furnace can be equipped with a tight sealing lid. A hydraulic cylinder tilts the unit for metal discharge.

The channel induction furnace (2.22) consists of a round vessel and an inductor flanged to its bottom. As in the well-known electrical transformer, an alternating electric current around an iron core induces also an alternating magnetic field. This in turn induces an electrical current in the secondary circuit. Liquid metal passing through a channel arranged around the primary system forms this secondary circuit. The induced current generates heat in the liquid aluminium passing through. It also forces the aluminium to flow, thus transferring the heat to the metal in the vessel. Due to its design principle, the channel induction furnace can only operate with a heel of liquid metal. The power generated is usually only sufficient to hold the metal at temperature. The furnace is usually closed with a tight lid and is tilted by a hydraulic cylinder to discharge the liquid metal. Also the standard furnace is equipped with one inductor. Channel induction furnaces can be designed with two or even more inductors as well.

Rotary drum furnaces (3.00)

These furnaces follow a design principle that differs very much from that of the hearth furnaces. It comprises a refractory-lined steel vessel that rotates around the central axis. Scrap is charged through one of the front ends. The only burner is also arranged at one front end. Traditionally the rotary drum furnace has a fixed axis (3.00) design. That means the center line of the rotation remains in the horizontal position. The material will be fed through one of the frond ends. The charging door also carries the burner. Flue gas leaves the furnace through an opening at the other front end. Liquid metal as well as flux is discharged through tapholes at the circumference of the drum. The tiltable rotary drum furnace (3.10) comprises a refractory-lined vessel that is arranged on a tilting frame. This facilitates tilting of the complete furnace within the design limits. Due to the tilted operation a large charging door can be arranged in the upper front end which facilitates easy charging as well as tapping of metal and discharging of slag. The opposite end of the furnace is closed so the furnace looks like a rotating crucible. A large burner, located at the charging end, provides the energy required for the operation.

2.1.3 Design criteria and application of furnaces

The basic technology for the fuel-heated furnace is the reverberatory furnace (**Fig. 2.2**). Its name originates from the kind of heat transfer. It is assumed that most of the energy is passed onto the metal by radiation following the principle of reverberation. In practise, this is, however, only the case when the furnace is filled with liquid metal. During melting heat exchange by convection will be the case with only a small portion of radiation (**Fig. 2.3**). During super-heating of the metal, the portion of radiant heat is much higher now.

Some essential design features, as covered in the first section in more detail, will apply for other furnaces as well and may help to understand the reasons for their specific design, their advantages and their possible shortcomings. Although a reverberatory furnace has a simple appearance, the physical conditions and the interaction of the individual components with each other as well as the specific operation conditions are quite complex. Such a furnace is a system but not just an agglomeration of individual parts such as refractory lining, steel shell, burners and process control with the metal burners interacting at ever changing operation conditions. Mistakes made right from the beginning in design and manufacture or selection of components can have a negative impact on the economy and quality requirements of the entire process, even if the individual elements of the system work perfectly, if regarded as isolated units.

2.1.3.1 Hearth furnace

The principal design of the hearth furnace comprises a closed refractory-lined rectangular box (**Fig. 2.4**). One or more burners provide energy while the flue gases escape through an opening in the roof or in the furnace wall. The furnace can have a stationery or tiltable design.

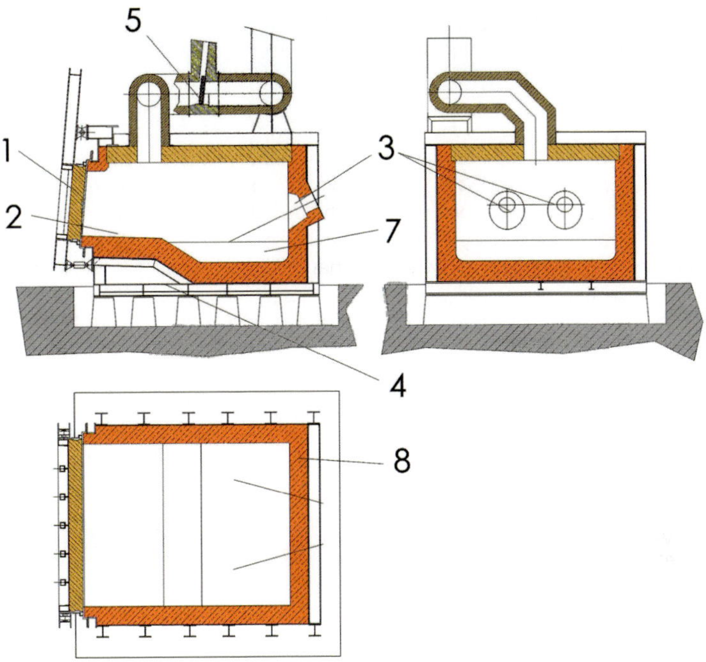

1 - charging and service door, 2 - preheating ramp, 3 - burners, 4 - steel structure, 5 - pressure control damper, 7 - metal bath, 8 - refractory lining

Fig. 2.2: Reverberatory furnace

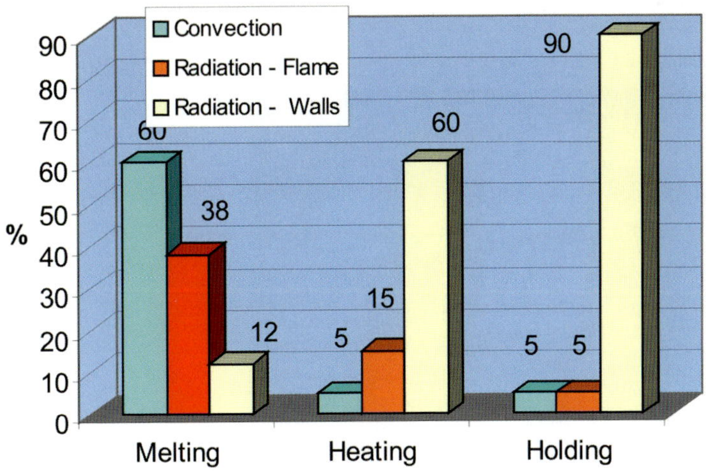

Fig. 2.3: Heat exchange in a reverberatory furnace

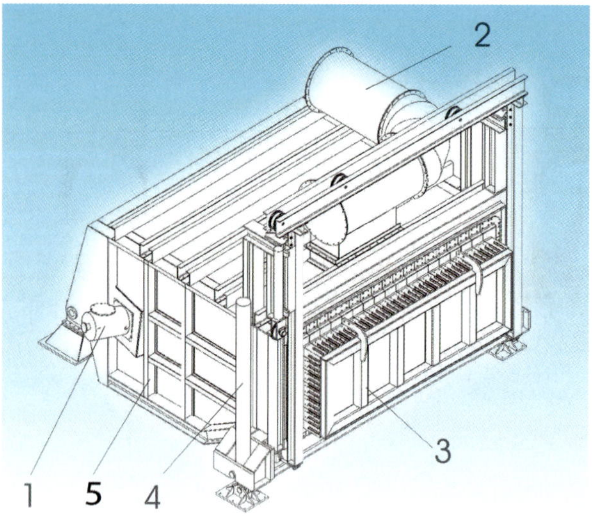

1 - Burner
2 - Flue gas duct
3 - Charging door
4 - Lifting/tilting cylinder
5 - Wall reinforcement

Fig. 2.4: Principle design of a reverberatory furnace

A shallow bath forming the furnace bottom holds the metal to be processed. The room above the bath enables good combustion and good heat exchange between combustion products and the material as well as the furnace body and, at the same time, is adequately large to receive a sufficient quantity of solid metal.

For ease of operation one large door is arranged at one of the long sides of the furnace. The door should open the entire sidewall of the furnace so that no edges obstruct the cleaning of the furnace walls (Fig. 2.2). In some cases it may be very convenient, regarding heat losses and easy operation, that the furnace door is a two part door with no center column (**Fig. 2.5**).

A charging machine or a forklift truck will load solid metal through this furnace door which does also allow skimming with the aid of heavy equipment. Liquid metal will be charged through a

Fig. 2.5: Tiltable reverberatory furnace, 60 tons capacity

Fig. 2.6: Charging of liquid metal

separate charging well by means of a crane or a tilting device (**Fig. 2.6**). Feeding of liquid metal through the large furnace door is also possible but it requires opening for some time period with resulting energy loss and additional dross generation. The furnace can be designed for top loading depending on the type of scrap to be processed.

In order to reduce heat losses through the furnace walls, the total exterior surface of the furnace should be at its minimum. This will also save cost due to reduced furnace weight. The ideal furnace is a square box with very low height. For operational reasons this design is not advisable. As already mentioned, the space above the bath must provide sufficient room for receiving solid metal. Furthermore, the furnace room must be large enough to allow efficient combustion and good flue gas turbulence of the combustion products to facilitate good heat transfer. A good value developed by experience is 1.8 m to 2.5 m above bath level independent of the furnace bath capacity.

For a good heat exchange the residual time of the combustion products should be as long as possible. The burners are arranged in such a way as to force a turbulent flow of the flue gas prior to leaving the furnace. To facilitate good preheating of large blocks of material placed onto the preheating well, the flue gas duct has its favorable location in the roof above the charging door. But some plant layouts may not permit such an arrangement. Thus, compromises have to be found frequently.

The depth of the furnace, i. e. the distance between the door and the opposite wall should be short enough to permit easy skimming of the bath surface. A large distance results in long skimming tools that are very difficult to handle, even with tools attached to a forklift or a special skimming vehicle.

The height of the bath needs some consideration. It is quite obvious that the metal bath fulfils the conditions for a hydrostatic equilibrium. This is the case if the vector of the density gradient is parallel to the gravity. Since the density of the aluminium bath is proportional to the temperature gradient, this means that no internal movement of the liquid metal will occur unless initiated by external forces. Energy to the metal bath is introduced from the top of the bath which results in decreasing temperature from top to bottom resulting in increasing density. Heat to the lower levels of the metal bath can only be transferred by means of convection. This may be described by the equation

$$t \cong \frac{h^2}{a} \tag{2.1}$$

indicating that the heating time is governed by the height of the metal bath.

The consequence is that the bath should be as shallow as possible.

An example may explain this. For liquid aluminium the value of temperature conductivity

$\alpha = \lambda/\rho \cdot cp$ expressed in m/s^2 is $37 \cdot 10^{-6}$ m^2/s at a temperature of 800 °C. Assuming a bath height of 800 mm, the characteristic time for the heat diffusion from top to bottom will be

$$t = h^2/a = 0.8^2/37 \cdot 10^{-6} = 17{,}300 \text{ s} = 288 \text{ min}$$

Reducing the bath level to 500 mm will reduce the diffusion time to 112 min.

This time is important for the temperature homogeneity and for the heat transfer for cooling or heating of the liquid metal. As a good standard the bath height of 500-600 mm has been established with some variations due to furnace design. But still the temperature difference between top and bottom is in the range of 23 to 25 °C.

In order to improve the temperature distribution some methods for stirring have been introduced. Most plants apply the practise of stirring with mechanical tools attached to a heavy forklift truck. Already within few minutes after stopping the action, the bath has settled in a stable condition again. For stirring the furnace door must be open resulting in additional dross generation.

Therefore, mechanical means for stirring have been developed.

A centrifugal pump could be installed in a side well that "communicates" with the metal in the furnace. It comprises a pump unit with impeller made of special carbon material (**Fig. 2.7**). Its drive consists of an electric motor or a compressed air motor that is connected to the impeller by means of a long shaft. The centrifugal pump is able to create a steady material flow at high velocity which may be in the range of up to 300 ms^{-1}. The resulting flow rate should be adapted to the furnace size. This ensures very good circulation of the metal in the furnace and comparatively low energy requirements. Operators have to take care that the impeller always operates in liquid metal. If the furnace is emptied totally or if the furnace is being tilted, the pump needs to be removed from the pump well and stopped. The pump should be kept in a special heating box to maintain its temperature thus avoiding damage to the impeller due to thermal shock when returned to the liquid metal. The impeller is a wear part so that annual cost in the range of the price for one complete pump must be considered. Investment cost comprises the cost for the pump, the pump well attached to the furnace, the preheating box and the lifting gear.

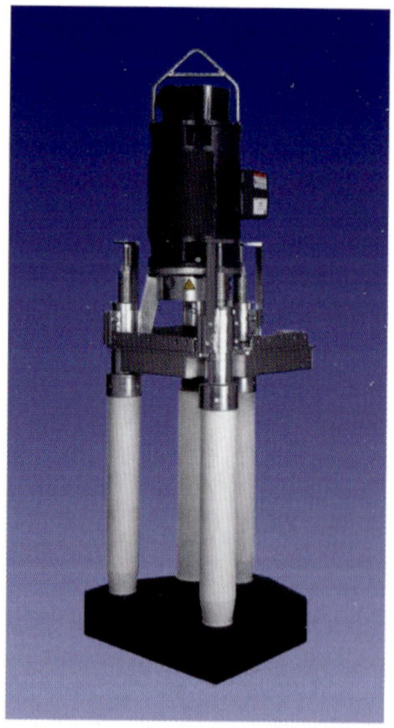

Fig. 2.7: Mechanical pump with impeller

The electro-magnetic pump (EMP) conveys metal through special tube channels. It is based on the so-called linear motor and can be described as a conventional cylindrical rotary motor which has been cut along an axial plane and then unrolled to give a flat structure. Whereas in a conventional motor the rotor

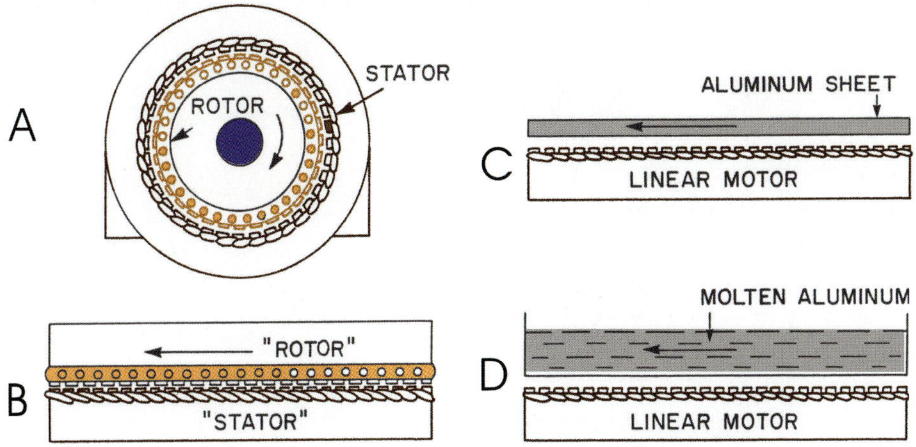

A- normal electrical motor comprising rotor and stator, B- normal motor unrolled, C- stator replaced by an aluminium sheet, D- instead of the aluminium sheet, liquid aluminium receives the horizontal thrust

Fig. 2.8: Linear motor principle

spins, in a linear motor the rotor is given a linear thrust. The rotor can be a sheet of aluminium or a batch of molten aluminium (**Fig. 2.8**). The electrical coil (stator) can also envelop a tube or a rectangular channel. The velocity of the liquid aluminium and the resulting flow rate is lower than the one obtained with the centrifugal pump but is still far above the actual requirements of the furnace system. The energy consumption is also moderate. The linear motor system can be flanged to the sidewall of a furnace. It can be designed as side channel installation or as system attached to a furnace wall. For the good efficiency of the electro-magnetic forces, the unit requires a comparatively thin high duty refractory layer between liquid metal and the magnetic system. The linear motor can be arranged around conveying tubes or rectangular channels. The strong flow of energy requires an efficient cooling system. There is also an increased danger of accidents due to the fragility of the refractories. However, contrary to the centrifugal pump, it is not necessary to have liquid metal in the furnace. Frequent monitoring of the channel conditions is required to keep them free of scaling. Due to dross generation in the furnace, this is sometimes difficult to achieve. Careful and frequent maintenance is required in any case.

The electro-magnetic stirrer (EMS) is attached to the bottom or to a sidewall of the furnace (**Fig. 2.9**). As in an induction furnace, it induces an alternating electro-magnetic field in the liquid aluminium thus causing it to rotate. The only condition required from the furnace builder is a non-magnetic steel plate in the area of activity of the EMS. The unit can operate through the full thickness of the refractory lining and is, therefore, very safe to operate with very little maintenance efforts. But the price to pay is the relatively high cost of electrical energy.

The system of gas injection bricks in the bottom of the furnace differs very much from the principles as described above. It is applicable in hearth furnaces and comprises a number of porous bricks installed in the refractory lining of the furnace. Each brick is connected to a feeding line for inert gases such as argon or nitrogen. The bubbles floating to the surface of the liquid metal bath create a stirring effect and hence ensure disturbance of the stable temperature distribution within the liquid metal. The efficiency of this system is limited to hearth furnaces since a strong horizontal movement of liquid metal cannot be generated. The system does not require much maintenance

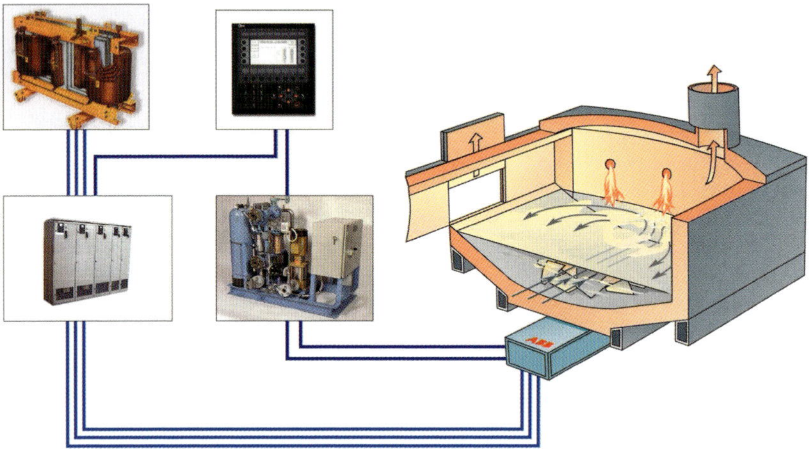

Fig. 2.9: Electro-magnetic stirrer (Source: ABB)

and energy cost is almost nil. But cost for gas has to be taken into account. The gas flow through the batch has a very convenient side effect. Due to lowering of the partial pressure in the melt, hydrogen is removed and oxides trapped in the melt are forced to float to the top of the liquid metal bath. However, both the centrifugal pump as well as the linear motor – with the exception of the electro-magnetic stirrer – can be equipped with a gas injection device and hence obtain a similar degassing effect.

All systems have been applied in the secondary aluminium industry. They permit a deeper bath which means furnaces with smaller dimensions and, for the operators at the furnace, less physical work in uncomfortable ambient conditions.

For standard operation a shallow bath with a depth of 500-600 mm seems to be the preferable solution for furnaces in the secondary aluminium industry since this design feature ensures trouble-free operation providing satisfactory results.

This is invariably true for casting furnaces as well. The temperature of the liquid metal bath needs to go up to casting temperature in most operations. Due to comparatively small surface area, the time to increase the temperature is very long and even gets longer with the depth of the bath, as shown before. On the other hand, a deep bath is helpful when metal is treated in the furnace. Reaction between gases and metal is improved when the passage of gas bubbles through the liquid metal bath takes more time. And this is certainly the case with greater bath depth. In such applications the installation of a stirrer is helpful. Circulation of the bath offers even more benefit. Since the bath surface can be maintained at a lower temperature level for good heat transfer, less dross will be generated. Furthermore, intensive bath movement suppresses demixing of the various alloying components.

The furnace bottom is sloped from the door opening to the lowest point of the hearth (Fig. 2.2). This eases skimming and cleaning of the furnace. Large blocks of solid material can be dropped directly into the bath or placed onto a bridge next to the door for preheating or drying (**Fig. 2.10**). This is quite important when charging 500 kg or 1,000 kg ingots, commonly called sows. Shortly before reaching melting temperature the operators push the block into the liquid metal bath. The procedure is very important when melting material with high moisture content. During transport and storage in the open, the ingots are exposed to rain and snow and may pick up a substantial

Fig. 2.10: Preheating ramp

amount of water, collected in the cracks and shrinkage cavities. As soon as this water gets in contact with liquid metal it evaporates instantaneously and the steam is suddenly heated to the metal temperature. One liter of water volume expands to 1,25 m^3 of steam in standard conditions and the rise in temperature almost triplicates this volume. Entrapped in liquid metal, the explosion upon expansion causes the hot metal to erupt thus causing damage to equipment and it may also cause heavy and sometimes lethal injuries to the operators. But not only water is able to cause accidents. Oil-bearing blocks are also critical. In such a case it is not only the evaporation. The oil will react instantaneously with oxygen generating combustion products. Only 1 kg of oil results in almost 11 m^3 in standard conditions of flue gas also expanding due to the reaction temperature to almost 50m^3. Assuming this reaction will occur in 1 second, the equivalent flow rate would reach more than 178.000 m^3/h from every kg of oil, a volume no waste gas cleaning system is able to handle. This is in addition to the fact that the hot combustion products will leave the furnace in a flash or, in case the furnace is closed, may cause substantial damage to the equipment. Thus, every melting furnace should have a preheating bridge and also most of the casting furnaces because, unavoidably, block material must be melted to adjust the alloy or incorporate blocks from previous no-go production. Some plants operate a pre-heating chamber in order to use the energy of the furnace's flue gas. During such an operation also moisture will escape and safe charging is assured. For monitoring the status of the material placed on the preheating bridge some plant operators equip the furnace with a TV camera. The interior of the furnace and the status of the material can be observed on a monitor without opening the large furnace door; an action that causes energy loss.

The preheating bridge offers advantages that help to improve the overall heat economy of the furnace. Bulky material, such as sows or bales, can be placed onto the bridge for preheating. Due to their small specific area, i. e. m^2 per weight unit, large blocks take much time for melting. The heat received by the surface flows through the material into the interior, thus increasing the total block temperature. Heat exchange with the exterior heating medium, as a function of surface area, temperature difference and heat transfer coefficient, can not keep pace with the heat transport within the block. Material with larger surface area is able to accept more heat and together with the short distance within the blocks the increase of temperature is improved. Heat transfer between hot combustion products is already quite favorable due to the higher temperature of the heating media and the better heat transfer coefficient. Pushing the block into the liquid metal bath decreases the heat transfer. There are various reasons for it. The heat transfer coefficient α is lower if energy flows from liquid metal to solid metal as it is for the transfer from furnace gas to solid metal. One ruling factor is the smaller temperature difference between the interacting media. Last but not least is the trivial fact that the temperature of the liquid metal lowers when it passes heat to the block. Thus, the liquid metal in contact with the block surface must permanently be replaced by fresh hot metal by natural draft or, even much more effective, by forced circulation.

Transferring the heat through the liquid metal to arrive at a homogeneous temperature distribution is one challenge but just as important is to solve the question how much energy can be transferred through the surface to the in liquid metal bath or, in other words, how much energy can the surface accept? This a fundamental question for the dimensioning of a furnace independent whether the furnace is used as melting furnace or as casting furnace and also independent of agitating the bath or not. While melting the solid metal is comparatively fast, the increase in liquid metal temperature takes much more time.

The surface area is calculated from the desired bath capacity and the selected bath depth. Now the question comes up about the size of the burner. If it is too big valuable energy is lost through the stack and if it is too small heating of metal takes too long for an economical production.

The heat transfer to the bath is calculated according the equation

$$Q = A \cdot \alpha \cdot (\vartheta_2 - \vartheta_1) \tag{2.2}$$

where Q is the energy transferred in Watt, A the surface area of the bath in m^2, ϑ_2 the temperature of the flue gas and ϑ_1 the temperature of the liquid metal, both in °C. The heat transfer coefficient α combines the heat transfer by convection between flue gases and metal and the radiation from flue gases and furnace walls. It is actually not a heat transfer coefficient in the sense of the laws of thermodynamics but a factor considering very different parameters. Different α-values are calculated and proven by experience for different phases of the furnace cycle. Depending on the furnace design, these values range from 80 to 200 W/m^2 · °C. The higher value is valid for an efficient furnace properly designed and equipment to improve heat exchange such as metal circulating systems. For more details refer to section 2.1.3.1 of this book.

Example:

A furnace is to have a bath surface area of 5 x 6 m that results in an area of 30 m^2. The metal temperature is 720 °C, the flue gas temperature 1,100 °C. It is a well designed furnace with optimal heat distribution characteristics. Thus, α is 150 W/m^2 °C. This is also on account of the high flue gas temperature. With lower gas temperature α will be lower, too.

The heat transferred is now calculated as

$$Q = 30 \cdot 150 \cdot (1100 - 720)$$

$$Q = 1{,}710{,}000 \text{ W or } 1{,}710 \text{ kW}$$

From this results a good rule of thumb value. With maximum flue gas temperature of 1,100 °C, the maximum heat that can be transferred to a bath surface amounts to 57 kW per m^2. Since we do not have to consider furnace losses through stack and wall we will be able to calculate sufficient accuracy by using the theoretical energy value for melting 1 t of aluminium amounting to 316 kWh. Considering a furnace, that receives all heat input through the liquid metal surface, as is the case with the twin chamber furnace, sufficient heat must be provided to obtain a melting rate of, let us say, 5 t/h. This will require a heat input of 316 kWh/t melting rate (refer also to chapter 3.1).

The required surface area will be

$$A = 5 \cdot 316/57 \cdot (1{,}100 - 720\,) = 27.7 m^2$$

for the hot metal section. Assuming a bath depth of 800 mm, we arrive at a bath capacity of 44 t. Considering the charging section of such furnace, we will come up with a total furnace capacity of approximately 60 t. This corresponds very well with practical experience for this type of furnace.

The heat transfer, as calculated before, can be used as basis to select the firing equipment. When sizing the burners, the losses through furnace walls, door and stack have to be taken into account. This is expressed by the overall efficiency of a furnace which is in the range of 40 to 60 %. The internal transport of heat, as discussed before, requires consideration, too. It is of no use to offer the maximum heat the surface would be able to receive if the heat cannot flow fast enough to the interior of the batch. The calculation only refers to a liquid bath of aluminium. Solid material absorbs much more heat at a faster rate for the internal heat distribution. Thus, the burner size is selected accordingly. This is the reason for down-rating the burner input as soon as all material is molten and the temperature of the bath must get the requested value for further processing, although substantial heat input would be desirable for shortening this part of the furnace cycle.

Bath capacity is a characteristic parameter for the size of the furnace. Having selected the bath depth, the surface area of the bath can be calculated very easily by the equation

$$A = \frac{G}{\gamma \cdot h} \left[m^2\right] \qquad (2.3)$$

whereby A is the bath surface in m^2, G the bath capacity or the weight of the metal in the furnace in tons, γ the specific gravity of the liquid aluminium in kg/m^3 and h the selected height of the liquid metal bath in mm.

For all calculations in connection with liquid aluminium a specific gravity of 2.3 is used. This is different to the one of solid aluminium which amounts to 2.7. A correction of the result will be required since the furnace will most definitely have a certain shape beneath the liquid level to facilitate cleaning and skimming and which also may be required for design purposes. Depending on plant layout and operation procedures, a side ratio length/depth is to be defined. A ratio ranging from 4/2 to 4/3 is very convenient. It gives sufficient length for the combustion products to transfer the heat but does not result in a furnace shaped too deep for easy skimming. When sizing the depth of the furnace the preheating bridge must be considered as well. With these parameters the furnace size is defined now. This refers to the inside dimensions significant for the process.

Besides their duty to envelop the furnace interior, the furnace walls have to be designed to fulfill other duties. Therefore, the furnace is lined with refractories.

The duty of the refractory lining is to avoid chemical and metallurgical reaction between the aluminium and the furnace body and to provide good heat insulation in order to reduce temperature losses through the walls. These different duties cannot be accomplished by one type of refractory material alone. For the contact with metal a dense and wear-resistant material will be required. This refractory material must have good mechanical properties and must be able to resist thermal shocks caused by the temperature difference between hot flame impingement and cooling by open doors as well as metal charge. But then, the insulation characteristics of such materials are not that good. Low density refractory material will provide excellent heat insulation but then the mechanical properties are rather poor. Therefore, the refractory lining of a furnace will comprise different layers. The layer inside the furnace, the “hot face“ will be made of highly dense material. Because of the required penetration and temperature resistance, the alumina content is in the range of 90 %. The layers towards the furnace shell will comprise insulating material.

Different areas within the furnace require specific considerations. For the bath area the industry favors two different philosophies. One is directed to maximum heat insulation, the other one to long service life.

It is due to the nature of the refractory material that thin cracks will occur within the material. Liquid aluminium, having a viscosity similar to that of water, will penetrate into those fissures until it solidifies. It is desirable that solidification will take place in the hot face layer which means that the 658 °C isotherm will be in the outer third of the front layer as viewed in the direction of the heat flow. This design feature assures that no aluminium can penetrate into one of the lower layers even after extensive wear of the hot face. Presence of aluminium underneath the front layer would result in deformation of the refractory materials allowing even more liquid metal to collect there. Finally, the refractory materials will be destroyed and the change of volume may result in extensive deformation (bulging) of the steel shell in the long run, even with a sturdily built furnace casing. However, shell temperature will be somewhat higher with this design resulting in a slight increase of the total wall losses of the furnace.

For the calculation of the heat flow through the furnace walls and the related losses please refer to section 3.5.

Fig. 2.11 shows the temperature drop for each layer assuming a 3-layer system. Two possibilities are shown for a total refractory materials lining thickness of 500 mm in both cases. If the condition must be fulfilled that the 658 °C -isotherm has to be within the hot face, this layer must be quite thick. Fortunately, another factor helps to arrive at reasonable lining thickness. The temperature gradient depends also on the specific heat loss at the furnace shell which is a combination of losses due to radiation and losses due to convection.

In case B (solidification within the hot face layer) the resulting shell temperature is above 100 °C which means higher heat losses in the range of 1290 W/m^2 compared to case A (solidification point within the insulating layer), where the shell temperature is only 60 °C and the resulting heat loss is 410 W/m^2. It is certainly possible to increase the total thickness of the refractory lining. However, this would result in much more weight with the consequence of heavier furnace shell design and larger hydraulic tilting cylinder, in total resulting in increased equipment cost. Upon evaluating these values, it becomes quite clear that the higher energy loss is very well compensated for by lower investment cost and much lower maintenance cost in the long run. It must also be considered that the higher energy loss is limited to the bath area which accounts for only 30 % of the total wall losses.

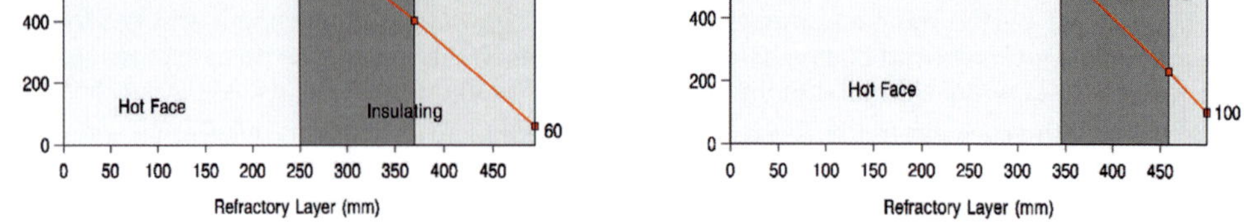

A - solidification will occur in the insulating layer, B - the 658 °C isotherm is in the hot face layer, resulting in increased shell temperature

Fig. 2.11: Heat flow through furnace walls

Liquid aluminium penetrates very easily through small cracks and fissures in the refractory lining. It is impossible to avoid this effect totally in spite of excellent refractory engineering. Aluminium passing through causes damage to the insulating layer and heats up the steel shell at so-called "hot spots". The local heat expansion causes the steel to expand locally and create deformation of the shell leading to even more cracks and damages. Generally, the refractory system is designed so that the penetrating aluminium solidifies at least in a sealing layer between hot face layer and first insulating layer. This, however, works only with small cracks and fissures. If the steel shell deforms due to lack of mechanical strength, the refractory lining cracks accordingly with the described results. This occurs particularly in tilting furnaces since the substantial forces of the furnace weight have to be lifted into position. In spite of the mechanical strength properties, the elastic deformation of the steel members must remain within the narrow limits the refractory material permits. A spring-like back force can by no means stabilize the refractory materials once aluminium has penetrated into the cracks. Furthermore, cracks having damaged the painfully established solid body of the refractory can not close again and establish the former properties.

Sealing of the furnace against penetration of cold air during the complete furnace cycle is a very important feature. Cold air not only reduces the energy efficiency but also increases the generation of dross inside the furnace. Therefore, all openings in the furnace, such as doors, tapping spout, openings for thermocouples and feed wells, are critical items when trying to maintain proper sealing of the furnace. As a consequence, the main door is the most difficult challenge for the furnace designer.

The full length door is exposed to the extremes of heat inside the furnace and the cold ambient air. The structure must be light enough for easy opening while sealing properly when closed at a positive pressure inside the furnace. One critical item of the entire door design is the door frame. Different methods have been developed in the past to maintain a distortion-free structure. A water-cooled design provides good stability but creates some problems with the cooling water system and the effective sealing of the door frame towards the hot metal. There is some potential for heavy accidents as well. The modern "dry door" avoids this problem. Its design comprises a sturdy frame of castings with a comparatively light but distortion-free and refractory-lined door. The door seal system, depicted in **Fig. 2.12**, utilizes a heat-resistant mineral wool rope while hydraulic cylinders press the entire door against the door frame to ensure a tight seal.

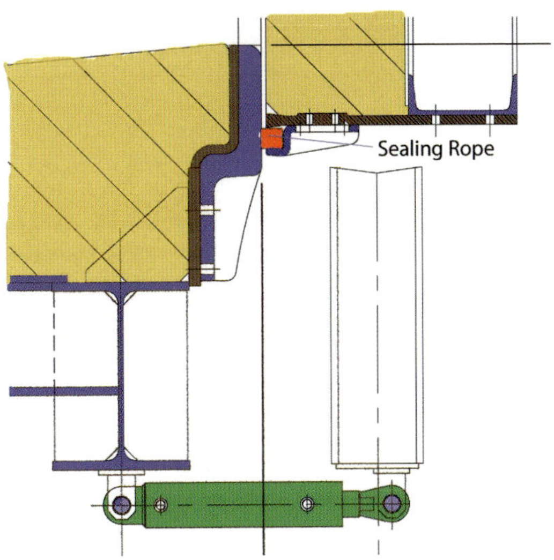

Fig. 2.12: Door-sealing system

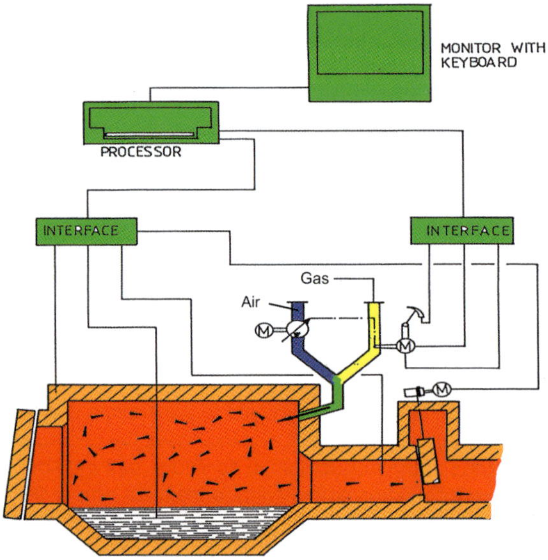

Fig. 2.13: Pressure control damper

Another measure to avoid incoming cold air is the furnace pressure control. The furnace is equipped with a pressure control damper for maintaining a slight positive pressure of 50-100 Pa during the complete furnace cycle (**Fig. 2.13**). Another desirable effect is that no heat is taken out of the furnace due to the negative pressure generated by the waste gas scrubber. The furnace pressure control is based on the differential pressure between the furnace interior and the ambience ouside the furnace. Considering the very low pressure in the furnace chamber, a fixed setting of the absolute pressure would have to be set at every change of the weather outside and its resulting alteration of the atmospheric pressure if the pressure control is not based on the differential pressure. The damper to control the furnace pressure is installed in the hot flue gas stream leaving the furnace. It has to operate safely in a hot and mostly dirty atmosphere.

A very robust and reliable method is the air curtain. It comprises a ring tube, manufactured of heat resistant steel, which carries a series of nozzle holes. Low pressure air is provided by the combustion air blower or by a separate fan providing the air curtain. The air flow is controlled depending on the pressure difference between the furnace interior and the atmosphere. It is convenient to use the combustion air blower for supplying the low pressure air since the air curtain must be quite firm during low fire and may be less when the full power of the burners provides ample overpressure in the furnace. But during low fire less combustion air is required and the excess air is available for the damper. The cold air introduced by the air curtain helps to lower the flue gas temperature as well.

The air curtain works fine in dusty and hot atmosphere. However, its control behavior is very inaccurate.

If the dust load in the furnace permits, a butterfly valve can be utilized. The damper blade is usually made of a heat-resistant steel casing and is operated by a step control motor or a controllable pneumatic positioning cylinder. The system operates safely in the rough conditions of the flue gas stream. The disadvantage of this design is its control characteristic. In the very low range, which is required for the small furnace pressure difference, only little change of angle of the butterfly damper results in remarkable change of gas flow i. e. furnace pressure.

A gate-type damper, as depicted in **Fig. 2.14**, offers best control characteristics. Similar to the butterfly damper, the dust load of the flue gas must not be too high. The system comprises a ceramic damper blade operating in a refractory-lined housing. A tube system within the damper blade provides cooling to the blade. The cooling air is provided by a small blower attached to the damper housing.

Independent of the design of the furnace pressure control, the positive pressure must be reduced to zero or slightly lower as soon as the large furnace door is opened. This is necessary to protect the furnace operator against a stream of hot flue gases leaving the furnace immediately through the open door.

The combination of furnace pressure control and furnace sealing, particularly the large furnace door, provides very effective measures to improve heat efficiency and, at the same time, helps to reduce dross generation.

If the furnace is used as casting furnace, another criteria is the controllability of the metal flow at constant temperature. By tapping metal from a stationary furnace through a taphole at the furnace bottom, the operator will have to face increasing temperature during casting since more and more metal from the upper layers will flow into the launder and the taphole. It is also quite difficult to achieve a constant metal flow through a taphole. Particularly at the beginning of the cast with the full metal head, turbulence will cause oxidation of aluminium because of the intensive contact with air during the turbulent flow condition. These oxides will be trapped in the metal as unwanted impurities. Even modern inline metal filters are not able to remove all of these impurities. No satisfactory solution for the taphole control has been developed up to date.

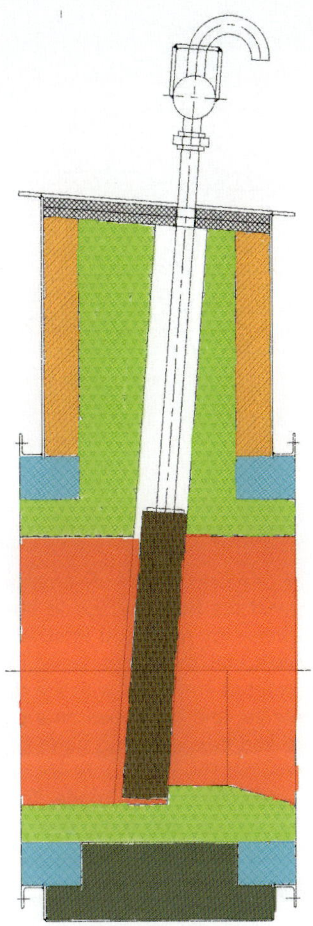

Fig. 2.14: Air-cooled pressure control damper

On the other hand, constant metal temperature and very well controlled metal quantity are essential for the operation of modern high-performance casting machines. Tiltable furnaces will meet these requirements. Due to accurate control, one or two hydraulic cylinders tilt the furnace automatically and precisely as required. Metal will flow through a spout into the launder system. (**Fig. 2.15**) shows a spout system whereby the sealing is assured by spring tension and ceramic rope. A ceramic cloth may also be used for connecting the furnace and the launder. The connection between spout and launder system is located in the swivel axis of the furnace, thus avoiding a metal cascade into the launder. Since the spout is arranged on the bath surface only metal from the top is poured. In this case the temperature can be maintained very easily since the heat transfer takes place in the comparatively thin layer on the top of the metal bath. Metal flows through the spout turbulence free under an oxide skin so that there is no danger of trapping dross particles within the aluminium. The furnace bottom is designed with the slope of the maximum tilting angle. This limits the required angle for complete emptying of the furnace to 30 to 35 degrees. The angle also creates a slight bath movement since a horizontal component of the gravity force will be induced. However, this flow is far too small to improve the temperature distribution.

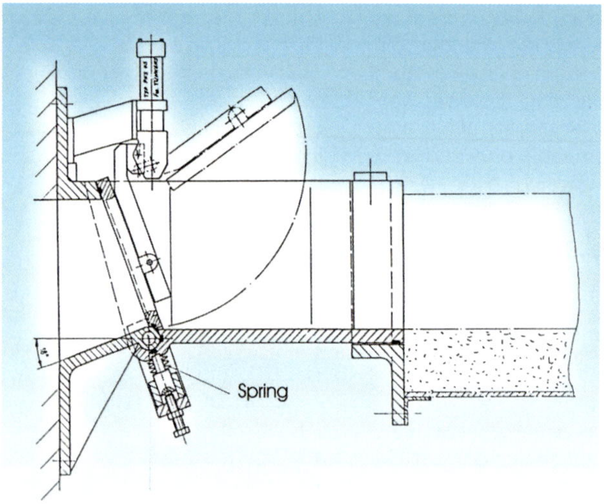

Fig. 2.15: Furnace spout for a tiltable furnace

One or more burners are arranged at the sidewalls of the furnace. Their center line is tilted slightly against the bath surface to obtain a good circulation of the combustion products. This is also obtained by the good internal circulation of flue gases of high-velocity burners. High-velocity means that the gases leave the burner nozzle with a velocity of 80 to 120 m/s. This velocity decreases inevitably with the downrating of the burner. At that point the combustion is not complete yet and continues on a passage through the furnace with the flue gas expanding at a fast rate and thus ensuring good heat transfer. Heating fuel may be natural gas, light fuel oil or even heavy oil. The basic concept of the firing equipment comprises one or more burners arranged in the sidewalls of the furnace. These burners cause the fuel to emit their energy in a time-dependent process. The ratio of fuel and combustion air is automatically controlled. But manual setting of the

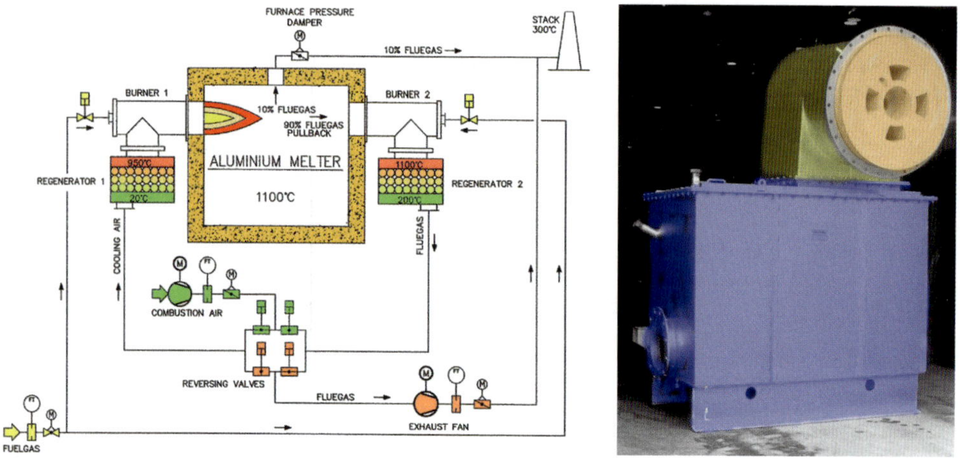

Fig. 2.16: Regenerative burners installed in a reverberatory furnace

basic parameters, e.g. excess air, permits to adjust the firing to specific requirements. This may be desired if, for instance, scrap with a high content of organics shall be processed. In such case it is favorable to burn these organics in the furnace. Therefore, sufficient quantity of excess air must be provided to allow the combustion of these contaminations. As described in a later chapter, the combustion air may arrive at the burner preheated in a recuperator in order to save some of the energy otherwise leaving the system unused. If the aluminium to be processed is clean, regenerative burners using some of the excess heat of the flue gases may be installed to improve the heat efficiency of the furnace. These burners use a regenerator instead of a recuperator. In such case heat is stored in a suitable medium and is then released when used. In a recuperative burner system such medium consists of a bed of tabular alumina balls or of ceramic tubes. The regenerative burner system (**Fig. 2.16**) comprises one or more pairs of burners. While one burner is firing, the additional burners take in the combustion products which are passed through the regenerator bed on their way to the stack. The energy is stored on the bed by increasing its temperature. After reaching a maximum temperature, the system switches to the other burner which now takes in the flue gases while the other burner is firing. Prior to combustion, the combustion air passes through the regenerator bed thus being heated. During the cycle the regenerator of the firing burner cools down while the other regenerator bed is heated up. Such a system is in used since centuries in blast furnaces for iron production. Here it is called hot blast stove. However, the size of such regenerator is much, much bigger than used in ore melting furnaces.

In a properly designed direct-charged reverberatory furnace, intelligent location of burners and flue gas discharge, proper selection of burner equipment and sophisticated control creates a very efficient fuel economy during all parts of the furnace cycle. The conditions in the furnace undergo permanent change during the cycle (**Fig. 2.17**). At the beginning, cold metal is charged to the hot furnace with the result that the furnace interior, i. e. the walls cool down to a notable temperature drop. Solid material charged to the furnace absorbs the heat very fast from the combustion products (POC), which in many cases impinge directly on the scrap pile (**Fig. 2.18**). The total surface area of the pile is also large. Thus, flue gases leaving the furnace are at relatively low temperature. This has an effect on the possible heat recovery system which can not operate at good efficiency during this period. As the temperature of the solid material increases, the heat exchange between combustion products and metal is reduced. The burners will reduce the energy input. Finally alu-

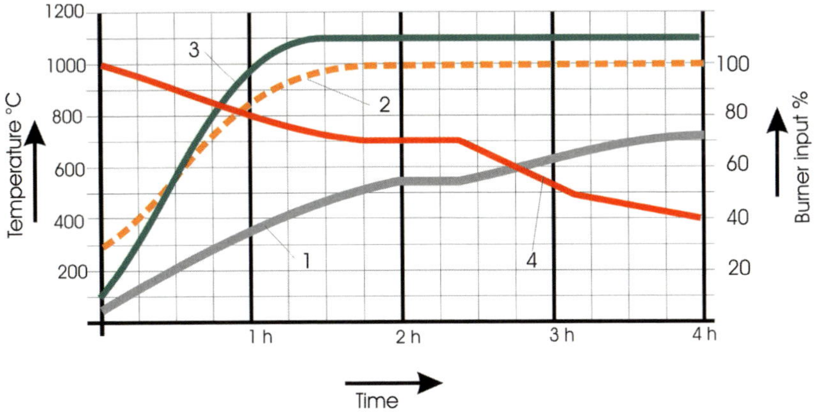

1 - metal temperature, 2 - furnace temperature, 3 - flue gas temperature, 4 - energy input

Fig. 2.17: Temperature and energy input during a furnace cycle

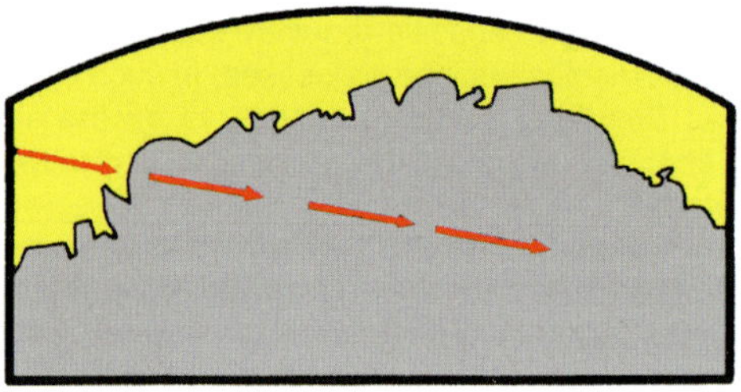

Fig. 2.18: Burner input with scrap

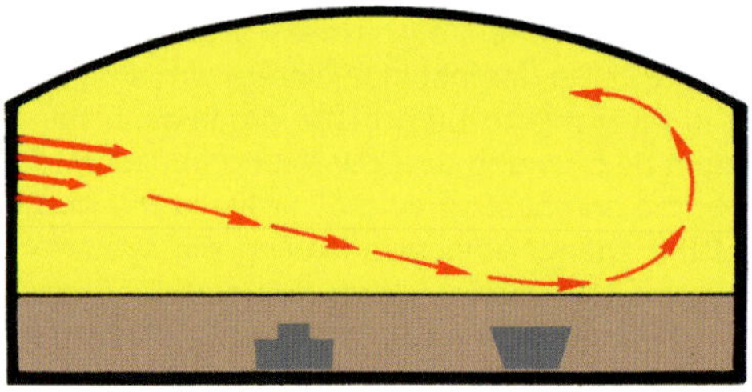

Fig. 2.19: Burner input with flat bath

minium starts melting and the charge begins to go flat (**Fig. 2.19**).The temperature of the exiting gases increases dramatically since temperature difference and metal surface area are reduced. As discussed before, the surface of the liquid metal bath is not able to absorb the heat offered fast enough. The temperature rise of the liquid metal is slow. As discussed before, the energy offered causes an overheating of the top liquid aluminium layer at very stable stratigraphy. Therefore, frequent stirring or forced circulation as described is essential. Since the flue gas temperature is high, measures for heat recovery are efficient during this phase of the furnace cycle.

Selection of firing units and also their control must be able to take care of these conditions. The flue gas leaves the furnace through an outlet arranged near the opposite wall. A rotary joint connects the flue gas discharge of the furnace to the waste gas scrubbing system of the plant. This arrangement depends on the actual plant situation and the furnace design must necessarily adapt these conditions. Depending on the position of door, burners and flue gas, discharge must be arranged accordingly to assure good heat distribution without hot spots and cold corners.

During the heating stage of the furnace cycle it is advisable to ensure heat transfer by convection. The solid material is directly in contact with the hot combustion products which creates a large temperature difference between the combustion products and the charge. This is obtained by a short mixing flame. This will be the result of the high gas velocity. The momentum of the flame leaving the burner will work as injector and draw flue gas from the furnace room into the jet of the flame. Studies have shown that the quantity of the gas circulated increases with the height of the furnace and the gas velocity in the jet. The use of high momentum burners, having a gas velocity of more then 100 m/s, forces high gas circulation in the furnace with the result that the turbulence of the gases is very high which improves the heat exchange by convection. The recirculation has another favorable effect. It lowers the flame temperature. This certainly decreases the heat transfer by radiation but it also reduces the generation of NO_x. This unwanted gas is formed at temperatures above 1,300 °C. Even at the high momentum burner, the first section of the combustion which is close to the burner nozzle, the flame temperature will be much higher and some NO_x will be generated but decreases rapidly as the hot combustion products are passing into the furnace room.

Once the bath is flat a different mode of heat transfer is required. Since temperature difference between reduced high level and the surface area is reduced to the furnace plane, radiate heat exchange becomes more effective. This can be achieved with the furnace roofs and the walls where the maximum permissible temperature is maintained by the burner input. The limiting parameter for the input of energy is the set by the refractory lining of the furnace that must not exceed 1,200 °C. Because of the design parameters of the burners, there is a lower limit for the turn down rate of the firing system. Thus, the burner input is approaching a more or less on – off control. The energy input is used to rise the bath temperature to the level wanted by production and to compensate for the wall losses of the furnace. As soon as the holding stage is reached, which means the metal has the required temperature and is just held on this level in anticipation of further processing, i. e. casting, only losses have to be compensated for. After the flat state of the bath is reached, a long luminous flame could be of advantage. This would require a dual capacity or an additional set of burners. It is common understanding that the long and soft luminous flame radiates more energy to the bath than the short, intensive mixing flame.

But is this really true?

Mostly, the expression "flame" is used for the shining flow of gases leaving the burner. As soon as the fuel terminates its reaction with oxygen, the combustion products do not emit light of visible wavelength' and the hot combustion products continue their flow through the furnace "flameless" still transferring heat by radiation and – if they are in contact with furnace walls or metal – by convection. The visible brightness of the flame is caused by carbon particles which are generated during the combustion process. This soot absorbs and, consequently, emits radiant energy. The conditions, such as soot concentration, reaction of the different components, temperature distribution, are difficult to obtain and still require extensive research work. Here again, the furnace designer has to rely on empirical data. Tests have shown a trend that the portion of heat transfer by soot radiation in the total radiation is approximately 10 % for natural gas and 35 % for oil firing. As consequence: If long luminous flames are able to cover the complete bath surface or at least a major section of it, the resulting heat transfer could shorten the heatup time of the liquid metal bath. Flame temperature is 100 to 200 °C higher than the permissible temperature of the refractory lining. It has to be considered as well that during the long flame period no heat can be received from the furnace walls since the flame will absorb the quantity of heat emitted by the walls as it would emit having wall temperature. For economical reasons, however, it may be of more advantage to except the on-off control of the burners and the perhaps slightly lower heat increase of the bath.

The mechanism of heat transfer in the furnace will be discussed in chapter 3.2.

Heat recovery during the cycle period of heating the liquid bath is much more effective than during heating of solid material. As a consequence, the furnace could be equipped with high velocity cold air burners and regenerative burners for the second heating step. **Fig. 2.20** shows a solution for such a multiple burner system. The regenerative burners are installed in the vertical furnace walls. There is always a pair of burners required with space-consuming gate valves, blowers and

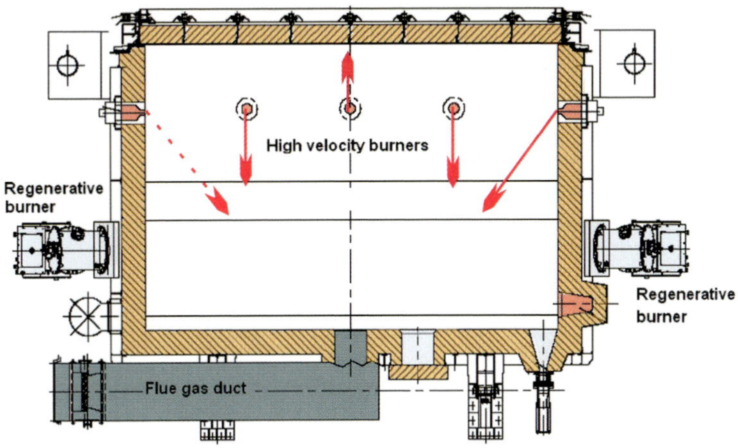

Fig. 2.20: Multiple burner installation in a reverberatory furnace (Source: Thermcon)

ducts as well as the voluminous regenerative beds. The cold air burners are, therefore, arranged in the furnace roof. This is no problem for the flow of the combustion products from these burners through the furnace since the high velocity burners generate a very turbulent flow and good internal circulation, thus ensuring good heat distribution in the furnace.

In general more than one burner is used in a furnace. It may be of advantage for design and cost reasons to have these burners located next to each other. This will definitely reduce pipework. On the other hand, burners located strategically at different points in the furnace may have the advantage of very good heat distribution within the furnace. In the lower range of the energy requirements, burners could be operated in cascade which means one burner fires for a certain period of time. Then the firing is switched to the next burner and so on until burner number one will be start firing again. Considering the "fly wheel" effect of refractory lining and metal charge, the system will evenly distribute the heat.

The plant requirements, which in fact mean the location of the door in relation to tapping spout or – as in the case of a tiltable furnace – the tilting axis, will influence the arrangement of the burners. The designer of the furnace must take care in any case to achieve the best routing for the combustion products on their way to the stack.

The flue gas leaves the furnace through an outlet favorably arranged next to the charging door. For a stationary furnace the flue gas duct is directly connected to the plant waste gas system. This is more difficult with the tiltable furnace. The flue gas duct is connected to the plant system by means

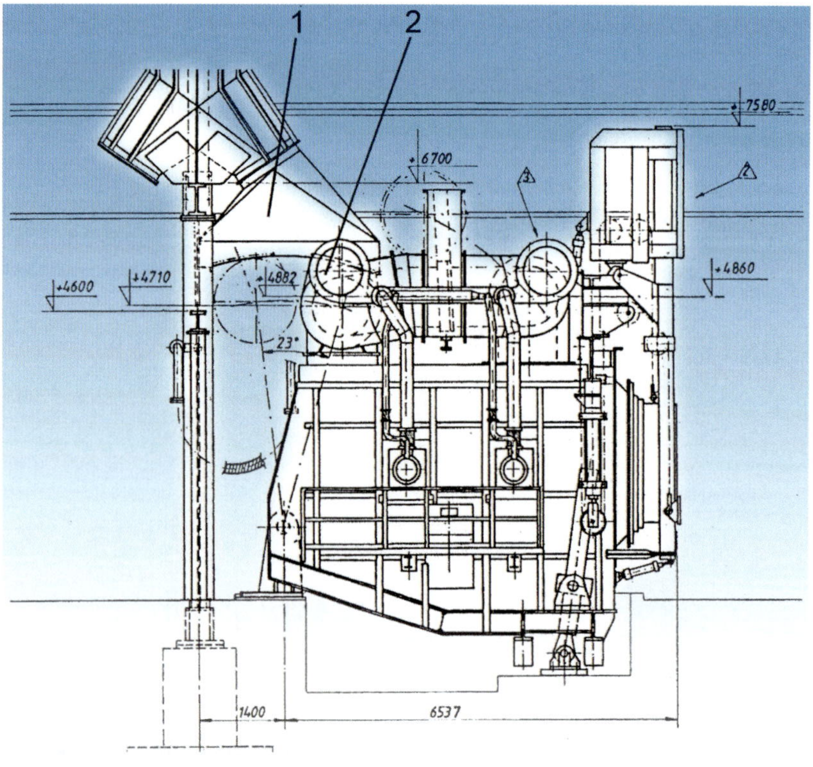

Fig. 2.21: Stack discharge of flue gas

of a rotary joint arranged in the swivel axis of the furnace. The flue gases may also be released via a collecting hood (**Fig. 2.21**). In this case the pressure control damper must be installed ahead of the flue gas outlet. State-of-the-art furnaces have a very efficient flue gas collecting system that does not even permit the escape of diffuse emissions to the furnace atmosphere.

Heat recovery either by regeneration or by recuperation will be covered in more detail in section 3.4.

Facing the challenge of more and more stringent environmental protection, one has to face the NO_X problem when installing heat recovery systems. This poisonous gas is a reaction product between the nitrogen of the combustion air and oxygen which takes place at high flame temperatures particularly near 1,300 °C. By recovering heat from the flue gases the flame temperature increases. Thus, heat recovery needs to be limited to an acceptable temperature of the preheated combustion air. Techniques have been developed to reduce this effect. They all have the target of reducing the flame temperature. This can be achieved by lowering the combustion air temperature to approximately 400 °C, multi-step combustion or flue gas recirculation and proper burner design to obtain an evenly distributed combustion temperature without temperature peaks. **Fig. 2.22** shows the permissible emission of NO_x depending on the temperature of the preheat temperature of the combustion air according to the German regulation TA-Luft. But standards may change very quickly. Thus, it might be advisable to consult the regulations valid for the location of the plant.

A sophisticated furnace control system will definitely contribute to an efficient furnace operation.

Burner control will have to take care of the requirements during the various cycles of the furnace operation sequence. Temperature control works in a closed loop circuit with fuel and combustion air supply to the burners considering a value preset by the operators. The automatic temperature control loops are based on continuous and steady state control, considering the necessary transfer functions such as PI or PID control. Safety functions need to be built into the control circuits with the necessary interlocking and alarms to protect operators and equipment. For more details on furnace control please refer to section 7.

If the furnace is used for casting only and the operation is focused on liquid metal treatment, a tiltable barrel type furnace may be selected. The limited purpose is also the reason that this type of furnace is still called converter. The increased depth of the liquid metal bath leads to a longer

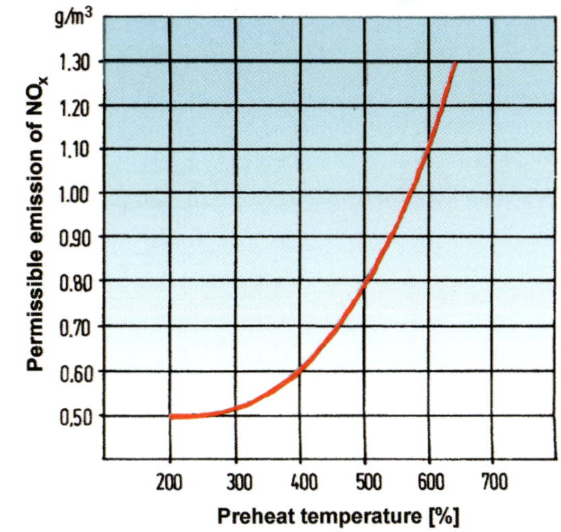

Fig. 2.22: Permissible NO_x concentration (TA-Luft)

passage of the treatment gases introduced by graphite lances from top or by porous bricks from the bottom. Since the mechanism of heat transfer and heat conduction are identical to those of the rectangular reverberatory furnace, the performance of the barrel type furnace is poor when the metal temperature needs to be increased or solid metal is to be melted. Due to the deeper metal bath, the size of the furnace is smaller than that of a rectangular furnace. Together with the fact that the round shape is statically simple and does not require many reinforcing members, the furnace is less expensive. Because of the round shape, refractory material is somewhat more complicated and that may eliminate some cost benefits. A short calculation may show the size difference between a hearth furnace and a converter.

Example:

A furnace with a bath capacity of 25 tons is be designed as converter or as rectangular furnace. Both furnaces will solely be used for metal treatment and casting.

Case 1: Converter:

The optimal degree of filling is 50 % which means the furnace is filled up to the horizontal center line. The length/diameter ratio is 2.5.

$$\frac{G}{\gamma} = \frac{1}{2} \cdot \frac{d^2 \pi}{4} \cdot 2{,}5\ d$$

Resulting from this

$$d = \sqrt[3]{\frac{8 \cdot G}{2.5 \cdot \pi \cdot \gamma}} = \sqrt[3]{\frac{8 \cdot 25}{2.5 \cdot \pi \cdot 2.3}} = 2.23\ \text{m}$$

The resulting length will be

$$l = 2.5 \cdot 2.23 = 5.57\ \text{m}$$

The furnace will have a diameter of 2.1 m and a length of 5.25 m. The dimensions are, of course, inside diameter and inside length.

Case 2: The furnace will have a bath depth like the maximum depth of that of the converter that is 1 m and a length/width ratio also identical to the converter of 2.5.

$$\frac{G}{\gamma} = h \cdot (2.5 \cdot W) \cdot W$$

Resulting from this

$$W = \sqrt{\frac{G}{2.5 \cdot \gamma \cdot h}} = \sqrt{\frac{25}{2.5 \cdot 2.3 \cdot 1.0}} = 2.01\text{m}$$

And the length is

$$2.5 \cdot 2.01 = 5.03\ \text{m}$$

Compared to the converter, the rectangular furnace having the same duty, i. e. casting and metal treatment only, is even smaller than the round furnace. The sizing will be different if the rectangular furnace is equipped with a preheating bridge of 1 m width. Then the total furnace depth will be increased accordingly to 3.00 m, an acceptable length to facilitate skimming. The converter could certainly also have a preheating bridge. The interior length becomes 6.60 m, although skimming is somewhat more difficult.

If the rectangular furnace is to be used as melting furnace as well, bath depth may be selected to be 500 mm only. The size of the furnace will then be

$$W = \sqrt{\frac{G}{2.5 \cdot \gamma \cdot h}} = \sqrt{\frac{25}{2.5 \cdot 2.3 \cdot 0.5}} = 2.95 \text{ m}$$

and the resulting length, having the same side ratio:

$$2.5 \cdot 2.95 = 7.37 \text{ m.}$$

The related skimming length of 4 m (plus 1m for a pre-heating bridge) is still very convenient.

Sometimes both front walls of the converter-type furnace have a service door. The available geometrical dimensions limit the size of the door. Apart from difficult skimming, the unavoidable edges in the furnace make cleaning of the furnace wall a difficult task. To get more space for the door, the furnace cross-section is expanded to form an oval shape (**Fig. 2.23**). The system for stress calculation is almost identical to the one for the round shape but the front door or doors can now be larger. There are still the inevitable edges next to the door. Also the conditions for heat exchange are similar, although improved, with the oval furnace. Burners are arranged on one front wall left and right of the door. They are slightly tilted towards the furnace bath and to the center of the furnace.

To allow treatment of the melt with aggressive gas, such as Cl_2, the burners can be pulled back and the burner tile can be protected by a protective shield.

In some plants, particularly in extrusion plants and rolling mills, a large quantity of long but thin scrap pieces are collected in tangled piles. This could be deformed no-go extrusion shapes or bent pieces from process start or cut from trimming rolled plates. To be able to process this scrap, it would be very convenient to have the material baled in large presses. Some plant operators do not consider having a benefit from this procedure and would like to charge the tangled

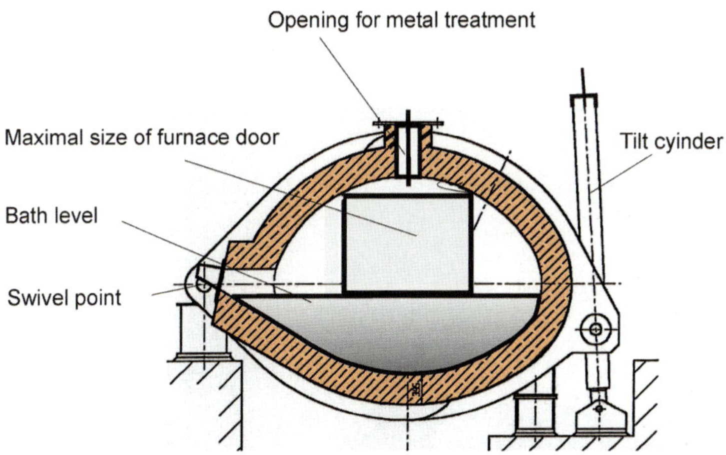

Fig. 2.23: Oval furnace

material directly into a melting furnace. The handling through a door, even a large door at one side of a furnace, would be too complicated and time-consuming. Therefore, they use a top loading furnace (**Fig. 2.24**). Usually the complete furnace roof can be moved to a side thus opening the furnace rectangle for charging by using a crane with grab. For opening, the lid is first lifted and then moved to one side on rollers and rail by a horizontal drive unit via chain or steel rope. A heat shield underneath the open door protects the mechanical structure as well as the personnel working underneath. During opening and charging, the furnace is wide open and will inevitably lose heat. Also fumes, that may be generated by rolling or extrusion lubricants or coating, will be released to the outside and have to be collected by a large hood arranged above the operating range of the charging crane. A service door is arranged at one side of the furnace which permits charging of other scrap, alloying elements as well as de-drossing and sampling. Similar to the standard reverberatory furnace, the burners are arranged in the sidewalls and will permit all methods of heat recovery. The furnace is always of the fixed type.

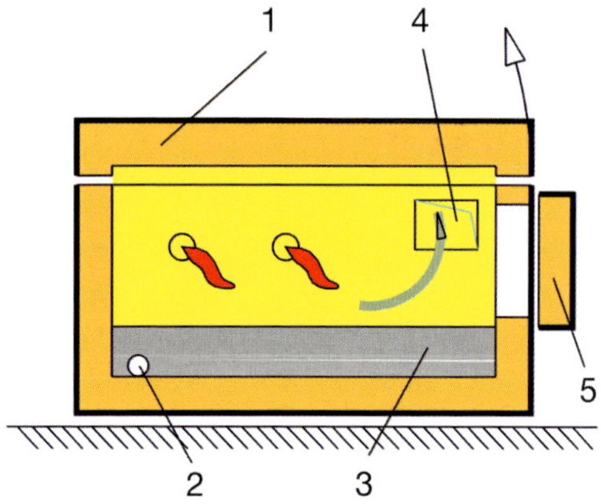

1 - removable lid, 2 - taphole, 3 - metal bath, 4 - flue gas discharge, 5 - furnace door

Fig. 2.24: Top-loading furnace

For larger capacity and for even more tangled material the furnace may have a round shape. Operation of the large lid and principle design is similar to the one of the rectangular top-loaded furnace (**Fig. 2.25**). A service door is also located in the sidewall. The burners are arranged on the circumference of the furnace and provide very good flow of combustion products for equal heat distribution. The furnace can be equipped with a preheating chamber that is very helpful if for instance complete coils of rolled aluminium need to be recycled (**Fig. 2.26**). Since flue gas

Fig. 2.25: Round top loading furnace (Source: Maerz Gautschi)

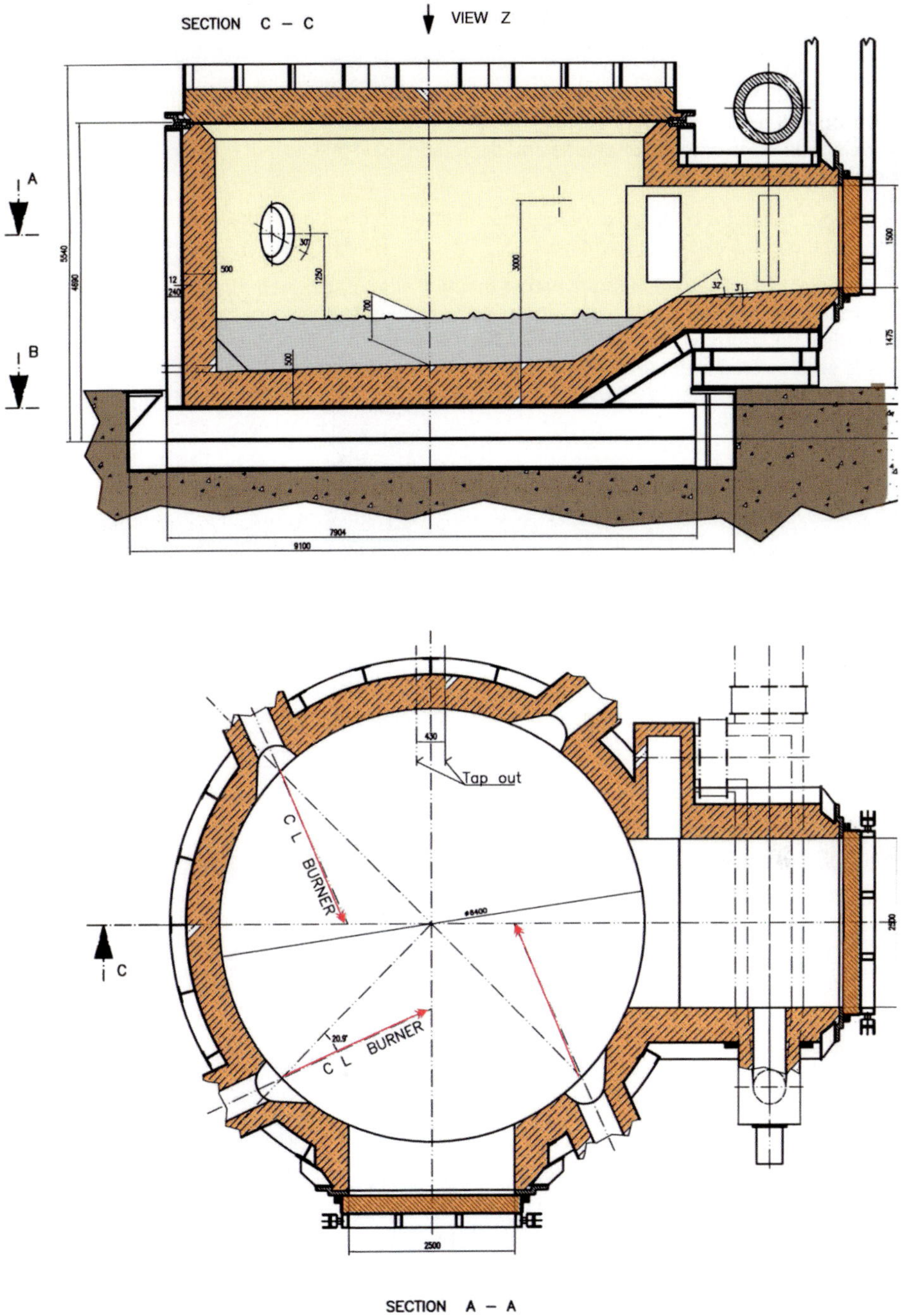

Fig. 2.26: Round top-loading furnace

from the furnace will heat this chamber, decoating of such material is possible with internal heat circulation, if required.

The round furnace can also be designed as tiltable furnace. However, this will then be a side-loaded unit since top-loading will not be possible or at least difficult to design.

Like the standard reverberatory furnace, all means of liquid metal circulation can be installed in a top-loading furnace and all conditions and design criteria – except, of course, the furnace lid – are identical or at least similar.

The flue gas outlet for both types of furnaces is fitted to the side of the unit and can then be connected to the plant waste gas system.

Reverberatory furnaces, no matter whether they are side or top-loaded, round, oval or barrel type, are suitable for melting large blocks of aluminium, ingots, bales or bundled shapes, crop end and the like. In short, all material with small specific surface area. If, however, material with large specific surface area is charged to the furnace, the oxygen contained in the flue gases reacts with the aluminium at high temperature. The result is formation of a substantial amount of aluminium oxide, i. e. high melting loss.

2.1.3.2 Twin chamber furnace

With already existing liquid metal bath and turned off burners, however, roof and walls transfer energy by radiation to the batch (**Fig. 2.27**). In the lower section of the furnace material dives into the bath and liquid metal will float around the individual particles. In the furnace a combination of melting under absence of air and melting by submerging into liquid occurs. Additionally, fresh liquid metal can be supplied to the metal particles by a forced bath movement generated by a metal pump or by an electromagnetic stirrer.

The heat transfer is, however, limited. If no energy is provided by an outside force only the heat capacity stored in the furnace body and the liquid metal can be effective for melting. This is possible only at a temperature that is remarkably above the melting temperature of aluminium i. e. 680 °C. Before charging of metal the furnace body and liquid metal should be brought to a temperature that is as high as possible. This should be approximately 1,200 °C for the furnace body and 800 °C for the metal.

The available heat can be expressed as

$$Q_B + Q_W = Q_M + Q_S + Q_F \tag{2.4}$$

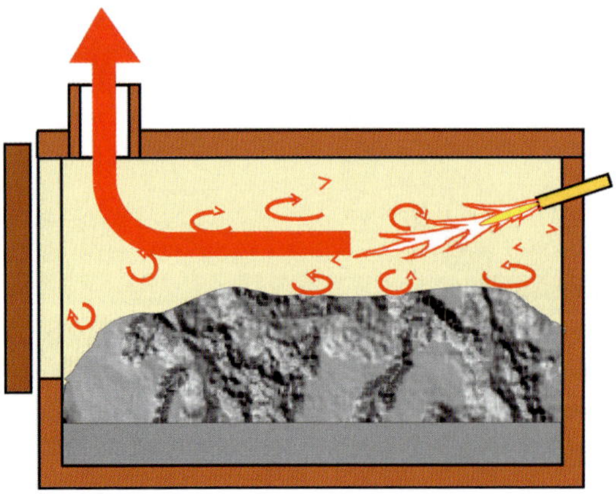

Fig. 2.27: Reverberatory furnace with all doors closed

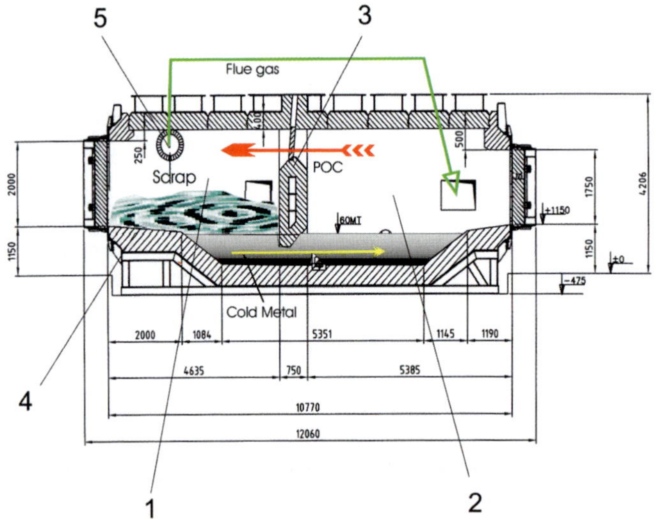

1 - melting chamber,
2 - liquid metal chamber,
3 - separating wall,
4 - preheating ramp,
5 - gas circulation blower

Fig. 2.28: Closed well furnace

Q_B is the available heat of the liquid metal, Q_W the available heat of the furnace body, Q_M heat required to reach the melting point, Q_S heat of fusion, Q_F heat required to heat up the liquid metal.

This heat balance limits the melting capacity. For additional energy input, the burners would have to be switched on. However, the effect of repeated melting, solidifying and melting again, as described above, will occur including all negative effects. It would be worth considering operating the furnace at reducing atmosphere. The poor heat transfer within a bulk of material will, however, prevent a successful melting process. And there is still a remarkable quantity of free oxygen available in the flue gas.

Sufficient energy transfer can be realized if the furnace is divided into two sections (**Fig. 2.28**). Heat will be supplied to the hot metal section while metal is charged into the melting section.

A dividing wall, that reaches below the liquid metal level, separates the furnace into two chambers. Scrap charged submerges into the metal bath in the melting chamber and will be melted. Natural convection provides a continuous flow of liquid metal (**Fig. 2.29**). A pumping system can improve the metal circulation remarkably. All of the energy is introduced into the hot metal section where the bath surface is only available for the heat transfer. Contrary to the large surface area generally available with a solid metal batch, this area is somewhat limited whereby the less favorable heat transfer coefficient I and the lower temperature difference need consideration as well. We have discussed this taking an example in the previous chapter. The solid metal poor heat conduction coefficient I can be improved to a large extent by intensive bath movement with a pumping system or an electro-magnetic stirrer. In order to obtain an economical melting rate, the bath surface and, in combination with that, the bath capacity must be quite large. Another disadvantage is that the furnace has to operate with metal heel: This complicates a change of alloy. In many recycling plants this is not necessary or at least only necessary at large intervals. The melting chamber of the twin chamber furnace can be used like a normal reverberatory furnace. It is equipped with a preheat bridge. Thus, ingots or other large pieces of metal can be charged and melted efficiently. This provides the required liquid metal heel for melting in the melting chamber. As soon as the required level of the heel is obtained, the melting chamber can be charged with loose scrap that may be contaminated by organics. Oxide contaminations or such caused by foreign metal and

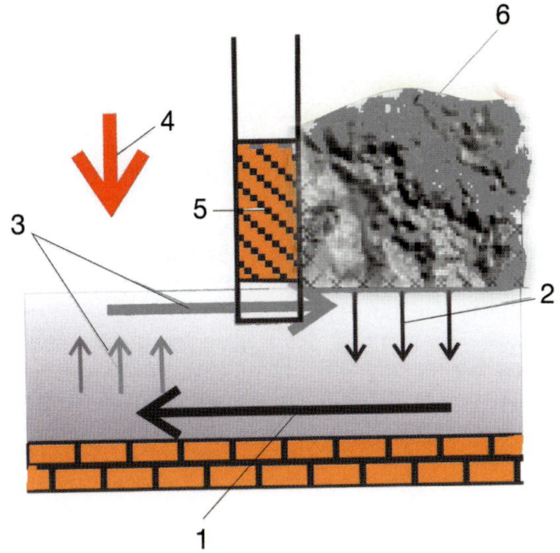

1 - cold metal returning to hot metal chamber,
2 - cold metal sinking to bottom,
3 - hot metal flowing to melting chamber,
4 - heat input,
5 - separating wall,
6 - scrap charge

Fig. 2.29: Metal flow in a twin chamber furnace

tramp iron are difficult to process. This material will collect at the bottom of the melting chamber and is difficult to remove. Due to the circulation of liquid metal, contaminations may also be distributed in the entire melt. Considering these limitations the furnace is able to operate efficiently.

The standard twin chamber furnace is fixed but tiltable furnaces are in operation as well. If metal of a defined analysis is used, this furnace is able to operate continuously whereby the quantity of metal discharged corresponds to the metal quantity charged.

When processing clean scrap of defined analysis, the furnace may be designed as tiltable furnace.

The furnace was developed from the side well furnace (**Fig. 2.30**) used in the US for a long period of time and the furnaces are still in use for specific applications. The units are small and the melting rate is low. Since the side well is open, energy loss is inevitable. The logical step was to close the side well and incorporate this chamber in the furnace and to establish the circulation of liquid metal by natural or forced draft in the melting chamber. The "communication" of the liquid metal

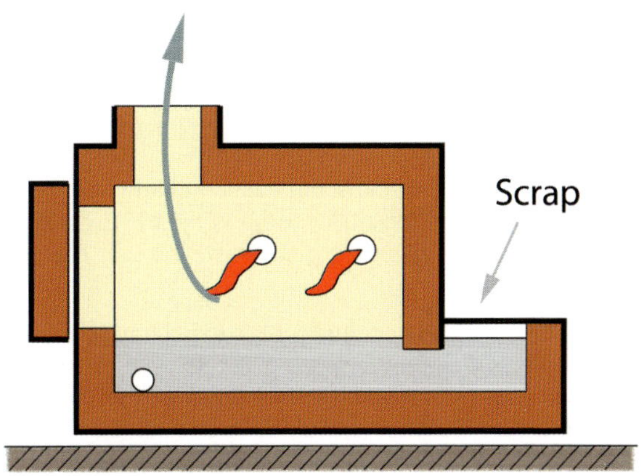

Fig. 2.30: Side well furnace

takes place via ports in the separating wall. It would be of advantage to have the ports above the liquid metal level to permit the free circulation of hot surface metal. But in order to avoid contamination of the metal in the heating chamber, this opening is preferably located below the liquid metal surface. No matter where the ports are located, the variation of liquid metal level when charging and tapping needs to be considered. The dividing hot wall also seals the upper part of the furnace against each other. Combustion can take place in the hot metal chamber without influence on the metal in the melting chamber. For removing the organics from the scrap charge, some combustion product is directed from the hot air chamber to the melting chamber via a small opening above the bath level. It is intended to start a pyrolytic process that means a process for removing organics under absence of oxygen. The hot metal chamber operates with some excess air to obtain good combustion. The flue gases entering the melting chamber will still have oxygen not having reacted. But since the oxygen quantity is low, it will be consumed rapidly by some organics. The pyrolytic gas leaving the melting chamber comprises combustion products from the melting chamber and not yet reacted, i. e. unburned organics. This gas cannot be released to a waste gas scrubbing plant before it passes an incinerator. The quantity of fresh air introduced at that point depends on the oxygen content of the gas supplied from the melting chamber. The oxygen content will be measured by a so-called lambda probe and the quantity of fresh air provided is adjusted accordingly. In the ideal case no oxygen is contained in the gas. But as the decoating commences, less reaction product is available. Eventually the material is clean and the lambda probe will measure the excess air content of the melting chamber. The pyrolytic gas from the melting chamber is recycled by hot air blowers. Generally, additional energy is required for the incinerator. Thus, the enthalpy of the flue gases leaving the incinerator is quite substantial.

The pyrolytic gases originating from the decoating process create a reducing atmosphere in the melting chamber which more or less suppresses the oxidation of the metal. An additional pilot burner installed in one of the walls of the melting chamber may support the decoating such as returned extrusion shapes. The gases developed consist mainly of organics and must, therefore, pass through a post combustion system before they can be released to the outside. The subsequent flue gases may also pass through the hot metal chamber, thus improving the overall efficiency of the furnace as mentioned above.

It is indispensable that all combustion products generated by combustion air and fuel have to leave the system eventually. Unfortunately, all of the enthalpy of these combustion products cannot be utilized. One reason is that part of the combustion products must leave the furnace to provide room for newly introduced flue gases as described above. The other reason is that the bath surface is not able to receive fully the enthalpy provided from the combustion products of the incinerator gases.

Since the basic concept does not differ for the various types of twin chamber furnaces, the actual differences are the methods of heat recovery. There may also be different layouts for the door arrangement or variations in the location of the pumping chamber: The basic process is identical. Variations refer to the arrangement of the pump well or the charging system. Some designs include a dry hearth section that is connected to either the melting chamber or to the hot metal chamber. Some furnaces may be designed as top loaders for bulk material such as swarf or shredded scrap arranged above the melting chamber with a more or less automated charging system. In another design, the bulky scrap is held in a decoating chamber above the melting section to allow the furnace gases to pass freely through the batch for decoating before it is dumped into the liquid metal.

Some means of heat recovery will be required to improve the efficiency of the system and, therefore, there is a common request for all heat recovery systems in the twin chamber furnace: All pyrolytic gases or all pyrolytic fumes generated by decoating or drying must be heated to an igniting temperature of at least 800 °C and must ignite. Hence, they need to pass through a zone with very high temperature which is easily obtained in the flame zone of a burner. Finally, they leave the system after passing through the furnace system without contact with fresh scrap and must not be able to pick up organics. All carbon must be burned to the permissible level but all NO_x generation must be kept within the limits stipulated in the rules and regulations for environmental protection such as TA-Luft in Germany. Most of the furnaces supplied are equipped with additional firing equipment in the melting chamber to ensure complete organics removal. The designer has to take caution when selecting size and arrangement of these burners. Direct impingement of the hot flame on metal must be avoided. Furthermore, the metal temperature of the solid metal charged must not exceed 350 °C to 400 °C since oxidation increases above this temperature.

One method for heat recovery is to install a recuperator or a regenerator for preheating the combustion air for the burners in the hot metal chamber. The balance of flue gas is then released to the waste gas scrubbing system together with the excess flue gas that is not passed to the melting chamber. This system has been practised and is in use since many years.

Heat recovery systems installed more recently still need to justify their economics. The waste gas treatment of the twin chamber melting furnace is quite complex particularly if the emissions during charging require treatment.

Generally, a treatment system comprises a recuperator, or a regenerator, a quenching chamber to remove furans and dioxins and the bag house. Furans and the extremely poisonous dioxins crack at a temperature above 900 °C. Unfortunately, they are generated again in the so-called "de-novo synthesis". Only rapid cooling suppresses this reformation. Therefore, the flue gases are passed through a quenching chamber to cool down to approximately 200 °C by distracting heat by a water spray before they are passed on to the bag house. Introducing water to the flue gas is somewhat difficult since the lowering of the dew point must be considered in order to avoid clogging of the filter bags. Therefore, dry systems are also in use which involve adding active carbon powder or other chemicals together with the lime to the gas stream ahead of the filter.

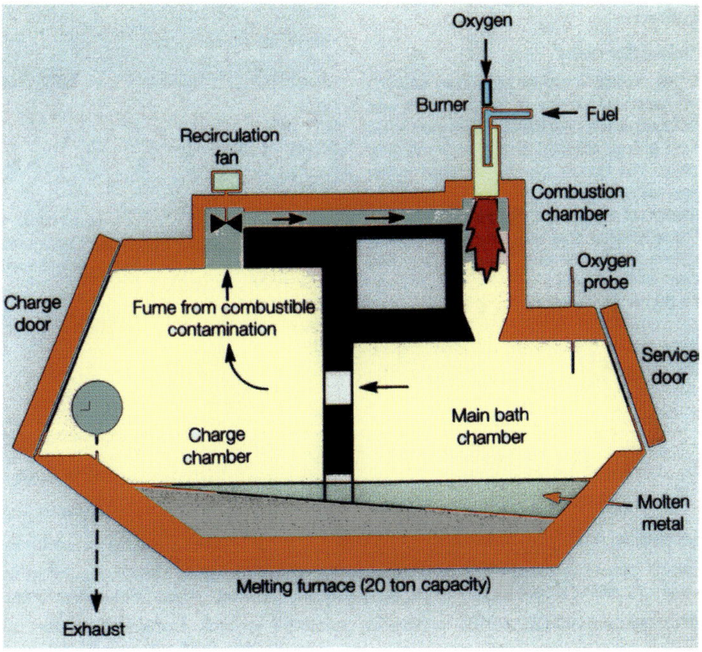

Fig. 2.31: RILEE-System (Source: Air Products)

The hot gases leaving the incinerator may be directed to the hot air chamber without passing through a heat recovery system. In this case the direct contact of the hot flue gases will transfer energy to the metal by radiation and even more by convection directly to the metal.

The hot metal chamber of the furnace can be used to burn the organics removed from the material in the melting chamber thus replacing a separate incinerator (**Fig. 2.31**). The process will be explained taking the example of the Rilee system as developed by Air Products. The pyrolytic gas is mixed intimately with the hot combustion products generated by the main burner to raise the temperature to the level required to ignite the pyrolytic gas. A small combustion chamber above the main bath chamber contains the burner. The recycled flue gas is injected around the burner flame and the combined gases are then directed downwards to the bath. The charging chamber contains an uptake flue duct where a hot gas circulation fan recirculates combustion products and pyrolytic gas to the small combustion chamber. The excess gas quantity leaves the charging chamber through the exhaust at a rate controlled by a hot gas damper. In order to reduce the total flue gas quantity – recirculated and generated – it is of advantage to use an oxyfuel system. This also improves the heat efficiency of the furnace since no nitrogen needs to be brought to flue gas temperature. All combustible material is progressively burned under controlled partial oxidation conditions. Therefore, the entire system requires sophisticated control which monitors the furnace temperature and oxygen level in the flue gas and adjusts the burner firing rate according to the demand by the metal. The control system also regulates the volume rate of the recycled exhaust gas by varying the fan speed to maintain the preset combustion chamber temperature. By increasing the fan speed, the gas recycling rate increases. Thus, the recycling rate of the combustion chamber decreases. Slowing the fan speed has the opposite effect; the temperature in the combustion chamber increases. The incineration of combustible pyrolytic fumes is monitored and adjusts the oxygen content in the furnace by setting the ratio of fuel and oxygen fed to the burner. As more combustible fumes are fed to the combustion chamber, the preset oxygen level will begin to reduce. To compensate for the lower oxygen level in the combustion chamber, the burner fuel rate is reduced, thus making more oxygen available to burn the combustible fumes. As the fumes are eliminated, oxygen concentration and fuel rate return to the perset ratio. Finally, flue gases are discharged through a stack in the melting chamber. This is certainly not the optimal solution since POCs (product of combustion) have to pass through freshly charged scrap and will pick up organics. The better solution would be to discharge these POCs through a flue in the hot metal chamber.

Since oxygen is used the generation of NO_x is very low. Only nitrogen available for air trapped in the scrap and taken in by the furnace, due to leaks still possible at the doors, is able to react with

Fig. 2.32: Regenerative burner arranged in a twin chamber furnace (Source: LOI Thermprocess)

oxygen to form NO_x. This is different from other firing systems and here in particular for regenerative burner systems.

These burner systems are also applied in twin chamber furnaces to improve the heat efficiency (**Fig. 2.32**). A pair of regenerative burners is installed in the hot metal section of the furnace. These burners operate in cycle; while one burner is firing, the hot combustion products from the furnace interior pass through the second burner and its generator bed comprising ceramic balls or tubes. As the regenerator is heated the flue gas cools down. This has the comfortable effect that the reformation of dioxins is suppressed by the rapid cooling and no downstream quenching chamber is required. After reaching the maximum temperature of the regenerator, the cycle is terminated and the second burner starts firing. Combustion air now passing the ceramic regenerator is heated before it participates in the reaction with fuel (**Fig. 2.33**). There is a difference to the application of regenerative burners in a standard reverberatory furnace. The pyrolytic gases from the melting chamber are recycled to the hot metal chamber. These are produced by additional burners since flue gas cannot pass from the hot metal chamber to the melting chamber due to the characteristics of the burners installed there. The combustion products, as generated by the regenerative burners in the hot metal chamber, need to be recycled within the pair of burners. Some excess gas discharged from the furnace, as practised in a reverberatory furnace, could be passed into the melting chamber but this would create complications in the gas balance of the complete furnace system. The pyrolytic gases from the melting chamber could be treated entirely different but for improving the heat balance they should be passed – in this case not recycled – to the hot metal section. The temperature maintained here is not sufficient to ignite the combustible fumes and must, consequently, mix with the fuel gas directed to the firing burner.

A pump chamber is integrated into the melting chamber to accommodate a centrifugal pump. Since a melting chamber can never be too large, it is more convenient to attach the pump chamber as separate section to the outside of the furnace thus providing more room for charging and receiving more spacious scrap. A small charging door is sometimes arranged at the other small

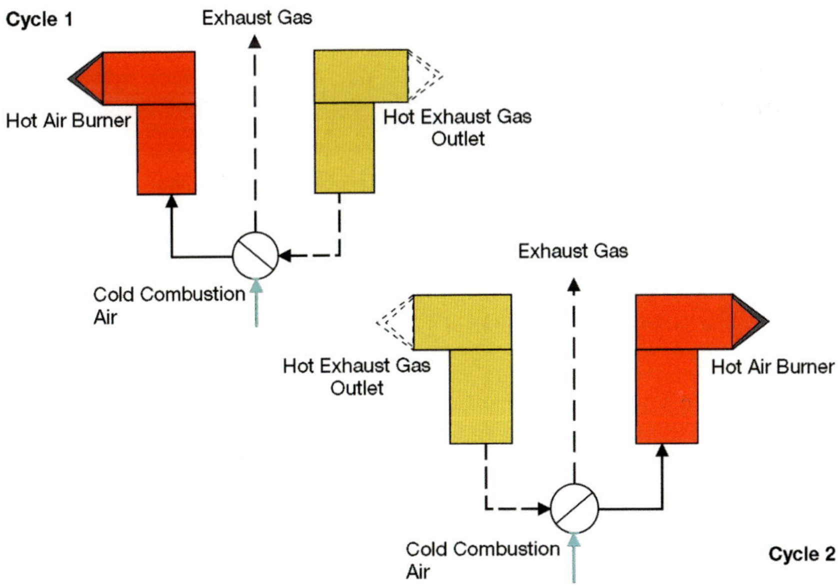

Fig. 2.33: Firing cycle of regenerative burners (Source: LOI Thermprocess)

end of the melting chamber. This creates a narrow and deep chamber shape that may be difficult to charge and to clean. In general, it is more convenient to have a wide charging door at short depth to facilitate ease of operation.

However, it is not always possible to choose the optimal solution for the furnace design. This is very simple in case of the linear motor or the electro-magnetic stirrer. Space conditions, particularly if new equipment is to be installed in an existing plant, shape and arrangement of the furnace have to be adjusted to match space requirements and existing operational conditions. This refers mainly to door arrangement but this in turn does influence the burner arrangement, too. Fortunately, hearth furnaces, such as the reverberatory furnace as well as the twin chamber furnace, can very well be adjusted to the plant conditions.

Compared to the standard reverb furnace, the concept of the two chamber furnace offers one more advantage. The reverberatory furnace can only process clean scrap. This may also be scrap with somewhat larger specific surface such as returned extrusion shapes. The melting chamber of the two chamber furnace can, however, also accept scrap, contaminated with plastic or coating. Organic components, such as paint or plastics originating from compound material, can be removed and burned efficiently thus improving the energy balance of the furnace. Melting becomes more difficult if the scrap to be processed is contaminated by non-aluminium components, tramp iron or other non-ferrous metals or oxides.

The twin chamber furnace is a very good choice for remelting plants with a fairly limited range of different alloys or large quantities for every type of alloy. The fact that a large furnace capacity is required to obtain a suitable melting rate may also be critical in some cases. The melting rate obtained in a well designed twin chamber furnace with forced metal circulation of 60 ton bath capacity is 5 to 6 t/h. In comparison, a normal reverberatory furnace of the same size could easily have a melting rate of more than 20 t/h. But its limitation is material with large specific surface area (i. e foil scrap and thin shapes) and heavy contamination with organics.

The example in the previous chapter shows that only a limited quantity of energy can be introduced to the liquid metal via the bath surface. This energy is required to melt the charged scrap. The calculation in the example was based on the assumption that all melting energy introduced into the liquid metal is required for melting. This assumption is not quite correct. Material charged is heated up to approximately 350 °C by the POC generated by the firing system and passed partly to the melting chamber. The heat required for this preheating need not to be provided by liquid metal; the energy input into the hot metal section and the liquid metal bath can be lower:

$$q_{met} = 1.070 \cdot (660 - 350) + 390 + 1{,}050 \cdot (720 - 660) = 785 \text{ kJ} \cdot \text{kg}^{-1}.$$

This is the energy required to increase the metal temperature from 350 to 660 °C, melt the metal and superheat the liquid metal to a temperature of 720 °C. (For details of the calculation please refer to section 3.1). This heat can be expressed as 218 kWh per ton of metal.

Coming back to the example, as calculated in the previous chapter, the area required to introduce this energy would be

$$A = 5 \cdot 218/76 = 14.5 \text{ m}^2$$

required for a melting rate of 5 t/h.

Table 2.2 gives some melting rates referring to furnace sizes.

Melting tests with UBC chips and swarf, however, have not been successful, neither in the reverberatory furnace nor in the two chamber furnace. Here special measures have to be taken to submerge the loose material rapidly underneath the liquid metal surface. There are different systems available.

Table 2.2: Typical melting rates and bath capacities for twin chamber furnaces (Source: Termcon)

Melt rate t/h	Area liquid metal chamber m^2	Area melting chamber m^2	Dorr liquid metal chamber mm	Inside dim. liquid metal chamber mm	Bath capacity tons	Heel capacity tons
2.50	10	8	3800	2800	16	10
5.00	20	16	4800	4200	36	20
7.50	30	24	6000	5000	60	30
10.00	40	32	7200	5600	80	40

One of the processes comprises a metal pump that creates a vortex in a liquid metal chamber attached to the furnace (**Fig. 2.34**). The bulk material is directly charged into this vortex and its vertical flow will draw it into the metal bath. Suitable pumps could be impeller pumps or electro magnetic pumps based on linear motor technology. The pump chamber is connected to the main furnace chamber by means of tubes or channels. These connections have to be cleaned in frequent intervals and they could be a source for some trouble. The vortex chamber attached to a twin chamber furnace or a reverberatory furnace must be removed from the furnace from time to time for cleaning of the vortex housing and cleaning or replacement of the connecting tubes to the furnace. The tubes are usually made of SiC. It is indispensable for the operation of the vortex that sufficient metal heel is maintained in the furnace. It is also necessary that the bulk material is charged to the vortex chamber as continuous flow. **Fig. 2.35** shows a system which is used to feed swarf from a surge bin via a screw conveyor to the top of the vortex chamber. A burner system keeps the temperature of the vortex chamber at the required level or is used to preheat the chamber.

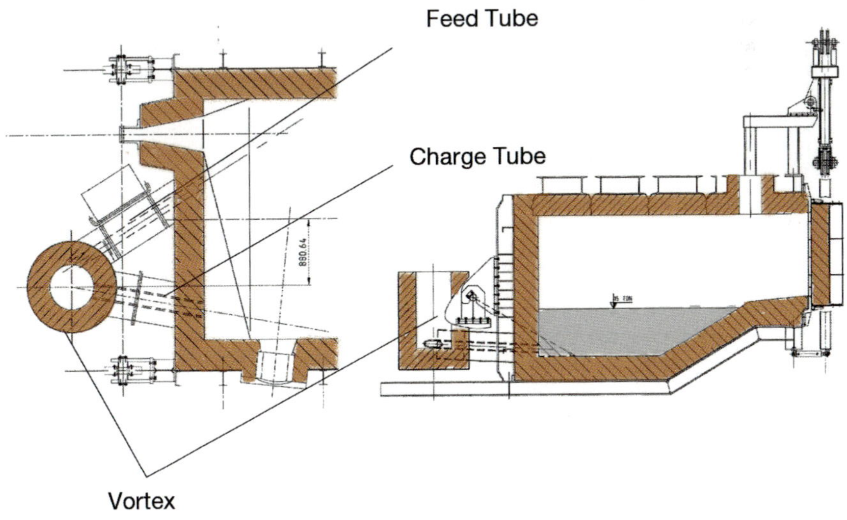

Fig. 2.34: Vortex system attached to a reverberatory furnace (Source: EMP/Alcutec)

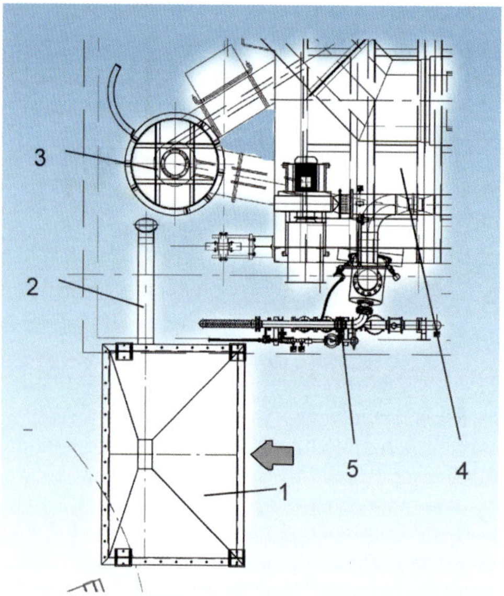

1 - storage bin,
2 - feed conveyor,
3 - vortex chamber,
4 - furnace,
5 - burner control ramp

Fig. 2.35: Feed system for small particles

A very robust system with linear motor was developed by Alcan (**Fig. 2.36**). An electro-magnetic linear motor is attached to one furnace wall. After switching on the electrical current an upward directed liquid metal flow is initiated by the electro-magnetic field that flows against a deflection

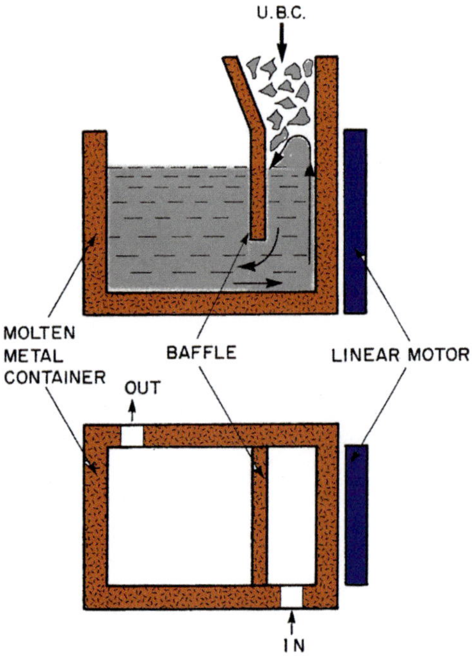

Fig. 2.36: Alcan-system to melt small aluminium particles

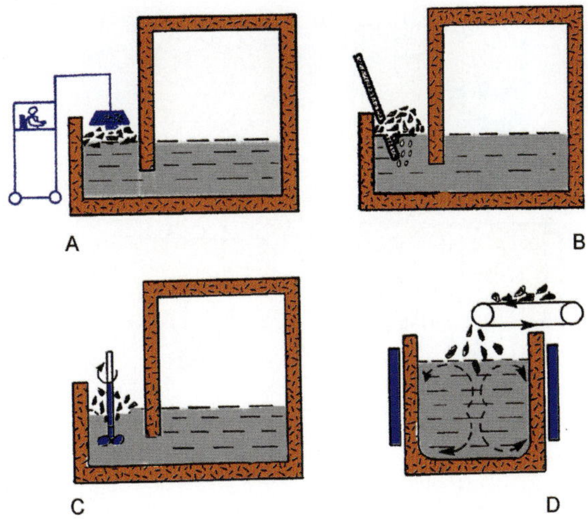

A - plunging,
B - gas lance,
C - vortex,
D - induction furnace

Fig. 2.37: Methods for submerging small particles in a liquid metal bath

plate and creates a kind of stationary wave into which the material is charged. The downwards directed metal flow draws the bulk material immediately underneath the metal surface. The system is almost maintenance-free since there is no contact with liquid metal. But until today there is still one problem. Only a very thin layer of refractory material can be tolerated between the linear motor and the liquid metal. This must be regarded as potential danger apart from the problem to provide sufficient removal of the heat passing through the thin refractory layer. The melting chamber can be integrated into the melting chamber of a two chamber furnace as well as attachment to a standard reverberatory furnace.

One other method is the application of air/nitrogen injection where a pile of loose scrap is charged in a similar fashion as described above. A nitrogen lance is introduced into the liquid metal underneath the pile of scrap (**Fig. 2.37 C**). Bubbles of gas create a "fountain" of metal which gradually erodes away the solid scrap within the pile. A similar system is created by using porous bricks installed in the furnace bottom for introducing nitrogen and thus generating the liquid metal fountain. To be frank, these systems have not been proven to be very efficient for melting metal. But they are very useful for metal treatment.

Both methods have in common that they can only operate if there is a sufficient metal heel. There is also one other problem. Due to the unavoidable intake of air trapped in the loose bulk of material, there is some extensive generation of dross. This dross must be skimmed off. In case of the vortex, dross needs to be skimmed from the entire furnace bath. The design of the Alcan system includes a dedrossing chamber of limited size.

There is also the method of plunging where a pile of loose or baled scrap is charged into an open well and then submerged using a mechanical plunger (**Fig. 2.37 A**). As discussed before, the open well furnace has its limitations and, therefore, the plunging system is not applied very frequently.

The chips must be charged to the system by a suitable conveyor with a continuous feeding rate. If economically justified, a drying plant can be arranged inline with the melting furnace so that the chips are directly charged to the furnace. Steel swarf needs to be separated prior to feeding the system. The flue gases leaving the furnace can be utilized for providing energy to the chip dryer. It is indispensable that the treatment of combustion products is considered as integral part of the entire system. There is a selection of chip dryer equipment which is well-suited for an integrated system. Please refer to part 2 for more information on chip drying.

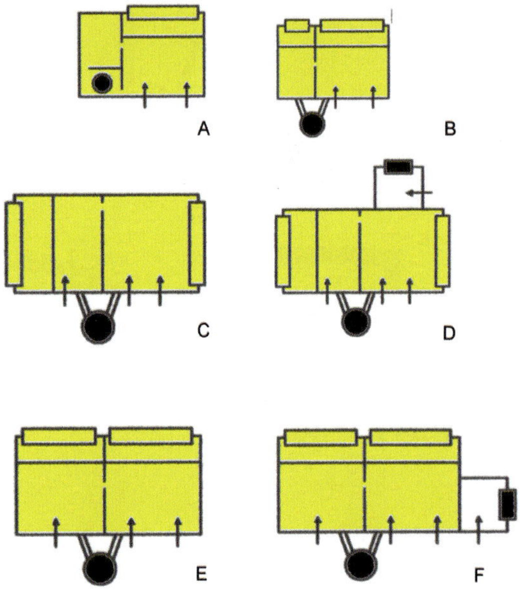

Fig. 2.38: Arrangement of twin chamber furnaces (Source: Thermcon)

Based on the principle of the twin chamber furnace, some special design features are possible to enable this furnace a wide range of application. **Fig. 2.38** shows possible arrangements based on the requirements of the production plant. Picture A) is the very basic design of the furnace. An open well with circulation pump is attached to a reverberatory furnace to provide metal circulation. Material is charged into the furnace by means of a large charging door. A twin chamber furnace is depicted in picture B). The melting chamber is used as de-drossing chamber and only required if material comprising small bulky particles such as swarf, for instance, is charged to the furnace by means of a vortex. The standard twin chamber furnace is shown in picture C). The furnace is equipped with a charging ramp to which material is fed and organic contaminations can be removed before the material is pushed into the liquid metal bath. The furnace can also be equipped with a dry hearth (picture D) as described in the next chapter for removing large non-aluminium particles such as bronze or iron components. However, this procedure results in increased melt losses and should be applied only if no other means of processing this kind of scrap are available. In the design, as shown so far, the hot metal section is of standard design without pre-heating bridge. Both sections of the furnace can be equipped with such a bridge as shown in picture E). Finally, a twin chamber furnace can be equipped with everything described so far including a hot metal section with preheating bridge, a melting chamber with preheating bridge and finally a dry hearth section. This solution is depicted in picture F).

Which design will be selected finally by a plant depends very much on the requirements of the operation. The concept of the twin chamber furnace offers good flexibility for a remelting plant. While thin material is melted with acceptable metal recovery, ingots and large inhouse scrap can be processed in the hot metal section. Parallel to that, contaminated scrap can be melted using the dry hearth section with limitations regarding pickup of non-aluminium metals and reduced metal loss.

The metal tapped from the twin chamber furnace is transferred to a casting furnace for final analysis correction and temperature adjustment.

2.1.3.3 Dry hearth furnace

The scrap supplied to the aluminium refining plant sometimes contains a substantial amount of metallic components which may be part of the previous life or may be collected with the scrap. This material should be removed during mechanical processing but this may not be always possible. Hence, the materials need to be separated the "hot" way. Also for this case the industry has developed a suitable furnace: The dry hearth furnace (**Fig. 2.39**). The furnace comprises a holding section for the liquid metal and a dry hearth section which is sloped towards the holding section. They are separated by a dividing hot wall. Both sections are equipped with burners. Scrap is placed in the dry hearth and energy is introduced. The dry hearth can be directly fired to heat up and finally melt the scrap. The separation of the different components of the scrap is based on different melting points. Generally aluminium melts first if there is no lead in the scrap and flows through the channel into the holding section of the furnace. Although heat exchange between solids and gas is very good, material will start melting at the surface of the pile. As soon as it is in the liquid state it will float down to the bottom of the dry hearth. On the way down liquid aluminium will flow over solid metal of somewhat lower temperature and may solidify again until this area of the pile has reached sufficient temperature. During all phases of the melting process, the scrap offers a large surface to the hot combustion products. Therefore, it is inevitable that melting in a dry hearth goes with additional oxidation of aluminium with the related metal loss. The situation can be improved if hot flue gas recycled from the holding section of the furnace is mixed with the new POC of the dry hearth chamber. The heat recovery system is then similar to that of the twin chamber furnace.

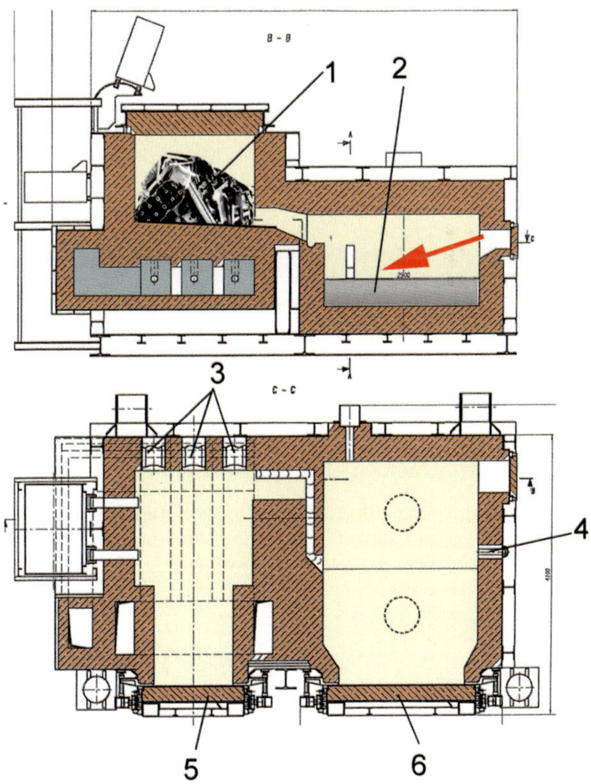

1 - contaminated aluminium scrap,
2 - holding section,
3 - burners,
4 - taphole,
5 - service and charging door,
6 - service door

Fig. 2.39: Dry hearth furnace

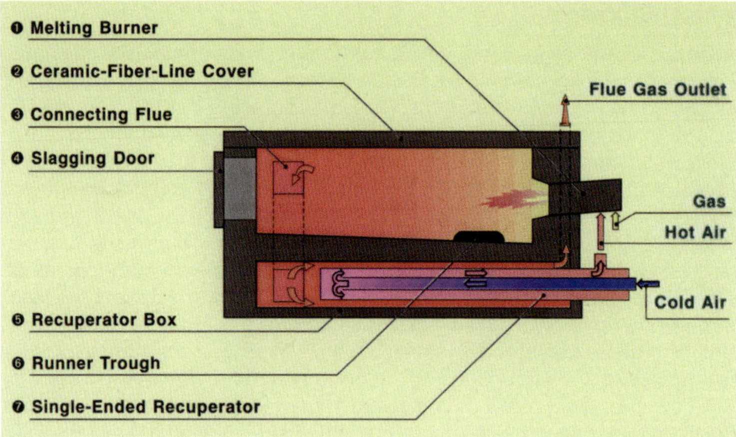

1 - top charging lid, 2 - service door, 3 - burner, 4 - connecting duct, 5 - furnace body, 6 - connection to heating section, 7 - recuperator

Fig. 2.40: Quick melting furnace (Source: Gas de France)

After melting the aluminium, the foreign components can be taken out of the dry hearth through the large door. Metal is tapped from the holding section by means of a fixed taphole with plug. The metal molten in the dry hearth furnace is passed to the casting furnace where final alloying and refining take place.

The difference of the dry hearth and the pre-heating bridge of a reverberatory furnace or a twin chamber furnace is quite clear: While on the pre-heating bridge the material is raised to a temperature just below melting point and it is pushed then into the liquid metal, the aluminium positioned in the dry hearth is melted and flows to the holding section of the furnace.

A variation of the dry hearth furnace, as shown in **Fig 2.40**, is the so-called quick-melting furnace. It also consists of a dry hearth section and a holding chamber. The dry hearth section can be loaded from top or through a charging door at the front end of the chamber. The dry hearth is also sloped towards the holding section which is arranged rectangular to the dry hearth. A burner is arranged at the wall next to the transfer point firing into the metal batch. The flue duct is at the side opposite the burner. POCs are directed from the furnace to a furnace chamber underneath the dry hearth, thus heating the bottom of the dry hearth. Hence, heat transfer occurs also from the bottom on to the batch with the result of fast heating up. A recuperator can be placed in the hot gas chamber underneath the dry hearth.

This furnace is very effective in foundries for melting ingots and inhouse scrap. Since its use for processing contaminated scrap and small particles is very limited, and together with low melting rate and small bath capacity, it is not used in the secondary aluminium industry.

2.1.3.4 Tower furnace

The tower furnace consists of the melting tower section and a furnace section (**Fig. 2.41**). Its basic idea is the recovery of energy usually released through the stack. Flue gas from the holding section passes through the aluminium placed in the flue gas duct, i. e. the “tower”. The material is pre-heated before it passes into the furnace section. In theory this sounds very good. The problem is how to transport the solid metal into the furnace. Tower furnaces represent a compro-

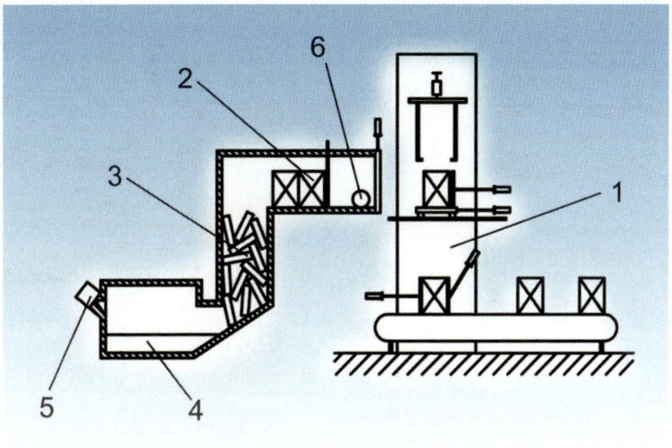

1 - charge feed system,
2 - stacks being preheated,
3 - metal being melted in the tower,
4 - holding section,
5 - burner,
6 - flue discharge

Fig. 2.41: Tower furnace

mise between pre-heating and melting. The aluminium is melted in the lower section of the tower while the POC flow upwards through the tower, thus preheating the newly charged material. One or more burners are arranged in the melting section of the tower to provide the required energy. Liquid metal flows via a duct into the furnace section that now works as holding and overheating chamber only that is equipped with the necessary high performance burners. As the charge is dropped into the furnace, some means is required to stop the material to build a pile at the bottom of the tower before melting: **Fig. 2.42** shows the tower section of a Striko tower furnace widely used in die-casting plants. Material is charged to the tower by means of a skip hoist and dumped into the tower. The material charged is held by a refractory bridge that permits only liquid metal to pass further down into the holding section. This section heats the metal by heat of radiation. It is, therefore, equipped with radiative roof burners.

Although applied very successfully in the copper industry, the application of the tower furnace within the aluminium industry is limited to foundries. There is, as in the case of the quick melting furnace, the restriction to melting ingots and returned scrap from the foundry. The high melt losses, when processing small and contaminated scrap and the complicated charging, may be the reason for the reluctance of plant operators to use this kind of furnace in the secondary aluminium industry.

2.1.3.5 Rotary drum furnace

The rotary drum furnace (**Fig. 2.43**) is designed to process bulk material consisting of small individual particles. Scrap of larger size is usually processed in reverb furnaces due to the heat transfer mechanism. Heat flows from the atmosphere within the furnace into the metal via the surface of the piece of metal and continuously into the block and hence increases its temperature up to the melting point. The heat transfer is a function of the heat transfer coefficient, the temperature difference between furnace atmosphere, the block surface area and the conductivity factor λ.

As described for bulk material, the heat transfer is to some extent more complex than processing large pieces of scrap. Although the specific surface area is large, the surface exposed to the furnace atmosphere is quite small. Furthermore, heat conductivity within the bath is extremely poor due to the numerous particle boarders and the space between the individual pieces of metal. In order to arrive at an acceptable heat transfer, the bulk has to be revolved. This is ideally achieved in a rotary drum furnace. For protecting the surface of the material and for other purposes the process operates with flux. Please refer for details to section 1.6 above.

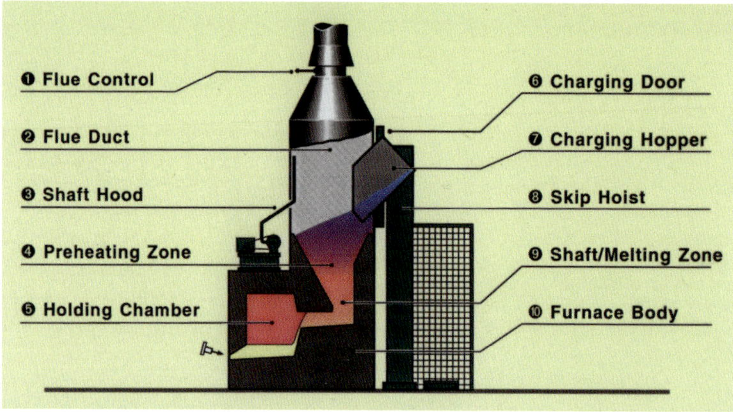

1 - flue gas damper, 2 - tower, 3 - shaft hood, 4 - material charge, 5 - holding section, 6 - charging door, 7 - charging hopper, 8 - skip hoist, 9 - melting zone, 10 - furnace body

Fig. 2.42: Striko tower furnace

Due to the rotation of the furnace drum, material is moved and the individual particles will be frequently exposed to the combustion products.

The heat exchange within the furnace comprises three components (**Fig. 2.44**):

The heat exchange by convection is from the hot flue gases to the wall of the furnace and to the batch of salt and metal.

The heat transfer coefficient α_g is the decisive factor for the heat transfer for the heat flow from the combustion products to the batch of scrap as well as to the furnace walls. We know this from pre-

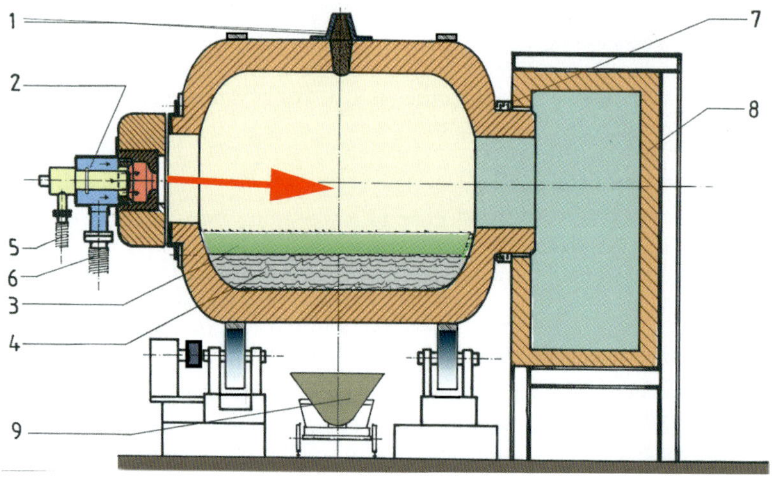

1 - slag discharge, 2 - burner, 3 - salt, 4 - liquid aluminium, 5 - gas connection, 6 - combustion air connection, 7 - rotary drum sealing, 8 - fume-collecting chamber, 9 - dross train

Fig. 2.43: Rotary drum furnace

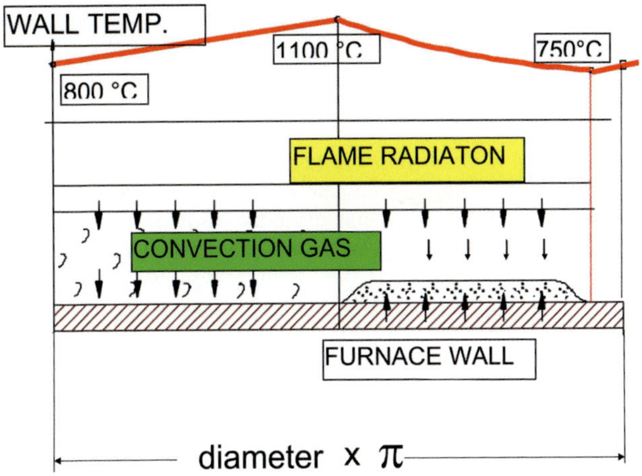

Fig. 2.44: Heat exchange in a rotary drum furnace

vious calculations. This factor depends on the surface conditions and on the materials exchanging heat and to the most extent on the gas velocity; with high gas velocity w_g increases. Thus, the velocity of the hot gases within the furnace and the turbulence, as result of flue gas velocity and flue gas volume, are very important design factors. Therefore, the burners selected will be high velocity burners. With a small quantity of flue gas the velocity of the gas within a defined furnace room is also small.

Radiation from the hot flue gases is emitted by water steam and carbon oxide only. Nitrogen does not emit energy but this content is important for heat exchange by convection. For the heat transmitted to the batch, in the furnace and to the furnace wall with a resulting heat reflection from the wall to the batch, the heat transfer depends on the temperature of the combustion products and the surface to receive the heat.

The surface area participating in the heat transfer, the emission coefficient ε, the partial pressure (or at atmospheric pressure the gas content), the flame thickness, the gas temperature and the batch temperature are the factors for the radiative heat exchange. The total heat transferred by gas radiation is roughly the sum of both fractions. The temperature is very important because the heat transmission increases with the power of 3.5.

There is a benefit of using oxygen instead of combustion air. With nitrogen being absent, the partial pressures of H_2O and CO_2 are higher and the portion of radiation is accordingly. The flame temperature is higher, too. This is even more effective since the heat transfer increases with the power of 3.5 assuming the identical batch temperature. The heat transfer by radiation increases very steep with increasing temperature but high flame temperature has some limitations. A hot flame results in very high radiation to the furnace wall, heating it up rapidly to a high temperature. Fig. 2.44 shows the development of the furnace drum at a certain position. The surface of the refractory lining is exposed to rapid temperature changes along the circumference of the drum. This, together with the grinding effect of the material, results in short service life of such a refractory lining. Design has to consider this fact very carefully.

The third component is the heat exchange between furnace wall and material covering the furnace wall. Due to the rotation of the furnace, part of the wall is always covered by material. Since the wall temperature is higher than the temperature of the bath, some heat is transferred from the wall to the material. This effect of heat transfer is generally overestimated. With the comparatively high number of revolutions of the furnace drum, time is too short for the heat to penetrate into the refractory lining very deep. This means there is not much heat stored. Hence, very little heat can be transferred from the furnace wall to the batch. We will come back to this later.

The radiation from the furnace wall to the batch can be neglected because of the comparatively low wall temperature and the absorption of heat by dust, smoke and the flame itself.

Generally, in a rotary drum furnace combustion products generated by the burner enter the furnace from the one side and flow through the drum to the flue gas exit (Fig. 2.43). During this passage the hot combustion products transmit their energy to the batch of material and to the furnace wall. There are always different conditions in the furnace along the furnace axis as well as on the circumference of the furnace drum. As the furnace heats up, these conditions will never be stationary until the end of the cycle. Since the heat exchange by convections represents the major portion of the energy transmission it is important to maintain the gas flow level as high as possible. If the energy input needs to be downrated, which may be the case once the maximum temperature of the refractory lining is reached, it appears to be a good technical solution to bring in a burner with a lower firing rate (**Fig. 2.45**).

The residual time of the flue gases in the furnace is also important for the heat transfer. In modern furnaces, particularly when equipped with oxygen burners, the combustion products are not leaving the furnace through a flue arranged at the furnace end opposite the burner side. This flue is closed and the combustion products are forced to flow counter-currently to the gases freshly introduced. This is always the case in a tiltable rotary drum furnace since it is one of the relevant design requirements that the back wall of the furnace is closed. The flue gases are reflected at the back wall and must leave the furnace on the burner side. This provides a fairly long residual time for the flue gases. However, the proper sizing of the furnace drum must consider that the back wall has to be kept hot in order to avoid scaling. Therefore, the combustion products require sufficient momentum

$$p = m \cdot v \tag{2.5}$$

to be able to reach the back wall during all firing phases. In this equation p is the momentum, m the mass of combustion products and v the gas velocity. During its flow from the burner to the back wall the jet of combustion products will take in some gas from the furnace atmosphere. Additionally, the gas flow is disturbed by the gas stream flowing counter-currently to the burner side to be able to escape through the flue gas system. Consequently, the burner jet has to be shaped to be very stable in the first section of the furnace and finally opens to provide sufficient area for heat exchange. Although the conditions of gas flow in the furnace are chaotic in a physical sense, it is

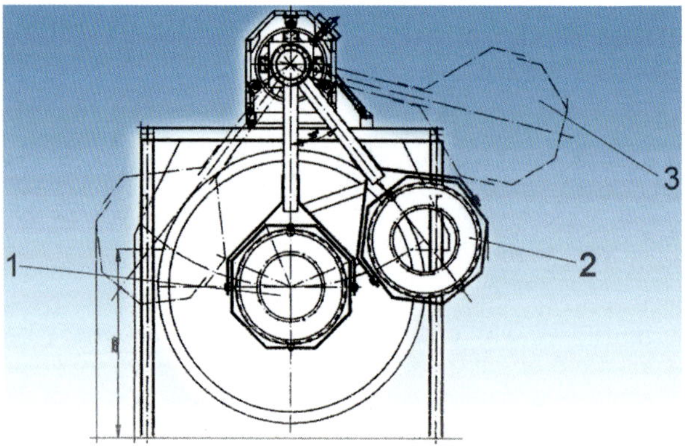

1 - low capacity burner,
2 - high capacity burner,
3 - charging position

Fig. 2.45: Dual burner arrangement

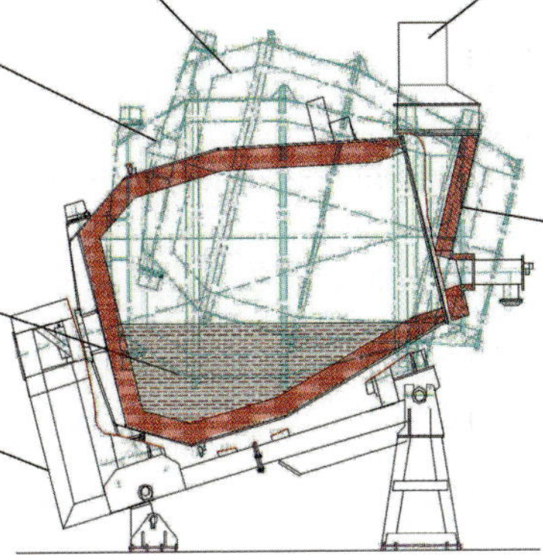

Fig. 2.46: Tiltable rotary drum furnace (Source: Alcutec)

fair to assume that widening the drum in the back section of the furnace will result in lowering the pressure to some extent thus assisting the gas flow to be stable for a longer time. This is realized in a well designed rotary drum furnace (**Fig. 2.46**).

The temperature setting in a rotary drum furnace is more difficult than in a hearth furnace. Due to the rotation of the drum, no temperature pick-up is possible in the batch or in the furnace wall. The only possible temperature reading can be obtained from the exiting flue gases. However, practise teaches that there is a definite temperature difference between the furnace wall and the flue gas. The flue gas temperature is always 100-150 °C higher than the average refractory material temperature. Therefore, this temperature, as only means of a temperature relation of the batch, is used to control the energy input.

Material charged to the furnace is usually highly contaminated with oxides and organic components. A de-coating phase to burn off the organics is, therefore, the very first stage of the furnace cycle. The energy input will be set to a low value sufficient for removing the organics but not too high to heat the metal to a level above 350 °C since oxidation increases rapidly above this temperature. Sufficient excess air must be provided by the firing system to burn the organic components which are ignited by contact with the hot burner flame. This is necessary to ensure that the content of organics in the waste gas does not exceed the level as set by the environmental regulations. If the organics are not burned totally, an incinerator may be required. A good solution is to incorporate such an incinerator in the waste gas ducting system of the rotary drum furnace. The main burner is installed in the flue gas hood or in the furnace door. It ignites the flue gases passing it and the downstream ducting provides sufficient residual time for the carbons and hydrocarbons to react. As discussed before, if the burner system is designed properly some of the heat generated by combustion of the organics may remain within the furnace thus improving the heat balance of the entire system. Unlike in the hearth furnaces, the energy of the organics once having left the furnace is difficult to recycle since the waste gases of the rotary drum furnaces are heavily contaminated with dust. This will clog heat recovery systems very quickly. This is not a feature typical for rotary drum furnaces but it is due to the type of material generally processed in those furnaces. As the organic content decreases, the excess air will be reduced. This is usually an automatic process. A lambda probe arranged in the flue gas stream measures the oxygen content of the combustion products. The signal is used to set the combustion air valve to the required volume. In principle, the system is very simple and quite reliable. Considering the contaminated material processed in a rotary drum furnace, the flue gases carry a high dust load. This could be a problem for the function of the lambda probe. The unit has to be cleaned frequently – and this cannot be assumed for a production plant in general. Thus, a pre-setting of the parameters seems to be more appropriate for a secondary smelter. But some units have been developed to overcome the cleaning problem. Refer also to section 7 "Process control".

The de-coating phase is followed by the melting stage. Now the flue gas temperature is set to the highest possible value which in general is 1,000 °C, considering some safety margin with respect to hot spots acting on the refractory lining. The burner input follows this control parameter automatically. Due to the high energy requirement of the batch, the actual flue gas temperature does not reach the set-point yet. As material is heated up, less energy is required and the energy input of the firing system is automatically reduced by the burner control loop. The flue gas temperature remains at the set-point. The automatic control also sets the rotation speed of the furnace drum as the friction between drum wall and material decreases. It is quite difficult to obtain this function automatically. It could be based on the power requirement of the drum drive. But this parameter depends too much on the material processed and thus the data available are not very reliable for automatic control but may provide an indication to the operator about the status of the melting phase. The next cycle stage is overheating of the melt, followed by metal tapping and slag discharge.

It may be necessary to charge the furnace two or three times if the nature of the scrap does not permit one charging only. Thus, de-coating and melting have to be repeated as well, whereby the next charging can be done when the batch in the furnace is molten or broken. Operators have to be careful when the next batch is selected. No humid or oily material may be charged to the furnace in order to avoid an explosion. The persons responsible for the batch calculation and batch preparation must carefully consider this when the total furnace batch is prepared.

The parameters for the different steps of the cycle depend very much on the material to be processed. Therefore, they are set prior to the start of the furnace cycle based on experience of the operating personnel with identical or similar material. The necessity to have more than one charging needs to be considered when defining the recipe for the specific material. The control system of the furnace is designed to permit the setting of various recepies which can be selected by the furnace operators.

After the last batch – the full load is in the furnace – the rotating speed of the furnace drum should increase gradually. As soon as the aluminium is melted, two different procedures are recommended:

Material having a high content of oxides should be processed at comparatively high speed. This helps to achieve good coagulation of molten metal particles. Since there is sufficient non-metallic material present, there is also satisfactory covering available to reduce the danger of melting loss due to contact of liquid metal within the furnace atmosphere.

The situation is different with material having low oxide content. Only a thin layer of protective skin is formed that should not be broken off. This is now similar to the conditions within a reverb furnace. In such a case, the speed of the drum should be set at 0.5 to 1 r.p.m.

Mixing of the material in the drum becomes more intensive with increasing rotation speed. Because of the friction at the drum, the material is lifted to a level depending on the rotating speed and cascades back onto the material below thus forming a new top layer. The mathematical model developed for ball mills is very helpful for understanding the mixing mechanism of the rotary drum

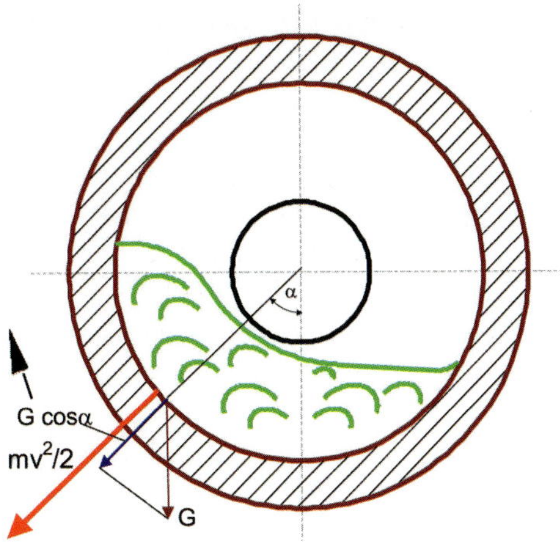

Fig. 2.47: Material movement in rotary drum furnaces

furnace. A particle located close to the drum wall (**Fig. 2.47**) is affected by the radial force N resulting from the centrifugal force and the radial vector of the weight.

$$N = \frac{m \cdot v^2}{r} - m \cdot g - \cos\alpha \tag{2.6}$$

During rotation, N will cause friction between the particle and the wall enabling it to adhere to the wall thus being lifted and then accelerated. The radial component of the weight becomes smaller with increasing angle α and will be negative after the particle passes through the horizontal axis ($\alpha = 90°$, $\cos\alpha = 0$). Depending on the speed, sooner or later the particle loses contact to the wall and enters into a ballistic curve upon falling down to the bottom: The particle collective forms a cascade.

At the critical speed the centrifugal force is equal to the weight of the particle. At this speed all material adheres to the wall:

$$\frac{m \cdot v^2}{r} = m \cdot g \tag{2.7}$$

With rotating velocity $v = 2r\,\pi\,n$ with n = number of revolutions per minute and r = the drum radius, the gravity factor 9.81 ms^{-2}

$$4 \cdot r \cdot \pi^2 \cdot n^2 = g$$

Resulting from that with D = 2r

$$n = \frac{42.3}{\sqrt{D}} \tag{2.8}$$

This equation offers a very convenient method to define and compare test results. Since the behavior of a batch is very different from the one for a single particle, the best speed has to be tested with relation to the degree of filling and material composition. Depending on the degree of filling, tests in rotary drum furnace showed best results at

$$n = \frac{10 \ldots 27}{\sqrt{D}} \qquad (2.9)$$

As the temperature of the batch increases and approaches the melting point, some liquid phase of material, that could be metal, salt or both, will be present, thus decreasing the friction coefficient m. The desired cascade movement can only be achieved by increasing the rotating speed of the drum. On the other hand, decreasing cascading tendency of the batch indicates that material is partly melted and the next batch needs to be charged. There are different means to force cascading of the material. By designing the refractory lining as polygon, material is forced to lift and the friction is supported. Unfortunately, the lining will wear off and we will end up with a round shape in the end.

The standard rotary drum furnace (Fig. 2.43) comprises a rotating drum with both ends closed during operation. The required energy is supplied by a burner arranged at one end of the furnace and the resulting combustion products pass through the furnace and leave it finally through a flue at the opposite end of the furnace connected to the waste gas scrubbing system of the refining plant via rotating joint or just being released and collected by means of an open hood. On its passage through the furnace, the flue gas transfers its energy to the material in the furnace. This will be charged through the charging door at the front end of the furnace. The burner is incorporated in this door and is swivelled with the door to be able to charge scrap. The furnace drum is supported by four roller stations via thrust rings attached to the furnace drum by an adjustable wedge to obtain smooth rotation of the furnace drum. The drive can be executed by means of a chain drive or by motor and gearbox acting to drive the one or more supporting rollers. Molten metal is tapped by means of a taphole at the circumference or at the flue gas side of the drum. During melting it is closed by a plug system. After melting, the liquid metal is tapped first. After this the slag, comprising oxides and salt, needs to be discharged from the furnace. This is difficult due to the high viscosity of the slag. For slag discharge a large taphole is arranged at the circumference of the drum. The slag is collected in slag bins arranged as slag train underneath the furnace. If the slag is still too viscous, additional salt has to be charged to the furnace to obtain better flow conditions for the slag. Tapping and slag discharge are normally quite time-consuming actions.

Material is charged to the furnace through an opening in the furnace axis. Due to the size of the taphole some time is required to drain the metal. Discharging of salt is also somewhat difficult since it has to be made sure that salt is always liquid enough to be tapped.

Fig. 2.48.1: Liquid metal discharge from a rotary drum furnace and transfer to a casting furnace (Source: INTEC)

Fig. 2.48.2: Slag discharge from rotary drum furnace (Melting solutions)

To be able to charge material very rapidly, the relevant opening needs to be quite spacious. However, increasing the size of the charging opening requires also that the drum diameter is also enlarged in order to obtain the same material quantity in the furnace.

The tiltable rotary drum furnace (Fig. 2.46) was developed to overcome these problems.

A very large charging opening is achieved by tilting the furnace back. For melting, the furnace rotates in the back-tilted position. To discharge metal the furnace is now positioned horizontally or slightly more and metal flows in a wide stream within a short time out into a launder system (**Fig 2.48.1**) connected to a holding furnace. After metal is tapped, slag is removed by tilting the furnace somewhat more and as the drum slowly rotates the slag is discharged into containers arranged underneath the furnace opening (**Fig. 2.48.2**). As in the standard rotary drum furnace, the burner is arranged at the charging side either in the swivel door or in the hood above the door.

There is another important design criterion for the tiltable rotary drum furnace.

The original concept of the tiltable rotary drum furnace is based on a cylindrical shape of the furnace drum. For smaller furnace capacity this design is very convenient since it offers the maximum size of the charging opening. As capacity increases, the cylindrical shape becomes critical. Larger size of the furnace opening does not result in faster charging and discharging anymore and energy losses through the large diameter get unacceptably high.

Increasing the capacity by extending the drum length is also not possible. There is a limitation arising from the burner momentum; a long furnace will have a cold back wall. Thus, the length/diameter ratio should be close to 1.5. There is a manufacturing problem as well if the furnace diameter gets too spacious. Large machine tools to turn supporting rings and drum are not readily available anymore. Thus, some solution must be found. One possibility is to design front and back of the drum as steep cone (**Fig. 2.49**). This will help with regard to charging opening and energy loss but inevitably leads to an excessively large tilting angle that may result in a dangerous situation at the end of the slag discharging operation due to uncontrolled sliding of hot slag. The tiltable rotary drum furnace discharges metal and slag very fast through a large door in the front of

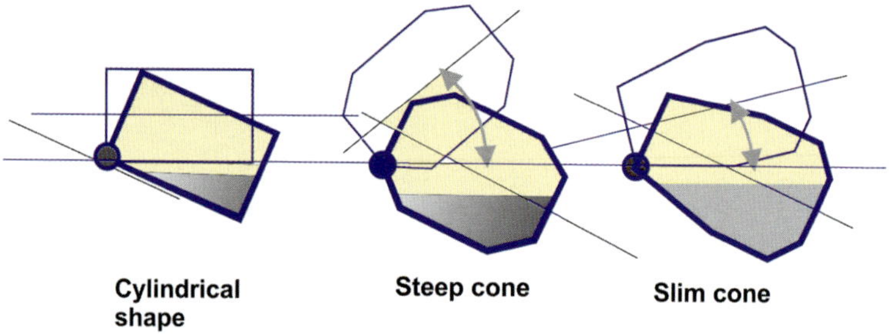

Fig. 2.49: Shapes of tiltable rotary drum furnaces

the furnace. Consistency of the slag is not critical for this step of the furnace cycle. Due to efficient heat transfer in the furnace and together with the advantage of reduced charging time and the very rapid discharge, the overall cycle time of the furnace is shortened remarkably. Typical time for one furnace cycle from tap to tap is four hours.

The rotary drum furnace is the only furnace that is able to process scrap with a high content of oxides or other contaminations. It can also process clean material, particularly material with a large specific surface area, i. e a large area per kilogram of weight. The rotary drum furnace process uses salt to protect and to clean the metal in order to get an optimal yield. This is covered in detail in section 1.6. There is no doubt that salt is a cost factor and the deposit of salt slag creates cost, too – at least in industrial countries. Thus, plant operators try to reduce the quantity of salt. On the other hand, manufacturers of equipment try to convince plant operators that their specific process is able to operate without salt. This may be true if one does not consider that operating without salt increases metal losses remarkably. It is also true that processing material with high content of oxides requires less salt for protection since oxides protect the metal very well. Furthermore, they contain a high percentage of salt originating from pervious processes. One solution to save salt is to operate in reducing atmosphere. If there is no oxygen available for reaction, no oxidation is possible.

One should consider, however, that no combustion can happen at a 100 % level. Consequently, a certain percentage of not yet reacted oxygen is still present in the combustion products. Establishing a reducing atmosphere in a furnace helps definitely to improve metal yield. The method has one disadvantage: The flue gases released from the furnace contain a high percentage of unburned organics. In order to fulfill the requirements of environmental protection, these organics must be converted into non-reactive products, i. e they must be burned. This is partly done directly in the furnace with the result that the reducing atmosphere does not exist anymore since excess oxygen needs to be provided for the combustion of the organics. But the energy of the organic contamination can still be utilized to improve the fuel economy as discussed before. The limiting factor is the ability of the furnace batch to absorb the energy provided. By controlling the combustion air the time for reaction may be controlled to a certain extent. Another time-delaying factor is that mostly bales of the organics containing scrap are to be melted. Since disintegration of such bales will certainly take some time the combustion can take place only once the organics are freed. But still the major portion of the heat generated is lost. **Table 2.3** summarizes one example. If this procedure is not desired or possible, an incinerator can be installed in the flue gas system also in the rotary drum process before the combustion products are released to a waste gas scrubbing plant (**Fig. 2.50**). It does not matter if the process uses normal air or oxy-fuel.

Oxygen is used in rotary drum furnaces already for a long time.

Table 2.3: Example 10 t scrap, 10 % organics

Type of organic	Polyethylen
Weight of organics	1,000 kg
Lower heating value	12.2 kWh/kg
Volume of heat Q_{org} generated	12,200 kW/h
Time combustion	30 min
Total heat required by aluminium (20 °C to 720 °C)	6,000 kWh
Heat absorbed in 30 min	2,000 kWh
Increase of heat efficiency	**16.4 %**

In past years operators have been reluctant to use oxygen burners since some problems were encountered. These include frequent burner water jacket leaks and on-going maintenance problems, severe refractory material damage in the burner tile and surrounding furnace areas due to poor burner location resulting in reflection of the flame back on the brickwork and reduced metal yield due to either poor burner position or overheating the metal. Severe damages were also encountered due to excessively high flame temperature resulting in heating the refractory lining to above its service limit temperature. In the meantime, most of these problems are solved either by better burner design, more intelligent burner arrangement as well as more reasonable application of existing technology.

It is stated that there are remarkable advantages when using oxy-fuel. Mainly these are less energy consumption, faster melting, lower emissions of pollutants and dust and lower flue gas volume.

We will have a closer look at these statements.

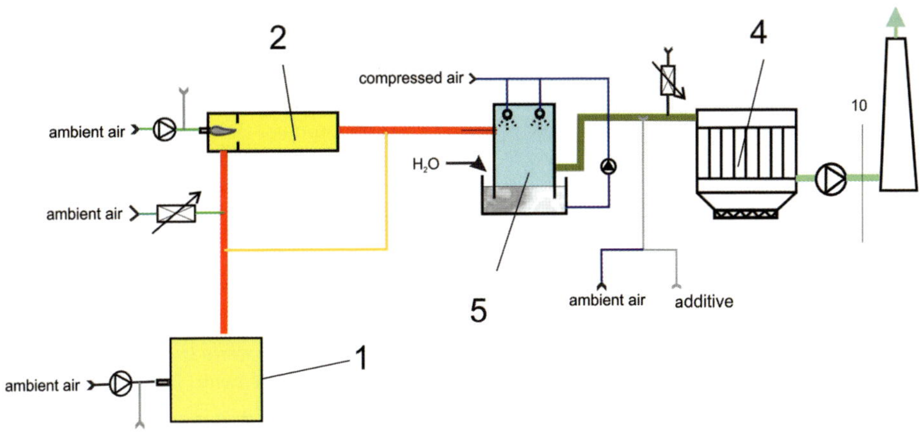

1 - rotary drum furnace, 2 - incinerator, 4 - bag house, 5 - quenching chamber. The small lines with numbers indicate the system for calculating the parameters of the waste gas system.

Fig. 2.50: Rotary drum furnace with incinerator

When firing fossil fuel, oxygen can reduce the nitrogen load of the process. Nitrogen is contained in the combustion air and is mainly inert to the aluminium melting process. However, when nitrogen passes through the furnace, it is heated up. Typical temperatures are 20 °C for nitrogen entering the furnace and about 1,000 °C exiting the process if there is no heat recovery (as is standard in the rotary drum furnace). Approximately 600 kg of nitrogen are heated up for 1,000 kg of melted aluminium and this takes away a significant amount of energy through sensible heat from the furnace. When nitrogen is passing through the burner flame it also cools the temperature of the exothermic reaction in the flame. For example, the adiabatic flame temperature of natural gas in air is 1,950 °C, whereas in oxygen it is 2,780 °C. Such high temperature is not in itself necessary for a low-melting metal such as aluminium. Radiant heat transfer, however, is proportional to the power of 3.5 (theoretically 4) of the temperature differential. $\left(T_{flame}^{3,5} - T_{metal}^{3,5}\right)$ If T_{metal} is 750 °C, for example, the oxy-gas flame provides more than four times the rate of radiant heat than the air/gas flame. Theoretically, this increase in heat transfer reduces melting time and energy input in some cases by approximately 60 %. The furnace also requires a smaller burner. Thus, for example, instead of a burner rate of 6 MW of an air/fuel burner only 4 MW are required for the oxy-fuel firing. Although the theoretical heat requirement should be down to 2.5 MW, the actual heat input must compensate for the other furnaces as well.

One condition must be considered when discussing radiant heat transfer in a rotary drum furnace. Particularly when processing contaminated scrap the dust load of the flue gas is quite high. Hence, the flue gas absorbs a large amount of the radiant heat at least during the de-coating stage. The temperature of the combustion products is also higher in case of the oxy-fuel burner resulting in a better heat exchange by convection. Here it must be taken into account that towards the end of the melting cycle the gas temperature reaching the set-point cannot be further increased in order to protect the refractory lining against overheating.

We can summarize these aspects, that is the reduction of energy consumption now. Since there is definitely less gas volume to be heated, the energy requirement is reduced with oxyfuel. During certain stages of the furnace cycle, the higher flame temperature is of advantage for a better radiant heat transfer to the furnace charge. Also the higher flue gas temperature may increase the heat transfer by convection. The effects of a better heat transfer mechanism are reduced in case of the radiation by the dusty furnace atmosphere and, in the case of convection, by the much lower gas quantity resulting in a lower flow rate of the gas.

The faster melting rate, resulting in higher furnace throughput, is directly linked to and result of the improved heat exchange. If the batch can reach the required temperature faster, then melting time is shorter and also less energy is required. Practical operation shows that one additional cycle per day is possible when using oxy-fuel burners instead of air/fuel burners. Since normally five to six cycles are run per day, this is an increase from six to seven cycles. The advantages of the oxy-fuel technology show much more improvements in case of revamping an existing rotary drum furnace than in a well designed tiltable rotary drum furnace of more recent design.

The statement that there are less emissions when using oxy-fuel refers more to other furnace technologies, e. g. to the twin chamber furnace. Practise did not show improvements in the rotary drum furnace.

But there is definitely less dust compared to the air/fuel burners. Since there is less gas volume, the gas velocity in the furnace is much lower. Thus, less dust is carried away. This is certainly of advantage since the cleaning interval for waste gas ducting is extended.

As mentioned before, the total volume of waste gas is lower since no nitrogen is involved in the combustion process, reducing the gas load on the waste gas filter. This is the case during combustion. When loading the furnace, a certain air volume is essential to provide sufficient air to take in the dust generated during charging. Mostly, the relevant air quantity is higher than the gas quantity during combustion. This dust-loaden air must pass the waste gas scrubbing system as well. Hence, the gas load arising from combustion is not the ruling factor but the total air intake. This must be considered when sizing a waste gas treatment system for a recycling plant.

As an option, the burner can be operated with a mixture of air and oxygen. This permits adjustment to very specific operating conditions. It is also a measure to adjust the flame temperature which may be desired when the furnace is operated at a low rotation speed. Another possibility is to have the furnace equipped with a dual burner system; one burner is used for oxy-fuel only while the other burner operates with air/fuel. Since each burner can be controlled independent of the other, best operating conditions can be obtained for the various phases of the furnace cycle. These parameters are set on the furnace control panel within the definition of the recipies for different types of scrap and product. A safety measure has to be considered when operating a dual burner system. Since both burners are sized for the maximum energy input when operating as single burner only, parallel operation may exceed the permissible energy input to the furnace. Therefore, an interlocking is required to limit the total energy.

Looking at the application of oxygen in a rotary drum furnace, and in particular at the application in a tiltable rotary drum furnace, we can fairly state that the use of oxy-fuel offers some advantages. It is imperative, however, that the firing system is designed properly to avoid excessive wear of the refractories and high maintenance efforts. The economics have to be taken in account very thoroughly. Saving fuel and possible increase in production must be compared with the cost of oxygen and gas (or oil). Combustion air is free, oxygen not. Also the available infrastructure for the supply of oxygen must be looked at.

2.1.3.6 Crucible furnaces

Crucible furnaces are not typically used in the aluminium recycling industry. Their major application is in small or midsize capacities, having a wide range of alloys. They are used as holding and casting furnaces in sand or die-casting plants. They are ideally suited for metal treatment. Due to their typical design, crucible furnaces are not the preferred type of furnace for melting although a melting rate of 300 to 400 kg/h can be obtained. For operation the crucible is filled with liquid

Fig. 2.51: Crucible furnace in tilting position (Source: Nabertherm GmbH)

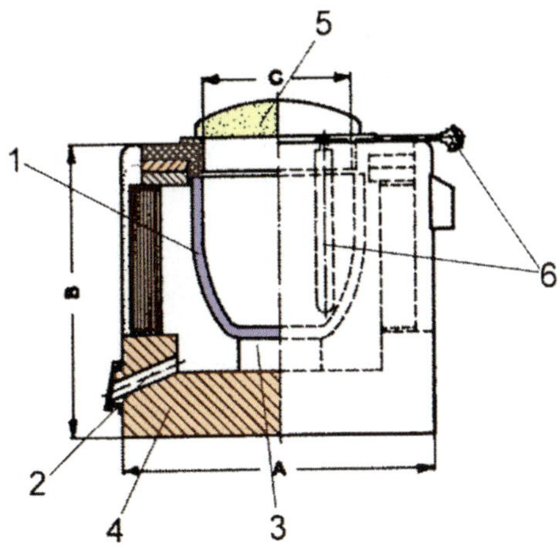

1 - crucible, 2 - burner, 3 - crucible support, 4 - furnace housing, 5 - lid

Fig. 2.52: Crucible furnace

metal followed by the required metal treatment and temperature adjustment. After thorough metal treatment, excellent metal quality is obtained.

The furnaces are designed for small capacities in the range of 100 to 800 kg. They comprise a crucible made of silicon carbide containing the aluminium that rests on the crucible base. Usually the furnace is equipped with a swivel lid (**Fig. 2.51** and **2.52**). Heating is by means of gas or oil burners that heat the metal in the crucible from outside. The combustion products do not have contact to the melt. The outer shell is designed as steel structure that is lined with refractories on the inside. The preferred refractory lining is out of monolithic material. The burner is arranged in the lower part of the furnace shell. The flue gases pass through the flue passageway to a flue duct. Since only clean material is processed the flue gases are released into the casthouse. In furnaces of the new generation the flue gases pass through a recuperator system for heat recovery. The furnace can be designed as tiltable unit for easy pouring.

Instead of heating with fossil energy, the crucible furnace can be heated by means of electrical resistance heating elements. This results in very good temperature distribution within the liquid metal bath. As the fossil heated furnaces, electrically heated furnaces are used as casting furnaces and are preferred as bale-out furnaces required for sand casting.

2.1.3.7 Crucible induction furnaces

A very specific type of a crucible furnace is the crucible induction furnace, also called coreless induction furnace. Instead of having burners or electrical resistors to heat the metal indirectly, in this furnace the induction principle is applied (**Fig. 2.53**). Induction heating uses the contact-less power transmission between a coil and electrically conductive charge. The charge may consist first of solid metal pieces that are finally melted. The induction furnace comprises a crucible made of refractory material that contains the metal. A water-cooled power coil surrounds this crucible and a number of magnetic yokes which concentrate the magnetic field established by the coil current. These yokes also eliminate the generation of a high magnetic scatter field and heating up of the furnace structure (**Fig. 2.54**). The power coil carries a large electric current which establishes an alternating magnetic field. The field induces electric currents, called eddy currents, in the metal bath. This eddy currents effect resistance heating of the bath.

There is always an ideal relationship between the size of a coreless furnace and its operating frequency. As a general rule, a small furnace gives best results at high to medium frequencies and large furnaces work best at the lower frequencies.

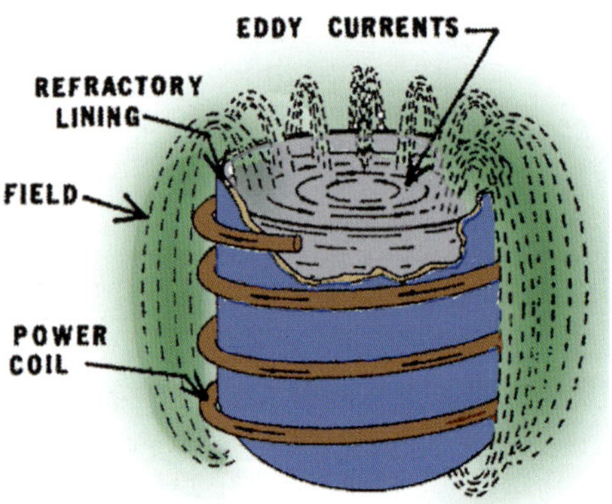

Fig. 2.53: Crucible (coreless) induction furnace

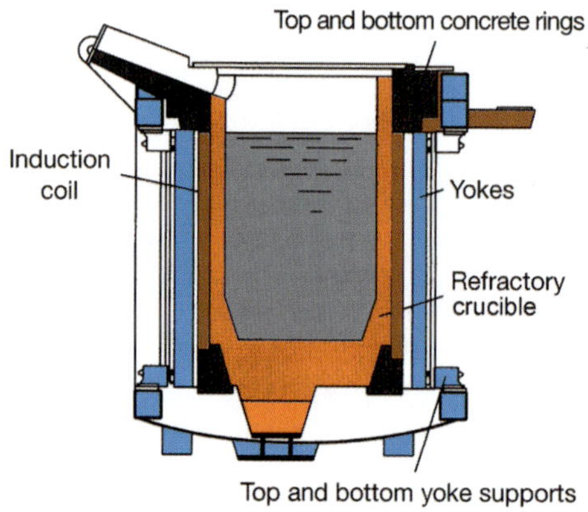

Fig. 2.54: Crucible induction furnace

The current density of the induced eddy current is highest in the refractory crucible and the surface of the metal batch and decreases to zero in the center of the melt. This effect is generally known as skin effect. The penetration depth and the bath movement increase with lower frequency while the energy input increases with higher frequency. Depending on the application, induction furnaces are used as line frequency furnaces or medium frequency furnaces with a frequency ranging from 110-1,000 Hz. The aluminium industry uses mainly line frequency furnaces operating with the grid frequency of 50-60 Hz. The technology of thyristor control permits adjustment to a frequency that may best suit the operating conditions. The eddy currents also cause a bath movement in the furnace that decreases with increasing frequency. This bath movement is characteristic for crucible induction furnaces. It ensures effective stirring of the bath with the resulting homogeneity of the melt. Well-designed crucible induction furnaces have the coil system separated into an upper section and a lower section. Depending on the actual melt condition and the degree of filling of the furnace, power can be directed to the lower or to the upper section of the coil system. The so called "power-focus-technology" permits optimal usage of the furnace power. Since the metal velocity depends on the frequency, the flow of material can be optimized and adjusted to the actual requirements by the "multi-frequency-technology" (Otto Junker), whereby upper and lower coil are switched to either parallel or inline operation. The combination of both technologies is, for instance, used for melting swarf. The process starts with the power focused on the lower coil. In order to avoid a bath movement that is too severe, the furnace operates in this phase with high frequency. As soon as the first charge is melted the frequency is lowered. The now strong bath movement creates a downwards directed metal flow which submerges the newly charged swarf immediately under the liquid metal bath thus avoiding oxidation.

For reduced power input, as it is the case for a holding furnace, the furnace may be only equipped with the lower induction coil (system Junker).

The coreless induction furnace is usually charged full and tapped empty, although on line frequency it may be necessary to retain a certain amount of metal in the furnace to continue the operation since it is difficult to start the furnace with small particles, such as turnings and borings, in a cold crucible. As a result, it is general practise to retain a heel in the furnace that may be as much as one-third of its molten metal volume. This problem can be avoided in furnaces of higher frequency, where startup can be performed with small size material charges without carrying a heel.

Crucible induction furnaces are particularly attractive for melting charges and alloys of known analysis; in essence, the operation becomes one of metal melting with rapidly absorbed electric heat without disturbing the metallurgical properties of the initial charge.

Crucible induction furnaces are supplied from a single-phase source. In order to obtain a balanced three-phase input, it is necessary to specifically design the electrical equipment for the inclusion of capacitors which are generally automatically switched (by inductance changes) during the operation in order to provide a reasonably high power factor. Power factors in such furnaces can be kept at or near unity. In high frequency coreless induction furnaces, high power factors are necessary to prevent overburdening the frequency control equipment.

To obtain a good throughput, coreless induction furnaces can be arranged in tandem. There is only one electrical power unit for both furnaces. The power directed to the one or the other furnace can be split in any ratio (**Fig. 2.55**). This allows optimal utilization of both furnaces but requires good organization of the casthouse process. It also requires a good process control system.

The efficiency of crucible induction furnaces is in the range of 60-70 %. Thus, for melting and overheating to 720 °C of metal, a specific energy consumption of 450-600 kWh/t is required. The melting rate of crucible induction furnace may reach 4 t/h for a large bath with a capacity of seven to eight tons. Smaller furnaces do have a lower melting rate of approximately 1.5-3 t/h. The

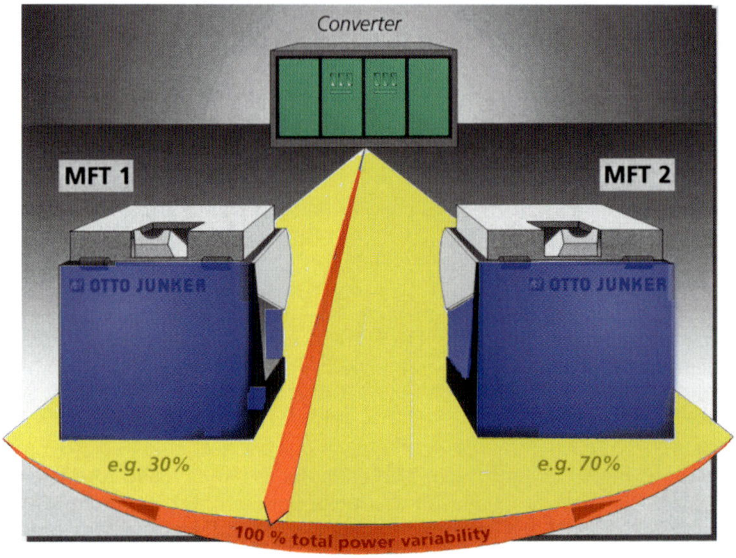

Fig. 2.55: Tandem arrangement for core-type induction furnaces (Source: Otto Junker)

furnace is housed in a steel structure with the necessary heat insulation. It can be tilted by means of hydraulic cylinders for discharging the liquid metal into transport crucibles.

As mentioned before, the furnace is charged from the top. Different charging systems are available. This can be a movable vibrating trough charging machine or crane type loading system with charging containers with bottom flap gates or crane-operated grips. For charging swarf a system comprising storage bins, discharging via vibrating troughs and a sophisticated conveyor system to feed the recalled material to the furnaces is very useful.

Alloying elements charged to the furnace are very well distributed through the melt. Usually the furnace is equipped with a lid that can be swivelled to one side permitting access to the bath for skimming, alloying addition and, of course, charging. If the lid is closed no air can penetrate with the result that no oxidation can occur.

The crucible induction furnace is ideally suited to melt scrap with high specific surface area since there is no contact with combustion products. Some air trapped in a bulk of loose material will cause some metal loss, but this is minimal. It requires, however, that the furnace crucible is cleaned after every furnace cycle.

In a secondary aluminium smelter, it is of no use to melt ingots or similar size material in an induction furnace. Thus, the application of the furnace is limited to melting swarf, chips or shredder. The material must be dry and de-coated. Let us have a look at processing swarf. This material has a bulk density of 0.6-0.8 t/m^3. A furnace with a bath capacity of 8 tons will have a useful volume of 8 / 2.3 = 3.5 m^3. This equals to a swarf weight of 2 to 2.8 tons. Consequently, 3 to 4 charges to fill the furnace are required. This requires mechanical charging equipment for efficient operation. The repeated opening of the furnace, required for fresh metal and the related introduction of air trapped in the material, will definitely spoil original benefits of low metal loss when melting in the crucible induction furnace. It can be fairly assumed, however, that the melt loss in the coreless induction furnace is lower than in other furnace systems. Not considering the original oxide skin, the melt loss should be in the range of 1-2 % only.

Although very useful in aluminium foundries, there are limitations for the application in the secondary aluminium industry. The handicap is the large production capacity required in the secondary plant. Crucible induction furnaces, as used in the aluminium industry, have been built with a bath capacity of eight tons. This is fairly small for a secondary smelter. A reasonable capacity would be at least 20 tons. But at this size the diameter of the furnace is already more than 2.2 meters. Since the furnace needs to be operated from the top, this size could be a handicap already for the operators but definitely for the designers. The related cost should also be considered. Thus, a bath capacity not exceeding 8 tons seems to be a reasonable size.

The coreless induction furnace is very useful for the production of high quality alloys. All means of treatment to clean the melt can be done in the furnace. But throughput, energy cost and investment cost limit the application in aluminium foundries. The application in the secondary aluminium industry is limited to very specific applications such as melting swarf or the production of master alloys.

Channel induction furnaces can be built in very large sizes since more than one inductor can be attached to the furnace bottom (**Fig. 2.56**).

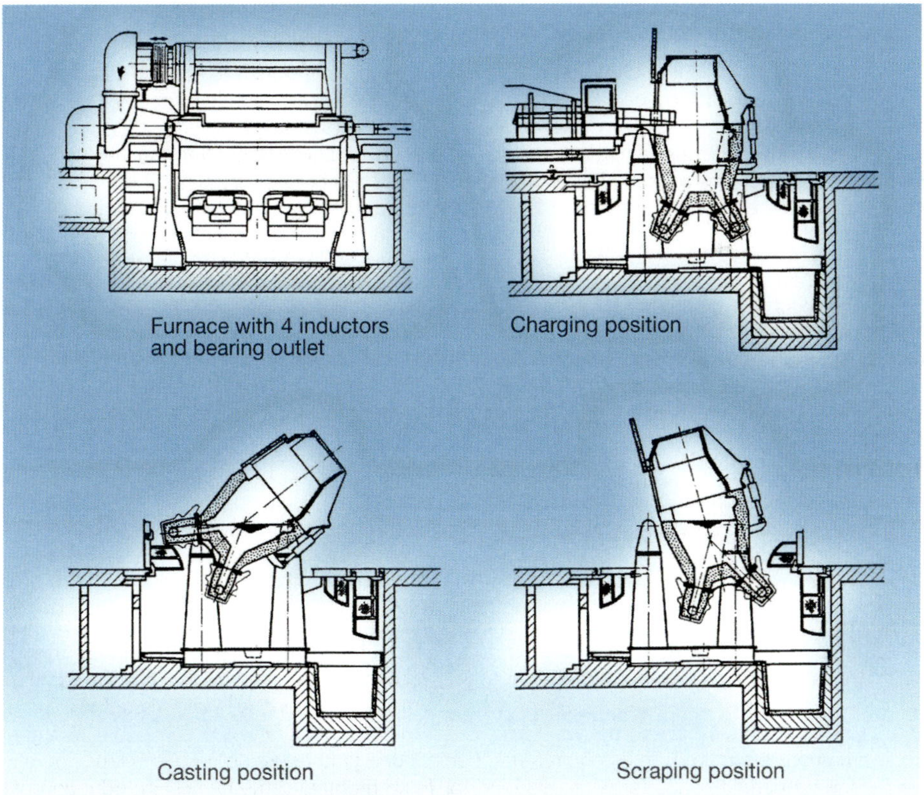

Fig. 2.56: Channel (core-type) induction furnace

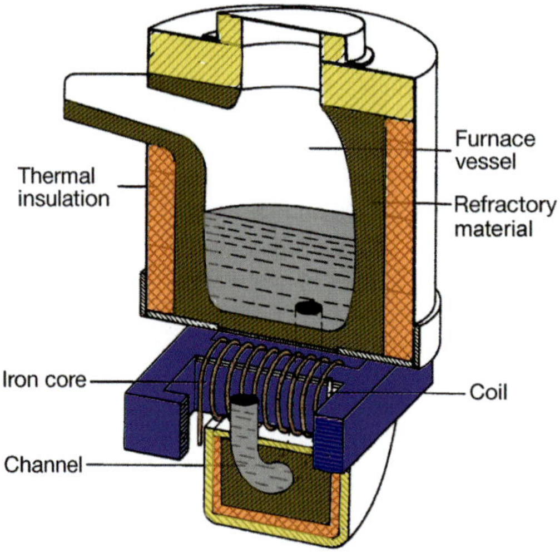

Fig. 2.57: Channel induction furnace with four inductors

2.1.3.8 Channel induction furnaces

In contrast to the crucible induction furnace, the principle of heat generation in the channel induction furnace (core-type furnace) is comparable to that of a transformer (**Fig. 2.57**). It actually conforms to a typical transformer design having an iron core and layers of wire acting as primary circuit. The melting channel acts as a ring short circuit around this transformer in the melting chamber. According to the desired melting capacity, one, two or even more such transformers (or inductors as they are called) may be added to the furnace shell. At all times, the channel must hold sufficient metal to maintain a short circuit around the transformer core. Air-cooling is used as required to prevent excessive heating of the inductor coils and the magnetic cores.

The melting output is controlled by varying the voltage supplied to the inductors with the aid of a variable-voltage transformers connected to the primary circuit of the supply. Core-type furnaces always use line frequency. These transformers are single-phase units and by using three such units a balanced three-phase input can be obtained. The current flowing trough the primary inductors by transformation causes a much larger current in the metal loop whose resistance generates heat for melting.

The core-type furnace is the most efficient type of induction furnace because its iron core concentrates magnetic flux in the area of the magnetic loop, ensuring maximum power transfer from primary to secondary. Efficiency in the use of power can be as high as 95 to 98 %.

The essential loop of metal must always be maintained in the channel induction furnace. If this loop is allowed to freeze by cooling, extreme care must be taken when trying to remelt because the loop may rupture and disrupt the circuit. Generally, the only means to rectify this is dismantling the coil and restoring the loop. This is connected with extensive work. To keep the downtime at a minimum, a spare inductor should always be kept in stock. Consequently, core-type furnaces are rarely permitted to cool. This makes alloy changes difficult because a heel of molten metal is always required.

The relatively narrow melting channel must be kept as clean as possible since a high metal temperature exists in this loop. Non-metallics or tramps in the charge tend to accumulate on the wall in the channel area, restricting the flow of metal and ultimately closing the passage.

The furnace is designed with a sturdy steel structure or with rectangular cross-section like a standard reverberatory furnace. This steel housing is lined with refractories with high alumina materials that is structured as multi-layer system as described in connection with the hearth furnace. Due to the very particular features of the inductors, the furnace designer is free to attach more than one inductor to the furnace bottom. This means almost any size of the furnace can be built. Fig. 2.56 shows a furnace with four inductors. Thus, the size of the batch can be very well adjusted to the size as marketing or operation require. The furnace is tilted by two hydraulic cylinders which allow precise control of the casting rate.

In spite of some limitations, from the mechanical and operational point of view the channel induction furnace is the ideal holding furnace. There is no contact with combustion products and with ambient air. The furnace can be closed completely and, therefore, no or only very little melt loss will occur. The eddy currents in the inductors cause a thorough stirring effect which creates a good mixture of all alloying components with the result of a homogeneous melt. Also very good temperature distribution is obtained over the entire bath which is easy to maintain during casting.

The maintenance problem in connection with the inductor requires that spare inductors are held as stock items. It has to be accepted that these items are a wear part and replacement within the framework of preventive maintenance should be scheduled. Alloy change is also more difficult than in a standard reverberatory furnace. Here careful production planning will help to overcome this problem. If a certain range of alloys can be planned over a lengthy period of time, alloy change should not create a difficult problem. For the change, the furnace can be switched off and emptied for a short time. If a crucible or a batch in a melting furnace is held as standby, liquid metal can be filled into the channel induction furnace and the new production campaign can start. Once the furnace contains a certain quantity of liquid metal, solid material can be melted at a good melting rate. Of course large blocks of metal, i. e. sows, may still present some problems.

The channel induction furnace is very suitable for remelters and, to some extent, also for refiners having a large plant capacity. Availability of low cost electrical energy may be decisive for the decision to install such equipment and the investment cost will also be a very important factor to be taken into consideration.

3. Furnace technology

Christoph Schmitz

3.1 Energy balance and efficiency

3.1.1 Basics of thermodynamics

Furnaces are the key equipment of a recycling plant. The purpose of these furnaces is to melt the scrap, remove contaminations still adhering to it and blend the liquid aluminium with alloying components in order to achieve the required alloy. To be able to do so, heat has to be introduced to the metal to arrive at the necessary state at the required temperature.

The exchange of energy within the furnace follows the rules of thermodynamics which generally deal with systems in equilibrium. The quantity of matter under consideration is called the system and everything else is referred to as the surroundings. With a closed system there is no interchange of matter between system and surroundings; with an open system there is such interchange. Any change that the system may undergo is viewed as a process. Any process or series, in which the system returns to its original condition or state, is called cycle.

3.1.1.1 Energy

Heat is the energy in transit from one mass to the other because of a temperature difference between the two. Whenever a force of any kind acts through a distance, work is done. Like heat, work is energy in transit.

Heat transfer can be direct as conduction within the material or heat convection between different matter such as gas and solid or liquid material or indirect as radiation. Energy can appear as kinetic energy or potential energy or as combination of both whereby each form of energy can be converted to the other form. This is more of interest for mechanical calculations. There are other forms of energy:

- Structural energy which is characterized by the internal strength of the material due to the crystal structure. It appears in characteristics such as resistance against deformation or elasticity.
- Chemically bound energy, characterized by the ability of a substance to react with other substances whereby energy is transmitted to or from the surrounding area.
- Nuclear energy which is the result of reactions within the nuclear structure of matter.
- Electrical energy created due to the movements of electrons or ions by electrical or magnetic fields.
- Internal energy is the energy of the movements of atoms and molecules within a substance. Its external expression is the temperature of matter and – in the case of gas – the pressure.

Each form of energy can be converted into mechanical work, expressed by the product force x displacement and heat. Again, heat can be converted in mechanical work as very well-known from the car engine or gas turbine for instance and mechanical work into heat as, for example, known from friction. There are many well-known conversions of one form of energy into another form. However, there are two basic laws of thermodynamics.

First law:

Energy may be neither created nor destroyed.

Second law:

Heat will not by itself flow from low to high temperatures.

For furnace technology the most important energy is heat. Therefore, we should have a brief excursion on the basics of the heating process.

3.1.1.2 Enthalpy

The enthalpy H is a thermodynamic characteristic of a substance and is defined by the internal energy U and the work W of volume p · V (pressure p x Volume V) that is

$$H = U + p \cdot V \tag{3.1}$$

This is a state that is variable and means that its value depends solely on state and quantity of a substance but not on their previous state.

The differential notation is written as

$$dU = \delta Q + \delta W \tag{3.2}$$

The expression of the differential small heat supply and the differential small work is δ since these are process parameters and not state parameters.

Most of the furnace processes run under constant pressure whereby the only mechanical work is a change of volume (W = – ∫p · dV. The change of enthalpy is the heat Q that is exchanged with the surrounding area

$$Q = H_{after} - H_{before} = (U_{after} + U_{before}) + p \cdot (V_{after} - V_{before}) \tag{3.3}$$

This follows the first law of thermodynamics stating that the change of internal energy U is equal to the amount of heat and work exchanged with the surrounding area. Equation (3.3) does not define absolute values of enthalpy. Those cannot be determined by methods of thermodynamics. However, all process calculations use relative values of enthalpy.

3.1.1.3 Heat capacity

If only heat is transmitted in a system, where no chemical reaction occurs and no change of phase takes place but volume work only, temperature T will change. The ratio between received heat and temperature increase is called heat capacity. For an isobaric process as per (3.3) the heat capacity C_p is

$$C_p = \left(\frac{\delta H}{\delta T}\right)_p \tag{3.4}$$

By integration the temperature dependency of the enthalpy of a substance as a function of C_p is obtained as

$$H_{T_2} = H_{T_1} + \int_{T_1}^{T_2} C_p dT \tag{3.5}$$

whereby T_1 is the temperature before and T_2 the temperature after the transit.

For practical calculations the heat capacity is related to the particular substance, i. e. aluminium or air or flue gas.

The specific heat capacity is related to one unit (1 kg or 1 m^3) and is defined as

$$c_p = \frac{C_p}{m} \tag{3.6}$$

m is the mass of the substance.

If heat is transferred in a system where the volume is constant, no work is performed ($p \cdot v = 0$). Therefore,

$$dU = dQ \tag{3.7}$$

with the heat capacity written as

$$C_v = \left(\frac{\delta U}{\delta T}\right)_v \tag{3.8}$$

and similar to (3.6) the result is

$$c_v = \frac{C_v}{m} \tag{3.9}$$

Within the operation range of aluminium melting furnaces the change of volume, due to temperature, is very low. Solid material to be heated and liquid metal do not change volume with temperature remarkably. The linear expansion coefficient is only 23.6 x 10^{-6} K^{-1}. Furthermore, it is very difficult to distinguish between c_v and c_p and in fact the difference between c_v and c_p is below the limits of measurements for these parameters. Consequently, it is meaningful to use the term specific heat capacity c_p in calculations. The change in enthalpy can be expressed as differential quotient

$$dh = c_p \cdot d\vartheta \tag{3.10}$$

and after integration

$$h_2 - h_1 = \int_{\vartheta_1}^{\vartheta_2} c_p \cdot d\vartheta \tag{3.11}$$

Fig. 3.1 shows the integral $\int_{\vartheta_1}^{\vartheta_2} c_p \cdot d\vartheta$ as shaded area. From this follows

$$\int_{\vartheta_1}^{\vartheta_2} c \cdot d\vartheta = \int_{0}^{\vartheta_2} c_p \cdot d\vartheta - \int_{0}^{\vartheta_1} c_p \cdot d\vartheta \tag{3.12}$$

This shows that the energy exchanged in a system is the difference in enthalpy of the starting state and the final state.

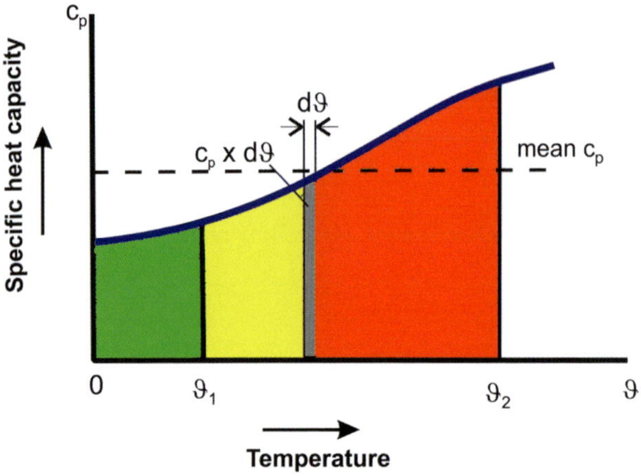

Fig 3.1: Specific heat capacity

The specific heat capacity cp is the energy required to raise 1 unit of a material 1 deg in temperature. The ratio of the amount of heat transferred to raise unit mass of material 1 deg to that required to raise unit mass of water 1 deg at some specified temperature is the specific heat of the material, sometimes called water value. For most engineering purposes, heat capacity may be assumed numerically equal to specific heat. For engineering calculations, it is very convenient to use a mean heat capacity that is valid for a certain range of temperature

$$c_{p_m} = \frac{1}{T_2 - T_1} \int_{T_1}^{T_2} c_p dT \tag{3.13}$$

The integral is of theoretical interest. For practical calculations linear relation is assumed. Following that, the mean specific heat capacity is

$$c_m = \frac{c_{T_2} + c_{T_1}}{2} \tag{3.14}$$

Example:

The specific heat of aluminium at a temperature of 20 °C or 293 K is c_{293} = 900 J/kgK. At 660 °C or 933 K of solid metal, the specific heat amounts to c_{873} = 1,240 J/kgK. According to (3.14)

$$c_m = \frac{1240 + 900}{2} = 1070\,[J/kgK]$$

Or, as a second example, the specific heat of liquid aluminium at a temperature of 660 °C or 933 K is c_{933} = 1,040 J/kgK. At 800 °C or 1,073 K of liquid metal, the specific heat amounts to c_{1073} = 1,060 J/kgK. Again according to (3.14)

$$c_m = \frac{1060 + 1040}{2} = 1050\,[J/kgK]$$

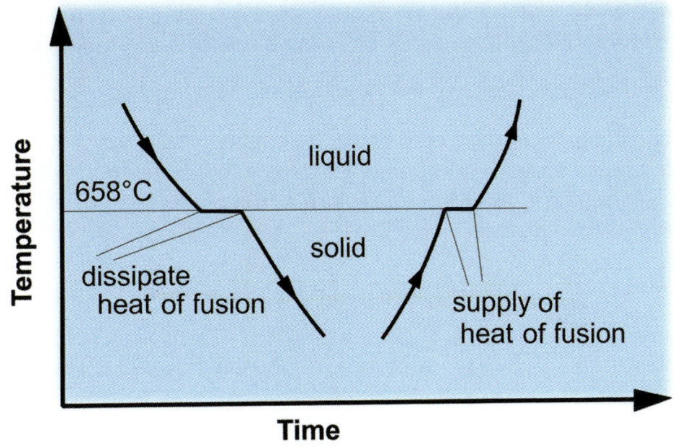

Fig. 3.2: Cooling and heating of aluminium

From the examples we notice that when melting metal or when metal solidifies a change of phase takes place. Solid phase and liquid phase of a substance have different enthalpies. The difference of these enthalpies is the melting enthalpy q_f. If energy is supplied to a system, the internal energy increases in this case resulting in a higher oscillation of the atoms around their position within the crystal grid. This is noticed as an increased temperature. But also the distance between the atoms increases slightly with resulting heat expansion. At a certain temperature some atoms leave their position and as more and more crystal defects occur, the crystal texture finally ends up in a crystallographic disorder, i. e the matter has changed into a liquid state. During this transition the temperature does not increase since all energy is used to cause the atoms to break free. **Fig. 3.2** shows the increase of temperature with time. As soon as the metal has reached melting temperature, we notice a so-called holding point that is commonly referred to as melting point. After all of the metal has reached the liquid state the temperature raises again. In the other direction, i. e. during cooling, all the metal has solidified before temperature continues to decrease. Alloying components change the melting point. The melting point, depending on the ratios of different components, may be shown in a diagram, the so-called phase diagram. At a certain ratio of the various components the melting point is at a minimum. This point is called eutectic. **Fig. 3.3**

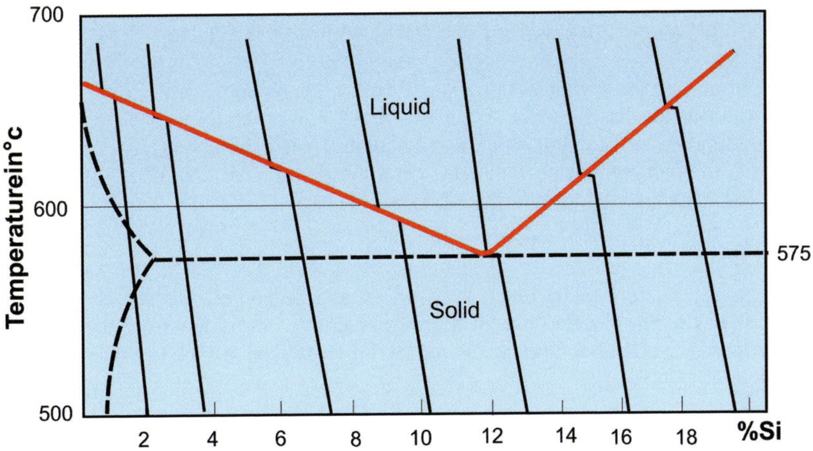

Fig. 3.3: Phase diagram Al – Si

shows an example for an aluminium-silica alloy. It shows that the melting point decreases with the Si content and then increases until it reaches the melting point of silica.

As mentioned before, the holding point also appears when metal solidifies.

For the following equations the conditions of the solid state of aggregation are marked with "while the liquid state of aggregation receives the marking". With T_u being the melting temperature or the melting point, the total change of enthalpy is

$$h_{T_2} - h_{T_1} = \int_{T_1}^{T_u} c_p' dT + \Delta h_f + \int_{T_u}^{T_2} c_p'' dT \tag{3.15}$$

Assuming linear conditions and with $h_{T2} - h_{T1} = q$ and $\Delta h_f = q_f$ equation (3.15) now reads

$$q = c_p'\left(T_u - T_1\right) + q_f + c_p''\left(T_2 - T_u\right) \tag{3.16}$$

By using specific units, q is expressed as kJ/ kg, cp' and cp" as kJ/kg · K and T as K, heat values can now be calculated.

Kindly note that T is the absolute temperature expressed in K (Kelvin) known as $\Delta\vartheta + 273$ and ϑ is the temperature in Centigrade (°C). For the temperature difference centigrade may be used as well, hence

$$q = c_p'\left(\vartheta_u - \vartheta_1\right) + q_f + c_p''\left(\vartheta_2 - \vartheta_u\right) \tag{3.17}$$

Example:

By applying (3.17) we are able to calculate the energy required to melt aluminium with a temperature of 20 °C and to increase the temperature to 720 °C. For the specific heat the values shall be as calculated. That is 1,070 J/kgK or 1.07 kJ/kgK for the solid metal and 1,050 J/kgK or 1.05 kJ/kgK for the liquid metal. The heat of fusion q_f of aluminium is 390 kJ/kg. Therefore,

$$q_{met} = 1{,}070\ (660 - 20) + 390 + 1{,}050\ (720 - 660) = 1137{,}8\ [\text{kJ/kg}]$$

This is a very important value that will be used in many calculations involving heat processing of aluminium and it will appear, therefore, in some of the following chapters. This value may also be expressed as 316 kWh/ton of metal. The value calculated is the theoretically required heat to raise the temperature of solid aluminium from 20 °C to a temperature of 720 °C of liquid aluminium. It does not consider the losses of the furnace process.

3.1.1.4 Heat flux

Heat exchange in a system takes place as soon as two bodies of different energy level are in contact. According to the second law of thermodynamics, heat flows from the higher level of energy to the lower level, depending on the temperature difference. The velocity of the heat flow is called heat flux

$$\dot{Q} = \frac{Q}{t} \tag{3.18}$$

The heat flux is heat Q divided by time t. However, the heat flow is not constant but depends on various parameters that change during a heat process, such as temperature difference or change of surface. Therefore, the heat flux should be written as the differential quotient in regard to time

$$\dot{Q} = \frac{dQ}{dt} \tag{3.19}$$

The unit of the heat flux is J/s being equal to W (Watt) that is equivalent to power, the work per unit of time. $\dot{q}$ describes the flux as specific to the surface area A, the heat flux density is obtained

$$\dot{q} = \frac{\dot{Q}}{A} \tag{3.20}$$

If the heat flux density is not constant over the surface being considered, the general form of the heat flux density should be written as

$$\dot{q} = \frac{d\dot{Q}}{dA} = \frac{d^2Q}{dt \cdot dA} \tag{3.21}$$

The dimension of the heat flux density is W/m^2.

Gas equation

In most of the aluminium melting furnaces heat is generally provided by combustion of fossil energy such as natural gas or oil. The combustion products are gases that change temperature, volume and pressure during the heat process. Three conditions characterize a gas process. At a constant volume (isochoric process), only temperature and pressure of the gas change. At a constant pressure (isobaric process), only temperature and volume and at constant temperature (isothermic process) volume and pressure change. If all of the parameters change, the heat process is called adiabatic process. The laws of Gay-Lussac describe the interaction of the characteristic parameters. At constant pressure the volume is proportional to the temperature T expressed in K.

$$V_T = \frac{V_0}{273} \cdot T \tag{3.22}$$

V_T is the volume at temperature T, V_0 the volume at 273 K.

At constant volume the pressure p_T is proportional to the temperature T

$$p_T = \frac{p_0}{273} \cdot T \tag{3.23}$$

with p_0 = the pressure at 273 K.

From (3.22) follows $V_1:V_2 = T_1:T_2$ and from (3.23) $p_1:p_2 = T_1:T_2$

According to the law of Boyle-Mariotte, the product of volume and pressure is constant:

$$V_1 \cdot p_1 = V_2 \cdot p_2 \tag{3.24}$$

If now the pressure is increased at constant temperature T_1, the volume obtained will be

$V_2' = \frac{V_1 \cdot p_1}{p_2}$ and now increasing temperature at constant pressure p1 we obtain

$\frac{V_2}{V_2'} = \frac{T_2}{T_1}$ and by substitution of V_2' we arrive at $\frac{T_1}{T_2} = \frac{V_1 \cdot p_1}{V_2 \cdot p_2}$ or

$$\frac{V_1 \cdot p_1}{T_1} = \frac{V_2 \cdot p_2}{T_2} \tag{3.25}$$

or written in a different format

$$\frac{V \cdot p}{T} = \text{const} \tag{3.26}$$

Equation (3.26) is called the gas equation. It is very useful when calculating gas flow in a furnace process. The flue gases of the furnace do have different temperatures and thus different volumes at various systems within the furnace atmosphere on their flow through the furnace with heat recovery system to the waste gas scrubber and finally to the stack.

The general gas equation was obtained by experiments. However, if the gas molecules are considered as round particles having different velocities, the general gas equation can be obtained by a theoretical calculation using the laws of mechanical physics. It is a requirement that the gas has certain characteristics to be a so-called ideal gas. The temperature is expressed as the velocity of the gases which impact the wall of a vessel due to their kinetic energy. The number of impacts on the wall at their specific velocity is a measure for the pressure. If the volume of the vessel is increased, the number of impacts is reduced and this means the pressure decreases. It is obvious that the number of particles is also of importance. The theory behind these calculations is called the kinetic theory of gases. The constant in equation (3.26) is different for each type of gas since each gas has a different number of molecules at an equal mass. This becomes different if we use the specific volume v instead of any random volume. This parameter is expressed as volume per kmole and it is now only a function of temperature and pressure. For mean conditions (273 K and 9.81 Nm^{-2}) the specific volume for all gases is 22.42 m^3kmol^{-1}. The constant in the equation is now identical for all gases and is called the gas constant R having the value 8.317 · 103 $JK^{-1}kmol^{-1}$.

The final form of equation (3.25) now becomes

$$v \cdot p = R \cdot T \tag{3.26}$$

This equation is called the universal gas equation.

The reciprocal value of v, that is 1/v, is the density r with the dimension kgm^{-3}.

For most of the engineering calculation, equation (3.25) will be applied since in most of the cases relative values are used.

Example:

Flue gas leaves a melting furnace at a temperature of $\vartheta 1$= 900 °C. The volume V_1 is 10,000 m^3/h. How much is the volume V_2 at the waste gas scrubbing plant if the temperature ϑ_2 is 170 °C ? The pressure within the entire degassing pipeline is constant.

$$\frac{V_1}{V_2} = \frac{T_1}{T_2} \text{ from that } V_2 = V_1 \cdot \frac{(\vartheta_2 + 273)}{(\vartheta_1 + 273)} = 10000 \cdot \frac{(170 + 273)}{(900 + 273)} = 3777 \text{ m}^3$$

This calculation is, for example, required to dimension pipe diameters.

3.1.2 Heat balance

We take a furnace that provides liquid aluminium for further processing by alloying, refining and casting. The material must finally arrive at the liquid state at the temperature required by the subsequent process steps. The heat introduced to the furnace needs to be sufficient to increase the enthalpy of the metal to reach this state.

In fuel-heated furnaces, the chemically bound energy of the fuel is first transformed into heat and then loaded on to the flue gases. From these flue gases the heat is directly or indirectly transferred to the batch as available heat. Unfortunately, not all of the energy will reach the material but a substantial portion is lost because of typical furnace features and physical condition. There are some complex mechanisms for transferring the heat within the furnace from the energy source to the material and to the furnace walls, from the walls to the material and through the walls to the furnace atmosphere. According to the second law of thermodynamics, inside the furnace the temperature of the flue gas cannot be lower than the temperature of the metal. Therefore, it will leave the furnace at a temperature that is far above the temperature of the surrounding area. The enthalpy of the flue gas is lost for the process unless it passes through an energy-recovery system.

The heat balance is a good indication of how the different portions of energy are distributed within a furnace. Its basis is the first law of thermodynamics, stating that no energy is lost and the total amount of the energy remains constant.

The chemically bound energy H_b, which is totally (perfect combustion) or only partly (incomplete combustion) transformed during combustion into heat, is the energy supplied to the furnace. The combustion products carry this energy through the furnace. Part of it is still contained in the flue gases when it leaves the furnace. Before that portion of the energy has found its way to different directions and, fortunately, also to the metal. The total energy is equal to the quantity introduced to the furnace.

The different energy flow distributed within the furnace and leaving the furnace is considered as loss except for the increase of enthalpy of the metal.

There are various losses in a furnace process. The following defines these losses. Some typical values are listed in **Table 3.1** Process adjustments or specific design methods to eliminate or to reduce the effect of these losses are described in separate chapters.

The largest quantity of energy is lost by the hot flue gas escaping through the stack. This is called stack loss and expressed as H_{st}. During heat transfer the temperature of the combustion products drops since energy flows from the hot flue gases to the surrounding area and to the metal. Their temperature must always be above the present metal temperature to maintain an energy difference. Considering the large flue gas quantity, a substantial amount of energy is transported as the enthalpy of the flue gas out of the furnace. Techniques have been developed to recycle some of the heat lost through the stack to the system thus improving the overall efficiency of the furnace.

It is obvious that the interior temperature of the furnace needs to be high enough to arrive at and to maintain a sufficient liquid metal temperature. The material in contact with liquid aluminium must be chemically resistant and must also withstand the temperature. But even the most efficient refractory lining cannot avoid that energy passes through it to the furnace shell and from there to the outside of the furnace. This energy flow is the second largest loss of a furnace. It is called

Table 3.1: Energy requirements of aluminium melting technologies

Process	Energy consumption [kwh/ton]	Efficiency [%]	Remarks
Reverberatory furnace, no recuperation	1,000 – 1,200	<32	
Reverberatory furnace with recuperation	600 – 650	<53	
Closed well furnace no recuperation	800 – 1,000	<40	
Closed well furnace with recuperation	600 - 650	<50	Data improved by burning organic components
Rotary drum furnace fixed axis	800 - 1,000	<40	Data improved by burning organic components and application of oxygen
Tiltable rotary drum furnace	530 - 600	<53	Data improved by burning organic components and application of oxygen

wall loss Q_w. Its quantity depends on the heat resistance of the furnace walls, the radiation from the walls to the outside and the convective heat flow by the air movement around the furnace. The wall losses include the losses of the furnace roof, the bottom and the flow of heat through the furnace door or doors lined with refractories. This loss is usually minimized by a very carefully designed multi-layer refractory lining system.

If the furnace is not in the state of equilibrium, which is the case during normal operation, the furnace walls are heated up. Energy is required to increase the enthalpy H_w of the furnace. During a cold start the available heat flows into the walls since the furnace is empty. At normal operation walls will pick up heat but energy is transferred back to the furnace during another process step. Every time the furnace doors open the burners will be in "off" or "low fire" state but still energy is lost mainly by radiation from the furnace walls. This means the wall temperature decreases and has to be brought to operation conditions as soon as the door closes again. This part of the energy loss contributes to the door losses as described in more detail below. In a rotary drum furnace, part of the wall will be covered by material at each revolution of the furnace drum. Thus, the walls will supply some energy to this material with the result that the wall temperature decreases. After the wall is free it will pick up heat again and the temperature increases. These examples show that this energy exchange takes place as part of the internal heat process during a furnace cycle and needs not to be considered as separate entry in the total heat balance. This with the exemption, of course, during heat-up and open door situation, which is considered, however, as part of the door losses.

A furnace requires doors for charging metal and for furnace operations such as skimming or stirring or taking samples. The door must be sufficiently large to simplify these activities. Thus, when opening the door, some energy flows to the outside by escaping furnace gases or by radiation from the walls, as already mentioned before. On the other hand, some cold air may enter the furnace. Even with the door closed some ingress of cold air must be noted. This results in energy losses that are called door loss Q_d. Some flue gas may escape through gaps at door or inspection openings. These are part of the flue gas losses and should not be handled separately. Here again there are design methods to minimize these losses. In general, these corrective features comprise proper door design with adequate door sealing and furnace pressure control maintaining a slight overpressure of the furnace interior against surrounding area.

One major loss is the heat of the dross or, as in the case of the rotary drum furnace, of the slag. Dross leaves the furnace at the temperature of the liquid metal. Its enthalpy H_{Sl} is beyond question lost for the heat process. The situation in the rotary drum furnace is more difficult. The heat of slag H_{sl} depends very much on the quantity of slag which is defined by the content of non-metallic components in the scrap and the amount of salt required for the process. Therefore, it is more realistic to relate the energy input to the quantity of scrap charged including contaminations and flux. Positive heat input to the furnace process arrives from the oxidation of metal. Some aluminium at the bath surface (and even at the solid metal surface) will react with the oxygen present in the furnace atmosphere to form aluminium oxide (Al_2O_3). This exothermic reaction even continues after dross is skimmed with the result that metal skimmed also burns off. Aluminium is without doubt an expensive fuel. Consequently, oxidation to create metal loss should be kept at an absolute minimum.

After having looked at the different losses of an aluminium melting furnace, the energy input requires a closer look. Energy is provided by fossil fuel or electrical energy. Since most of the furnaces are fired with fossil fuel, this technology is discussed in more detail. The input with electrical energy is discussed in a separate chapter. Together with the fuel, the energy of the mass of fuel as well as the enthalpy of the air required for combustion is introduced into the furnace. When part of the flue gas enthalpy is recycled to the furnace this part becomes quite substantial. For engineering calculations, the enthalpy introduced by the fuel quantity is generally not considered because its magnitude is comparatively low. Therefore, the energy introduced comprises the chemically bound energy of the fuel H_b, the enthalpy of the air introduced H_{air} and the recycled energy H_{rec}. The enthalpy of the combustion air H_{air} is considered to be small compared to the other energy input. It is generally not considered in aluminium furnaces. However, looking at the total energy consumption of a furnace, the electrical energy required for the blower is definitely an energy input to the furnace. Upon setting the system boundary to the furnace wall, only the enthalpy of the combustion air is to be considered.

The amount of heat shall be calculated using the following example.

A melting furnace is to have a burner input of 6,000 kW. The required combustion air will be supplied by a blower having a motor of 25 kW power. Even considering the energy losses of the blower to be part of the furnace system, the ratio $P_{blower} / P_{burner} = 25/6{,}000 = 0.00417$ or 0.417 %. Also looking at the tolerances we have when calculating energy values, it is obvious why this part of the energy input need not to be considered.

The total heat balance is shown in a so-called Sankey diagram. This diagram gives the different energy flows as bars where the thickness of each bar is a measure of the quantity. **Fig. 3.4** illustrates the heat flow in a melting furnace. The energy introduced is split in different branches for heat utilized or heat losses.

The heat balance appears as follows

$$H_b = Q_d + Q_w + H_k + H_{sl} + H_{met} + H_{st} \tag{3.27}$$

When summarizing the heat balance it is essential to use the same unit for all heat values. If, for instance, H_b is expressed as heat per hour, all the other values must be expressed as value per hour as well, or if H_b is expressed as power in kW all other values must read kW, too.

We have already discussed that the energy input by the combustion air is negligible. However, another heat input needs to be considered. In many cases the plant has to process contaminated

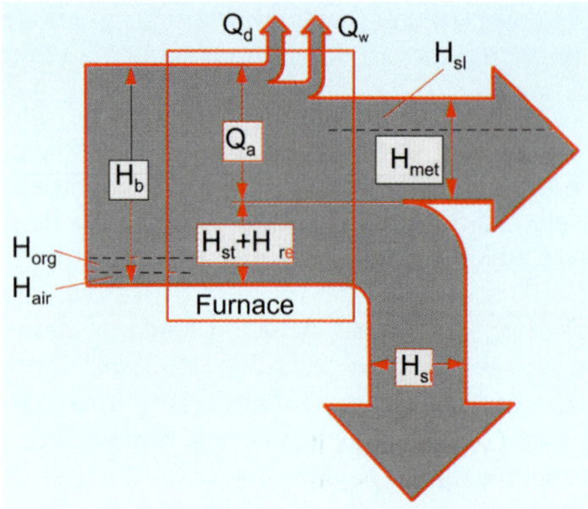

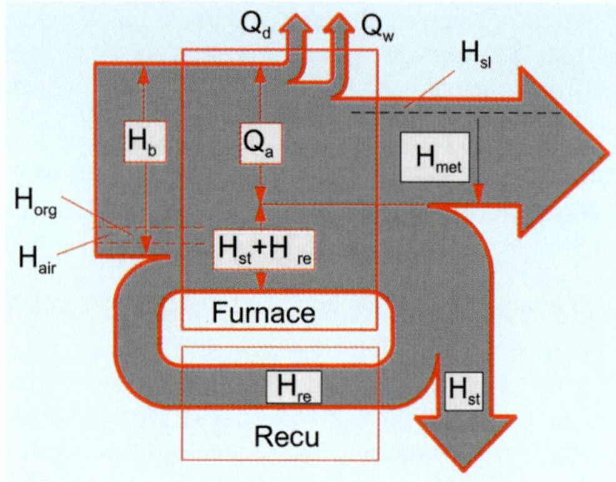

H_b = Heat introduced into the furnace, Q_a = total heat utilized, Q_d = door losses; Q_w = wall losses
H_{sl} = enthalpy of slag; H_{met} = enthalpy of metal; H_{st} = enthalpy of flue gas leaving the furnace through the stack;
H_{re} = enthalpy of flue gas that could be recycled; H_{org} = enthalpy of combustible organics;
H_{air} = enthalpy of air introduced

Fig. 3.4: Sankey diagram of a furnace without energy recycling

Fig. 3.5: Sankey diagram of a furnace with heat recycling

scrap whereby most of the contaminations comprise organic material such as plastics, coating or even oil. Undoubtedly this material is flammable within the temperature range of a recycling furnace. Hence, the chemically bound energy H_{org} of such organics adds to the positive side of the heat balance.

$$H_b + H_{org} = Q_d + Q_w + H_k + H_{sl} + H_{met} + H_{st} \tag{3.28}$$

When processing scrap with organic contaminations it may be necessary to establish an incomplete combustion in the furnace in order to protect the metal from contact with oxygen contained in the furnace atmosphere. This is vital to avoid excessive metal losses. The resulting combustion products still carry chemically bound energy and must not be released to the atmosphere. This is avoided by passing these gases through a combustion chamber. In most cases the temperature of the flue gas is too low for self-ignition. Hence, the temperature needs to be increased by introducing additional energy H_{pc}. In general, the post-combustion burner is considered to be part of the furnace system. In this case the energy input of the relevant burner adds to the energy input to the furnace system. The heat balance now reads

$$H_b + H_{org} + H_{pc} = Q_d + Q_w + H_k + H_{sl} + H_{met} + H_{st} \tag{3.29}$$

In many cases it is possible to recover heat from the flue gases by using a recuperator or a regenerator. As shown in **Fig. 3.5**, this heat is recycled to the heat input thus reducing the heat required from fuel. Unlike the input of the energy of the combustion air, the gas recycled carries a substantial amount of energy. In this case equation (3.29) is extended as

$$H_b + H_{org} + H_{pc}\ H_{re} = Q_d + Q_w + H_k + H_{sl} + H_{met} + H_{st} \tag{3.30}$$

Comparing equation (3.29) to (3.30) it is obvious that the overall efficiency of the process is improved due to the recycled energy from the flue gases and it also demonstrates that the fuel input of two identical furnaces operating under identical conditions is reduced if a recuperator or a regenerator is installed.

It is quite time-consuming to calculate the values for the heat balance in every case. During operation and commissioning of different plants quite a number of data have been collected. It is general practise to state typical consumption figures for specific techniques expressed as heat per weight unit or melting rate. Data for aluminium melting techniques range from 500 to 1,200 kWh per ton of melting rate. Table 3.1 shows such data for typical melting processes.

The efficiency $\eta_{tot,}$ as indicated in the table, will be explained in the following section.

3.1.3 Efficiencies

The second law of the thermodynamic provides the basis for the definition of efficiency factors for furnaces. This law states that heat can only flow from a body at higher temperature to a body having a lower temperature. With regard to furnace technology this means that heat at higher temperature is more valuable than heat at lower temperature.

As per definition the efficiency expressed as h is the ratio of output to input

efficiency = energy utilized / energy input

In the case of the furnaces it means the ratio between energy utilized and energy supplied. Different efficiency factors h depend on the definition or scope of the expression of heat utilized. If the furnace is regarded as closed system, the heat utilized is the total heat consumed in the furnace including all losses. In this case the efficiency factor η_{th} compares the enthalpy of the flue gases leaving the furnace to the energy introduced to the furnace. This theoretical value is helpful in some cases but explains only little about the quality of the furnace design. To obtain a better indication the efficiency factor $_0$ compares the heat transferred to the metal to the total energy input. The efficiency factors give a good indication about furnaces of similar design. The subsequent data are based on tests and experience. Some typical technologies are compared in Table 3.1.

3.1.3.1 Combustion efficiency

The combustion efficiency will be better if the temperature of combustion products will be high at the point where they are generated and low when they leave the furnace. Thus, the combustion efficiency indicates which portion of the energy input will be used in the furnace for useful or not so useful purposes.

If we summarize the heat balance as per equation (3.29) we have

$$H_b = Q_d + Q_w + H_k + H_{sl} + H_{met} + H_{st}$$

Similarly, we simplify the equation by combining the heat requirements to $Q_b + Q_w + H_{sl} + H_{met}$ to the available heat Q_a (see also Fig. 3.4). From that $H_b = Q_a + H_{st}$

As per definition for the combustion efficiency we obtain

$$\eta_{th} = \frac{Q_a}{H_b} \tag{3.31}$$

With high stack losses Q_a is comparatively small and the efficiency drops. The smaller the stack losses are, the more favorable the efficiency is. The equation may be written in a different format by using the stack losses as factor:

$$\eta_{th} = \frac{H_b - H_{st}}{H_b} \text{ or}$$

$$\eta_{th} = 1 - \frac{H_{st}}{H_b} \tag{3.32}$$

This is a very convenient equation which permits to use parameters that can be easily obtained.

The calculation of η_{th} is even made simpler when the enthalpies are based on 1 m^3 of combustion products. Consequently, instead of H_b the enthalpy of 1 m^3 of combustion products includes process air that may be introduced within the system. Thus, instead of H_b the enthalpy of 1 m^3 of the preheated combustion products V_{st} is used. V_{st} is the quantity of combustion products leaving the furnace through the stack. Therefore, the expression flue gas is also commonly used. In general, the quantity of combustion products leaving the furnace through the stack is identical to the total quantity of combustion products since all gases entering the furnace have to leave it eventually. For the equation we obtain

$$h_1 = \frac{H_u}{V_{st}}$$

with Vst taken at standard conditions. H_u is the lower heating value. It is the chemically bound energy of one unit of fuel. This should not be confused with the energy of 1 m^3 of combustion products as we calculated for better reference. By burning this fuel, which may be either oil or natural gas, the volume Vc is generated. Please note that the volume must be expressed in m^3 at standard conditions since the actual value of gases change with the temperature. This volume is to be calculated by the combustion equations. It also includes a certain amount of excess air expressed by the factor λ that is required to provide a sufficient amount of oxygen for perfect combustion. In some processes a so-called reducing furnace atmosphere is required. In such a case λ becomes <1. **Fig. 3.6** and **3.7** show the combustion efficiency of natural gas for different

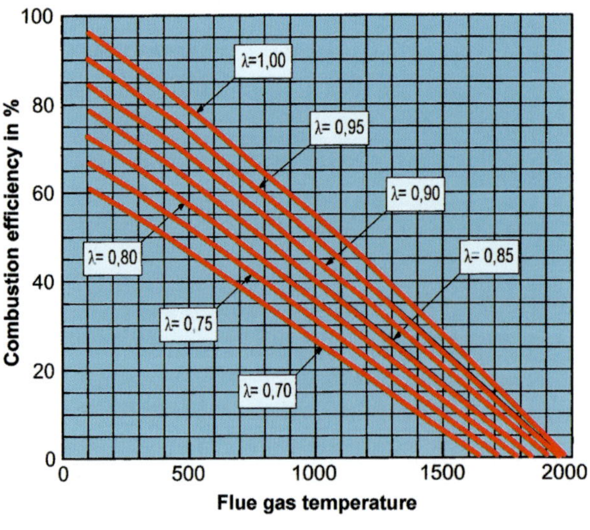

Fig. 3.6: Combustion efficiency for natural gas (according to Woelk)

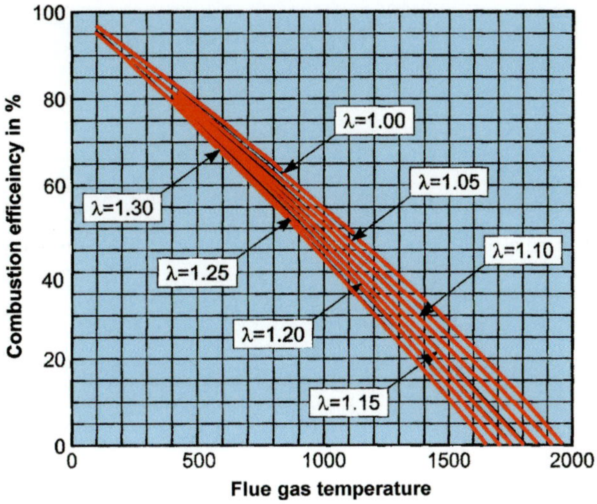

Fig 3.7: Combustion efficiency for natural gas (according to Woelk)

values of λ. Additionally, it may be necessary to introduce air at the volume Vat into the system that is finally part of the gases leaving the stack. Thus, the total flue gas is $V_{st} = V_c + V_{at}$. Please remember that the combustion efficiency compares only the total energy introduced to the furnace system to the total energy leaving the furnace system. All flow of energy or heat transfer within the system has no effect on this comparison.

In the same way we refer the enthalpy H_{st} of the flue gas leaving through the stack to 1 m^3 of combustion products:

$$h_2 = \frac{H_{st}}{V_{st}}$$

From this follows

$$\eta_{th} = 1 - \frac{h_2}{h_1} \tag{3.33}$$

But how is the situation if a portion of the energy from the flue gases is recycled to the furnace system? The energy recycled remains within the furnace system. Hence, the energy introduced into the system is the burner energy H_b only. Thus, equation (3.33) still applies. But there is a remarkable difference. The temperature of the flue gases leaving through the stack is now much lower since energy has been used to preheat combustion air. The gas has lost heat resulting in a decline of temperature. Consequently, the combustion efficiency is far better now.

A more detailed discussion may be helpful to determine fractions of energy input and recycled energy. We look at the energy h_2 that is provided to the regenerator or the recuperator. Within a system without energy recycling this is identical to the specific heat h_{st} leaving the system through the stack. Now we take a fraction h_{rec} of this energy to be utilized to preheat combustion air for recycling that share within the system. Since $h_2 - h_{rec} = h_{st}$ equation (3.33) now reads

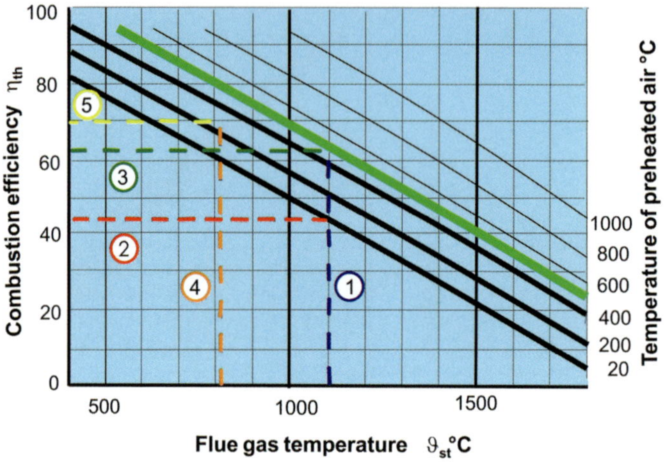

Fig. 3.8: Combustion efficiency with combustion air preheating

$$\eta_{th} = 1 - \frac{\left(h_2 - h_{rec}\right)}{h_1} \tag{3.34}$$

or

$$\eta_{th} = 1 - \frac{h_2}{h_1} + \frac{h_{rec}}{h_1} \tag{3.35}$$

From equation (3.33) we are able to determine the quantity of air for preheating the equations of the kinetic theory of gases.

The increase of efficiency is expressed in **Fig. 3.8**. At a flue gas temperature of 1,100 °C and the air temperature of 20 °C – that means no preheating of combustion air – the combustion efficiency amounts to 44 % (line 1 – 2). The picture changes if the combustion air is preheated to 500 °C at the same (line 1 – 3) flue gas temperature. The result is a combustion efficiency of now 62 %. During a melting process, conditions in the furnace change continuously. High flue gas temperature (in our assumption 1,100 °C) will occur towards the end of the cycle. At the beginning the temperature of the newly charged metal is low resulting in a high temperature difference between flame and metal. Also the surface area of the metal not yet melted is larger than the surface of the liquid metal bath. These conditions mean high heat transfer resulting in a larger temperature drop, i. e. lower temperature of the combustion products. As shown in Fig. 3.8, the combustion efficiency (line 4 – 3) with no air-preheating is almost identical to that at 1,110 °C flue gas temperature and air-preheating to 500 °C.

But we still have the air heat exchanger for combustion air in operation!

Since this unit is designed to provide optimal results at high flue gas temperatures, the efficiency at lesser flue gas temperatures is lower, too. Thus, in that case the combustion air temperature may reach only 300 °C. But still there is a remarkable increase of the combustion efficiency which is slightly above 70 % (line 4 – 5). Compared to the efficiency at 1,100 °C flue gas temperature, this is an improvement from 62 to 70 %.

The discussion demonstrates that the efficiency is better the more energy is used within the furnace. This is quite obvious. It also shows that there are remarkable benefits if the combustion air is preheated.

However, there are limitations. If applied when contaminated scrap needs to be processed, maintenance of the heat exchanger or the recuperator becomes excessively intensive. Some other limitations are even more stringent. The working mechanism of the heat recycling is the increase in flame temperature. Similar to the application of oxy-fuel, the refractory lining limits the temperature of the furnace interior. Too high flame temperature inevitably results in severe damage to the refractory material or at a minimum reduces its service life. Another limit is set by environmental conditions. High flame temperature leads to the generation of NO_x above the acceptable limits. Thus, considering economical as well as environmental values, an acceptable balance has to be found between reducing consumption and CO_2 generation and the production of gaseous matter that is harmful to the environment. Environmental regulations limit the temperature of the preheated air to 450 – 500 °C.

The problem is covered in more detail in the chapter 7 “environmental control”.

In a system with post-combustion, additional energy input is required if the flue gas temperature leaving the furnace is too low to ignite the not burned organics contained in the gas. The energy introduced now comprises the chemically bound energy of both the main burner and the post-combustion burner. Additionally, the chemically bound energy of the organics must be added to the heat balance. The volume of the combustion products may become quite high depending on the quantity of organics and the capacity of the post-combustion burner. Generally, the quantity of the organics charged with the scrap cannot be defined accurately. Consequently, a safety margin resulting in a comparatively high volume of excess air is introduced in order to make sure that all organics – and unreacted fuel – are burned in the post-combustion process. It is very inefficient to pass the now large volume of flue gas at high temperature directly to the stack. The combustion efficiency will be quite low in such a case. Hence, it is much better to use part of the energy for preheating combustion air for both the main burner as well as for the post-combustion burner or as part of the excess air directed to the post-combustor.

There is another method to improve the energy efficiency of the entire system. A part of the hot flue gases is directed to the furnace heating chamber where now the post-combustion takes place. This application depends very much on the melting process since the whole furnace interior must maintain the maximum temperature required for the post-combustion. It is also imperative that the combustion products generated by combustion leave the furnace since mass going in must go out as well. Thus, the energy balance of such a furnace system must be analyzed very carefully. In most cases preheating of combustion air seems to be the more favorable procedure since it permits to control temperatures as well as gas and air flow very precisely.

3.1.3.2 Furnace efficiency

The combustion efficiency compares the energy leaving the system to the energy introduced into the system. There is no indication for what the energy is utilized within the furnace. Major fractions may be used to compensate for wall losses or disappearance through the door. But it does not give any indication how much of the energy is effectively utilized for its intended purpose that is to melt and heat metal. To get to a better indication we have to use a different basis for comparison. Hence, we take the heat utilized in the furnace and compare it to the useful heat for melting and heating of the metal.

$$\eta_0 = 1 - \frac{h_{met}}{h_1} \tag{3.36}$$

which is called the furnace quality factor or, in short, the quality factor.

The total heat supplied to the furnace by the combustion products starting at the time they enter the furnace up to the time when leaving it is Vc · (h_1 - h_2). Provided for melting and overheating of G kg of metal, a quantity of B m^3 or kg of fuel is required. The energy transmitted in the furnace amounts to B · V_{st} · (h_1 - h_2)

The useful heat, i. e. the heat passed within the furnace to the metal, is G · qmet whereby q is the heat of 1 kg of metal expressed in kJ/kg. As explained before, it is the heat required for heating the metal from the ambient temperature to the melting point, the heat of fusion and the heat required to heat the metal from melting temperature to the final process temperature as calculated by using equation (3.17). From this follows for the quality factor

$$\eta_0 = \frac{G \cdot q_{met}}{B \cdot V_r(h_1 - h_2)} \quad (3.37)$$

Remember, h_1 is the enthalpy of 1 m^3 of combustion products immediately after combustion and h_2 the enthalpy of 1 m^3 combustion products at the point of leaving the furnace. If the furnace is equipped with a recuperator or a regenerator system, the heat recycled to the system needs to be considered and thus we arrive at

$$\eta_0 = \frac{G \cdot q_{met}}{B \cdot V_r(h_1 - h_2 + h_{rec})} \quad (3.38)$$

3.1.3.3 Total efficiency

The combustion efficiency η_{th} and the furnace efficiency η_0 provide a good look at the quality of design or a specific technology of a furnace. This may be very helpful for the designer or the plant manager. But a production manager may ask the simple question as to much of the energy put into the furnace is used for melting. In other words, what is the ratio of energy input to metal quantity processed? Thus, the metal heat has to be compared to the energy introduced to define the total efficiency of a furnace.

$$\eta_w = \frac{G \cdot q_{met}}{B \cdot V_r h_1} \quad (3.39)$$

Since $V_c \cdot h_1 = H_u$, equation (3.39) can also be written as

$$\eta_w = \frac{G \cdot q_{met}}{B \cdot H_u} \quad (3.40)$$

The equation may be extended

$$\eta_w = \frac{G \cdot q_{met}}{B \cdot H_u} \cdot \frac{B \cdot V_r \cdot (h_1 - h_2)}{B \cdot V_r \cdot (h_1 - h_2)} = \frac{G \cdot q_{met}}{B \cdot V_r \cdot (h_1 - h_2)} \cdot \frac{B \cdot V_r \cdot (h_1 - h_2)}{B \cdot H_u}$$

This can also be expressed as total efficiency

$$\eta_w = \eta_{th}\, \eta_0 \quad (3.41)$$

For furnaces operated with identical heat process, the performance can be monitored by applying equation (3.39). As mentioned before, heat efficiency η_{th} and furnace efficiency h_0 refer only to a very specific phase of the furnace process. During melting, the flue gases transmit energy to the material. Thus, the temperature drops depending on the heat transfer mechanism and the temperature difference between metal and flue gas. As metal temperature increases, the temperature difference decreases. Hence, less energy can be transferred resulting in a higher flue gas temperature. This in turn changes the enthalpy of the flue gas leaving through the stack with the result that the efficiency is now different. On the other hand, energy input would be reduced which changes the enthalpy of the combustion products. Considering these facts, η_{th} and η_0 can only be regarded as snapshot at a very specific moment during the furnace cycle.

The total efficiency is different. Parameters applied can be measured and accumulated during a complete furnace cycle.

For aluminium melting furnaces the heat requirement q_{met} is known. We have calculated it before at 1,138 kJ/kg or 316 kWh/t. The lower heating value H_u is provided by the gas or fuel oil supplier. Hence, production has to register the quantity of aluminium produced and the quantity of fuel used. The efficiency can now be calculated and monitored. If it is noticed that the efficiency is getting worse (in normal cases it does not get better) problems should be investigated. It could be that the door sealing is damaged and the door does not close properly anymore or the refractory lining is damaged or worn out. Usually the efficiency declines very slowly over a lengthy period of time. In most of the cases the production personnel will notice problems around the furnace. Sometimes a thorough investigation based on regular measurements may be required.

3.1.4 Specific energy consumption

By using equation (3.7) we have calculated that the heat required to melt solid metal from 20 °C and to obtain liquid metal at a temperature of 720 °C amounts to 1,138 kJ/kg. It is also obvious that a defined throughput of a furnace is required in a commercially operating plant. Since this means that the heat must be provided within a certain period of time, the heat flux must be 1,138 kJ/kg·per hour for each kg of metal melted within one hour. As defined above, the heat flux is expressed as a power expressed in W. Since 1 kWh is equivalent to 3,600 kJ the calculated heat flux will be

$$1{,}138/3{,}600 = 0.316 \text{ kWh/kg or } 316 \text{ kWh/t}$$

The expression in kWh is quite convenient since the suppliers of burners always define the power of the burner in kW.

The quantity G of metal melted within one hour is defined as melting rate m (to be precise in a mathematical sense the melting rate should be G' or dG/dt but we will use the common expression)

$$m = \frac{G}{t} \text{ expressed in t/h} \tag{3.42}$$

Besides the furnace capacity, which means the metal quantity that can be charged into a furnace, the melting rate is a very important figure characterizing a furnace. As we will see in a later chapter, the melting rate is linked to the bath capacity that defines the size of a furnace and the process.

Based on the melting rate m, it is quite convenient to use the specific energy consumption to define the main feature of a furnace. This factor includes the efficiency of the furnace since it indicates the actual quantity of energy required to melt one ton of metal per hour. From this the energy required, i. e. the burner capacity, can be calculated by using q_S expressed in kW/t (kW per tons) and m as t/h (tons per hour)

$$P_b = q_s \cdot m \tag{3.43}$$

The application of energy recycling within the furnace results in a lower energy consumption.

Table 3.1 gives some typical values for the specific energy consumption and the total efficiency for different types of furnace processes.

3.1.5 Introduction of heat

In the previous chapter we have calculated that the increase in enthalpy for converting solid aluminium having a temperature of 20 °C to liquid aluminium with a temperature of 720 °C is 748 kJ/kg (**Fig. 3.9**). But we noticed that this energy input will not be sufficient since losses of the furnace need to be considered. Somehow we have to get the required energy into the furnace. This can be done by electrical energy or by fossil fuel. The equipment converting the chemically bound energy or fossil energy – this is certainly not new to us – is a burner. The fuel is mixed with air to provide pure oxygen and by igniting the mixture the spontaneous reaction starts and releases the chemical energy h_u of the fuel. This energy should now be equal to the increase of enthalpy of the products of combustion hb. Now we should be able to calculate the temperature of combustion. Since $h_b = \vartheta p \cdot c_p$, we obtain from this

$$\vartheta = \frac{h_b}{c_p} \tag{3.44}$$

If we now assume that we burn a natural gas with a heating value of 38,000kJ/m^3 and a specific heat capacity of the flue gas of 1.52kJ/m^3 at a flue gas volume of 10.6 m^3 or flue gas per m^3 of natural gas, we obtain a combustion temperature of

$$\vartheta = \frac{38000}{11.3 \cdot 1.52} = 2{,}212\ ^\circ\mathrm{C}$$

This temperature is called calorimetric or adiabatic combustion temperature and cannot be observed in practice. The expression adiabatic means that there is no heat exchange in such a process. There is neither heat passed to the surrounding area nor is any heat absorbed from

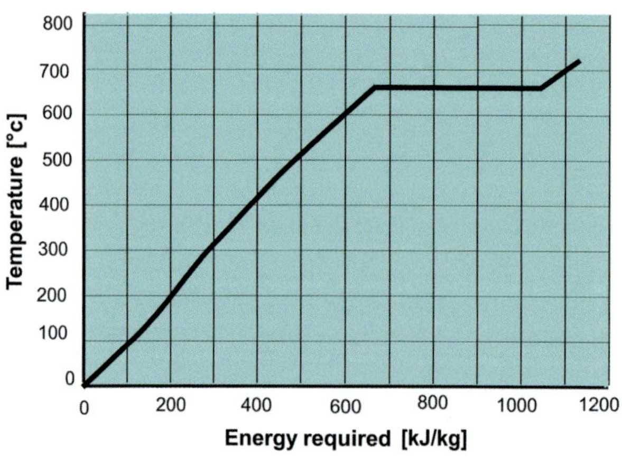

Fig. 3.9: Energy consumption for melting aluminium

the surrounding area. The actual combustion temperature is far below this value. The reason for this discrepancy is that the combustion takes time and part of the heat is already passed to the surrounding area during this process. Flue gas and oxygen have to be brought into contact for reaction. The faster this process will be, the higher the combustion temperature is. This can be observed when oxy-fuel is used for combustion. In this case no nitrogen is present and the contact with the fuel is more rapid. There is also less quantity of flue gas due to the missing nitrogen. Finally, the reaction of the gas with oxygen to carbon oxide and water cannot be complete since it will result in equilibrium, i. e. a certain ratio between the combustion products.

We have calculated the increase in enthalpy. The components participating in the combustion do have a certain enthalpy since air and fuel are supplied at ambient temperature. Therefore, we have expressed the adiabatic temperature in °C. If excess air is provided, the increase of enthalpy refers to a larger volume and the flame temperature will be lower. If a smaller quantity of air than required for total combustion is supplied, the flame temperature will also be lowered since the full chemical energy of the fuel cannot be released. If additional energy is provided by pre-heated combustion air, the flame temperature will be higher according to the additional heat supply.

A high actual combustion temperature results in good heat exchange between flue gas and material, thus reducing melting time. But the progression of combustion must be controlled in order to avoid overheating of the refractory lining and the extended oxidation of metal. For the understanding of the conditions in the furnace, it may be worth time to have a short look at the progress of combustion in relation to time, e.g. the distance from the burner nozzle, expressed by the equation

$$\frac{dh_{bt}}{dt} = k \cdot \left(h_0 \cdot h_{bt}\right) \text{ kJ/m}^3\text{s}$$

with η_{bt} = sensible heat of 1 m^3 of flue gas at mean conditions η_b = in sensible heat converted heating value, t = time and k = constant factor and dh_{bt}/dt the combustion velocity. The combustion velocity means the time-dependent change of the enthalpy, the value η_0-η_b the distance from the final condition. The equation is resolved to

$$\eta_{bt} = \eta_b\,(1 - \exp - kt) \tag{3.45}$$

The flue gas temperature can now be determined with

$$\vartheta_t = \frac{\eta_{bt}}{c_p} \tag{3.46}$$

This, however, under the condition that factor k is known. This factor depends very much on the design of a furnace or a burner since it includes factors such as mixing of the components for the combustion, shape of the combustion chamber or furnace, turbulence of the gases, gas pressure as well as heat losses during the progress of combustion. The evaluation is very interesting for the design of low NO_x burners. In order to reduce the formation of this poisonous gas, the combustion temperature must not exceed a certain value. Also for particular designs of furnaces, such as a tiltable rotary drum furnace, the combustion in a certain position in the furnace is interesting to know. **Fig. 3.10** shows an example of the temperature distribution in a furnace.

Follow-up of the progress of combustion is a good tool for the designer of burners and furnaces. As in many other cases, it is good to know about the physical principles, but data tested at production units or at least pilot plants are essential to be able to use the theoretical equation for calculation or computer-modelling of the process.

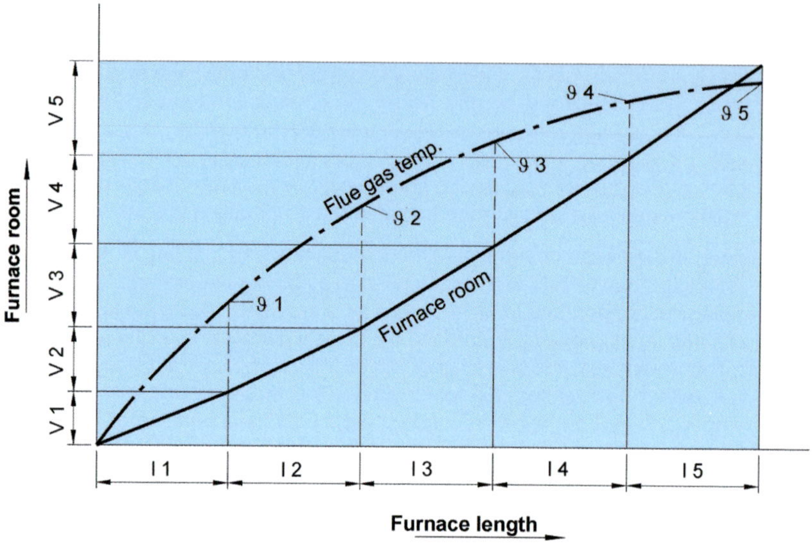

Fig. 3.10: Flame and furnace temperature

3.1.6 Fuels

The secondary aluminium industry uses mainly mineral oil and natural gas as fuel for their furnaces. Electrical energy for heating is applied in specific cases. LPG (liquefied petroleum gas) might be required for heating launders or where other means of heating are too difficult to handle. That might be the case if a plant operates with fuel oil for the furnaces. In such a case the ignition of the oil burners may also require the use of LPG.

Natural gas

Natural gas is a fuel gas stored naturally in subterranean deposits, comprising mainly methane (CH_4). Other components are propane and others out of the paraffin series as well as small contents of hydrogen, sulfur, carbon oxide and nitrogen. Since natural gas is a natural resource – as indicated by the name – its actual composition varies with the place of origin. Major deposits are in the countries of the former Soviet Union, North America, Iran, Algeria, Saudi-Arabia, Canada, Nigeria and Great Britain. To supply the European market, deposits in the North Sea have become important. There are deposits in other countries but their total is less than in the countries listed above. Natural gas appears in combination with oil or independently. Gas is supplied to countries without own deposits by tankers carrying LNG (liquefied natural gas).

If the natural gas does not contain volatile condensable components, it is called dry natural gas. If these volatiles are present, it is called wet natural gas or petroleum gas. The gas is refined if it cannot be used as it comes from the gas field and does not fulfil the quality requirements.

Gas is very well distributed in industrialized countries and it is, therefore, the most convenient fuel in secondary aluminium plants. It is easy to handle and can be used in almost all areas where heating is required and no specific storage is required on site. The most important figure for the plant operator is the heating value of the gas. It depends on the place of origin.

The gas is provided by the supplier up to the battery limits of the plant. At the incoming station the gas pressure is generally reduced to the plant requirements for the different firing equipment.

Maintaining such a pressure in the mains reduces the need for pressure control at the individual burners.

Fuel oil

Oil is also a natural inflammable resource appearing in subterranean deposits. Its color ranges from black, brown, greenish to amber. As it comes from the underground, it is called crude oil or petroleum. It also contains some elements such as sulfur or some heavy metals which are not really wanted in the final products. Gas is a mixture of different hydrocarbons of the paraffin and olefin series as well as other hydrocarbons. Due to its contents, a wide range of products are made out of the crude oil.

More than 50 % of the crude oil deposits are located in the Middle East (half of them in Saudi-Arabia), followed by Russia with approx. 12 %, North and South America with 10 % and Africa with 9 %. Europe and the Far East can be neglected.

With exemption of heavy oil, used for steam boilers and sometimes for furnaces as well, crude oil has to be refined to be used as heating oil. By fractionated distillation the different components are separated starting with the very volatile gasoline, followed by benzene (car fuel), light fuel oil and finally lubricating oils. The remaining waste fraction is asphalt.

If a plant has no access to natural gas, the preferred fuel is light fuel oil (Diesel).

The different types of fuel oil also have different heating values as shown in **Table 3.2**.

Table 3.2: Different heating values of different types of fuel oil

Type of oil	Lower heating value H_u (kJ/kg)	Lower heating value H_u (kWh/kg)
Light fuel oil EL	42,000	11.7
Light fuel oil L	41,100	11.4
Fuel oil M	40,300	11.2
Heavy fuel oil S	39,500	12.0

The table shows that only very little difference exists in the heating value of the different types of oil. Therefore, a designer uses a figure of 42,000kJ/kg, equal to 11.7 kWh/kg of oil.

Oil is provided by tank trucks and stored in a tank. It is distributed to the consumers by means of a ring pipe. The pipe diameter must be adequately sized and sufficient pressure is required to ensure that all tapping points are able to deliver the required oil quantity without disturbing the downstream furnaces. A continuous circulating flow provided by the circulating pump station ensures good distribution and prompt availability of the fuel.

If heavy fuel oil is used, the storage tank must be heated to maintain an oil temperature of at least 80 °C. This is necessary to obtain a viscosity that is sufficiently low for pumping. The ring pipe must also be heated. This can be achieved by an electrical line heating system.

Electrical energy

Secondary aluminium plants do not use electrical energy for their furnaces, except if there are some induction furnaces installed for a certain reason. However, a substantial amount of electrical energy is used for their auxiliary equipment. The main single “consumer” is the large blower for the plant’s waste gas treatment system. All drive motors of the equipment used in the plant, such as conveyors, compressed air and water circulation pumps, combustion air blowers at the furnaces, casting machine drives, hydraulic pumps and the like, require electrical energy that may add up to a large consumption. Smaller quantities are used for illumination of production plant and yard. And there are finally the requirements for the electrical control systems and the offices.

Electrical energy is generated by converting the chemical energy of oil, coal, and natural gas or by nuclear power plants or hydro-power plants and fed into a public grid with sometimes international interlocking. From this grid electrical energy is supplied to the plant as high voltage power. In the plant-owned transformer station, it is converted to the voltage required in the plant and distributed by the plant power system. The power supplied by the electrical company is recorded and registered at the transformer station according to the agreed contract conditions, including perhaps arrangement for reactive power compensation. For the individual consumers in the plant, power will be registered individually and recorded by the process monitoring system to be able to get a picture of the power consumption of individual sections and fluctuations over time.

At this point, we should have a brief look to the future. It is common knowledge meanwhile that we all have a problem with the greenhouse effect. Electrical power generated by converting the chemical energy generates – besides other poisonous gases – a huge quantity of carbon oxide, which is the main reason for the greenhouse effect. Programs are underway to separate the carbon oxide from the flue gases and to store them in geologically safe structures underground or in the oceans. The separation technology is known and the technology for storing the carbon oxide as well. But good efficiency can only be obtained by large size plants. This is certainly the case for the large power plants. But the situation is different for the many individual consumers and this is also true for our aluminium industry. It is very difficult for the production plants to avoid fossil fuel altogether or to install individual carbon oxide separation plants. More use of electrical power would shift the problems to the large power suppliers which are equipped to handle it. But using electrical furnaces is not possible in many cases in our production sector although their use could be extended. The resulting higher cost will occur anyhow with the extension and improvements of the waste gas treatment systems. Another solution would be to use oxy-fuel. The resulting flue gas would contain only water and carbon oxide (and some traces of other reaction products and contaminations), which are easy to separate. The carbon oxide could then be collected in special containers and transported to the central station responsible for deposal.

In the end, production cost will increase no doubt. But if all nations would join the program for environmental protection – and there is no doubt that they will eventually have to – cost increases will hit all production plants and the competition will be ruled by best equipment at optimal operation and access to inexpensive raw material. But this is already the case today.

3.1.7 Reaction of combustion

The fossil fuel has to react with oxygen to get access to its chemically bound energy. The oxidation of the hydrocarbons of the fuel is an exothermic reaction – the combustion – that is used to provide heat.

The oxygen required for the combustion is provided by air having an oxygen content of 21 %. This means the remaining 79 %, comprising mainly nitrogen and some traces of inert gases, does not participate in the reaction but is present with the combustion products. The combustible contents of the fuel react with the oxygen to form carbon oxide, water and sulfuric acid if sulfur is contained in the fuel. Natural gas comprises hydrocarbons according to the structure $H_n + H_{2n+2}$. Thus, only carbon oxide and water are generated. Liquid fuel comprises hydrocarbons but contains some sulfur. Therefore, besides water and carbon oxide, a certain quantity of sulfuric acid is produced.

For the combustion the following reactions take place

carbon $$C + O_2 \rightarrow CO_2 \tag{3.47}$$

hydrogen $2H_2 + O_2 \rightarrow 2H_2O$ (3.48)

sulphur $S + O_2 = SO_2$ (3.49)

There is the reaction of carbon monoxide to carbon dioxide that may be interesting for the incineration and for furnaces with flue gas recycling

carbon monoxide $2CO + O_2 = 2CO_2$

Hydrocarbons have a more complex structure. Thus, the general reaction equation applies:

$$C_xH_y + \left(x + \frac{y}{4}\right) O_2 \rightarrow xCO_2 + \left(\frac{y}{2}\right) H_2O \tag{3.50}$$

For example, considering the reaction of propane having the structure of C_3H_8, the combustion equation would be

$$C_3H_8 + \left(3 + \frac{8}{4}\right) O_2 \rightarrow 3CO_2 + \left(\frac{8}{2}\right) H_2O$$

or $C_3H_8 + (5)\ O_2 \rightarrow 3CO_2 + (4)\ H_2O$

We would obtain the same result if we would calculate the percentage of carbon and hydrogen and use the reaction equation for the individual components. This way of calculation is of interest when looking at the combustion of oil.

When looking at the emissions, the quality of the combustion is important. It is expressed by the combustion ratio h_v:

$$\eta_v = \frac{CO_2}{CO + CO_2} \cdot 100\ (\%) \tag{3.51}$$

The factor is a good indication of the heat performance of a furnace. If $h_v < 100$ the combustion is not complete and not burned organics are in the waste gas.

For the calculation of the combustion products the reaction equations are sufficient. They do not permit an analysis of the reaction velocity as discussed before. In fact, the reaction is a complex process that depends on many factors and intermediate steps. To start the reaction, the energy of activation is required, as we know from one of the previous chapters for the reaction of aluminium with oxygen. To raise the energy to the required level, it is required to have a high temperature at any place in the mixture of gas and oxygen. The hydrogen molecules are activated and react. The exothermic energy of this local reaction must continue and activate new hydrogen molecules. After ignition, a chain reaction starts, whereby intermediate products are formed and split again. Hydrogen atoms may be deactivated by passing their energy to the furnace wall, for instance, and will be activated again. The recirculation of combustion products into the flame will have an influence, too. These complex reactions determine the reaction velocity. There is another factor as well. The air must contain a minimum content of gas to be ignitable. But also with a gas content that is too high, no ignition will take place. Any gas mixture between these limits may explode at just one spark. Different gases do have different limits (**Table 3.3**).

Table 3.3: Limits of different gases

Gas	Lower limit (%)	Upper limit (%)
Hydrogen	4.0	74.2
Carbon monoxide	12.5	74.2
Methane	5.1	13.5
Propane	2.1	9.3
Butane	1.8	8.4

Natural gas, as supplied from Russia, consists to more than 90 % of methane. The methane content of North Sea gas is lower but still amounts to more than 80 %.

The combustion of liquid fuel can be started either by evaporating the liquid, mixing with air and igniting. The other possibility is to atomize the liquid during mixing with air, followed by ignition. The size of the atomized droplets is important, as it may also be known from the diesel engine or the fuel injection in an automotive engine. If the droplet size is too big, not burned carbon will remain in the waste gas. The relative velocity between combustion air and oil is also a criterion since the oil vapor enveloping every oil particle must be disturbed to ensure access of the air to the droplet. Atomizing by air appears the most appropriate way of atomizing in a burner for our aluminium furnaces.

3.1.8 Combustion air requirements

Based on the chemical equations we can now calculate the combustion air requirement for our burners.

We will first look at oil burners. The fuel oil, as used in our furnaces, has the following approximate contents

hydrogen	13 % by weight
carbon	85 % by weight
sulfur	1.0 % by weight

the balance will be water and ash.

Hydrogen will react with oxygen to form water:

$$2H_2 + O_2 \rightarrow 2H_2O$$

or with mole values: $2 \cdot 2g + 1 \cdot 32\ g = 2 \cdot 18\ g$

This shows that 2 moles of hydrogen plus 1 mole of oxygen will result in 2 moles of water. Please note that water will be present as gas (steam). According to Avogadro's law, every mole of a gas will have a volume of 22.4 liters at mean conditions. Therefore, we get: 44.8 l of hydrogen + 22.4 l of oxygen will produce 44.8 l of water. In case of oil, we calculate with weight and thus can define that 4g of hydrogen and 32 g of oxygen will produce 36 g of water vapor. In other words, to burn 4 g of hydrogen we require 22.4 l of oxygen.

Now we will perform the same calculation for carbon with the following reaction:

$$C + O_2 \rightarrow CO_2$$

or again with mole values

$$12\ g + 1 \cdot 32\ g = 44\ g$$

From that we take that 32 g of oxygen are required to get 44 g of carbon oxide. Furthermore, to burn 12 g of carbon we require 22.4 l of oxygen.

Finally we burn the sulfur as well:

$$S + O_2 \rightarrow SO_2$$

or $$32\ g + 1 \cdot 32\ g = 64\ g$$

If we refer to kmole, the volume will be in m^3 at mean conditions. Thus, we can now summarize:

We require for

1 kg of hydrogen	–	8 kg or 22.4/4	= 5.60 m^3 oxygen
1 kg of carbon	–	2.67 kg or 22.4/12	= 1.87 m^3 oxygen
1 kg of sulfur	–	1.00 kg or 22.4/32	= 0.70 m^3 oxygen

On the basis of the contents of different elements in the oil, as we have listed above, we obtain the consumption oxygen for burning 1 kg of oil:

hydrogen	0.13 x 8.0 = 1.04 kg O_2
carbon	0.85 x 1.87 = 2.27 kg O_2
sulfur	0.10 x 1.00 = 0.10 kg O_2

Thus, for each kg of oil, 3.4 kg of oxygen or 2.28 m^3 at mean conditions are required. We will come back to this figure in the chapter discussing oxy-fuel. We have to calculate on the basis of weight since the contents of the elements are given in % of weight, too. Assuming that the oxygen content in the air is 21 %, we obtain an air volume of 2.28/0.21 = 10.86 m^3 or approximately 11 m^3 for each kg of oil. This is a figure we should remember as a good approximation when calculating air requirements of a furnace.

Let us now have a look at the combustion of natural gas.

The main component of natural gas is methane. Consequently, the combustion follows the equation

$$CH_4 + 2O_2 \rightarrow CO_2 + 2H_2O \tag{3.52}$$

which shows that 44.8 m^3 (1 kmole) of oxygen is required for each 22.4 m^3 (1 kmole) of methane. Consequently, each m^3 of methane requires 2 m^3 of oxygen or 2/0.21 = 9.52 m^3 of air. We now have one more approximation to remember: 1 m^3 of natural gas requires 10 m^3 of combustion air.

Natural gas comprises gas mixtures of different hydrocarbons and may also contain nitrogen. Thus, we will first have a look at a different hydrocarbon and then prepare a table for one typical gas. Propane has the structure C_3H_8 with the combustion equation

$$C_3H_8 + 5O_2 \rightarrow 3CO_2 + 4H_2O \tag{3.53}$$

showing that 112 m^3 (5 kmole) of oxygen are required for the combustion of 22.4 m^3 (1 kmole) of propane. Table 3.3 shows the summary for a typical European natural gas and a Russian natural gas.

The calculated oxygen/gas ratio can be calculated very simply on the basis of the combustion equation (3.50). Having a hydrocarbon with the structure $C_x + H_y$, the oxygen ratio will be

Table 3.4: Gas data, air consumption and flue gas generation

Sources		Gas contents (vol%)						
		CH_4	C_2H_6	C_3H_8	N_2	CO_2	Traces	Totals
Russian	vol %	93.00	3.30				1.50	100
	m^3	0.93	0.03	0.00	0.00	0.02	0.02	1.00
	m^3 O_2	1.86	0.12	0.00	0.00	0.00	0.00	1.98
	m^3 air	8.86	0.55	0.00	0.00	0.00	0.00	9.41
	m^3 CO_2	1.86	0.07	0.00	0.00	0.02		1.95
	m^3 H_2O	1.86	0.10	0.00				1.96
	m^3 N_2	6.99	0.43	0.00	0.00			7.43
	m^3 flue gas							11.33
European	vol %	81.30	2.90	0.40		14.40	0.50	100
	m^3	0.81	0.03	0.00	0.00	0.14	0.01	1.00
	m^3 O_2	1.63	0.10	0.02	0.00	0.00	0.00	1.75
	m^3 air	7.74	0.48	0.10	0.00	0.00	0.00	8.32
	m^3 CO_2	1.63	0.06	0.01	0.00	0.14		1.84
	m^3 H_2O	1.63	0.09	0.02				1.73
	m^3 N_2	6.11	0.38	0.08	0.00			6.57
	m^3 flue gas							10.14

$$V_{O_2} = \left(x + \frac{y}{4}\right) \cdot V_{gas} \tag{3.54}$$

Table 3.4 gives a summary of oxygen required for natural gas provided from Russia and from Europe. The content of methane in the European gas is less than that of Russian gas and thus the quantity of oxygen required for combustion. The air in our atmosphere has an oxygen content of 21 %. With that figure it is easy to get the volume of required air.

$$V_{air} = \frac{V_{O2}}{0.21} = 4.76 \cdot V_{O_2}$$

Air and oxygen contents calculated so far are the theoretical values, i. e. all of the gas reacts during the combustion without any combustible material remaining in the flue gas. In practice, such ideal combustion is not possible. The mixing of fuel and air can never be total and the time for mixing may not be sufficient as well during passage of the flame through the furnace. Therefore, a certain amount of excess oxygen needs to be offered to the process. This quantity is expressed by the excess air coefficient l, the theoretical air volume V_{air} and the blower air volume V_b

$$\lambda = \frac{V_v}{V_{air}} \tag{3.55}$$

For complete combustion l must be >1. This differs depending on burner design and type of fuel. Generally, λ will be 1.1-1.2 for light fuel oil and 1.05-1.1 for natural gas. In the secondary aluminium industry there is another reason for increasing the excess air. With the material charged to the furnace, a sometimes large quantity of combustible contaminations, i. e organics, get into the furnace. These should be burned for many reasons. For their combustion sufficient oxygen must be provided by setting the excess air to a higher value. After the organics are burned off, I is set back to the value as optimal for the burner operation.

Excess air must be set to the actual value required since additional air increases the waste gas value without contributing to the energy supply. This could result in a notable decrease of furnace efficiency.

3.1.9 Combustion products

We have calculated the air required for combustion and, together with the fuel volume, we now have a good picture of what is going into the furnace. But which volumes leave the furnace finally through the stack?

With the combustion equations, as discussed before, we are in position to verify the quantity of combustion products.

We have calculated that the combustion of fuel oil with the contents, as listed above, will result in the generation of carbon oxide and water vapor.

1 kg of hydrogen will produce	36/4 = 9.00 kg H_2O or 11.2 m^3
1 kg of carbon will produce	44/12 = 3.67 kg CO_2 or 1.87 m^3
1 kg of sulfur will produce	64/32 = 2.00 kg SO_2 or 0.47 m^3

These figures have to be related to the actual proportions of the elements in the oil. The combustion of 1 kg of oil produces from

hydrogen	0.13 x 11.2 = 0.37 kg H_2O or 1.56 m^3
carbon	0.85 x 3.67 = 3.12 kg CO_2 or 1.59 m^3
sulfur	0.01 x 2.00 = 0.02 kg SO_2 or 0.05 m^3

If only oxygen is used, as in the case of oxy-fuel, the total flue gas discharge would be 3.51 kg or, expressed as gas volume, 3.2 m^3 at mean conditions. The nitrogen content of the combustion air does not react. It will be discharged as it enters the process, but, of course, at higher temperature. As we have calculated before, a volume of 10.86 m^3 is required. This includes 10.86 x 0.79 = 7.93 m^3 of nitrogen which has to be added to the volume of combustion products. As a total volume of flue gas leaving the furnace for each kg of oil burned, we thus obtain 3.2 + 7.93 = 11.2 m^3.

This results in very useful and easy to remember approximation figures for calculating. For each kg of oil burned, a quantity of approximately 11 m^3 of combustion air is required and approximately 11 m^3 of flue gas will be generated. Both volumes are at mean conditions.

If the excess air is to be considered, the combustion air has to be increased, as discussed before. This will also result in a slightly larger flue gas volume. Having a λ of 1.2, the combustion air volume will be (using our approximation) 11 x 1.2 = 13.2 m^3 per kg of oil. The flue gas quantity, now increased, is not that much since only 79 % of the excess air is nitrogen: (1.2 x 0.79) 11 = 1.16 x 11= 12.7 m^3. For most of the practical calculation we must consider safety margins. The small difference can be neglected.

The percentage of the oxygen in flue gases is measured for evaluating the efficiency of combustion. The theoretical value that can be obtained as maximum:

$$\frac{1.59}{11.2} \cdot 100 = 14.2\,\%$$

Sometimes the so-called "dry flue gas" is considered. The temperature of the flue gas is lowered to surrounding temperature. Thus, the water contained in the flue gas condenses:

$$\frac{1.59}{11.2-1.56}=16.5\ \%$$

The value referring to dry flue gas appears to be of academic interest only. For sizing ducting and stack, the total quantity – including water vapor – at the actual temperature is required. The carbon oxide ratio may be of interest in case of oxy-fuel. Then the "dry ratio" comes down to the actual carbon oxide content, giving an indication how much carbon oxide has to be collected and stored – but this consideration is something for the future.

The basic combustion equations lead to the amount of combustion products if gas firing is applied. Using the example for burning methane only, we notice that burning of 1 kmole or 22.4 m^3 of methane will produce 1 kmole (22.4 m^3) of CO_2 and 2 kmole (44.8) of water vapor. Using this ratio, we can state that burning 1 m^3 of methane will produce 1 m^3 of carbon oxide and 2 m^3 of water vapor. The total of combustion products is then 3 m^3. We have also calculated above that 2 m^3 of oxygen or 9.52 m^3 of combustion air is required per m^3 of gas. The content of nitrogen in the air is 9.52 x 0.79 = 7.52 m^3. The total of flue gas comes to 3 + 7.52 = 10.52 m^3 which is approximately 11 m^3.

This figure is very convenient since it gives the same approximation for calculating the flue gas quantity:

The flue gas quantity of burning 1 kg of oil or 1 m^3 of gas is approximately 11 m^3.

Considering the required excess air, the same calculation applies as used above in the case of oil burning. Natural gas does not consist of pure methane. Thus, we apply Table 3.4 again. The Russian natural gas, with a methane content of 93 %, produces 11.33 m^3 flue gas. The North Sea gas has a methane content of 81.3 %. However, it has a CO_2 content of 14 %. This difference should be reflected by a lower heating value. Before we look into that we want to calculate the CO_2 content in the flue gas as in the case of fuel oil.

For the Russian gas we have a total flue gas quantity of 11.33 m^3 for one m3 natural gas. The CO_2 content is 1.95 m^3. Thus we get

$$\frac{1.95}{11.3}\cdot 100=17.3\ \%$$

and for the European gas

$$\frac{1.84}{10.14}\cdot 100=18.1\%$$

The quantity of CO_2 already present in the natural gas contributes to this high percentage.

The content of CO_2 in the flue gases can be monitored by test equipment permanently installed in the stack. The data are then transferred to the plant computer system. Sometimes it will be required to make spot-checks. For this reason an accessible platform with sampling points in the stack is installed at a defined height of the stack.

Table 3.5: Heating values of typical fuels

Fuel	Symbol	kWh/m³	H_u MJ/m³	MJ/kg	Density kg/m³
Carbon oxide	CO	0.00	0.00	10.12	1.25
Hydrogen	H_2	0.00	0.00	120.00	0.09
Methane	CH_4	0.00	0.00	49.94	0.72
Ethane	C_2H_6	0.00	0.00	47.47	1.36
Propane	C_3H_8	0.00	0.00	46.36	2.02
Carbon	C			34.12	12.00
Sulfur	S			9.07	32.00

Heating value of fuel

During combustion the chemically bound energy is set free and passed to the products of combustion which are generally gases. They contain also nitrogen from the combustion air and excess air. The flue gas mix carries the heat defined by the temperature of these gases after completion of the combustion. The gross heating value H_0 gives the heat set free by the fuel if the temperature falls to ambient temperature. Usually this is not the case in the furnace processes. Therefore, the lower heating value H_u, simply called heating value, is used for calculations. The difference between H_u and H_0 is the deletion of the condensation energy of the water vapor contained in the products of combustion to obtain the heating value H_u. **Table 3.5** gives the heating value H_u for some components of fuel. With these values the heating value of a specific type of fuel can be calculated.

For the oil, as defined above, the heat can be calculated by using Table 3.5 as follows

carbon	0.85 x 34,120	=	29,000 kJ/kg
hydrogen	0.13 x 120,000	=	14,000 kJ/kg
sulfur	0.01 x 9,070	=	907 kJ/kg
	total		43,907 kJ/kg

Expressed as kW/kg, we arrive at 43907/3600 = 12.2 kWh/kg of fuel oil.

The calculation for the different types of natural gas is summarized in Table 3.5. The different heating values are related to their actual content in the gas and added up to arrive at the heating value for 1 m³ of gas. Please remember that all values are related to mean conditions at = 0 °C or 273 K (to be precise, it is actually 273.15 K) and at an air pressure of 1.013 bar (1,013 hPa). If different temperatures have to be considered, volume-related values have to be referred to the different temperature using the equations as per section 3.1.1.

For standard calculations a detailed analysis of the heating value is not required. The data can be taken from generally available tables. For the calculations in this book we use the following standard figures:

Natural gas	**Lower heating value H_u (kJ/m³)**	**Lower heating value H_u**
Natural gas Europe	29,400 – 42,400 kJ/m³	8.2 –11.8 kWh/m³
Natural gas Russia	33,600 – 54,600 kJ/m³	9.3 – 15.2 kWh/m³
Natural gas average	36,000 kJ/m³	10 kWh/m³
Fuel oil	43.200 kJ/kg	12 kWh/kg

Table 3.6: Data for typical European and Russian natural gas

Sources		CH_4	C_2H_6	N_2	CO_2	H_2O	Totals
Russian	Hu (kJ/m³)	33313	1932				35245
	Cpm (kJ/m³ K)*			1.37	1.97	1.65	1.52
	m³ flue gas			7.43	1.95	1.96	11.34
European	Hu (kJ/m³)	29122	1868				30989
	Cpm (kJ/m³ K)*			1.36	1.97	1.65	1.52
	m³ flue gas			6.57	1.84	1.73	10.14

* average for a temperature range 0-1400 °C

The specific heat depends on the gas temperatures. For the practical calculation it is convenient to use c_{pm}, which is the arithmetical average between the value for 1,400 °C and ambient temperature which is generally assumed to be 20 °C. The values for c_{pm} for the combustion of gas are summarized in **Table 3.6**. The specific heat value for the different components is related to the content of these components in the flue gas and summarized for the total of the flue gas value:

$$c_{pmflue} = \frac{\Sigma c_{pmc} \cdot r}{V_{flue}} \tag{3.56}$$

with c_{pmflue} = mean specific heat of flue gas, c_{pmc} = specific heat of the individual gas component (N_2, CO_2, H_2O), r = the proportion of the individual component and V_{flue} = flue gas volume.

The specific heat c_p of the flue gas is required when calculating furnace efficiencies and flue gas systems.

Oxy-fuel

The required oxygen for the combustion of fossil fuel is generally provided by air. Air is a mixture of

nitrogen (N_2)	78.09 % by volume
oxygen (O_2)	20.94 % by volume
carbon oxide (CO_2)	0.03 % by volume
argon (Ar)	0.93 % by volume
other noble gases	0.01 % by volume

Only oxygen is required for the combustion. The other components of air, mainly nitrogen (the content of the other components is too low and can be neglected), do not participate in the reaction. But the volume of nitrogen must be heated up and will finally discharge a substantial amount of energy through the stack. Increasing the content of oxygen in the combustion air or providing pure oxygen will reduce the flue gas volume thus increasing the efficiency of the furnace. There is one disadvantage connected to the reduction of gas volume. Particularly in the lower temperature range, convection is the most effective means of heat transfer. Here the gas velocity is the major factor. Reducing the gas volume means also reducing the heat transfer coefficient. But this is partly compensated by the higher temperature difference between gas and material.

The temperature of the flame and products of combustion is increased if the combustion air is enriched with oxygen since less of the non-reactive nitrogen needs to be brought to flame temperature and the sequence of combustion is less disturbed by an inert component. A maximum is reached by using pure oxygen. Due to the absence of nitrogen the theoretical combustion temperature is very high

$$\vartheta = \frac{38{,}000}{3.9 \cdot 1.5} = 6{,}496\ °C$$

But although the reaction in an oxy-fuel flame is much more intensive, it is not possible to reach this temperature and the resulting flame temperature is much lower.

Due to the higher flame temperature, the radiative heat transfer to the furnace batch – but also to the furnace walls – is increased. The heat transfer increases with the power of 4. Thus, a small increase of temperature causes much higher heat transfer.

At high temperature the combustion products CO_2 and H_2O start to dissociate. This process is endodermic. Thus, the gas temperature is lowered. As soon as the dissociation products get in contact with a cooler surface, they recombine again and pass the energy to the material. However, this process takes place at temperatures above 2,000 °C and is, therefore, not of interest for furnaces in the secondary aluminium industry sector.

High flame temperature improves the heat transfer to the furnace charge, which is of advantage, and also to the furnace walls, which is of disadvantage. The service temperature of the refractory lining limits the radiative heat transfer. Well-designed furnaces with adequate burner systems are able to reduce this problem. There are different types of burners available, mainly from suppliers of technical gases. They are designed for pure oxygen as well as for a mixture of oxygen and air. Due to the high temperature of the combustion products, most of the burners require cooling. More convenient designs are realized with internal cooling with the oxygen and do not require water-cooling. The impact of the very hot combustion products is softened by shielding the oxygen flame with air (**Fig. 3.11**). This system can only be realized if a burner is only used for oxygen enrichment. The mixture of air/oxygen is adjusted as the process requires. The radiation from the burner flame can pass the enveloping air undisturbed since air is transparent and does not absorb or reflect energy. The cooling provided by the shielding will reduce the impact on the refractory lining. In theory, the system sounds good but it must be investigated if the quantity of shielding air is less than the volume of air that should be provided as combustion air.

The high flame temperature supports the reaction of oxygen and nitrogen to form poisonous NO_x. With pure oxygen burners, this should not create a problem, since no nitrogen is present in the combustion products - at least theoretically. In practice, there are always leaks and air also intrudes with the material. Oxygen-enriched combustion takes place under presence of air. Thus, increased NO_x formation is to be expected. In this respect, it appears less favorable to operate with a mixture of air and gas. There is still the advantage of improving the efficiency of the furnace. A proportion of 30 % of oxygen in the combustion air is already equal to the pre-heating of combustion air to 600 °C (which is also not permissible by the environmental regulations due to the increase of NO_x generation).

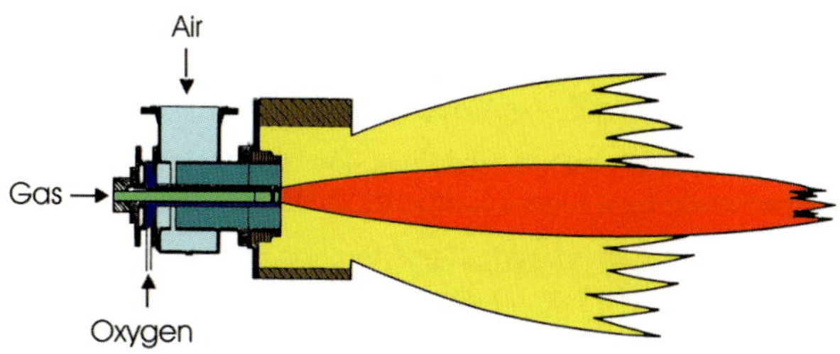

Fig. 3.11: Oxy-fuel flame shielded by an air curtain

Another effect of the combustion with pure oxygen or oxygen-enrichment requires attention. No combustion process reacts with the result of complete combustion. The higher oxygen content could improve the contact of metal with oxygen resulting in higher metal loss. This requires careful burner arrangement and adjustment of the burner flame (**Fig. 3.12**) to the actual conditions in the furnace. The burner, as shown in Fig. 3.12, allows shaping the flame by changing the nozzle position of the burner. The control of the oxygen content of the flue gas also helps to adjust the burner setting to the requirement of the specific process step in the furnace sequence.

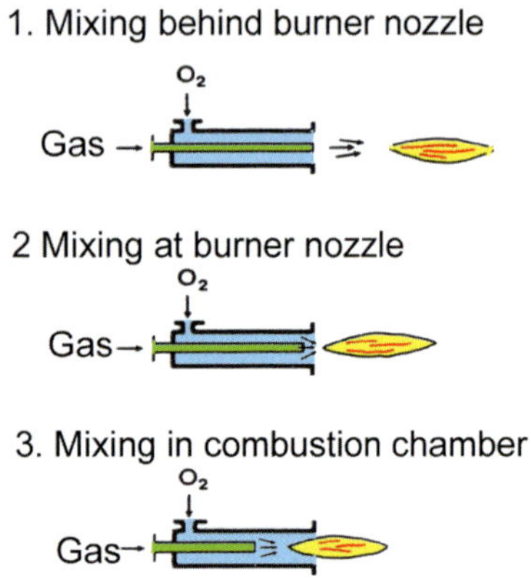

Fig. 3.12: Shaping of oxy fuel flames

In the aluminium industry, oxygen burners are preferably used in rotary drum furnaces, particularly for improving the performance of existing furnaces. But also in new furnaces the throughput can be increased. The hot flame temperatures in a rotary drum furnace do not have the same negative impact on the furnace lining as in stationary furnaces. Due to the rotation, the hot surface of the drum passes underneath the material in the furnace with the result of a cooling effect on the lining. The application in reverb furnaces seems to be less practicable. Some advantages can be noted by using oxy-fuel in twin chamber furnaces.

The results achieved after the installation of oxygen burners in a tiltable rotary drum furnace may help to understand the implication of oxygen usage.

The furnace was equipped with a low-NO_x, high velocity burner with a maximum firing capacity of 6 MW. An oxy-fuel burner was additionally installed to permit direct comparison of the performance of both systems. Fuel consumption was reduced from 560 kWh per ton material charged to 400 kWh. This is a fuel savings of close to 30 %. The most effective usage of the oxy-fuel system was achieved during heating-up of the batch. This is quite understandable since during this phase metal is cold and a very intensive heat exchange takes place that is proven by the low flue gas temperature. This expected increase in melting performance, by trying to improve the furnace cycle from 5 cycles/day to 6 cycles/day, could, however, not be achieved. One side effect was that the amount of dust discharged from the furnace by the large volume of flue gases was reduced due to the lower volume when using oxy-fuel.

The implications of using oxy-fuel can be summarized as follows:

- Due to the absence of nitrogen in the combustion products less gas needs to be heated up resulting in less fuel consumption. However, air is for free, oxygen not, thus the cost for oxygen has to be integrated in an economical comparison. Each m^3 of gas requires 2 m^3 of oxygen. Thus, a very simple condition must be met, using V_{gs} = gas volume saved, V_g = volume of gas used with oxygen, p_g price of gas/m^3, p_{ox} = price of oxygen/m^3:

  $$V_{gs} \cdot p_g \leq 2 \cdot V_g \cdot p_{ox}$$

- Better heat exchange may contribute to the saving of fuel but mainly increases the throughput rate. This is to be regarded as economical benefit.
- Less heat exchange by convection, due to lower velocity of the combustion products, is compensated by a higher temperature difference.
- If pure oxygen is used, the generation of NO_x is lowered. The unavoidable presence of air entering from leaks may cause an increase of NO_x generation in this small fraction of air. Enrichment of combustion air by oxygen increases the generation of NO_x.

- The firing with oxygen or enriched combustion air must be controlled carefully to get the optimal conditions for the usage of oxy-fuel during the furnace cycle. This can best be achieved by independent burner systems installed on the furnace.
- Control of the furnace atmosphere is required to avoid additional melt losses.

The quantity required in the standard aluminium recycling plant generally does not economically justify the installation of an in-house oxygen generation plant. Therefore, gas companies supply oxygen to a tank installed at site. The plant either installs its own tank installation or pays rent to the gas company which, of course, then has to be included in the economical evaluation.

3.1.10 Induction heating

Inductive heating is used for either crucible induction furnaces or channel induction furnaces.

The principle is that electrical energy is transformed into heat by inducing eddy current in the melt. The penetration of the eddy current depends on the frequency of the alternating current.

A characteristic feature is the frequency of the alternating current. It is important for the penetration depth. The penetration depth δ is expressed by

$$\delta = \frac{K}{\sqrt{f}} \tag{3.57}$$

with f = frequency and K a material-dependent factor. With increasing frequency the melting capacity and, consequently, the power requirements of the furnace increase. Apart from the type of the furnace, i. e. coreless or core-type, the furnaces are classified as main frequency or medium frequency furnaces. Main frequency furnaces operate with the frequency of the main power grid. The medium frequency ranges from 150 to 10,000 Hz; however, for practical purposes in operating plants, the highest frequency is 4,000 Hz. In former times the required frequency was produced by a speed-controlled motor / generator set. A modern system comprises an oscillating circuit (**Fig. 3.13**) which converts the mains frequency into the frequency required for the furnace. Main frequency induction furnaces do not require costly equipment for converting the energy. The pen-

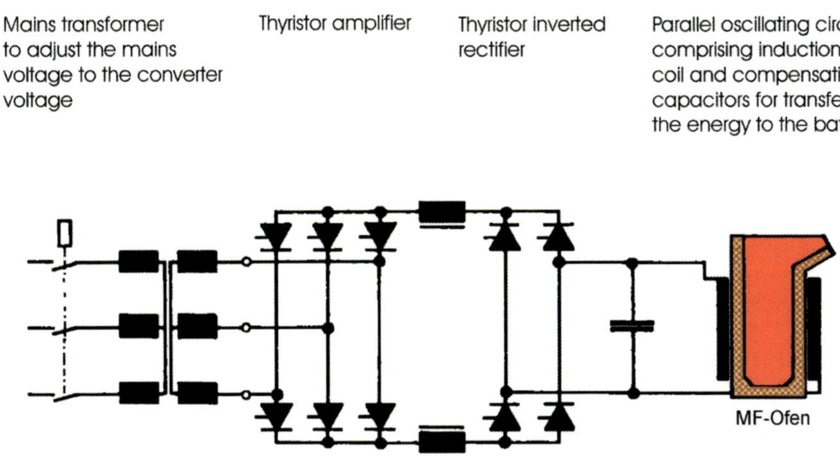

Fig. 3.13: Frequency converter for an induction furnace

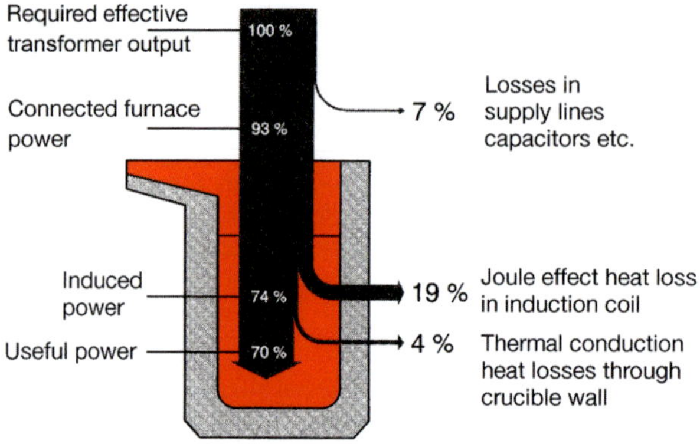

Fig. 3.14: Energy losses in a crucible induction furnace

etration depth of main frequency furnaces is very good but the intensive bath movement going in hand with it requires, however, a minimal size of the furnace which should not be smaller than 800 kg in case of the aluminium furnaces. The intensive bath movement also requires some care in operation.

In general, the efficiency of an induction furnace is expressed by

$$\eta_e = \frac{P_i}{P_i + P_v} \tag{3.58}$$

with P_i = the induced power in the melt and P_v = electric losses. The energy balance of a core-type induction furnace is shown in **Fig. 3.14**. Approximately 19 % of the energy is lost as Joule effect of the induction coil. This heat is dissipated into the cooling water requiring a closed loop cooling system. The efficiency for crucible induction furnaces used for aluminium is in the range of 65 % resulting in a specific energy consumption of 530 kWh/t. This is in the range of a good reverb furnace with heat recovery. The efficiency of a channel induction furnace is in the range of 80 %, which is a better value than the one for a reverberatory furnace.

The generation of electromagnetic fields requires reactive power. For the part of the magnetic flux flowing through the charge, the necessary reactive power is equal to the active power. The bigger demand of reactive power is due to the scatter field between the charge and the coil. The distance between coil and charge determines the requirement for the reactive power. As discussed in the case of the linear motor, the wall thickness between material in the furnace and the coil should be as low as possible to guarantee safe operation at best possible efficiency. This refers also to the inductor of the channel induction furnace. The reactive power increases with higher frequency. The ratio of the active power to the apparent power is characterized by the power factor cos φ.

$$P_W = U \cdot I \cdot \cos\varphi \tag{3.59}$$

and the reactive power

$$P_B = U \cdot I \cdot \sin\varphi \tag{3.60}$$

with P_W = active power, P_B = reactive power, U = voltage, I = current.

The power factor cos φ is compensated by a series of capacitors to the value of 1 with the target that the power supply only delivers active power.

3.1.11 Resistor-heating

Resistor-heating for furnaces is of very limited use in the aluminium industry. In former days, when electric power was available at the location of primary aluminium smelters, electrically heated reverb furnaces had been used as holding/casting furnaces. Today electric heating is used for crucible furnaces and heating lids for filters and in-line de-gassing units as well as for sampling furnaces. Also in homogenizing furnaces of the down-stream facilities, electric heating is quite common. Because of the very limited use in the secondary aluminium industry we will only touch the subject briefly. The situation may change in future if furnaces can be developed, perhaps on the basis of forced-air circulation to obtain a convective heat transfer because of more stringent environment protection regulations.

Mainly the generation of heat, using the Joule effect, is the basis for resistance-heated furnaces. An electrical current flowing through a resistor – in this case an electric heat conductor – will generate heat. According to Ohm's law, the electric current I (A) is proportional to the voltage U (V) with a proportionality factor R (W) which is called resistor.

$$U = R \cdot I \text{ (V)} \tag{3.61}$$

The electric power P is

$$P = U \cdot I \text{ (W)} \tag{3.62}$$

And from these equations follows

$$P = R \cdot I^2 \text{ (W)} \tag{3.63}$$

This law shows that the power introduced by the electric heat conductor depends on the resistance of this conductor. The specific resistance of the conductor material must be as high as possible to avoid excessively high current in the electric loop. The specific resistance should not change very much with changing temperature. The conductors should also not corrode and should be resistant to the furnace atmosphere. Different basic alloys have been developed. They consist of Cr-Ni containing ferritic alloys or Cr-Al containing austenitic alloys. Depending on their content of Ni or Cr, the service temperature ranges from 950 to 1,250 °C. The furnace temperature should be 50 °C lower than the permissible temperature of the conductor. There is also a limitation in regard to the permissible electric load in W/cm^2. For some austenitic conductor material the highest load can be accepted at a temperature of 500 °C with 3.4 W/cm^2 which comes down to 0.5 W/cm^2 at 1,000 °C.

The heat-conducting elements may comprise wire, shapes, strip or rods. For higher temperature ranges ceramic conductor rods made of silicon carbide are applied. Their service temperature my go up to 1,600 °C depending on the furnace atmosphere.

The temperature of the furnace is controlled by varying the power infinitely by a thyristor or on – off control. For the design of a furnace a specific load of 150 to 350 W/dm^2 furnace wall areas and 0.6 to 1.2 W/cm conductors can be calculated. From this figure the dimension of the heat conductor can be obtained, considering the service temperature. The heat conductors should be accommodated in refractory channels with a minimum Al_2O_3 content of 45 % together with very little Fe_2O. **Fig. 3.15** shows an example of a conductor arrangement in the roof of a reverberatory furnace of recent design.

The electric heat conductor must be monitored very carefully. Unbalanced operation must be avoided since this may result in short service life of individual conductors. The resistor heating systems can be operated in air or a protective atmosphere. The typical features, i. e. resistance

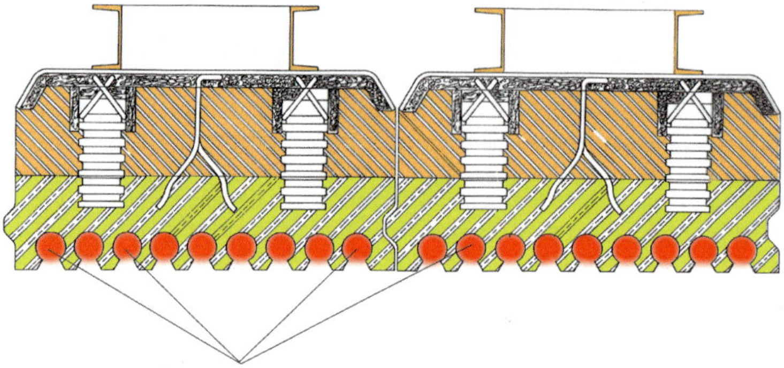

Fig. 3.15: Arrangement of electrical conductors in the roof of a reverberatory furnace

to gases, of the material need to be considered. In all cases where indirect heating is required, electric resistor heating is favored. The heat transfer in a resistor-heated furnace is generally by means of radiation.

Conductive heat transfer can be obtained with circulating air as executed in homogenizing furnaces in down-stream aluminium plants.

3.2 Transmission of heat

Christoph Schmitz

For the function of furnaces the energy exchange between different bodies is the basis for the design. This can be either by convection or radiation. Convection and conduction require that the two bodies are in contact. Radiation means that the heat transfer takes place without contact of the different bodies. Most often we will have a combination of the different kinds of heat transfer.

In a furnace the combustion products flowing through the furnace will touch the walls and, of course, the metal. According to the second law of thermodynamics, energy flows from the higher temperature to the lower temperature level. It is also obvious that the quantity of energy depends on the surface area and on the time. The heat transmitted can be defined with the equation

$$Q = A \cdot \alpha \cdot (T_1 - T_2) \qquad (3.64)$$

With Q the heat transferred in W, A = surface participating in the transfer in m^2, T_1 and T_2 = higher and lower temperature in K, the heat transfer coefficient α will be expressed as W/K m^2 h. The equation is very convenient for calculations and easy to use if α is known. And here it becomes more difficult since α depends on many factors such as type of surface condition, type of heat transfer, gas velocity and, particularly in the case of radiant heat transfer, on temperature. Values are obtained by tests under simplified conditions and practical experience, permitting recalculation of α. The surface area participating in the heat transfer is generally the contact area of the bodies or the areas opposite to one another. For bodies of a defined shape the contact surfaces are defined. With smaller pieces the area participating in the heat transfer is not clearly defined anymore. Particularly with bulk material of small particle size only part of the surface is exposed. Therefore, α refers to the volume of the bulk expressed in W/m^3· K· h. If a gas flows at least partly

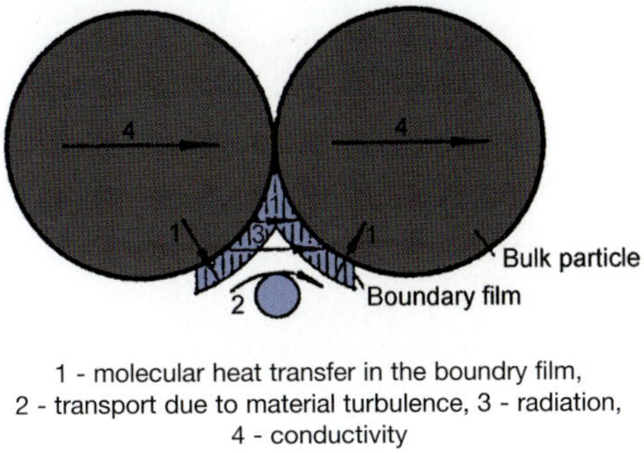

Fig. 3.16: Model for heat transfer in bulk material

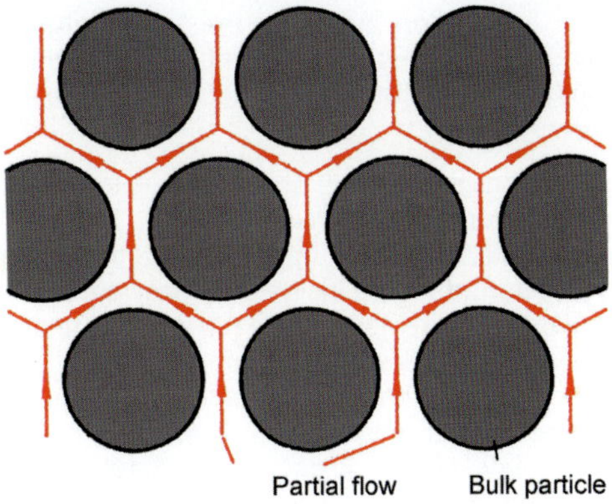

Fig. 3.17: Partial flow model in bulk material

through a bulk of material, as it is the case in a rotary drum furnace and in a chip dryer, the heat exchange will take place with different mechanisms (**Fig. 3.16** and **Fig. 3.17**):

- Molecular conductivity within the gas, acting in the boundary film of the particles
- Transport of energy between the boundary film due to material turbulence
- Radiation between the different particles
- Conduction within the particles.

Other material characteristics are of influence such as the real density of the material particle, its conductivity, the ratio of free space in the bulk, particle size, flow rate and velocity of the gas, temperature and the like. The mechanism of heat transfer in bulk material has been studied and models have been developed for a fluid bed where the gas flows through the bulk of particles. The possibility to apply these for our furnace processes is very limited. With the movement of particles, the radiation to the changing top layer and the contact of particles with the furnace wall do add some difficulties to the definition of the α-value in the rotary drum furnace. Thus, we have to rely on experience. α will be between 150 and 300 $W/m^3 \cdot K \cdot h$ for the rotary drum furnace.

3.2.1 Heat conduction

Heat will flow within a body from a point having higher temperature to a point with lower temperature. Experience tells us that the quantity Q of energy flowing from the area of higher temperature to the lower temperature section is proportional to the time t, the cross-section A available for the heat transfer and the temperature difference. The heat flow is, however, inversely proportional to the distance of the areas considered. The application of an equation developed on the basis of these considerations can only refer to a very small distance dx of these points of the body. Consequently, the temperature difference $d\vartheta$ must also be very small. Considering a proportionality factor λ, the equation reads

$$Q = -\lambda \cdot t \cdot A \cdot \frac{d\vartheta}{dx} \text{ (W)} \tag{3.65}$$

The expression $\frac{d\vartheta}{dx}$ indicates the temperature gradient between the different points of the body; it is negative in the direction of the heat flow.

The expression indicates the temperature gradient between the different points of the body; it is negative in the direction of the heat flow.

Some basic principals shall be explained by using a furnace wall as an example. The furnace shall be heated up already so that the temperature of the inside wall has reached the operating temperature. From the inside of the wall, heat flows now from the interior of the furnace to the outside wall. We have separated the wall into very small sections having a very small thickness δx (**Fig. 3.18**). Heat entering the section dx is used to heat up the wall element and to transfer heat according to the temperature difference δϑ. The quantity of energy entering section 2 in the very short time interval ϑ at the temperature difference dϑ between section and 2 will be

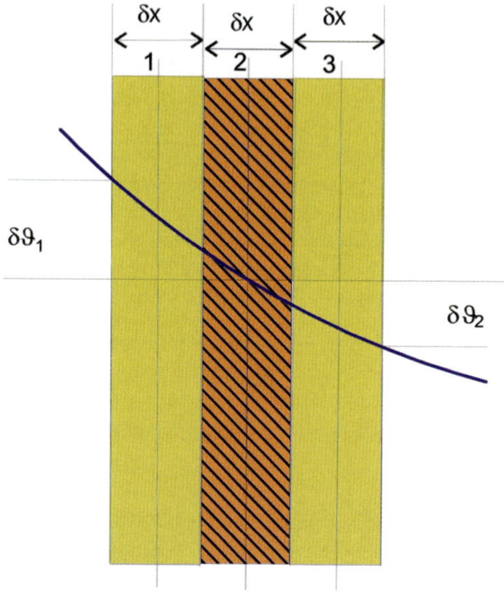

Fig. 3.18: Heat flow through a wall

$$dQ = -\lambda \cdot \delta t \cdot A \cdot \frac{\delta\vartheta_1}{\delta x}.$$

The heat remaining in section 2 results from the weight of this element, the specific heat and the temperature increase at

$$\delta Q_1 = -A \cdot \delta x \cdot \rho \cdot c \cdot \delta\vartheta_1.$$

Heat transferred to the adjacent section 3 is now:

$$\delta Q_2 = -\lambda \cdot \delta t \cdot A \cdot \frac{\delta\vartheta_2}{\delta x}.$$

According to the heat balance

$$\delta Q = \delta Q_1 + \delta Q_2$$

resulting in

$$c \cdot \rho \cdot \delta x \cdot \lambda \cdot \frac{\delta\vartheta_1 - \delta\vartheta_2}{\delta x}.$$

In the mathematical expression, the temperature difference $d\vartheta_1 - d\vartheta_2$ is expressed as $d^2\vartheta$ and thus the equation reads

$$\frac{\delta\vartheta}{\delta t} = \frac{\lambda}{c \cdot \rho} \cdot \frac{\delta^2\vartheta}{\delta x^2} \qquad (3.66).$$

The value $\frac{d\vartheta}{dx}$ gives the temperature gradient in the vicinity of the cross-sections, while $\frac{\delta^2\vartheta}{\delta x^2}$ indicates the change of this temperature gradient.

Time-dependent temperature changes will occur in case of different temperature gradients. Considering our furnace wall, we note a large temperature gradient at the beginning of the heating-up cycle. Near the furnace inside, the temperature difference is very large due to the hot furnace temperature. Near the outside, the gradient is small or even zero since wall temperature remains close to the ambient temperature. The wall temperatures near the furnace interior change rapidly due to the temperature increase in the furnace. Slowly, the temperature in the wall section will rise to the point where no temperature increase is possible due to the temperature gradient and all energy is passed to the next section. No temperature increase will occur in the section. If this process has passed through the entire wall, there is steady state condition now. All energy introduced to the inside of the wall is now transferred to the outside wall and from there to the surrounding area. The temperature gradient is now identical for all wall sections. Additional heat will not be stored in the wall. We can now apply the equation to the entire wall. Since the temperature gradient $\frac{d\vartheta}{dx}$ is equal in all wall sections, the infinitesimal temperature difference $\delta\vartheta$ can be replaced by the difference between interior wall surface and outside wall surface $\vartheta i - \vartheta o$, and dx can be replaced by the wall thickness s. If, furthermore, the wall area A =1 m^2 and t = 1 h, we obtain the specific value for heat flow from the inside to the outside as

$$q = \frac{\lambda}{s}\left(\vartheta_i - \vartheta_o\right)\left(W/m^2h\right). \tag{3.67}$$

This equation is applicable only if the wall is in a steady state condition. For changing conditions a different procedure using more complex equations is required. In our furnaces this would occur during preheating of a new lining or a start-up after a shutdown of the furnace. Different temperatures during the operation phases or the operation of a rotary drum furnace are also not steady state conditions. However, detailed studies of the wall conditions are only of academic interest in our case.

The thermal conduction coefficient λ is of much greater interest. It is characteristic for different materials but depends also very much on temperature. There is a very large fluctuation of this factor within the different refractory material depending on type and consistency of the material as well as on the temperature. **Table 3.7** gives some data showing the major differences for selected materials. A listing of the different materials can be obtained from refractory suppliers.

Table 3.7: Lambda-values for different materials

	Density kg/m³	λ in W/m °C at a temperature of °C						
		0	200	400	600	800	1100	1200
Fireclay	1850	0.87	0.96	1.05	1.14	1.23	1.33	1.42
High-alumina refractory	2400	1.71	1.63	1.58	1.54	1.51	1.49	1.47
Insulating firebrick	1100	0.36	0.46	0.44	0.48	0.52	0.6	
Insulating refractory	775	0.17	0.21	0.24	0.29	0.34	0.38	
Aluminium	2700	229	229	233		0.094		
Steel	7850	45	45	38	35			
Copper	8930	387	372	360	350			
Slag wool	130	0.32	0.062					
Vermiculite	100	0.056						

We have deducted our equations from the assumption that the furnace wall consists of one homogeneous material. This is not correct since the furnace wall is built from different materials for various reasons. The steel shell, to keep the shape of the furnace, is not so much of interest with regard to heat conduction since the heat passes through the steel much faster than through the refractory material. But the refractory walls at the furnace interior generally consist of very dense material with good mechanical strength but often not so optimum heat-insulation properties, backed by one, two or more layers providing the required heat insulation. The more the temperature decreases towards the outside of the furnace, the less dense material with better insulation characteristics can be installed. Each layer presents a resistance to the flow of heat to the outside of the furnace equivalent to the flow of electrical energy through a series of resistors.

$$R_{tot} = R_1 + R_2 + R_3$$

According to Joule's law, the electrical current depends on the difference of voltage, thus

$$U = R_{tot} \cdot I$$

and from that

$$U_{tot} = \Delta U_1 + \Delta U_2 + \Delta U_3$$

$$U_{tot} = R_1 \cdot I + R_2 \cdot I + R_3 \cdot I = I \cdot (R_1 + R_2 + R_3)$$

The similarity to our problem is obvious: I is equivalent to our flow of heat q, ΔU to the temperature difference between inside and outside the furnace ϑ_i - ϑ_o. But what about the resistance R ?

We can define

$$\vartheta_1 - \delta_3 = q \cdot \left(\frac{s_1}{\lambda_1} + \frac{s_2}{\lambda_2} + \frac{s_3}{\lambda_3} \right),$$

This is deducted from

$$\vartheta_1 - \vartheta_2 = \frac{s_1}{\lambda_1} \text{and } \vartheta_2 - \vartheta_3 = \frac{s_2}{\lambda_2}$$

We have some more resistances to consider. The wall temperature of the furnace wall is not equal to the temperature of the furnace interior but has to be calculated using the equation

$$q = \alpha_i \cdot (\vartheta_i - \vartheta_1)$$

and for the flow of heat to the outside of the furnace:

$$q = \alpha_a \, (\vartheta_3 - \vartheta_o)$$

Using the equivalence to the flow of electrical energy, we replace the total resistance by a factor k. The heat flow through a multi-layer wall is calculated according to the equation

$$\dot{q} = k \cdot (\vartheta_i - \vartheta_o) \tag{3.68}$$

The factor k is called the heat resistance coefficient and is calculated according to

$$k = \frac{1}{\frac{1}{\alpha_i} + \frac{s_1}{\lambda_1} + \frac{s_2}{\lambda_2} + \ldots \frac{1}{\alpha_o}} \tag{3.69}$$

The heat transferred to each individual refractory layer will be

$$q = \frac{\lambda_n}{s_n} \cdot \left(t_{ni} - t_{no}\right)$$

or

$$\frac{q}{\left(\lambda_n\right)} = \frac{\Delta t_n}{s_n}$$

λ depends on the refractory material. But we should have a look at the heat transfer coefficients α_i and α_o. Both are a combination of transmissions by radiation and by convection. The α-value for radiation is calculated according to the equation

$$\alpha_{1,2} = \frac{\varepsilon \cdot 5.67}{\vartheta_1 - \vartheta_2}\left[\left(\frac{T_1}{100}\right)^4 - \left(\frac{T_2}{100}\right)^4\right] \tag{3.70}$$

There are different equations used for the α_k-value which is the heat transfer coefficient for convection. The equation according to literature is

$$\alpha_k = 1.83 \cdot \left(T_2 - T_1\right)^{\frac{1}{3}} \tag{3.71}$$

This value refers to the outside wall only since the α-value for convection depends very much on the gas velocity which is close to zero in case of the outer furnace. The index 1 expresses the shell temperature and index 2 the outside temperature. For the heat exchange of the furnace interior, please refer to sections 3.2.2 and 3.2.3. The α-value for the furnace interior α_i is high compared to the value of the other data so that some uncertainty in the actual value applied will not upset the result of the calculation.

Different areas within the furnace require specific considerations. For the bath area the industry favors two different philosophies. One focuses on maximum heat insulation, the other one on long service life. If long service life is required, the solidification zone of the aluminium must be in the hot phase to avoid that liquid metal penetrates into the insulating layers.

Fig. 3.19 shows the temperature drop for each layer for the two different philosophies. In both cases, a 3-layer system is assumed. Two possibilities are shown for a total refractory material thickness of 500 mm in both cases. If the condition must be fulfilled that the 658 °C-isotherm has to be within the hot face, this layer must be quite thick. Fortunately, another factor helps to arrive at a reasonable lining thickness.

In case 1 (solidification within the hot face layer) the resulting shell temperature is above 100 °C which means higher heat losses in the range of 1,290 W/m^2 compared to case 2 (solidification point within the insulating layer) where the shell temperature is only 60 °C and the resulting heat loss is 410 W/m^2. It is certainly possible to increase the total thickness of the refractory lining. However, this would result in much more weight with the consequence of heavier furnace shell design and larger hydraulic cylinders, in total resulting in increased equipment cost. Evaluating

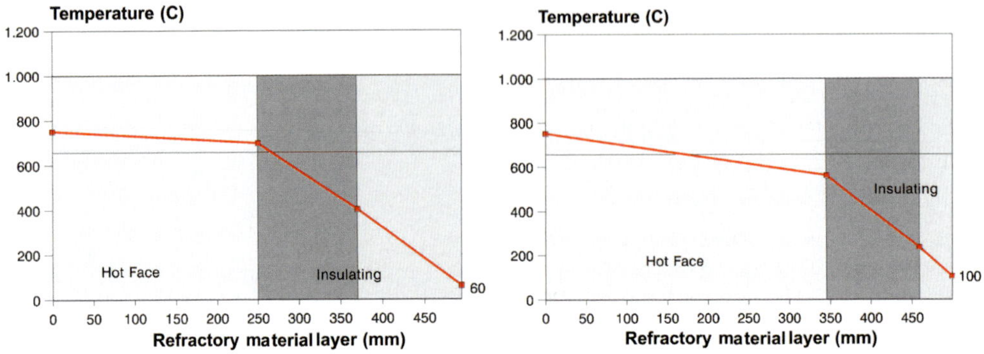

Fig. 3.19: Heat flow through refractory material layers

these values, it becomes quite clear that the higher energy loss is very well compensated for by lower investment cost and much lower maintenance cost in the long run. It must also be considered that the higher energy loss is limited to the bath area which accounts for only 30 % of the total wall losses.

The refractory companies offer the service of providing the calculation of wall temperature and losses through the furnace walls, roof and bath area. They have developed computer programs which enable them to verify the design with the selected refractory materials.

But still, it may be worthwhile to look into some details of the heat transfer calculation. The condition of the refractory material layers can be calculated manually by iteration. For such calculation a certain design and basic parameters are defined first and will then be checked by this calculation. This refers mainly to the temperature of the outside wall and the heat flow through the wall.

For an example we will use the following equations:

- Heat transfer inside

$$\dot{q} = \alpha \cdot \left(t_i - t_o\right) \text{ and } t_1 = t_i - \frac{\dot{q}}{\alpha_1} \tag{3.72}$$

Heat transfer in the first layer

$$\dot{q} = \frac{\lambda_1}{s_1} \cdot \left(t_1 - t_2\right) \text{ and } t_2 = t_1 - \dot{q} \cdot \frac{s_1}{\lambda_1} \tag{3.73}$$

- Heat transfer in the second layer

$$\dot{q} = \frac{\lambda_2}{s_2} \cdot \left(t_1 - t_2\right) \text{ and } t_3 = t_2 - \dot{q} \cdot \frac{s_2}{\lambda_2} \tag{3.74}$$

- Heat transfer outside

$$\dot{q} = \alpha_o \cdot \left(t_3 - t_o\right) \tag{3.75}$$

Since the flow of energy is the same for all layers and, consequently, the four equations, we can say

$$\dot{q} = \frac{(t_i - t_o)}{\frac{1}{\alpha_i} + \frac{s_1}{\lambda_1} + \frac{s_2}{\lambda_2} + \frac{1}{\alpha_o}} \qquad (3.76)$$

The refractory lining shall consist of a two-layer system (**Fig. 3.20**). The interior temperature t_i of the furnace shall be 1,200 °C, the outside temperature t_o shall be 20 °C and the shell temperature shall not exceed 160 °C. Furthermore,

α_i = 100 W/m²K, s_1 = 200 mm

αo = 12 W/m²K, s_2 = 80 mm

For the calculation we use the method of iteration. We estimate values for t_1, t_2 and t_3 and by using equation (3.76) we calculate the heat flow. With equation (3.72) we now calculate t_1 and with (3.73) we obtain t_2. Finally, we use equation (3.74) and get t_3.

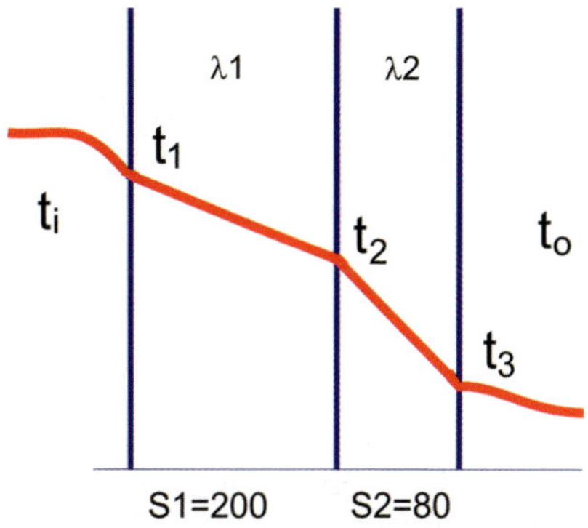

Fig. 3.20: Temperature distribution in a two-layer system

If the difference between the estimated values and the calculated values is too high, new estimates are made and the calculation is repeated until the calculated values are within an acceptable limit with the estimated values.

For the calculation we use a spread sheet program such as Excel with the equations in the subsequent columns. **Table 3.8** shows such a spreadsheet with the results of the iteration.

Iteration 1: t_1 = 1,180 °C, t_2 = 800 °C, t_3 = 150 °C, from a data sheet, we find for λ_1 = 2.25 W/m K and λ_2 = 0.12 W/m K

Using the spread sheet we obtain:

t_1 = 1,186 °C, t_2 = 1067 °C and t_3 = 136 °C.

We notice that the estimated value for t_2 differs very much from the now calculated temperature. Therefore, we make a new estimate and again that may result in differences that are too large.

t_1 = 1,183 °C, t_2 = 1,047 °C and t_3 = 158 °C.

Table 3.8: Spread sheet calculation for a furnace wall

Values	Inside (t)	Layer 1 (t2)	Layer 2 (t3)	Outside
Temperature	1200			20
Thickness (m)		0.20	0.08	
Alpha	100		12.00	
Temperature (°C)	1180	800	150	
Lambda		2.35	0.12	
Mean temperature	990	475		
Spec.heat flow	1396			
T1 calculated	1186			
T2 calculated	1067			
T3 calculated	136			
Diff. t2 estimated				-14
Temperature (°C)	1183	1047	158	
Lambda		2.43	0.15	
Mean temperature	1115	603		
Spec.heat flow	1664			
T1 calculated	1183			
T2 calculated	1046			
T3 calculated	159			
Diff. t2 estimated				1

We finally end up with the results as calculated with the spreadsheet in Table 3.8:

Heat loss through the wall = 1664 W/m² K h, t_2 is with 1,046 °C very close and the outside wall temperature of 159 °C (which is close enough to our pre-condition of 160 °C).

3.2.2 Convective heat transfer

If a gas passes along a surface it will transmit heat to this surface by convection according to the general equation:

$$q_c = \alpha_c \cdot \left(\vartheta_g - \vartheta_a\right) (W/m^2\ h) \qquad (3.77)$$

or for the total energy submitted during a time t and the temperatures ϑ_g for the gas and ϑ_a for the surface to the surface A in m²:

$$Q = A \cdot t \cdot \alpha_c \cdot \left(\vartheta_g - \vartheta_a\right) (W) \qquad (3.78)$$

Here again, the difficulty is to determine the heat transfer coefficient αc. The heat transfer by convection is due to the contact between the gas and the surface. The molecules of the gas pass their energy to the wall depending on gas velocity, kind of gas flow, viscosity and temperature of the gas. As we know that the temperature of a gas is determined by the velocity of the gas molecules within the gas volume, it is essential that the molecules are reflected by the wall thus passing some of their energy to this wall. If the conditions of the gas flow are such that the entire molecules move in the direction of the general gas flow, they remain in their position relative to each other due to the viscosity of the gas. Looking at the cross-section of a tube, the different gas molecules do not have the same velocity. Near the walls their velocity is lower due to the friction between wall and gas. It increases with growing distance from the wall and will reach its maximum in the center of the tube. This type of gas flow is called laminar or steady flow. Particles at the inside of the stream must pass their energy by conduction to the walls and, since gases are poor heat conductors, the flow of energy is very limited.

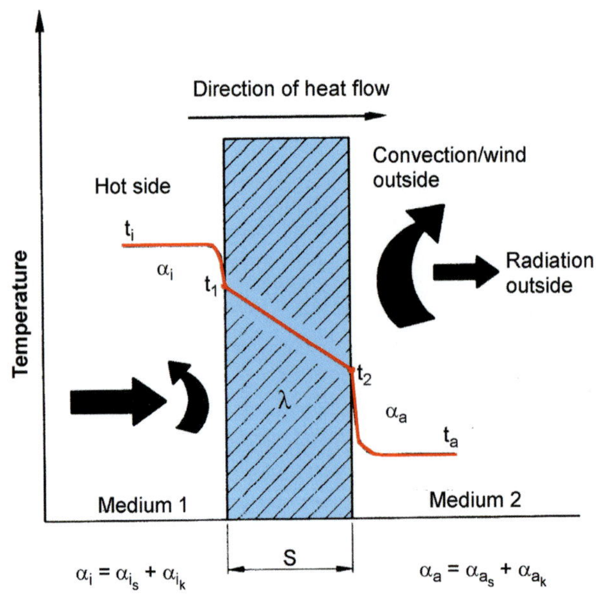

Fig. 3.21: Heat transfer from gas to a wall

As soon as the velocity of the gas increases above a critical value, turbulences will occur which are characterized by movement of particles rectangular to the flow direction and the gas particles can impact on the wall thus passing their energy directly. With increasing velocity turbulence will also grow. However, near the wall a thin film will remain in the laminar condition where heat exchange is disturbed (**Fig. 3.21**). The resulting temperature drop between gas and wall is characterized by the coefficient a. Not only high velocity can result in turbulences but also disturbances concerning the way the gas flows. In furnaces and ducts the velocity will mostly be turbulent.

We can notice that the heat transfer coefficient will increase with increasing velocity. A large cross-section will, however, reduce the effect of the turbulence since the individual gas particles have to pass a larger distance before they reach the wall. This is the case in our furnaces. Thus, we have to make sure that the gas flow remains turbulent and that the flow of the gas leaving the burners is directed within the furnace for maximum heat transfer.

The heat transfer coefficient can only be obtained by experiments by defining simplified or ideal conditions. One of the empirical equations for a forced flow parallel to a wall is (according to Jürges)

- for $w < 5$ m/s

$$\alpha_c = 6.2 + 4.2\ w \tag{3.79}$$

- for $w > 5$ m/s

$$\alpha_c = 7.52\ w^{0.78} \tag{3.80}$$

For w we have to use the value of

$$w = w_T \frac{293}{T_G}$$

with w_T = velocity at gas temperature T_G

If we assume the velocity of the combustion products circulating in the furnace at a temperature of 900 °C and having a velocity of 20 m/s we obtain

$$w = 20 \cdot \frac{293}{273 + 900} = 5\ m/s$$

and

$$\alpha_c = 7.52 \cdot 3.5^{0.78} = 26.4\ W/m^2\ K$$

This comes close to a value we know from comparable existing furnaces. But how good is our estimate of 20 m/s? This value can not be calculated from the quantity of flue gases passing through the furnace. The shape of the material batch also restricts the flow of the gases through the furnace. Considering high velocity burners, which create heavy turbulence in the furnace and also very intensive recirculation of these gases into the burner flame, an average between 25 and 40 W/m² K h seems to be a very realistic assumption.

For flue gas ducts the empiric equation

$$\alpha_c = \left[4.13 + 023 \frac{t}{100} - 0.0077 \cdot \left(\frac{t}{100}\right)^2\right] \cdot \frac{w_0^{0.75}}{d^{0.25}} \left(W/m^2K\right)$$

can be applied. For rectangular channels d is to be replaced by the equivalent diameter $d_a = \frac{4A}{U}$ with cross-section A and circumference U.

Another correction factor must be considered if the duct diameter is < 200 mm. Also for short distances some correction factors should be applied. This refers also to the inlet length of a ducting for different types of inlets.

3.2.3 Radiant heat transfer

As the name "reverberatory furnace" implies, radiation of energy is a characteristic feature of such furnaces – at least during certain phases of the furnace cycle. Thus, it may be worthwhile to have a look at the principles of radiant heat transfer.

If two bodies with different temperatures are located opposite to each other, the body with the higher temperature will radiate heat to the one with lower temperature. The energy received will not be absorbed in total. Part of it will be reflected and there may be also a portion that is just passing through that body. This will be the case if this body is transparent as, for instance, with atomic gases or glass. The ratio between the radiation absorbed radiation A_λ and the reflected radiation $A_{\lambda s}$ is called absorption ratio ε. Therefore,

$$\varepsilon = \frac{A_\lambda}{A_{\lambda s}} \tag{3.81}$$

A body that absorbs all energy without any portion passing through or being reflected is called "black body". A body not only absorbs energy but also emits energy depending on its temperature. The amount is equal to the energy which this body is able to absorb. Since the absorption of the black body is highest possible, it also emits the highest possible amount of energy. In practise, no absolute black body exists.

For the radiation of the black body the Stefan Boltzmann law applies:

$$Q = C \cdot 10^{-8} \cdot T \tag{3.82}$$

or the expression used in practise

$$Q = C \cdot \left(\frac{T}{100}\right)^4$$

whereby C is the radiation coefficient of the absolutely black body of 5.67 W/m^2K^4, Q the heat transferred and T the temperature in K. The gradient $\frac{T}{100}$ was used by engineers in the slide rule age in order to be able to handle smaller numbers. It still is convenient when doing manual calculations, even with pocket calculators. Therefore, we will use it in our considerations.

All materials in a furnace, including the metal to be processed, do not absorb all energy radiated to them and they are, therefore, called "grey bodies". They are characterized by an emission factor ε > 1 which is valid for all wavelengths. The equation used for the black body can now be applied if C is replaced by the factor $C_1 = \varepsilon \cdot C$. **Table 3.9** gives some values for ε which are of interest for our aluminium furnaces. Some bodies radiate at specific wavelengths only. At other wavelengths they let energy just pass and do not absorb or emit energy. Air is an example. Within gases, absorption takes place through the whole of the gas body since radiation must meet a sufficient number of

Table 3.9: Emission factors for selected materials

Material	Temperature [°C]	ε	$\varepsilon \cdot C$ [W/m² · K⁴]
Refractories, used	600 – 1000	0.75 – 0.80	4.3 – 4.6
Aluminium, molten	900	0.06	0.4
Aluminium with oxide skin	900	0.2 – 0.4	1.2

molecules before absorption is complete. This explains why the flame thickness of CO_2 and H_2O has a substantial influence on the radiation of these gases.

The heat transfer between roof and bath surface can be calculated according to the Stefan Boltzmann law

$$Q_{rC} = A \cdot C_{1,2} \cdot \left[\left(\frac{T_C}{100} \right)^4 - \left(\frac{T_B}{100} \right)^4 \right] \tag{3.83}$$

$C_{1,2}$ is a factor calculated as

$$C_{1,2} = \frac{C}{\frac{1}{\varepsilon_1} + \frac{1}{\varepsilon_2} - 1} \tag{3.84}$$

C_1 and C_2 depend on the material and calculated as c = 5.67 whereby 5.67 is the radiation coefficient of the black body. The reflection ratio ε for the refractory lining is 0.8. For the metal bath the factor is different for a surface with thin dross skin ($\varepsilon = 0.06$) or a thick dross layer ($\varepsilon = 0.3$) since a shiny surface will reflect a substantial amount of energy.

Example:

A furnace with a bath capacity of 50 tons shall have a bath surface of 8 x 5 m. The reflection ratio of the refractory furnace walls is $\varepsilon = 0.8$. The bath shall have a dross layer with $\varepsilon = 0.3$ and shall than be skimmed obtaining an ε of 0.06. How much energy is transferred from the roof having a temperature of 1,000 °C to the bath having a temperature of 700 °C?

According to (3.84)

$$C_{1,2} = \frac{5.67}{\frac{1}{0.8} + \frac{1}{0.3} - 1} = 1.6$$

The heat transferred between roof and bath is calculated with (3.83)

$$Q = 8 \cdot 5 \cdot 1.6 \cdot \left[\left(\frac{1273}{100} \right)^4 - \left(\frac{973}{100} \right)^4 \right] = 1107 \text{ kW.}$$

After the bath has been skimmed:

$$C_{1,2} = \frac{5.67}{\frac{1}{0.8} + \frac{1}{0.06} - 1} = 0.33$$

With this figure the energy transferred is now only 228 kW.

For our 50 t furnace, with the bath dimensions 8 m x 5 m and dross layer, the energy transferred by radiation from the roof to the bath amounts to 1,107 kW. A skimmed bath would receive only 228 kW! That gives a clear indication that the bath should be skimmed only after the holding period.

The sidewalls of the furnaces also contribute some energy to the bath. It has to be considered that the radiation will reach the bath surface from the vertical walls at an angle depending on the distance of the surface element from that wall. According to Lambert`s law, radiation decreases with the cos of the angle j between the perpendicular of the bath surface and the direction of radiation. For practical calculations a factor φ_j considers the influence of the sidewall.

$$Q_W = A_W \cdot \varphi \cdot \varepsilon_1 \cdot \varepsilon_2 \cdot C \cdot \left[\left(\frac{T_W}{100} \right)^4 - \left(\frac{T_B}{100} \right)^4 \right] \tag{3.85}$$

The factor φ depends on the design of the walls (**Fig. 3.22**). For the example above, the furnace height will be 2.5 m. From that we get

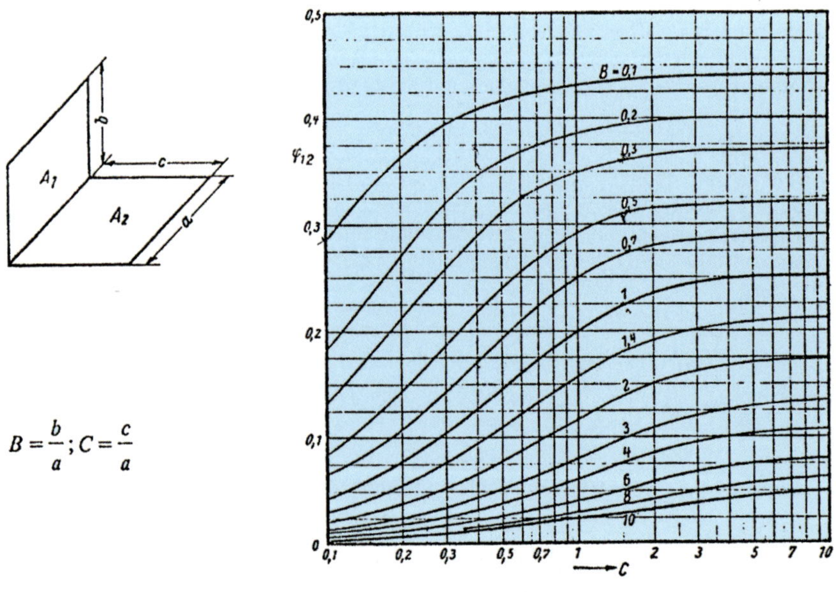

Fig. 3.22: Design factor for radiation from the furnace walls to the bath

B = 0.31

C = 0.63 for the long sides and from Fig. 3.22

$\varphi = 0.325$

Now we have all data to calculate the amount of heat transferred.

$$Q_W = 8 \cdot 5 \cdot 0.325 \cdot 0.8 \cdot 0.3 \cdot 5.676 \cdot \left[\left(\frac{1273}{100} \right)^4 - \left(\frac{973}{100} \right)^4 \right] = 306 \text{ kWh for 1 long side.}$$

For calculating the heat transfer from the short wall, we have to swap the values and get for B = 0.5 and C = 1.6. By using Fig. 3.22, we obtain $\varphi = 0.31$ and for the heat transfer we now have 292 kW. The total for two long sides and two small sides will be 2 (306 + 292) = 1196 kW.

There is not much of a difference between the two sides due to the size ratio. If, for instance, the ratio of the two sides would be 4 x 10 m to obtain the same surface area of 40 m^2, the amount of heat would be 358 kW for the long side and 245 kW for one short side adding up to a total of 1,206 kW. The difference is very little in the case of our aluminium furnace. This shows that the side ratios, as a result of the furnace sizes, do not have a major influence on the heat transfer from the walls to the bath. The condition of the bath surface is a direct proportional influence. Thus, for general calculation we can use $\varphi = 0.32$ with sufficient accuracy. The height above the bath is required for operating purposes. It is always in the range of 2.0 to 2.5 m and since the emission ratio of the refractory material will always be 0.8 we can simplify equation (3.85) to

$$Q_W = A \cdot \varepsilon_B \cdot f \left[\left(\frac{T_W}{100} \right)^4 - \left(\frac{T_B}{100} \right)^4 \right] \tag{3.86}$$

(f = 0.325 · 0.8 · 0.325 · 4 = 5.89)

to obtain Q_w for all four sides.

For the total heat transfer of walls and roof we have to add the value calculated for the energy transfer from the roof as we have calculated above.

$$1{,}107 + 1{,}206 \approx 2{,}300 \text{ kW}$$

Thus, we can fairly assume that the heat transfer from the roof is equal to the heat transfer from the walls provided the bath area is not skimmed. The ratio is disturbed when considering the heat loss through an open door. In such a case, the opposite wall will radiate its full energy to the surrounding area and quite a substantial amount of heat is lost.

If we go for specific values we can obtain the specific heat transfer per m^2 of bath area. In the case of our example this will be 2,300/40 = 57.5 kW/m^2. How does this value change when changing the temperature relation? If we increase the wall and roof temperature to 100 °C = 1,373 K, the heat transferred per m^2 will be

$$q = 0.3 \cdot 5.8 \cdot \left[\left(\frac{1373}{100} \right)^4 - \left(\frac{973}{100} \right)^4 \right] = 46.24 \text{ kW/m}^2 \text{ or } 92.5 \text{ kW/m}^2$$

including the ceiling. The little increase of 100 °C in roof temperature results in an improvement of the heat transfer by 60 %. Increase of the metal temperature by 100 °C will reduce the heat transfer to 2 x 22.63 = 45.26 kW/m², a decrease of only 21 %. As consequence, for operating a twin-chamber furnace it would be quite effective to increase the wall temperature by 100 °C and have a higher temperature of metal for circulation.

If the furnace has a preheating ramp the area of the bath surface has to be taken into account, not the total roof area. The remaining area will certainly contribute to the heat transfer but the difference will be in the range of accuracy of the assumed preconditions for the calculation.

One more case should be evaluated. Assuming that cold metal could be evenly distributed on the bath surface and wall temperature has come down during charging to 300 °C, the freshly charged metal shall have a temperature of 20 °C. Then the heat transfer would be

$$q = 0.3 \cdot 5.9 \cdot \left[\left(\frac{573}{100} \right)^4 - \left(\frac{293}{100} \right)^4 \right] = 1.78 \text{ kWh/m}^2$$

We notice that the radiation in a furnace at low temperatures is negligible.

The theoretical energy required to melt 1 kg of aluminium and superheat it to 750 °C is 208 kW. The energy of 2,300 kW/h, available from the walls, would be sufficient to melt 11 tons of metal/hour. In practise, this is not possible for various reasons. The main factor is that the surface of the 11 tons of solid metal is much smaller than the bath area. This illustrates the handicap of electrical resistor heating. The radiation can only be increased if the temperature of the roof increases, too. This is not possible due to the limitations of the refractory material. In case of the twin-chamber furnace, the temperature difference between walls and metal must be maintained. Hot metal must be moved to other furnace areas where the enthalpy can be utilized.

Now we have to consider the radiation from the flue gas. When firing natural gas, the combustion products mainly comprise N_2, H_2O and CO_2, perfect combustion assumed. N_2 does not emit energy by radiation; this is the case only with H_2O and CO_2. When firing fuel oil some unburned C could also radiate some energy.

The radiation of the flue gases to the wall or the bath, respectively, can be expressed for CO_2 as:

$$q_{gCO_2} = 4.1 \cdot \varepsilon \cdot \sqrt[3]{p \cdot s} \cdot \left[\left(\frac{T_G}{100} \right)^{3.5} - \left(\frac{T_B}{100} \right)^{3.5} \right] \quad (3.87)$$

and for H_2O

$$q_{gH_2O} = 41 \cdot \varepsilon \cdot p^{0.8} \cdot s^{0.6} \cdot \left[\left(\frac{T_G}{100} \right)^3 - \left(\frac{T_B}{100} \right)^3 \right] \quad (3.88)$$

Herein = q_{H2O} specific heat of H_2O, q_{CO2} specific heat of CO_2 (W/m²h), ε = emission ratio (W/K⁴ m²) h, p = partial pressure, s = flame thickness (m), T_G flue gas temperature (K), T_B = bath or wall temperature (K). To obtain the total heat transfer by radiation from the gas fractions, the two values have to be added. This does not consider that the combustion products are a mixture of gases and that there will be an influence on each other. This influence is very small and can be neglected for practical purposes.

There are more accurate equations developed on empirical basis which include a correction factor for the radiation of a different wavelength. However, these tests refer to furnaces whose design differs from that of aluminium furnaces. It is obvious that the minor adjustments are of theoretical interest only and may not be of interest for practical application. Besides, at least one factor needs some discussion. This is the flame thickness expressed as s in the above equations. For long furnaces s will be identical with the height above bath. For shorter furnaces s should be calculated according to Heiligenstedt by the equation

$$s = \frac{2}{3} \cdot \sqrt[3]{a \cdot b \cdot c} \tag{3.89}$$

with a, b and c being the dimensions of the furnace. This comes very close to the height above the bath. The calculations according to Brunklaus and the VDI Wärmeatlas result in a higher flame thickness. These figures for s are not logical in our case. With the very good turbulence in the aluminium furnaces, the flame thickness is identical to the height of the furnace room above the bath level, i. e. 2.5 m. The factor p is the partial pressure of H_2O or CO_2, respectively. Considering the only light positive pressure maintained above the ambience in the furnace, p is equal to the contents of these gases in the combustion products.

Now we come back to our example and try to get a picture of the amount of heat radiated by the flue gas. It would be convenient to be able to use similar equations for the heat transfer by convection. Therefore, we calculate values for different temperatures for bath and walls, respectively, and the combustion products. Doing so, we can simplify the equations (3.87) and (3.88) for our aluminium furnaces. The flame thickness is typically 2.5 m. The partial ratios of CO_2 and H_2O are also typical when firing natural gas and they are both approximately 0.2 %. Now we can define a factor for

$$CO_2:\ 4.1 \cdot \sqrt[3]{0.2 \div 2.5} = 3.25 \text{ and for } H_2O:\ 41 \cdot 0.2^{0.8} \cdot 2.5^{0.6} = 19.6$$

With these factors our equations now read

$$q_{CO_2} = \varepsilon_{B,C} \cdot 3.25 \cdot \left[\left(\frac{T_G}{100} \right)^{3.5} - \left(\frac{T_{B,C}}{100} \right)^{3.5} \right] \tag{3.90}$$

and

$$q_{H_2O} = \varepsilon_{B,C} \cdot 19.6 \cdot \left[\left(\frac{T_G}{100} \right)^{3} - \left(\frac{T_{B,C}}{100} \right)^{3} \right] \tag{3.91}$$

The total energy by gas radiation is

$$Q_{rG} = Q_{CO_2} + Q_{H_2O}\ (W/m^2 \cdot K \cdot h) \tag{3.92}$$

To obtain an equation that is similar to the principle equation using the difference in temperatures and a factor α, we divide the factors by $(T_G - T_{B,C})$ and define

Table 3.10: Alpha values for gas radiation

Bath/wall temperature	Gas temperature (K) 900	1000	1100	1200	1300	1400	1500	1600
300	34	41	49	58	67	78	89	101
400	39	46	55	64	74	85	96	109
500	45	52	61	71	81	92	104	117
600	51	59	68	78	89	100	113	126
700	58	67	76	86	97	109	122	136
800	66	75	85	95	107	119	133	147
900		84	94	105	117	130	144	159
1000			104	116	128	142	156	171
1100				127	140	154	169	184
1200					153	167	182	198
1300						182	197	214
1400							213	230
1500								246

$$\alpha = \frac{3.25\left[\left(\frac{T_G}{100}\right)^{3.5} - \left(\frac{T_{B,C}}{100}\right)^{3.5}\right] - 19.8\left[\left(\frac{T_G}{100}\right)^{3} - \left(\frac{T_{B,C}}{100}\right)^{3}\right]}{T_G - T_{B,C}}$$

As result we get

$$q_{rG} = \varepsilon_{B,C}\, \alpha \cdot \left(T_G - T_{B,C}\right) \tag{3.93}$$

The calculated results for different temperatures of gas at TG or TB are listed in **Table 3.10** and shown in **Fig. 3.23**. In the lower section of the table no values are listed. Since wall or bath temperatures are higher than the gas temperature there will be no heat flow from the gas but to the gas. In this case the gas will not heat up but cool down. The α-value increases with higher gas temperature at constant wall or bath temperature. With higher wall temperature, the level of the α-value will be higher, too. The way the α–value was calculated explains this. At first look not a logical tendency. This shall be explained by an example.

Gas temperature shall be 1,200 K, wall temperature = 800 K, ε = 0.8.

We obtain from Table 3.10 the α-value of 98. The resulting heat will be

$$q_{rG} = \varepsilon_W \cdot \alpha \cdot (T_G - T_W)$$

$$q_{rG} = 0.8 \cdot 98 \cdot (1{,}200 - 800) = 31{,}360 \text{ W or } 31.4 \text{ kW/m}^2$$

Now the wall temperature shall be 1,000 K, α from the table = 119:

$$q_{rG} = 0.8 \cdot 119 \cdot (1{,}200 - 1{,}000) = 19{,}040 \text{ W/m}^2 \text{ or } 19 \text{ kW/m}^2.$$

The table is very convenient and simplifies the calculation. We will now use it for our furnace example. The bath area is 8 m x 5 m = 40 m^2, wall temperature was defined to be 1,000 °C or 1,273 K and bath temperature 700 °C or 973 K. From the table we take the α-value closest to our data which will be 132.

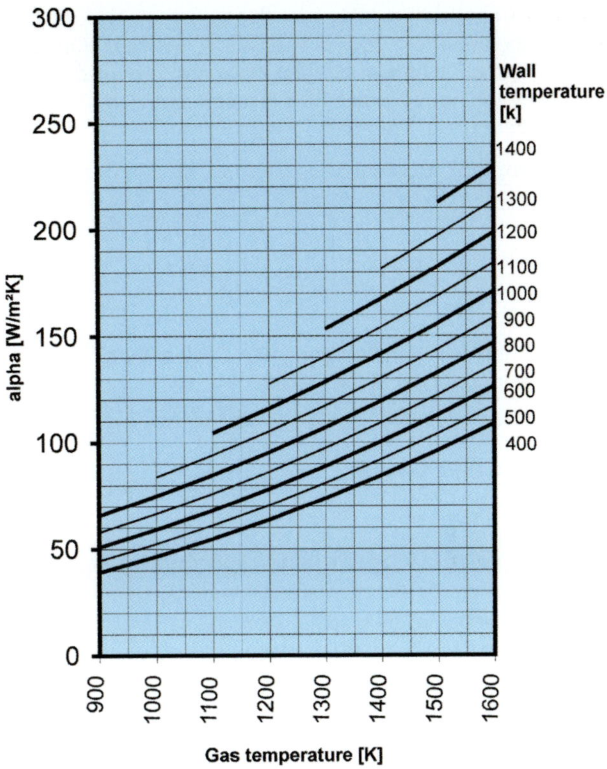

Fig. 3.23: Alpha-values for different wall and gas temperatures

For the heat transfer to the roof we now get

$$Q_{rG} = 40 \cdot 0.8 \cdot 132 \cdot (1{,}273 - 973) = 1{,}267{,}200 \text{ W or } 1{,}267 \text{ kW}$$

The total of energy supplied to the walls will be based on the specific value and the simplified equation. We will accept that the flame thickness may not be more than for the roof. But if we also assume that two burners are installed, the flame thickness will be identical.

$$A = 2 \cdot (2.5 \cdot 5 + 2.5 \cdot 8) = 65 \text{ m}^2$$

$$Q_{rG} = 65 \cdot 0.8 \cdot 132 \cdot (1{,}273 - 973) = 2{,}058{,}000 \text{ W or } 2{,}060 \text{ kW.}$$

The total of energy radiated to walls and roof adds up to 2,060 + 1,267 = 3,327 kW.

This heat flows to the furnace walls to compensate for walls losses and to heat up the refractory lining.

But how much energy flows to the bath having also a temperature of 700 °C or 973 K? There is less heat transferred since the emission factor ε will be 0.3 for the metal bath instead of 0.8 for the refractory walls and the surface area is only 40 m^2.

$$Q_{rG} = 40 \cdot 0.3 \cdot 132 \cdot (1{,}273 - 973) = 475{,}200 \text{ W or } 475 \text{ kW.}$$

We have assumed a fairly low flue gas temperature of 1,000 °C. If we would increase the gas temperature to 1,200 °C or 1,473 K, the heat transfer obtained then would be increased to 876 kW which is almost double. Walls and roof also transfer heat to the bath. The combustion products will, however, absorb the amount of energy that the gas would emit to the bath if it had the temperature of the walls.

For our example the heat transmitted to the bath from the walls amounts to

$$Q_{tot} = A \cdot \varepsilon_{W,B} \cdot 5.9 \cdot \left[\left(\frac{T_G}{100}\right)^4 - \left(\frac{T_B}{100}\right)^4\right] - A \cdot \varepsilon_B \, \alpha \left(T_G - T_B\right)$$

Normally the wall temperature is 100 °C lower than the gas temperature and thus we get for a gas temperature of 1,000 °C or 1,273 K

$$Q_{tot} = 40 \cdot 0.3 \cdot 5.9 \left[\left(\frac{1273-100}{100}\right)^4 - \left(\frac{973}{100}\right)^4\right] - 40 \cdot 0.3 \cdot 145\,(1273-100-973)$$

$$= 705{,}794 \text{ W} - 348{,}000 \text{ W} = 357{,}794 \text{ W or } 358 \text{ kW}$$

The heat transferred by radiation from the walls to the bath of our furnace is now much lower than the heat transfer without combustion products present in the furnace. If we compare the value for radiation 706 kW, we now notice the much lower value of 358 kW. The presence of the combustion products has reduced the radiation from the wall to the bath by close to 50 %. We now can obtain the α-value for these conditions, metal temperature 973 K (700 °C), wall temperature 1,273 K (1,000 °C):

$$\alpha_{rW} = \frac{358}{0.3 \cdot 40 \cdot (1273-973)} = 0.099 \left(\text{kW/m}^2\text{K}\right)$$

3.2.4 Total heat transfer

The total heat transfer in the furnace (**Fig. 3.24**) depends on many factors which are not available in detail. For the calculations important data have to be estimated on the basis of experience. An exact calculation of the heat flow in furnaces, particularly in melting furnaces, is not possible and this will be the case even after more research in detail and use of sophisticated computer models. Also for these models certain assumptions and simplifications have to be made. For practical purposes a calculation based on data available from experience and looking into specific conditions during a furnace cycle will be by far close enough to design and operate a furnace efficiently and economically.

However, to know about the basics of heat transfer it is very important to avoid mistakes and misunderstandings in design and operation. Therefore, we will look at heat transfer mechanisms during some significant stages in a furnace cycle.

During the holding phase the metal temperature must be maintained and the furnace waits for the casting. Metal and wall temperatures are in a steady state, doors and damper are closed. Assuming proper design of the door sealing, the only heat losses are wall losses through the furnace bottom, the upper walls and the roof. These losses are compensated by radiation from the walls and the roof which would slowly cool down. Therefore, heat is introduced by short pulses of the

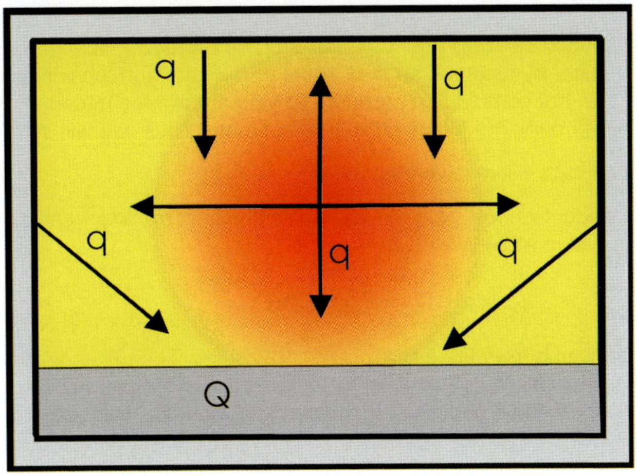

Fig. 3.24: Heat transfer in a furnace

burners which are characterized by long "off"-times and short "on"-times. The controls are set to maintain a roof temperature of 1,100 °C and bath temperature of 750 °C. This would result in an average bath temperature of 735 °C after stirring at the end of the holding period.

Before this stage is reached, solid metal must be charged, heated, melted and superheated. Let us have a closer look at these stages:

Stage 1: The doors are open for charging. During this period the burners are "off" and the walls will cool down to approximately 500 °C (773 K). Metal has the ambient temperature of 20 °C (293 K). After charging, the doors are closed and the burners start firing, thus heating wall and material. The temperature of the combustion products leaving the furnace will be 873 K (600 °C). The combustion products generated by the burner have a much higher temperature close to the burner nozzle. By transferring their energy to batch and walls they cool down. For the calculation, the arithmetic means between high temperature and flue will be used. Therefore, gas temperature = 0.5 x (1,373 + 873) = 1,123 K.

Stage 2: Metal has reached a temperature of 773 K (500 °C), burners operate at full power and gas temperature has reached 1,373 K (1,100 °C), roof temperature is 1273 K (1,000 °C).

Stage 3: All metal is molten and has a temperature of 973 K (700 °C). Gas temperature is at its maximum of 1473 K (1,200 °C), roof temperature has also reached it set-point of 1,373 K (1,100 °C).

We need to discuss the active surface of the batch of material. After charging, a large pile of material will fill the furnace. The surface area of this material will, without doubt, be larger than the bath area. Looking at a block of material having a side length of 1 m, 5 sides will be exposed to the furnace area with a total of 5 m^2. After melting, the material will be totally submerged in the metal bath having a depth of, let us say, about 0.5 m. The area of the former block, which is exposed to the furnace area, will now be about 3 m^2 resulting in poorer heat transfer from the furnace (we will not consider the heat transfer from the liquid metal in this consideration). Bulk material with a bulk density of 800 kg/m^3 will have three times the volume of solid material. This volume will decrease during melting. But the surface exposed to the furnace area is not in relation to the difference in density. Also piles of ingots will not expose their total surface to the furnace area. Considering this, we can fairly assume that the surface area to be taken into account is twice the bath surface as long as the batch is solid. After melting has started, the surface area will decrease gradually. This does not refer to radiation. While the combustion gases flow around the material in the furnace, radiation can only affect the parts of the furnace which are facing the roof. Therefore, for the heat transfer by radiation an area equal to the bath surface will have to be considered. For good order sake, hot combustion products flowing around the material will also radiate to the individual parts of the bulk. But this is difficult to estimate. Thus, we will stick to the bath surface area.

The bath will receive (Fig. 3.24):

Conductive heat from the flue gases (the heat transfer coefficient α for the convective heat remains constant at 35 W/m²K·h)

$$q_{cF} = \alpha_c \cdot (T_G - T_B) \text{ (W/m}^2\text{)}$$

Radiation from the walls

$$q_W = \varepsilon_B \cdot 5.9 \left[\left(\frac{T_W}{100} \right)^4 - \left(\frac{T_B}{100} \right)^4 \right] - \varepsilon \cdot \alpha_{WB} \left(T_W - T_B \right)$$

Radiation from the combustion products

$$q_{rG} = \alpha_G \varepsilon_B \, (T_G - T_B)$$

With these equations the heat transfer during the different stages can be calculated. By adding the specific heat value, we obtain the total heat transferred by convection, wall radiation and gas radiation. From this we can obtain a total α-value by dividing the total specific heat by

$$\alpha_{tot} = \frac{q_{tot}}{\left(\vartheta_G - \vartheta_B \right)}.$$

Of course α_{tot} depends very much on the temperature. For the bath, covered with a thin oxide layer, we have used the emission factor $\varepsilon = 0.3$. For a skimmed shiny bath the α-value must be adjusted. The emission factor for the furnace walls of $\varepsilon = 0.8$ does not change for the aluminium furnaces.

Now the different heat quantities can be calculated. During melting more than 63 % of the heat transfer will be by convection between flue gas and metal and between flue gas and wall. The resulting 55 – 143 W/m²·K for convection and radiation between flue gas and solid metal is very favorable in an aluminium melting furnace. The value of 110 W/m²·K in stage 2 matches the data obtained by experience very well. As discussed before, solid material exposes an area to the flue gases that is much larger than the bath area. With reference to the bath area, the heat transfer factor should be increased by a factor 2 in the lower temperature range where convection is the major means of heat transfer. As radiation increases, only the top surface is exposed and the calculated -value should be applied. The metal charge will fill some space in the furnace which will decrease the height of the furnace and the flame thickness. As soon as the metal starts melting, its surface will decrease steadily until only liquid aluminium is present. The α_c-value for convection comes down to a very low value. But since the temperature level is higher, there will be some compensation for the lower convective heat exchange. The poor heat exchange is partly due to the poor heat conductivity in the liquid metal as discussed before. Depending on the dross skin, more heat will be transferred by radiation. Nonetheless, there will be some heat exchange by convection to the roof and to the metal surface, but even at lower range the burner input has to be cut to avoid overheating of the refractory material. The heat transferred to the refractory lining is important when selecting refractory material and also with regard to the wall losses. The total heat input is restricted to the radiation from the flame and to convection. Since the emission ratio of the refractory material is 0.8, the heat transfer by gas radiation is increased by the factor 0.8/0.3. If only one wall is considered, the influence of the other walls and the bath surface is similar to that discussed for the bath area. For obtaining data for the selection and design of the refractory lining, it is common practice to consider one wall as independent unit.

Looking at Fig. 3.9 we notice that almost 90 % of the energy is required during stages one and two of the furnace. This means this energy is required to obtain liquid metal at melting temperature of aluminium. The remaining 10 % are required to bring the metal to the temperature required for operation. As consequence, we note that most of the heat transfer will take place in the lower temperature range of furnace and combustion products. As mentioned above, the bath is not able to accept the energy offered to it due to the poor conductive heat transport in the liquid metal. Thus, we understand the practical recognition that the time for superheating of the material to the required temperature is very long.

The heat transfer mechanisms in a rotary drum furnace are more complex than for a reverbaratory furnace. In addition to the influence of the burners, the walls do not remain in a static position; they rotate at varying speed. The direct influence of the combustion products on the furnace wall is limited in time. But after part of the wall has reached a position underneath the material batch, heat will flow from the wall to the material. Due to the short time this will only affect refractory layers near the surface. The outside temperature of the steel shell will not change during this period. However, during a lengthy period of operation with constant degree of filling at constant speed some temperature fluctuation of the steel shell may occur. However, there is no steady state condition in the furnace. Even with constant rotating speed and no additional charging, the temperature in the furnace will steadily increase and the steel shell may follow this trend with delayed reaction.

It is hardly possible and very time-consuming to calculate the heat flow and heat transfer in an aluminium furnace for every phase of the furnace cycle. But from our discussion, some consequences for the operation are obvious:

- The set-point for wall and roof temperature should be as high as permissible for the refractory material in order to improve radiation.
- The temperature of the combustion products should also be as high as possible. This will improve heat transfer of the gas to the bath by radiation and convection. However, metallurgical requirements define a lower temperature. High combustion gas temperature can be achieved by energy recycling from the flue gas or by using oxygen-enriched combustion air. A limitation is given by the pollution control (NO_x generation).
- High velocity burners shall be installed in order to improve the heat transfer by convection. The high velocity of the combustion products should be maintained at down-rated burner input by either operating only a reduced number of burners, pulsing the burners (on-off operation) or by installation of smaller burners for the down-rated input.
- Open door situation must be very short in order to avoid cooling of the furnace interior. This can be achieved by reducing charging time by installation of a charging machine and reducing skimming time by also using mechanical tools.
- Improve heat transfer to the liquid metal bath by disturbing the stable stratigraphy within the metal bath by using adequate stirring systems (liquid metal pumps or electro-magnetic stirrers).

Some of these methods are simple to achieve by just changing operation methods and training of the furnace operation personnel. Others require some investment which definitely will pay off, particularly if one considers other benefits such as less metal loss.

3.3 Burner technology

Josef Domagala

Electrical energy or fossil energy is used In the furnaces of the aluminum industry.

In non-electrical heated furnaces the source of heat is usually a burner firing a gas or liquid fuel. The main task of the burner is to change the chemical energy of the fuel in the heat content of the combustion gases and an efficient transfer of a possible large part of this heat to the charge. A proper choice and design of burners for the process, correct estimation of their input as well as the optimal burner location, depending on the furnace design,, have a substantial impact on the furnace capacity and its efficiency.

When designing combustion systems for aluminium furnaces, there are many factors that need to be considered.

a. The function of the furnace
b. Available fuels
c. For melting furnaces there is an optimum between achievable melting rate and a specific fuel consumption. The design of the combustion system for a maximal possible melting rate increases the specific fuel consumption.
d. For the aluminium holder the burner input should ensure heating of melted metal to a required temperature in proper time. For some holders also a certain melting rate is required
e. Fuel efficiency should be looked at: a proper system for a possible efficient furnace operation should be chosen: recuperative, regenerative or oxy-fuel combustion systems generate fuel savings.
f. Type of material to be charged at present and whether that will change in the future. Metal loss is the major factor in a profitable operation. The type of burners and their location should take in account the type of charge to avoid metal losses and to maintain optimal heat transfer to the charge.
g. Environmental regulations. What are the existing limits and what is the potential for tighter limits in the future. What are the taxes for CO, CO_2 and NO_x emissions.
h. Future maintenance costs.
i. Total installation costs.

The final system design should consider all of these points along with the particular site requirements.

3.3.1 Fuels

In the aluminium furnaces mainly rich gaseous or liquid fuels are used. Of all gas fuels the mostly used one is natural gas or LPG (mixture of propane and butane). The most used liquid fuel is fuel oil. In some plants still heavy oil is used.

The fuel characteristic numbers important for the heat balance of furnaces are:

- LHV (low heating value),
- indicates the chemical energy content of the fuel unit with the assumption that the water vapor in the waste gases stays in that form (no condensation),
- HHV (high heating value),
- is the sum of the LHV and the latent heat of water vapor in the waste gas.
- However, this latent heat is lost at practical flue gas temperatures so that LHV is the value used in practice in heat balances of aluminium furnaces.

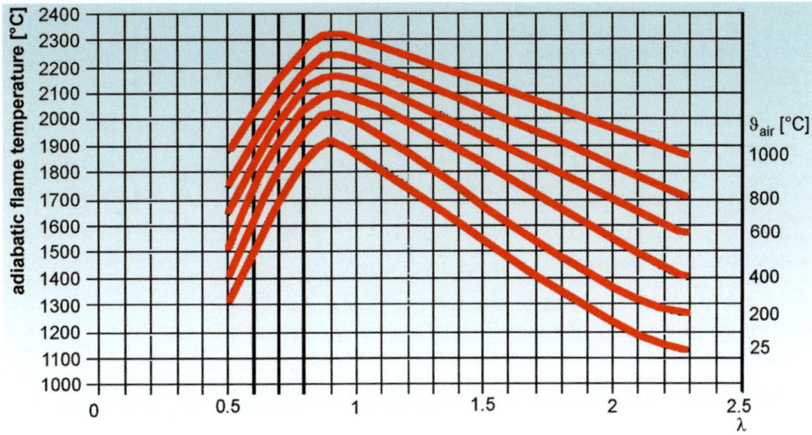

Fig. 3.25: Adiabatic combustion temperature of natural gas

Table 3.11: Properties of different natural gases

Natural gas	Unit	North-Sea	Natural gas	Russia	Netherlands	Natural gas
Components			Mixture H	Natural gas H	Natural gas L	Mixture L
CH_4	Vol.%	85.3	84.2	93.1	82.8	82.3
C_2H_6	Vol.%	8.17	5.63	2.85	3.31	3.08
C_2H_4	Vol.%	-	0.02	—		0.04
C_3H_8	Vol.%	2.84	1.76	0.76	0.64	0.66
C_4H_{10}	Vol.%	0.85	0.55	0.21	0.2	0.21
C_5H_{12}	Vol.%	0.18	0.13	0.08	0.06	0.06
C_nH_n	Vol.%	0.06	0.06	0.08	0.05	0.04
H_2	Vol.%		0.65			1.30
CO	Vol.%		0.05			0.12
CO_2	Vol.%	1.91	1.52	0.28	1.07	1.25
N_2	Vol.%	0.69	5.41	2.64	11.87	10.91
O_2	Vol.%		0.02			0.03
HHV	kWh/Nm^3	12.268	11.236	11.234	10.11	10.074
LHV	kWh/Nm^3	11.104	10.157	11.104	10.157	10.139
Rel. Density	Air =1	0.662	0.647	0.597	0.643	0.635
Air demand (L=1)	Nm^3/Nm^3	10.45	9.58	9.6	8.64	8.6
CO_2 max. (dry)	%	12.2	12.05	11.95	11.7	11.75
Max. Flame Temp. (w/o dis.)	°C	2050	2050	2050	2025	2025

- adiabatic combustion temperature,
- is the maximal temperature which could be achieved theoretically in a fully insulated combustion chamber at a given gas temperature, air temperature and combustion air excess.

Fig. 3.25 shows the combustion temperature of natural gas as a function of the combustion air preheat and the air excess factor.

Table 3.11 shows the characteristic average values for different kinds of the natural gas.

Table 3.12 shows the characteristic values for LPG (propane/butane). The fuel oil characteristic numbers can be seen in **Table 3.13**.

The gaseous fuels are available at the furnace gas supply point with a pressure of 4-6 bar. Gas trains with safety valves and pressure regulating valves are used to reduce the pressure below 200 mbar which is a typical application limit of gas armatures.

The fuel oil is normally available in the ring supply pipeline of the plant with constant pressure so that only a pressure regulator in the line to the furnace burner system is needed.

In other cases an oil tank and a proper oil pump with pressure regulating valve has to be installed.

The combustion of a mixture of propane/butane is slower than natural gas; however, with a higher combustion temperature. This results in higher NO_x-emission of the burners operated with this gas. Propane and butane can crack at a too high temperature in the gas nozzle. With hot air systems the burner gas nozzles should be insulated to avoid the cracking of the gas.

Using heavy oil as a fuel requires the most effort. The heavy oil in a tank has to be preheated to a temperature of approx. 60-70 °C to make the oil ready for pumping.

A proper oil pump with the pressure regulator keeps the pressure in the ring supply pipe constant. The pump has to have a capacity approx. two times higher than the capacity of the installed burn-

Table 3.12: Properties of propane/butane gas

		Propane	i-butane	Butane
LHV	kWh/Nm^3	25.99	33.78	34.31
LHV	kWh/kg	12.87	12.66	12.70
Lo	Nm^3 air/Nm^3	23.90	31.10	31.91
Lo	Nm^3 air/kg	12.15	12.02	12.02
Rel. Density	Air = 1	1.562	2.064	2.091

Table 3.13: Properties of the oil fuels

		E Lextra light	M middle	S heavy
LHV	kWh/kg	11.86	11.40	11.16
Density	Kg/dcm^3	0.86	0.91	0.96
C	%	85.50	85.10	84.70
H_2	%	14.00	12.10	11.50
S	%	0.50	2.50	3.10
Lo	Nm^3 air/kg	11.32	10.83	10.64

ers. The heavy oil has to be heated up to 130 °C at the burner to decrease its viscosity below the limit required by the oil atomizer. Most of the oil burners require a hot atomizing agent which can be the water vapor or the compressed air preheated to approx. 180 °C.

Table 3.13 shows the elementary analysis and the characteristic numbers for different types of oil. It is important to know the nitrogen content of oil and because it is a source of fuel-bounded NO_X emission (more than 95 % of nitrogen content in the fuel is turned into NO_x).

3.3.2 Burners

There are different classifications of the burners depending on the principle of operation, on

a stabilization mechanism or on an application and others. Generally based on the principle of operation, burners can be divided in two groups:

- premix burners

  are burners with mixing of the gas and air before they reach the outlet of the burner nozzle. In a typical premix burner a gas injector generates suction and transports a proper amount of air in a mixing zone. These types of burners have low capacity range and are used in the aluminium industry for refractory drying systems and heating of casting channels or metal filters.

- nozzle mixing burners

  are burners where the mixing of the fuel and air occurs after leaving the gas/air nozzle. The design of the gas and air nozzle allows different ratios of gas/air output velocity, different distances between gas and air stream and the direction of the gas and air streams. The proper use of these parameters enables the design of a burner which optimally meets the process needs.

Because most of the burners used in aluminum furnaces are nozzle-mixed burners, these burners only will be further considered.

Generally, all burners for aluminum furnaces should have the following properties:

- stable flame at each furnace operation stage
- proper characteristic of the flame (length, diameter, impulse, emissivity)
- high turn-down (range of decrease of the burner capacity with straight-forwarded flame)
- low NO_x and CO emissions.

3.3.2.1 Gas burners

The gas burners are operated with cold or preheated combustion air (up to 600 °C in recuperative systems and up to approx. 1,000 °C in regenerative burner systems).

Most of the burners used in the aluminium furnaces are parallel stream burners.

The gas and the combustion air come parallel to the burner axis in the burner port block (tile) where the mixing process is initiated. Especially the burners with ceramic nozzle are well suited for the melting and holding furnaces of the aluminium industry.

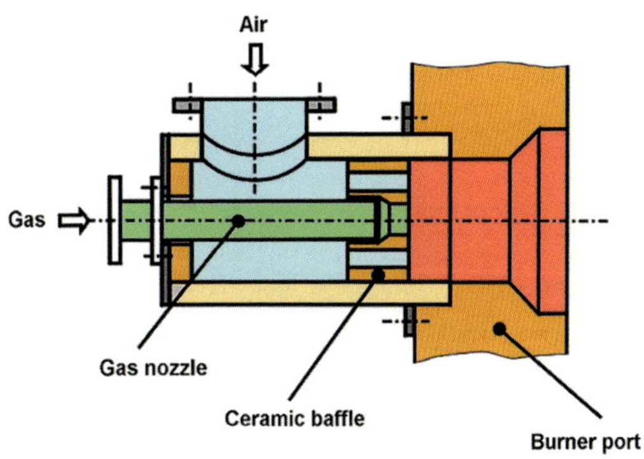

Fig. 3.26: Gas burner with a ceramic air nozzle (Source: Bloom)

The main components of such burners are shown in **Fig. 3.26**. The ceramic air nozzle (baffle) provides support to the gas tube and is a radiation shield between the flame and the internal burner parts. The baffle holes and the port designs determine the flame characteristics, such a shape, luminosity, impulse and the emissions of the flame.

The baffle passages are essentially nozzles which create a jet effect on the exit side. A jet exiting a nozzle creates a recirculation zone and a low pressure area at the exit. The jet phenomenon anchors the flame in the port, ensures thorough mixing of the air and fuel and provides the energy to recirculate the combustion gases in the port. The design of the air nozzle determines mainly the flame properties because the air amount going through the burner is approx. ten times higher than the amount of the gas.

The following factors have influence on the achieved flame properties:

- direction of the media leaving the nozzle in the combustion area: straight on or with a spin.
- A spin of combustion air shortens the flame: for example, the combustion air spin of 25 ° can result in a shorter flame by up to 50 %.
- output velocity of gas and air. The high output velocity generates high impulse of the flame which causes stronger recirculation of the furnace gases and high convective heat exchange. The strong recirculation of furnace gases in the flame also lowers NO_x emission.
- gas/air output velocity ratio and the distance between gas and air openings control the mixing process: these parameters have an impact on the flame length and the level of NO_x emission
- gas or air staging which means dividing of gas/air into two streams: primary and secondary one. This method of a combustion NO_x emission suppression is state-of-the-art today.

Fig. 3.27 shows the principle of the air staging in Low-NO_x baffle burners. The combustion air is injected in the air nozzle in two locations (air staging in primary and secondary air) while all of the fuel is injected through a single connection. In the first stage only the primary air is involved in the mixing process, in the second stage the other part of the air becomes involved in the mixing process. The purpose is to minimize locally the flame temperature and control the chemical environment within the flame, thus reducing NO_x formation. The air and fuel output velocities from the baffle have sufficient energy to create a jet effect which stabilizes the flame in the burner port and recirculates furnace gases into the flame while also producing good air/fuel mixing characteristics.

A very important part of a burner is the burner port block. Its length and shape have a large influence on the properties of the flame. Normally, the burner tiles are not supplied with the burners. They are fabricated (casted or rammed) from high heat-resistant (refractory) material directly at site in the furnace wall (according to the drawing of the burner supplier).

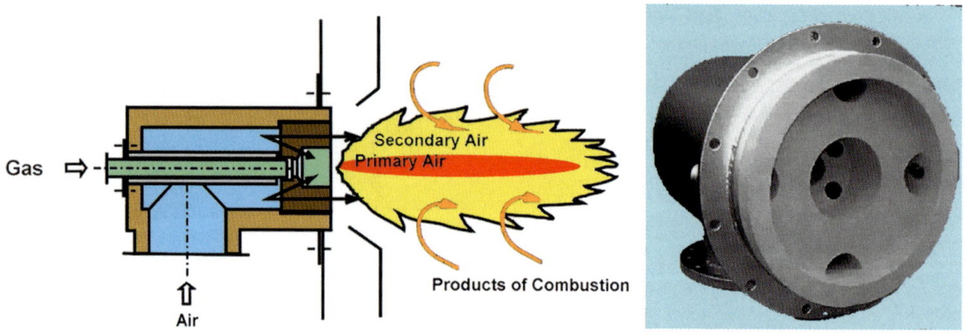

Fig. 3.27: Low-NO_x gas burner with air staging (Source: Bloom)

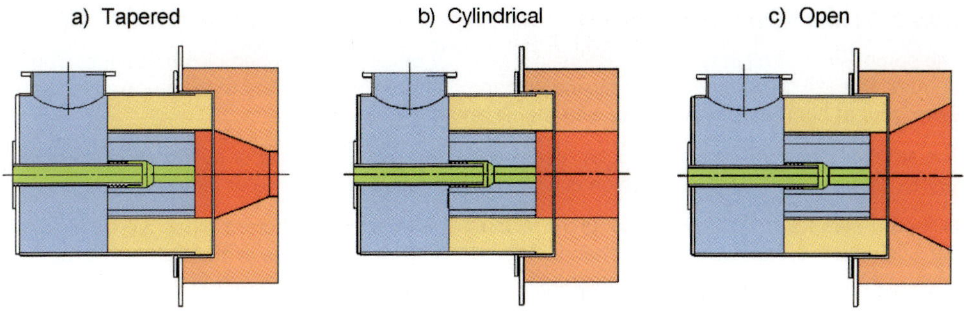

Fig. 3.28: Different shapes of burner port block

Fig. 3.28 shows different shapes of the burner tiles. In most burners today tapered or open tiles are used. The tapered tile increases the output velocity of the combustion gases and causes very strong recirculation of the furnace gases in the flame. However, they generate higher NO_X emission than burners with an open burner block. The NO_x emission can be suppressed by using gas staging.

The open tile allows the recirculation of the furnace gases up to the root of the flame at the surface of the air nozzle. With such port block the outlet velocity of the combustion air from the air slots of the nozzle is very high which ensures enough impulse of the combustion gases in spite of the open tile. The NO_x emission with such port blocks is lower than that of burners with a tapered port block.

Especially the ratio of a port diameter to its length (D/L) has a strong impact on the generated NO_x-emissions. Generally, in Low-NO_x burners this ratio is below 1.

Continuous falling below the valid limits for NO_x emission forced the burner designer to develop new Ultra Low-NO_x burners which are able to maintain the extremely low NO_x limits also in applications in regenerative burner systems (such burners are described in chapter 3.3.4.1).

Fig. 3.29 shows a burner suitable for smaller melters and aluminium holding furnaces. It is equipped with a metal gas and air nozzle. The burner can be used with port blocks of different shape. It has large turndown, good flame shape and emissions characteristics. The burner is also equipped with direct electrical ignition and ionization detection of the flame.

Fig. 3.29: Burner for small melters and aluminium holding furnaces
(Source: G. Kromschroeder AG)

3.3.2.2 Oil and dual fuel burners

The combustion of oil in furnaces occurs in a gas that envelope small particles of the fuel. The oil burners atomize the fuel before its mixing with the combustion air in very small drops which, after vaporizing at high temperature, mix with the air and combust.

The atomization of the oil can occur by means of a mechanical atomizer or by means of special oil lances with an atomizing agent which is normally compressed air or water vapor.

In the burners for the aluminium industry oil lances (atomizer) of different types are used which are built in the burners. A special mechanism for fixing of the lances in the burner allows their dismounting for maintenance or changing the fuel modus to gas in a very short time. In baffle-type burners it is possible to use the same burner equipped in a proper nozzle and an oil lance as an oil or dual fuel burner (oil or gas, oil and gas).

Fig. 3.30a shows a burner in a high pressure oil lance of a quick disconnect version. The burner can be used as a gas, oil or as dual fuel burner.

Oil burners can be equipped with low or high pressure atomizer. Low pressure atomizers are used for light oil (oil pressure of 2 bar and air pressure of 120 mbar are required).

A high pressure atomizer, using compressed air or water vapor with pressure of 5 bar as atomizing agent, can be applied for each type of oil. For light oil an atomizing agent pressure of 2-2.5 bar and oil pressure of 2 bar is required. Heavy oil burners need 4.5-5.0 bar oil pressure and the same pressure of an atomizing agent. The heavy oil has to be preheated to approx. 130 °C for the burner. The atomizing agent has to have the same pressure for the burner as oil.

If compressed air is used as atomizing agent, a preheating to approx. 180 °C is required.

It is of advantage to use oil burners with a quick retractable oil lance as shown in Fig. 3.30. The simple mechanism allows to dismount the oil lance and close the oil and atomizing air opening for maintenance or for changing to gas operation modus in a few minutes only (the oil and atomizing

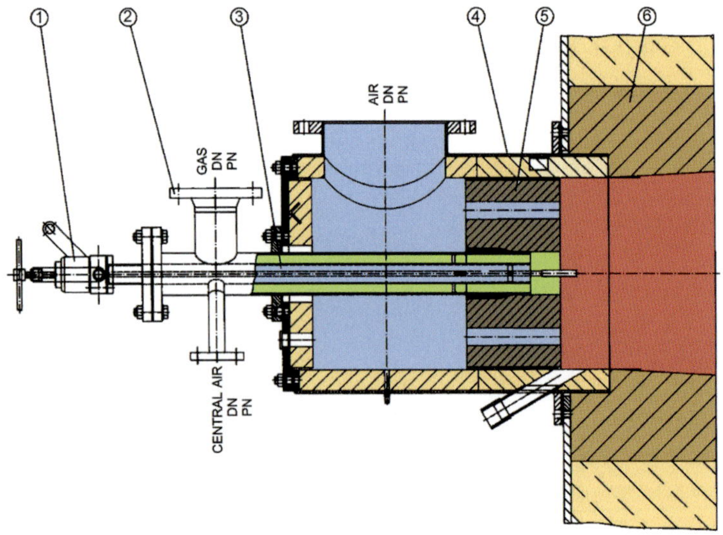

1 - Quick retractable oil lance, 2 - gas connection, 3 - central air, 4 - housing, 5 - baffle, 6 - port block

Fig. 3.30a: Dual fuel burner (Source: Bloom Engineering)

agent are connected to the lance head fixed to the burner and there is no need to disconnected them).

3.3.3 Combustion systems

There are different methods for optimizing combustion systems with the aim to achieve possible high capacity with a low energy consumption. The systems with cold combustion air can be improved by using recuperators which preheat air to 450-600 °C or using regenerators which make possible a preheat of the combustion air to temperatures only 150-250 °C lower than the temperature of furnace gases.

3.3.3.1 Efficiency of combustion systems

The combustion efficiency and the furnace thermal efficiency are considered in detail in chapter 3.1.3.1. Generally the combustion efficiency indicates which part of the total heat available stays in the furnace. The furnace thermal efficiency indicates which part of the total heat available is transferred to the charge.

For a rough comparison of different combustion systems it is sufficient to consider their combustion efficiencies.

The value of energy savings with a recuperative burner system versus a cold air system can be estimated approximately by applying the formula:

$$\text{Savings (recu\_sys versus cold\_air \_sys)} = \left(1 - \frac{\text{ETA cold air}}{\text{ETA recu}}\right) \cdot 100 \quad [\%] \tag{3.94}$$

An important issue is the choice of an "additional reserve energy input" which is added on the calculated furnace fuel input value. This factor can be between 1.1 to 1.35 and should be carefully chosen depending on the type of furnace and the planned furnace operation, because it is the decisive factor for the achievable furnace capacity and the specific fuel consumption.

A too high furnace input can result in high fuel consumption.

A too low furnace input is of advantage for the fuel consumption but it can be the reason for not achieving the planned capacity of the furnace.

3.3.3.2 Comparison of the efficiency of different combustion systems

Depending on the process, the furnace and the energy consumption requirements, different types of combustion systems are used:

- aluminum holding furnaces, metal filters, casting channels equipped with small capacity burners are operated with cold combustion air.
- with larger units, such as aluminum melting furnaces, recuperative or regenerative burner systems are used where the waste gas energy from the furnace is used for the preheating of the combustion air. The recuperative burner systems (air temperature of 450 °C) are able to save approx. 25 % of energy compared to cold air systems, whereas the regenerative burner system can achieve further savings higher than 20 %, compared to the recuperative systems and more than 40 % compared to the cold air systems.

Recuperative systems

Radiation or tube-convection can be applied for the aluminum furnaces.

The radiation recuperators ensure long service life in difficult conditions in melting furnaces. If chlorination is not used in a furnace, also central tube-convection recuperators can be used

(**Fig. 3.30b**) without any problems. With a realization of the processing concept and recuperator protective measures, an equally high heat recovery can be obtained throughout the entire melting process cycle.

A periodic cleaning of the heating surfaces during operation is necessary to avoid the formation of dust during the melting down of aluminium scrap. In addition, the recuperator can be designed so that it can be rolled in and out for cleaning purposes. This simplifies disassembly and assembly considerably. The tube-convection recuperators can also be installed where chlorination is performed using a bypass which makes it possible to bypass the recuperator during this phase.

From the point of efficiency, a tube-convection recuperator is more useful since air preheating temperatures can be obtained throughout the entire melting process.

Fig. 3.30b: Recuperator for an aluminium melting furnace (Source: Peiler)

Regenerative systems

Fuel-savings are still the main justification for a furnace conversion to regenerative firing.

Energy savings by higher air preheating are quite evident. **Fig. 3.31** shows the savings compared to recuperative and cold air firing.

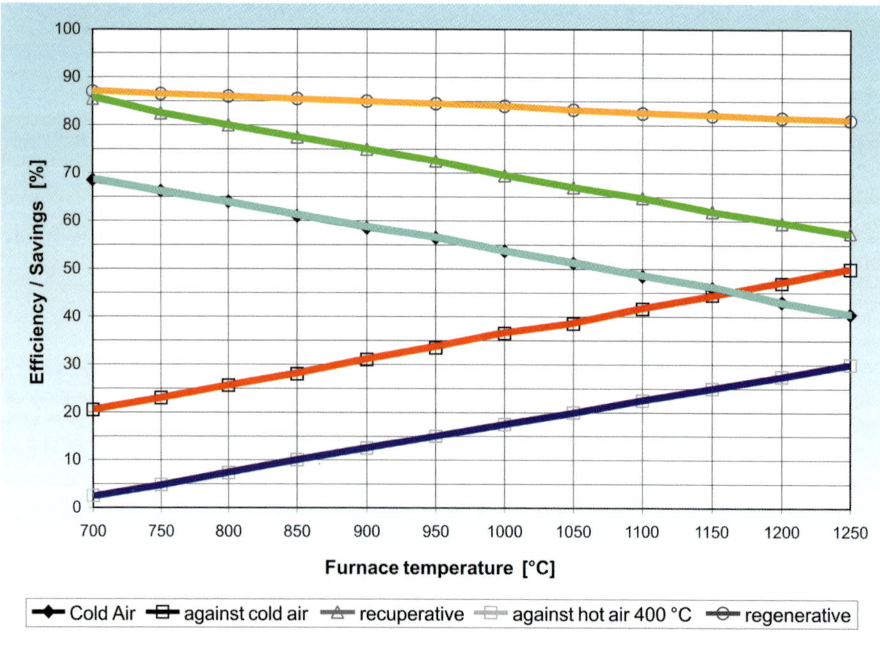

Fig. 3.31: Comparison of system efficiencies

A payback calculation, however, cannot just be based on a single point of the furnace operation but must take the entire cycle into consideration. Fuel savings are the highest at max. furnace temperature. A normal dry hearth scrap melter, however, especially after charging, works up to 25 % of the melting cycle at lower temperature. In this stage also air preheating is lower and fuel savings are lower. Fuel savings thus must be averaged and weighted over the entire melting cycle.

A different situation exists in two chamber furnaces for contaminated scrap.

If a regenerative burner system is used in the heating chamber, which is operated at the max. temperature, the total fuel savings in the chamber can be significantly higher than for the conventional melter (where the whole melting cycle from low to high furnace temperature has to be considered).

Real money-saving potential, when converting existing furnaces, is much higher than one can expect from theoretical calculations of efficiency improvement only because when rebuilding to regenerative firing usually also the furnace control system is improved.

Thus, after improving the poor air/fuel ratio control and poor furnace pressure control, high additional energy savings can be expected.

Many customers have reported fuel savings in excess of 50 % after a rebuild.

Table 3.14: Comparison of heat input for different firing systems

		FIRING SYSTEM		
		cold air	recuperative	regenerarive
Heat input	**MW**	**5,800**	**4,300**	**3,400**
Furnace temperature	°C	1,100	1,100	1,100
Air preheat	°C	20	450	950
Waste gas temperature enter. recu/regen	°C	-	900	1,100
Waste gas flow from combustion	Nm^3/h	6,400	4,800	3,800
Cooling air flow	Nm^3/h	-	1,300	no need
Total waste gas flow after recu/regen	Nm^3/h	6,400	6,100	3,800
Waste gas temperature before chimney/filter	°C	1,100	480	330
Total waste gas flow at temperature	m^3/h @°C	32 000 (1100°C)	17 000 (480°C)	8 400 (330°C)
FILTER				
Max. filter entry temperature	°C	200	200	200
Filter cooling air	Nm^3/h	42,000	9,700	2,750
Total waste gas flow into filter	Nm^3/h	48,400	15,800	6,550
Emissions				
NO_x concentration	mg/Nm^3 (2 $\%O_2$)	250	300	350
NO_x freight	kg/h	1.6	1.44	1.33
CO_2 freight	kg/h	1,140	850	670

The important factor in selecting a combustion system is the total amount and the temperature of flue gases to a chimney or to a filter system. Cold air-fired furnaces exhaust waste gases with

~ 1,050-1,100 °C and hot air-fired furnaces still have ~ 500 °C flue gas temperature exiting the recuperator. The waste gas temperature behind the regenerative burner system is normally below 300 °C. Because of the lower amount of cooling air with regenerative burner systems in the amount of 30 – 35 % compared to a cold air system (filter entry temperature normally is limited to ~ 200 °C), the filter can be smaller and substantially cheaper. **Table 3.14** shows the difference between various firing systems.

A rather new and even better for regenerative conversion is the expectation of a CO_2 emission tax. CO_2 emissions drop in the same percentage as fuel reduction occurs, i.e. a 30 % reduction in fuel consumption also means a 30 % reduction in CO_2 emissions and NO_x freight.

Part of the fuel cost reduction is offset by higher electricity cost due to the necessary flue gas suction fan and a higher air fan pressure. Some additional cost occurs also from the consumption of compressed air for the operation of the cycle valves but this is negligible compared to electricity costs. Regenerative burners, due to their excellent efficiency, can achieve with new aluminium furnaces a specific fuel consumption lower than 550 kWh/t aluminium.

3.3.4 Oxy-fuel burner systems

Thomas Niehoff

Oxy-fuel burners, which use pure oxygen instead of combustion air, are used in the aluminium furnaces especially in the case of dirty charge, for example in rotary furnaces for contaminated scrap or for the utilization of dross. The oxy-fuel burners have a fuel savings potential comparable to that of regenerative burners. However, they generate the added cost of purchased oxygen. For this reason it is of importance to consider the total savings in money not regarding fuel when a decision should be made in favor of an oxy-fuel combustion system.

There are the following considerations before applying it to production furnaces

a. A very high flame temperature of more than 2,700 °C can be of advantage for heat transfer. However, it also can cause hot spots in the aluminium bath (dross generation can increase)
b. In the initial part of the melt cycle in direct charged melters, convective heat transfer through the pile is of great importance. The transfer rate is based on the volume, temperature and velocity of the gases. With oxy-fuel, the volume of gases is considerably reduced.
c. It must be considered that the money-savings depend on the ratio "fuel price/oxygen price" and the ratio of "new productivity/old productivity". If the price of oxygen is too high, the fuel still can be saved but a money loss is generated
d. No NO_x generation (due to lack of N_2) is impractical due to the air leakage around doors etc. which includes nitrogen (some NO_x will be produced).

However, the oxy-fuel combustion systems have also some important advantages such as low investment cost and with a low oxygen price, cost-efficient operation of the furnaces where the installation of recuperative or regenerative would be not possible due to the type of charge. Please also see chapter 3.1.12.

3.3.4.1 Oxy-fuel burner principle

Oxy-fuel burners bring together the fuel and the oxidant. Oxy-fuel burners are connected to flow control and combustion control equipment to ensure safe and controlled operation. Oxy-fuel burners are typically burner tip mixing burners. Fuel and oxidant are mixed at the tip and the flame

Fig. 3.32: Oxy-fuel burner firing – blue part of oxy-fuel flame (Source: Linde AG)

is anchored at the burner pipe. Pre-mixed oxy-fuel burners are not common. When pre-mixing oxygen and fuel, the hot oxy-fuel flame can develop in the mixing part of the burner with a high potential to melt the metal of the burner body. Oxy-fuel burners often consist of several tubes (two or more) that bring the oxygen and fuel to the tip. Oxy-fuel burners can have several nozzles or lances to add fuel or oxygen at different gas velocities and different locations to create special effects such as flat flames, flameless flames …etc.

Oxy-fuel burners are designed for specific purposes and characteristics. The flame length of an oxy-fuel burner can be adjusted to a combustion space or furnace dimension by balancing the gas velocities with the burner designs. The characteristics of an oxy-fuel flame will be adjusted to process specifics. When firing in a furnace with powdery or dusty charge materials a low gas velocity (v = 10 to 50 m/s) range will be chosen to minimize the dust carryover into the flue gas system. Soothing flames with local reducing and oxidizing sections are designed to create special heat transfer effects. High momentum flames (v = 100 to 200 m/s) are created to achieve sharp and focussed flames to ensure directed heating. Burners with staged flames introduce oxidant and fuel at different locations and at different velocities to achieve a delayed mixing of the reactive gases. This will result in lower flame temperatures, extended flame volumes and reduced emissions (NO_x).

The oxy-fuel flame is brighter than an air fuel flame. An air fuel flame is typically extended in volume and diluted with nitrogen. Oxy-fuel flame is often blue at the anchor (**Fig. 3.32**) and develops a bright and shiny flame with sharp contours. High momentum flames are bluer in color and low momentum flames are more yellow to white in color.

3.3.4.2 Flameless oxy-fuel burner

Standard pipe in pipe oxy-fuel burners can create a local and hot flame. In a large furnace with one or more small oxy-fuel burners local overheating can occur. When melting aluminium in a reverberatory furnace without bath movement a local hot spot can cause aluminium oxidation and overheating of the refractory materials. In order to overcome local temperature issues, flameless oxy-fuel flames have been developed to create a more even heating and melting and to minimize metal losses and emissions.

Linde Gas developed the "Low Temperature Oxy-fuel Combustion Technologies" (LTOF) for melting aluminium (**Fig. 3.33**). It is a further development of the conventional oxy-fuel combustion technology which has been used in aluminium rotary furnaces.

In Low Temperature Oxy-fuel Combustion Technologies the flame temperature is reduced by staging the combustion process (**Fig. 3.34**). The flame is no longer seen by the human eye. The flame extends in volume and shape.

The dispersed flame still contains the same amount of energy and covers a larger volume. LTOF provides a more uniform heating and melting, avoiding hot spots and dross formation during melting. The flame temperatures achieved with LTOF combustion are comparable to air fuel flame temperatures. LTOF Combustion Technologies allow adjusting any flame condition between bright oxy-fuel flame and flameless oxy-fuel flame (**Fig. 3.35**).

Fig. 3.33: Low temperature oxy-fuel burner (Source: Linde AG)

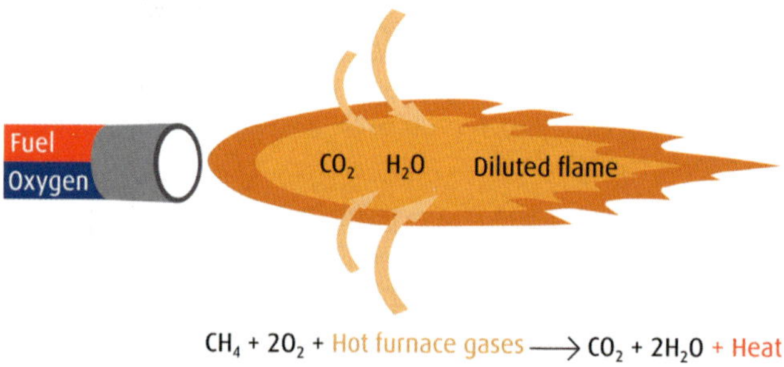

Fig. 3.34: Principle of staged combustion and flame dilution

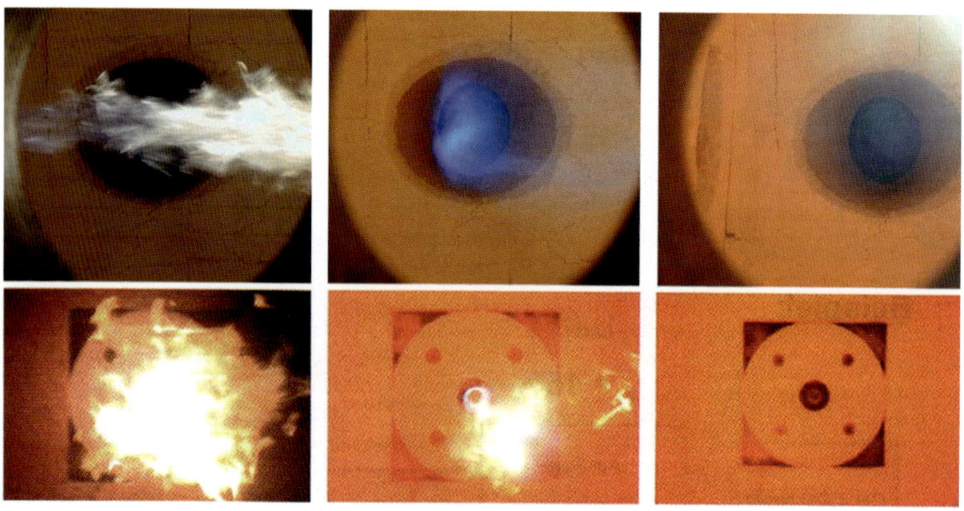

Fig. 3.35: Transition of oxy-fuel flame (left) to flameless (right) (Source: Linde AG)

3.3.4.3 Aluminium melting with oxy-fuel

Oxy-fuel combustion systems can be specifically tailored to achieve the best results in terms of energy efficiency, emissions and melting cost. Flames of oxy-fuel combustion systems can vary widely to give best results for your melting operation. Oxy-fuel flames can be bright and luminous, round, flat, long, short, flameless (Fig. 3.35), low temperature and high temperature.

Air-oxy-fuel combustion systems are flexible in their way to adjust to specific needs of a melting process or furnace. The flame length, luminosity and convective heat can be adjusted by the air to oxygen ratio, as desired. An extra oxygen lance can be used to burn out combustibles in the furnace chamber.

Bright and radiant oxy-fuel combustion systems are optimum for fast melting in rotary furnaces. There they achieve economic and efficient melt results. Oxy-fuel combustion systems can be equipped with an extra lance to support and enable combustion of charged organics.

The Linde combustion process incorporates oxygen lancing technology with burner combustion management to effectively combust emissions (i.e. volatile organic compounds) inside the furnace. By monitoring flue gas compositions, the combustion process burns off carbon-based materials in the charge and uses this energy to reduce overall burner heat input to the furnace.

Low temperature oxy-fuel (LTOF) combustion technology is designed to boost melting capacity within a converter or reverberatory "box-type" and rotary drum furnace. The uniquely designed burner allows for the recirculation of flue gases into the burner mixing zone, which dilutes the oxygen concentration and slows down the combustion reaction, thus creating a lower flame temperature comparable to air-fuel technology. The LTOF flame characteristics result in uniform furnace temperatures, elimination of furnace hot spots, reduced fuel consumption, reduced emissions and improved metal yields.

Rotary drum furnaces for melting aluminium can be regarded as ideal oxy-fuel burner operation (**Fig. 3.36**). Oxy-fuel burners in the traditional fixed axis rotary furnaces can be installed in two ways. A "single pass" operation is when the oxy-fuel burner is installed in the furnace door. Then the burner is firing towards to flue gas end. The flame and the heat pass the furnace volume one single time before exiting through the flue end. This installation has the advantage that most of the heat is provided at the furnace door where the aluminium is charged. Another option is to install the burner at the flue gas end. Then the burner will fire towards the door at an angle and the combustion gases will pass the furnace volume twice – "double pass". The advantages of a burner installation in the flue end of the furnace: during charging the burner can stay lit on low or medium fire and the combustion gases have the ability to transfer heat while taking the double pass turn through the furnace. Air-fuel burners could not be installed in the flue end of a rotary furnace due to higher flue gas volumes and larger space required for installation. When converting a fixed axis rotary furnace into oxy-fuel several process modifications are required to provide the best environ-

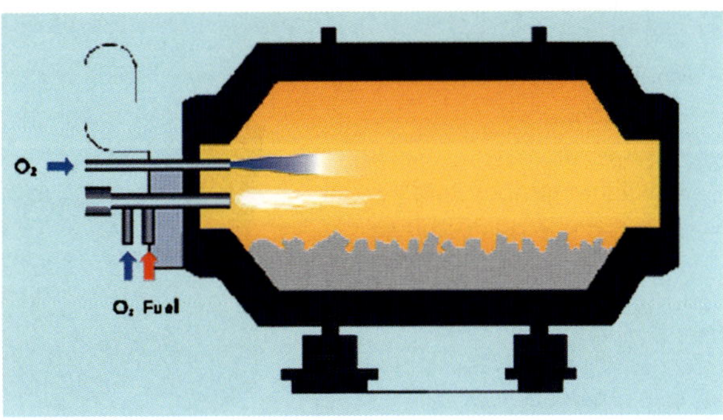

Fig. 3.36: Typical rotary drum furnace, stationary design (Source: Linde AG)

ment and to maximize the oxy-fuel benefits. The furnace door needs to be sealed and the pressure conditions inside the furnace have to be optimized in a way that the drastically reduced flue gas does not result in large air infiltration into the combustion chamber. In case cold air is pulled through the process it can destroy the benefits that oxy-fuel combustion would deliver. In case the Linde combustion process to burn off organics from the charge should be used, then a flue gas sensor and automatic and manual process control should be installed together with a control logic in the burner control software. The flue gas sensors used are often optical or optic-acoustic and sometimes laser systems. Laser systems can be expensive and require regular service. The sensor indicates when combustible gases exit the furnace and triggers the oxygen flow through oxygen lances and/or oxy-fuel burner.

Today the fixed axis rotary drum furnace is a tiltable rotary furnace (TRF) as described in detail in section 2.1.3.5. A tiltable furnace has only one opening where the charge door is located. A burner and/or oxygen lances can only be installed on that side of the furnace. The benefits, that oxy fuel operation offer, are well understood and documented. A payback time of less than one year of the oxy fuel combustion equipment can be achieved depending, however, on the type of scrap to be processed.

Linde Gas has even started to equip TRF's with flameless oxy-fuel burners to maximize the benefits of oxy-fuel operation. The idea here is fuelled by the desire to minimize metal oxidation and thermal efficiency even further by employing a spacious and low temperature flame. Oxy-fuel combustion in a TRF can also be paired with the Linde WASTOX® combustion process. In such a case, a sensor is installed in the flue gas system and the signal is used to set the oxygen requirement and timing to burn off the addtional organics from the charge. Very good results are achieved with a comparatively small content of (3 to 7%) organics in the charged material. An organic content of more than 10 % is often difficult to handle since the energy contained in the feedstock material is too high.

Reverberatory furnaces (**Fig. 3.37**) are often used for clean scrap with low specific surface area and often are fired with regenerative air fuel combustion systems. The heat is mainly transferred via radiation (once the metal is molten) from the burner flames and the hot walls. Bath movement enhances metal yield and energy efficiency. When installing oxy-fuel burners in a box type furnace it is very important to avoid local overheating. LTOF combustion Technologies has demonstrated to be a good concept for melting aluminium in a reverberatory furnace (**Fig. 3.38**).

Flameless oxy-fuel burners can be installed in reverberatory furnaces together with air fuel or air-oxy-fuel fired burners. Flameless oxy-fuel burners in reverberatory furnaces have the following features:

- high energy density in furnace
- high energy efficiency
- small wall temperature gradients
- avoiding of hot spots
- low NO_x emissions
- improved metal bath heat transfer
- conditions
- reduced aluminium oxidation
- low flame temperatures
- reduced flue gas volumes

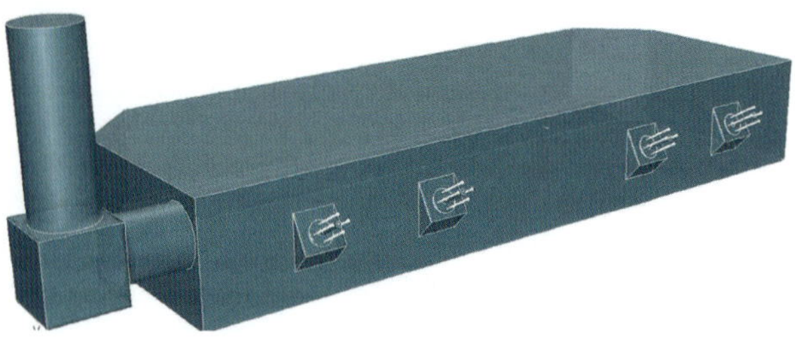

Fig. 3.37: Principal burner arrangement for a reverberatory furnace (Source: Linde AG)

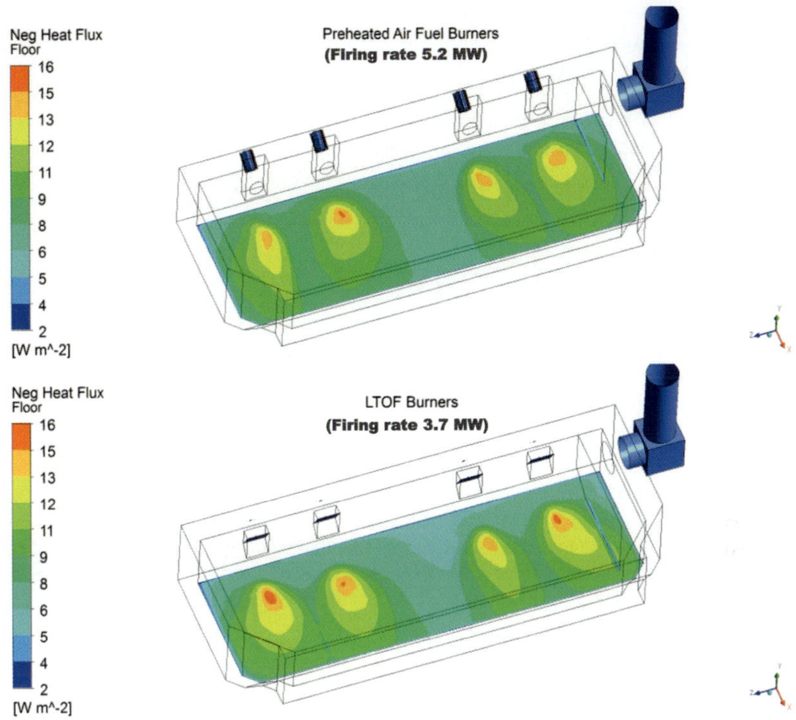

Fig. 3.38: Heat flux contours of preheated air fuel burners and oxy-fuel burners (Source: Linde AG)

3.3.4.4 Emissions

Energy savings of 30 % and in some cases even more can be realized when changing from air-fuel to oxy-fuel. The nitrogen content in the air is close to 79 %. In air-fuel combustion the inert nitrogen gets heated up from ambient to furnace temperature and then leaves the process with the flue gases. This takes up a substantial part of the energy. CO_2 emissions are directly related to the quantity of fuel needed and the composition thereof. For instance hydrogen (H_2) could be used as a fuel. The flue gas from hydrogen burned with oxygen ($2H_2 + O_2 \rightarrow 2H_2O$) would only be water vapor (H_2O) and no CO_2. Hence, the amount of carbon in the fuel is directly related to the CO_2 emissions. When the natural gas/ energy per ton of metals is reduced then also the CO_2 emissions are reduced, since they are directly related to the fuel usage. Gas or oil with a high carbon content and low in volatiles will mainly generate CO_2 when burned. Subject to the hydro-carbon ratio of the fuel used and fuel mix used, there will be a very specific CO_2 emission in [kg of CO_2 per volume or mass of fuel]. Per kg carbon 3.67 kg CO_2 are generated and per kg methane (CH_4) 2.75 kg of CO_2 are generated. Please refer also to section 3.1. When fossil fuel energy is saved this is then directly linked to the CO_2 emissions as shown. Approximately 331 kg per t of aluminum can be saved when converting an air fuel fired combustion process with an overall thermal efficiency of 15 % to oxy-fuel with an overall thermal efficiency of 50%.

The CO_2 emissions of a process are directly linked to the thermal heating or melting efficiency when firing fuels containing carbon. **Fig. 3.39** shows how the CO_2 emissions are being reduced when improving the thermal efficiency of combustion methane (CH_4). Oxy-fuel processes can reach thermal efficiencies of more than 60 %.

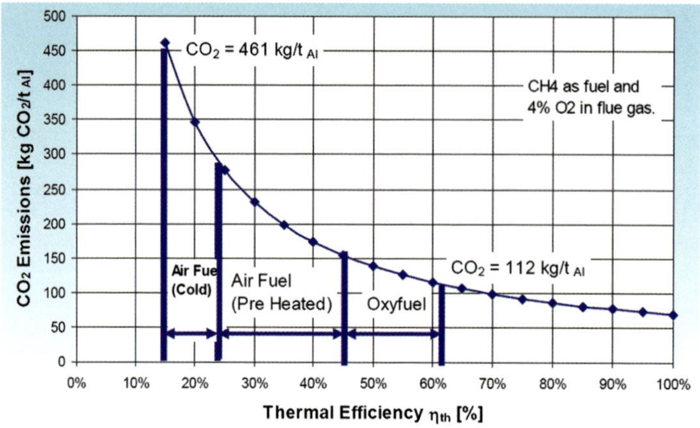

Fig. 3.39: Thermal efficiency of different firing systems

Emissions of CO typically occur when combustion is incomplete, i.e. too much fuel or not enough oxygen. Combustion processes use a ratio of fuel to oxidizer (air or oxygen) to ensure complete combustion and no CO emission. Oxy-fuel increases the retention time of combustibles in the hot zones and the reactivity of fuel and oxidizer because of missing nitrogen and reduced gas volumes. CO emissions are reduced when converting from air/fuel to oxy-fuel.

NO_X is generated in flames where nitrogen and oxygen is present. At higher temperatures (above 1,000 °C/2,000° F) the NO_X generation is higher due to thermal NO_X. Fuels can contain nitrogen. Nitrogen-free fuel combusted with oxygen will burn without NO_X. Industrial furnace conditions are not ideal. Air leaks and nitrogen-containing fuels are very common. Low NO_X burners are used to lower the NO_X emissions. Low NO_X burners stage the combustion process and reduce the flame temperature.

NO_X generation is affected by many parameters. In **Fig. 3.40** the influences of different firing technologies are shown. The more oxygen is available in the flue gas, the more NO_X can be formed. Regenerative air-fuel burners and conventional oxy-fuel burners show similar NO_X counts in Fig. 3.40. The flameless oxy-fuel technology clearly generates less NO_X compared to the other burners. NO_X also depends upon furnace temperature and very specific furnace conditions.

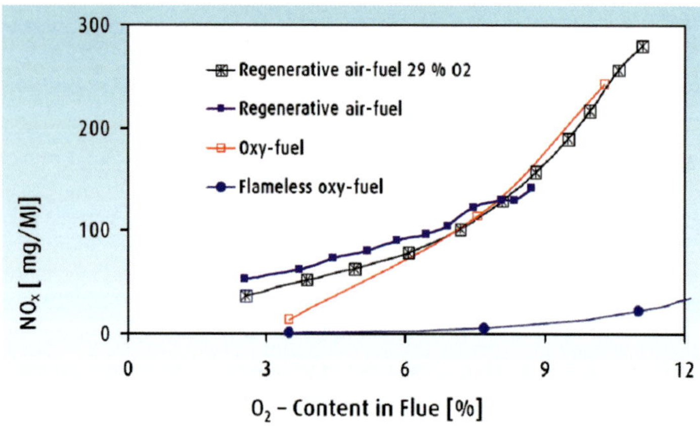

Fig. 3.40: Content of NO_X in flue gas for different firing systems

3.3.5 Regenerative burner systems

Josef Domagala

Regenerative burner systems are state-of-the-art today in aluminum melting furnaces with a low energy consumption. A lot of new aluminum melting furnaces are equipped with regenerative firing. Regenerative burner systems make it possible to reduce the heat consumption of a furnace by more than 40 % against cold air firing and more than 20 % against recuperative (450 °C) air preheat system.

3.3.5.1 System principle

The high air preheat and, consequently, low flue gas exit temperature of a regenerative system is due to direct heat exchange between the hot waste gas (respectively the cold combustion air) and an exchange medium. Contrary to recuperative heat recovery, air and waste gas are not separated by tubes. **Fig. 3.41** shows the basic design of such a burner.

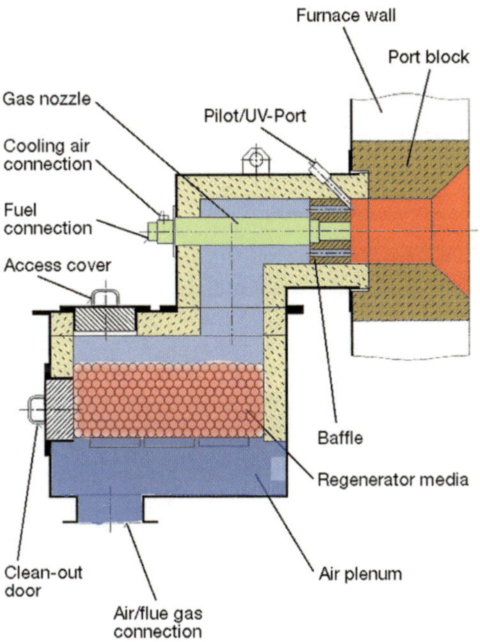

Fig. 3.41: Design of a regenerative burner (Source: Bloom)

The most commonly used system requires burners to operate in pairs, connected as pairs by piping or logically by the control system (usually an ON/OFF control).

When burner A fires, the waste gas is exhausted through burner B by passing through the burner body into a media case (the regenerator) containing the ceramic heat exchange media.

The exhaust gases heat the media, thus recovering and storing energy from the flue products. When the media bed is fully heated, burner A turns off and begins exhausting the flue products. Burner B, having a hot media bed, begins firing and combustion air passes through the media bed B where it is heated by the hot refractory material - see **Fig. 3.42** for the operating principle. The two burners cycle approximately every 60 to 90 seconds to heat the combustion air to approx. 150 °C below furnace temperature and cool down waste gas to below 200 °C (the regenerator valves switch faster if the temperature switch behind the regenerator shows a temperature of the waste gas higher than 200 °C).

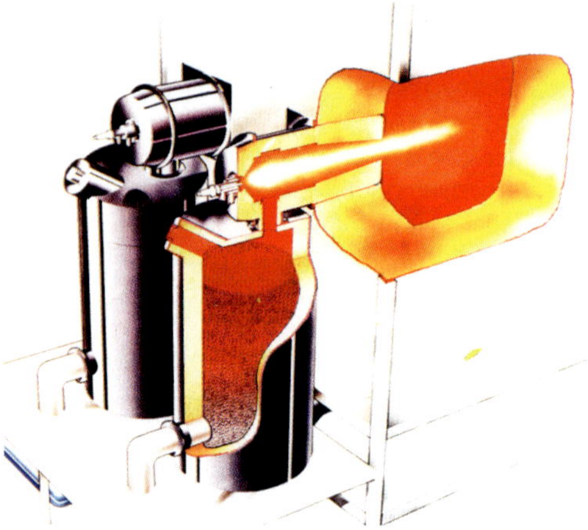

Fig. 3.42: Regenerative burner principle

Fig. 3.43 shows a regenerative burners system with two separate controlled burner pairs. The gas, combustion air and waste gas flow are measured by orifices with P-transmitters.

The "pull back", which is defined as the amount of the waste gas going through regenerators, is controlled parallel to the burner capacity and changes in range 80-90 % of the generated waste gas amount. With a pull back of 90 %, the temperature of waste gas to chimney (or to filter system) is below 300 °C. The system is controlled by means of PLC systems with standard burner control devices for ignition and supervision of the main (UV) and pilot (ionization) flames. The pilot burners (not shown) are normally operated continuously.

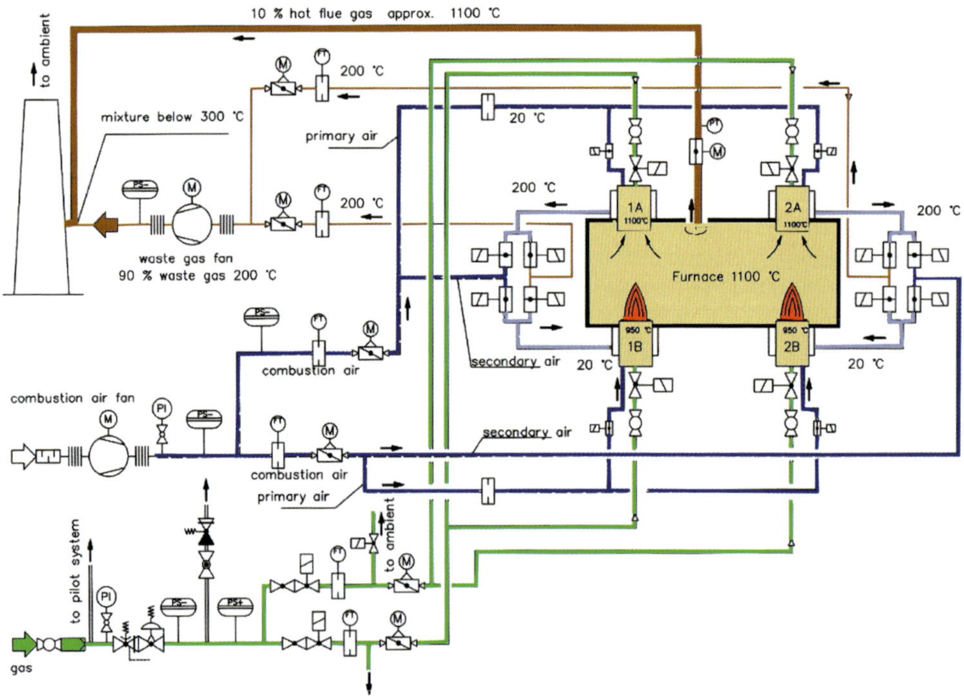

Fig. 3.43: Regenerative burner system with two pairs of burners (modulating control in two temperature zones)

3.3.5.2 Special ultra low-NO_x burners for regenerative systems

According to the valid limits, the NO_x emission of combustion systems with aluminum furnaces has to be not higher than 350 mg/Nm^3 flue gas measured under normal condition without O_2 correction. To maintain this value at a very high air preheat temperature, special ultra low-NO_x burners have to be used for regenerative burner systems. Special nozzles with gas and air staging allow the regenerative burner system to be operated with very low-NO_x emission. Through optimization of shape, exit velocities and location of the burner's gas and air ducts, it is assured that the individual gas and air streams, while exiting from the burner, can recirculate large amounts of flue gas before they meet and mix. This reduces the O_2 concentration in the mixing zone and generates the required low NO_x levels.

Fig. 3.44 shows the principle of such a burner. The flame developed is of large area, extremely luminous and directed forward.

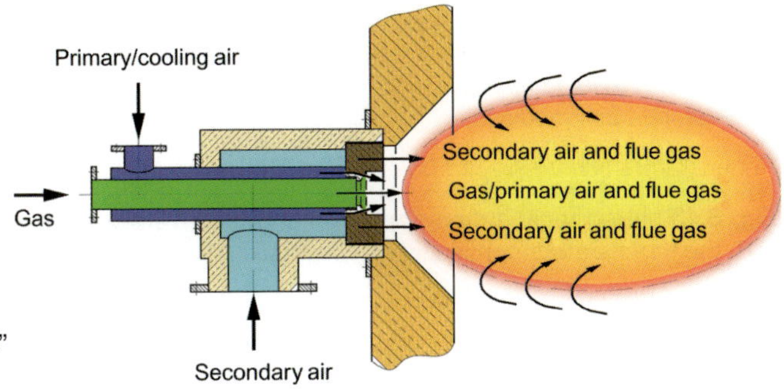

Fig. 3.44: Principle of "lumiflame" burners (Source: Bloom)

For cold furnace start-up and times of lower temperatures (immediately after charging or in standby operation), part of the combustion air is guided into an inner port where it directly reacts with the gas. In this stage a sharper flame for higher convective heat transfer is produced. At higher furnace temperature the burner is operated with the combustion air going mostly through the outside baffle channels. The high exit velocity of the gas causesd an additional staging in the mixing process. The NO_x emission in such burners is extremely low.

The other method of suppressing NO_x emission is gas staging by means of gas jet nozzles placed outside the burner. Such a burner is shown in **Fig. 3.45**.

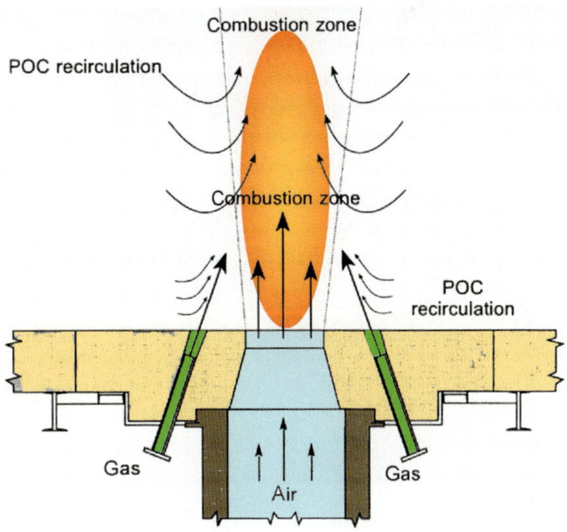

Fig. 3.45: Low-NO_x burner with gas injection (Source: North American)

The burner is operated as a conventional high velocity burner when furnace temperature is below approx. 800 °C. At higher furnace temperature the burner switches to low-NO_x mode: air continues to flow through the center port of the burner while the gas is switched to strategically located outboard injectors. The mixing of gas and air and combustion is relocated from the burner tile to the space in front of the burner. A strong recirculation of inert gases from the combustion chamber in the air and gas stream takes place. This slows the combustion process, reduces hot spots in the flame and thus substantially reduces NO_x-emission of the burner.

3.3.5.3 Regenerators

As heat exchange media for regenerators in the aluminium industry, ceramic balls with high content Al_2O_3 material or honey comb modules from the same material are used. The ceramic ball regenerator bed is easy to handle and resistant than mechanical stress.

The honeycomb regenerator media has a lower pressure drop to the ball media. However, it requires much larger dimensions of regenerators with the same cycle time (the volume weight of ball media is much higher than that of a honeycomb media). For this reason and the easier maintenance of the regenerator media, the high capacity regenerative burners are usually equipped with ceramic ball media.

Regenerators can be placed below or above the burners.

Fig. 3.46 shows regenerators (fast exchangeable version) installed on a tiltable melting furnace with the regenerators conventionally arranged below the burners.

By means of a moveable platform, the regenerators can be removed and installed in < 1 hour time.

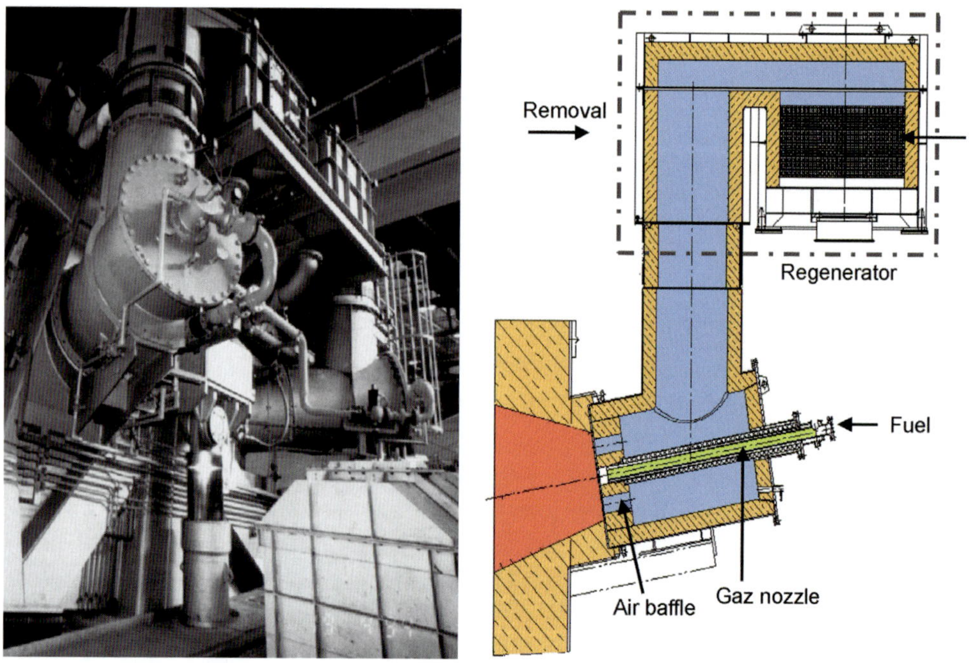

Fig. 3.46: Quick exchangeable regenerators installed above burners (Source: Bloom)

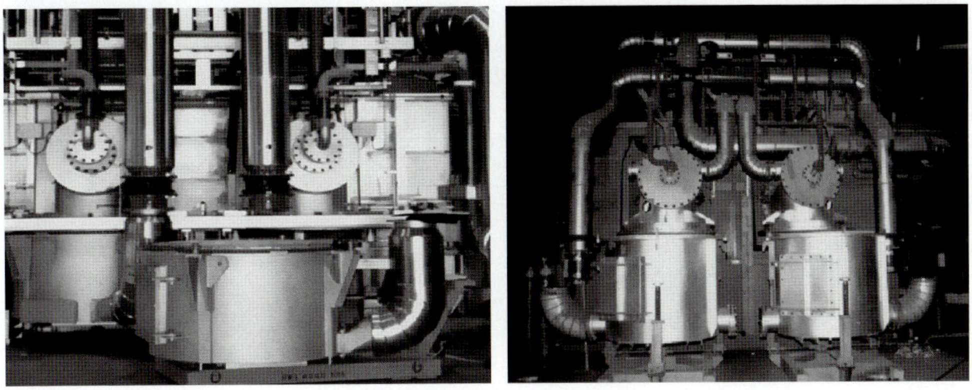

Fig. 3.47: Quick exchangeable regenerators installed below burners (Source: Bloom)

Fig. 3.47 shows a special space-saving arrangement of regenerators above the burners. For the exchange of regenerators only an overhead crane is required and there is no need for loosening any bolts or clamping equipment.

Regenerative burner system with rotating regenerator

Another regenerative burner system is the system with rotating regenerator.

The rotating regenerator consists of fixed upper and lower parts and a movable regenerator suspended between these parts. It has two chambers which are passed by air and waste gas (**Fig. 3.48**). The ceramic material of the regenerator (honeycomb ceramic material) is throughout the chambers and transports the heat from the waste gas to the combustion air.

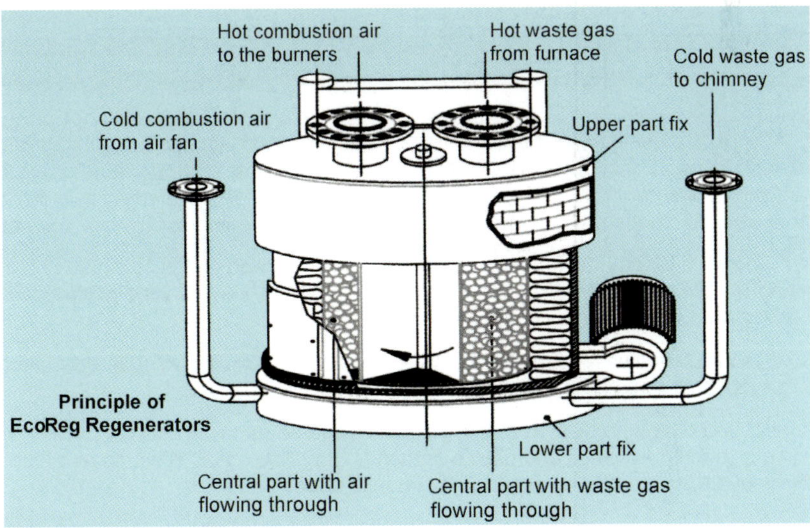

Fig. 3.48: Rotating regenerator (Source: Jasper)

Fig. 3.49: Burner system with rotating regenerator for a two chamber furnace (Source: Jasper)

The waste gases and the combustion air flow continuously through the regenerator (the thermal balance of regenerator is kept by means of partial blow-out of the combustion air).

The waste gases are sucked out of the furnace through a separate opening (they do not go through the burner). The system does not need cycled valves and because of no switching over the furnace, pressure control is easier. The special system ensures the tightness of the system regenerator/regenerator housing. The regenerator positioned beside the furnace has to be connected with the burners and the suction opening by piping with high temperature air or waste gas (up to 1,000 °C) piping. The burner for such a system has to be designed with a low air pressure drop to keep the pressure difference between air and waste gas regenerator chambers low. The waste gases going from the combustion chamber do not pass the burners as in standard regenerative burner systems. The system with a rotating regenerator is used very often in two chamber furnaces for contaminated scrap (**Fig. 3.49**).

3.3.5.4 Maintenance of regenerative burner systems

The maintenance of the regenerative systems is more intensive than that of a conventional combustion system because of a waste gas system, cycling valves and regenerators.

However, it is reasonable to take the following into account considering the energy savings achievable with this system.

- The maintenance expenditure for the additionally required waste gas fans is comparable to that of the combustion air fan. The waste gas fan works normally at a temperature around 200 °C and can shortly work with a waste gas temperature of 350 °C. It is protected by a temperature switch (a valve for suction of cooling air is opened upon exceeding of the max. temperature).
- The cycling valves for combustion air and waste gas are installed on the cold side of the system (operating temperature 200 °C, max. 250-300 °C) and have service lives of more than three years with nearly no maintenance or breakdown.
- Solenoid valves have service lives of more than two million cycles which is equivalent to 4-5 years of system operation.
- The control and operation system of a regenerative heated furnace functions similar to that of conventional heated furnaces and uses commercially available PLC systems as proven and in use for many other operations in an aluminium melt shop.
- Regenerative burner gas nozzles are usually insulated and air-cooled to avoid early destruction by the hot waste gases passing the nozzle.

Fig. 3.50: Low maintenance burner with sealed hot face

- Maintenance work on the burner itself is comparable to that of conventional furnaces, i.e. occasional removal of possible dross or aluminum deposits on the port block and burner face. A burner design (**Fig. 3.50**) with a rather sealed hot face leaves less open space for splashes extending into the burner body.

  A reverse-tapered port block is much more difficult to clean than a port block flaring out. All this experience, however, has also been gathered over years with standard burners.

  A properly designed regenerative burner can run > 5 years in 3 shift operations without major repair unless careless operation (splashing while charging, careless removal of dross etc.) occurs.

- The additional maintenance expenditure to be considered for regenerative systems is the need for an occasional cleaning of the regenerator media. This normally consists of high percentage Al_2O_3 balls in which dust and dross deposit during the exhaust cycle.

  The frequency of cleaning mainly depends on the operation conditions in the plant. When charging thin profiles, which may have plastic or finishing adhesions, dross is higher as in furnaces with clean charge such as ingots, sows or butts. Chlorine and salt are of less danger for the regenerator material than for a recuperative system (which needs an expensive bypass system), but they shorten cleaning intervals.

  A regenerator of a 4 MW burner with 2.5 t media fill can achieve cleaning intervals of 6 months if ingots or the like are charged only whereas extremely contaminated scrap may need monthly cleaning intervals.

  The time needed for cleaning a regenerator can be drastically reduced and simplified by the use of quick exchangeable media cases. Such a design permits replacing the entire regenerator with a new clean one in a short time without waiting for the media to cool down. The dirty regenerator can then be cleaned at a later time and at a more accessible location.

  Firing all regenerative burners as cold air burners during the bath treatment period can protect them against the penetration of waste gases with vapors of salt in the burners of the regenerator and make the cleaning intervals longer.

  Regenerative burners have proven to work reliably with minimum maintenance or repair in three shift operation on melters having up to 120 t capacity with > 25 t/h melting rate (**Fig. 3.51**).

  Quick removable regenerator (see also chapter 3.3.5.3) designs minimize the time and work needed for cleaning.

Fig. 3.51: Regenerative burners installed for a 120 t tiltable two chamber furnace (Source: GKI/OFU)

With rotary regenerators maintenance of the regenerator drive and a tight system between the middle, top and bottom part of the regenerator is needed. If the regenerator media is honeycomb refractory material, the exchange of damaged modules can be necessary.

3.3.6 Oxy-fuel technology

Thomas Niehoff

Air fuel burners and combustion theory have been described in chapters 3.1.7 and following chapters. Here the objective is to describe the specifics of combustion with technical oxygen rather than air. Oxygen is contained in air amongst other gases. A split of the gases contained in air can be seen in **Table 3.15**.

Table 3.15: Air gases in vol.-%

Gases	Volume %
N_2	78.1 %
O_2	20.9 %
Ar	0.9 %
H_2, N_e, H_e, CO_2,.....others	< 0.1 %

Air only contains about 20.9 % oxygen by volume. The other gases are present but do not contribute to combustion or oxidation processes. A combustion system on air only uses 20.9 % of the gases for combustion. When burning fossil fuels (hydrocarbons C_nH_m) the oxidizer is used to provide the oxygen to the oxidation/combustion of hydrogen or carbon:

$$2H_2 + O_2 \rightarrow 2H_2O$$

$$C + O_2 \rightarrow CO_2$$

Any nitrogen in the process will not add to the combustion but will have impacts such as cooling of the flame or generate other oxidation products like:

$$N_2 + O_2 \rightarrow 2NO$$

$$N_2 + 2O_2 \rightarrow 2NO_2$$

Nitrous oxides (NO_x) are pollutants and there are several different ways and technologies how to reduce NO_x generation or to minimize NO_x emissions once they are produced.

Any other fuel elements such as sulfur or even the metal can be oxidized as well during heating and melting.

$$S + O_2 \rightarrow SO_2$$

$$2Al + {}^{3}/_{2}\, O_2 \rightarrow Al_2 O_3$$

The missing nitrogen has the following major consequences for the combustion process:

- brighter, more luminous flame
- increased flame temperatures
- reduction of flue gases
- change of furnace atmosphere gases
- change of flue gas composition
- changes of heat transfer from flame to furnace and goods to be heated/melted
- reduced flue gas heat losses
- increased combustion reactivity
- higher flame speed
- impact on gaseous missions (CO, NO_x, CO_2, dust,...etc)

When looking at the changes it is obvious that when changing from air-fuel operation to oxy-fuel operation a simple burner switch will not deliver expected results.

In the following chapters an attempt is made to discuss the differences between air-fuel and oxy-fuel combustion processes.

3.3.6.1 Oxy-fuel combustion

Oxy-fuel combustion can be characterized as fossil fuel combustion in the absence of nitrogen in the oxidizing gas. Air is the best known oxidizing gas. There are several ways to use extra oxygen in air as oxidizing gas. When mixing technical oxygen with air the natural oxygen volume concentration of 20.9 % will increase. Some examples are given below (**Table 3.16**):

About 20 m^3 of oxygen can replace 100 m^3 of air. In order to maintain the oxygen at balance with the fuel for every 20 m^3 of oxygen added about 100 m^3 of air need to be reduced. This way the nitrogen in the combustion gas is cut back.

Oxygen enrichment is a very efficient and very basic way to use oxygen for combustion. Typically, a diffuser is installed in an air pipe or duct. This diffuser injects oxygen in an air stream which

Table 3.16: Impact on oxygen concentration

Air [Nm^3]	O_2 [Nm^3]	Total Volume [Nm^3]	O_2 – conc. [vol.-%]
100	0	100	20.9
80	20	100	36.7
60	40	100	52.5
40	60	100	68.4
20	80	100	84.2
0	100	100	100

mixes in the ducts which lead to the burner. For oxygen enrichment it is very important to adhere to strict safety standards to avoid other components from overheating or even ignition.

Oxygen lancing is the next option as to how to use oxygen in industrial production processes. For oxygen lancing specially designed and engineered oxygen lances are installed in furnaces under or above flames of air burners or they are integrated into air burners. Oxygen lancing opens up a wider range of safe oxygen usage for combustion. All the materials that come into contact with oxygen are selected and defined for their use. Oxygen lancing can allow staging of combustion in a way to reduce emissions (e.g. NO_x) and it can be beneficial for the heat transfer and temperature profile of furnace and heated goods.

Air-oxy-fuel burners are another option to use oxygen for combustion. Air-oxy-fuel burners are burners that use an oxygen stream and an air stream in parallel. Different flow legs control the flow rates of air and oxygen and calculate the ratio of oxygen to the fuel and adjust according to stoichiometric requirements. Air-oxy-fuel burners can be retrofitted air burners, where an oxygen pipe or oxy-fuel burner is integrated into an existing air burner. Alternatively, a specially designed burner may be used. While operating an air-oxy-fuel burner the flow rates of air and oxygen can be altered to adjust to different melting process steps or flame characteristics.

Oxy-fuel burners are burners that run on fuel and technical oxygen (close to 100 % by volume). Here the building size is extremely small when compared to air-fuel burners. Consequently, the mounting hole in the furnace will be small and the air blower and air ducting may not be needed any longer. Flow controls for oxy-fuel burner systems are small compared to air-fuel combustion systems. Flue gas ducts need to be resized and pressure conditions in furnaces need to be reviewed. Oxy-fuel burners have the potential to realize the largest benefits in terms of energy savings and productivity increases for industrial processes. The missing nitrogen can have the impact of very low NO_x formation if the fuel is nitrogen-free and the furnace is operated in a way to avoid air filtration.

3.3.6.1.1 Stoichiometry and reaction of combustion

The stoichiometry of combustion describes the ratio between fuel and oxidizer. Stoichiometry means in its original meaning (Greek) the assessment of amounts. In chemistry it is used to calculate chemical reactions – that means the quantitative description of chemical reactions.

The stoichiometric reaction of methane (CH_4) with air and oxygen is described below to give an example.

Equation by volumes for air:

$$CH_4 + O_2 + 7.52\ N_2 \rightarrow CO_2 + 2H_2O + 7.52\ N_2,\ \Delta H = -35{,}883\ kJ$$

Equation by mass for air:

$$1\ kg\ CH_4 + 4\ kg\ O_2 + 13.2\ kg\ N_2 \rightarrow 2.75\ kg\ CO_2 + 2.25\ kg\ H_2O + 13.2\ kg\ N_2,\ \Delta H = -49{,}610\ kJ$$

Equation by volumes for oxygen:

$$CH_4 + O_2 \rightarrow CO_2 + 2H_2O,\ \Delta H = -35{,}883\ kJ$$

Equation by mass for oxygen:

$$1\ kg\ CH_4 + 4\ kg\ O_2 \rightarrow 2.75\ kg\ CO_2 + 2.25\ kg\ H_2O\ \Delta H = -49{,}610\ kJ$$

The examples of air show a significant volume and mass contribution from nitrogen. In conventional combustion and furnace systems the nitrogen enters the process cold at ambient temperatures or pre-heated to 300 or 600 °C. Higher pre-heat temperatures are rare. Thus, in general the nitrogen is pulling heat away from the melting or heating process into the flue gas.

The nitrogen will be passing through the flame and reaction zones and will act as a coolant for the flame and will dilute the concentrations of the reactive gases and slow down the combustion reactivity.

3.3.6.1.2 Temperatures

Adiabatic flame temperatures for flames in air and oxygen have been investigated and are known to a large extent (see **Table 3.17**).

Table 3.17: Adiabatic flame temperatures of various fuels

	Air		Oxygen	
Fuel	**[°C]**	**[°F]**	**[°C]**	**[°F]**
H_2	2097	3807	2806	5083
CH_2	1950	3542	2780	5036
C_2H_2	2262	4104	3069	5556
CO	2108	3826	2705	4901

Non-adiabatic flame temperatures in hot furnace conditions are about 300 Kelvin lower as compared to adiabatic flame temperatures. Burner flame temperatures can be altered by staging combustion. Combustion can be staged by staging the oxidizer or fuel or both. Staged combustion uses the furnace atmosphere gases and mixes them with the combustion gases of oxidizer and fuel. This way the reaction is slower and in the reaction volume is increased. Hence, the flame temperature is being reduced.

Flame temperatures are influenced by various parameters. Higher carbon to hydrogen ratios (C/H–ratio) increase the flame temperature. Oil flames are hotter than gas flames and coal or coke dust flames are hotter than oil flames. Oxidizer preheating has a similar effect on flame temperature as oxygen use (see **Fig. 3.52**).

3.3.6.1.3 Flue gases

Flue gases from combustion with air will be larger than from combustion with oxygen. As quantified in the stoichiometry equations from chapter 3.3.6.1.1, the nitrogen takes a large volume of the flue gases. When combusting natural gas with air the wet flue gases are about 10.7 m^3 per 1 m^3 of natural gas. Using oxygen instead of air eliminates the nitrogen and reduces the flue gases down to 3 m^3 per 1 m^3 of natural gas. This is a volume reduction of approx. 72 %. Combustion gases enter a combustion process either cold or pre-heated through the burner and get heated in the furnace before they leave through the stack exit. These flue gases take away heat from the furnace and reduce the available heat for melting and heating of the goods in the furnace. It can

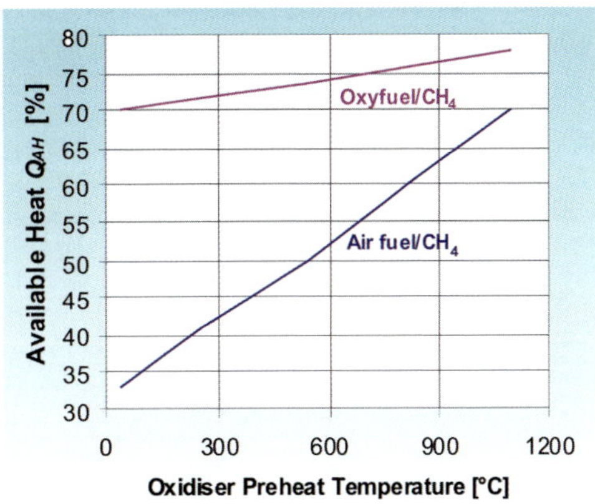

Fig. 3.52: Available heat over pre-heat temperature (Source: Linde AG)

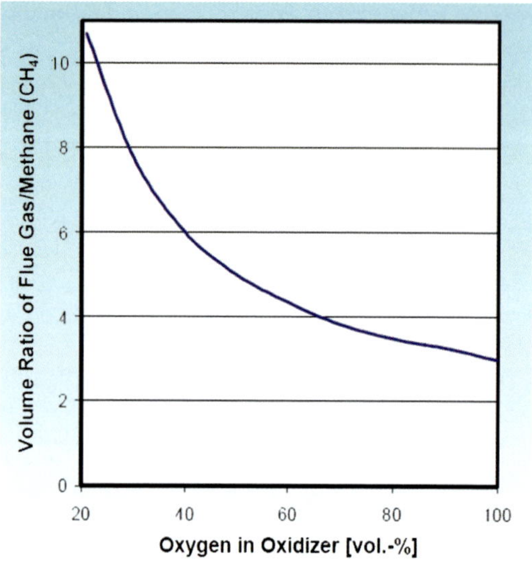

Fig. 3.53: Oxygen concentration over flue gas volume (Source: Linde AG)

be assumed that the furnace wall temperatures remain unchanged for air fuel and oxy-fuel. As a consequence, the wall losses are comparable, too.

Fig. 3.53 shows the effect oxygen use has on flue gas volumes. The figure shows that the flue gas reduction is not a linear function of the oxygen concentration in the oxidizer. Coming from air (20.9 % of oxygen) to 45 % of oxygen the curve is very steep. Consequently, oxygen use in this area has the largest effect on the flue gases.

Using oxygen instead of air as an oxidizer reduces the flue gas volumes by more than 70 %. The energy, which is taken out of the process by the flue gases, can be significantly reduced (see **Fig. 3.54**).

Applying oxygen in industrial combustion processes can save 30-60 % of the fuel input compared to air-fuel operation.

Combustion kinetics and flame temperatures are greatly increased using oxygen instead of air. When nitrogen passes through the flame it cools the combustion process and dilutes the reactants in a way that it slows down the combustion reactions in the flame. The flame is less intense and

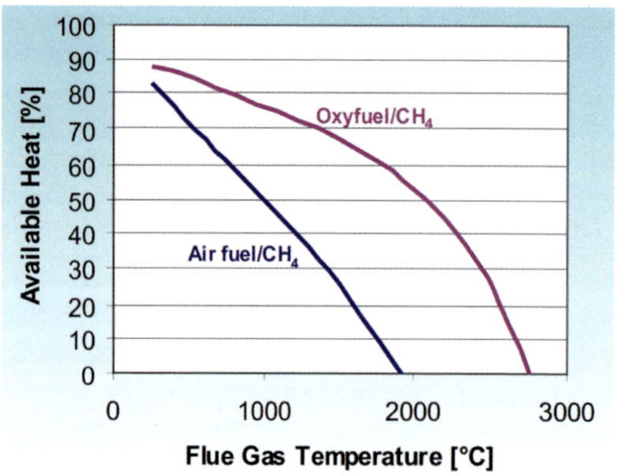

Fig. 3.54: Flue gas temperature for air-fuel and oxy-fuel over available heat (Source: Linde AG)

colder as compared to an oxy-fuel flame. Heat transfer at temperatures above 800 °C (1,600 °F) mainly occurs by radiation. The following equation describes the heat transfer by radiation.

$$q_s = \left[\left(\frac{T_A}{100} \right)^4 - \left(\frac{T_B}{100} \right)^4 \right] \cdot c_s$$

Where: q_S = radiation heat transferred

T_A – TB = temperature gradient

c_S = radiation coefficient

The increased flame temperature of oxy-fuel flames greatly enhances the heat transfer by radiation. The temperature affects the heat transfer rate by the power of four. This explains why melting and heating with oxy-fuel flames can be so much faster as compared to air-fuel. The combination of optimized heat transfer and greatly reduced flue gas losses contributes to dramatic process energy savings.

The second fundamental consequence is the increased concentration of the highly radiating products of combustion CO_2 and H_2O in the furnace atmosphere. Calculations according to VDI Wärmeatlas (**Fig. 3.55**) indicate that the emissivity of oxy-fuel burner exhaust gases is 30 to 60% higher than air-fuel burner exhaust gases in the temperature range of 400 to 1,200 °C. Therefore, the heat transfer through radiation increases when firing oxy-fuel instead of air fuel.[1)]

As the "radiative component of the heat flux dominates the heat transfer between gas and metal by far"[2)] this leads to a more efficient heating and melting process.

1) Gripenberg, H. et al.

2) Buchholz and Rødseth

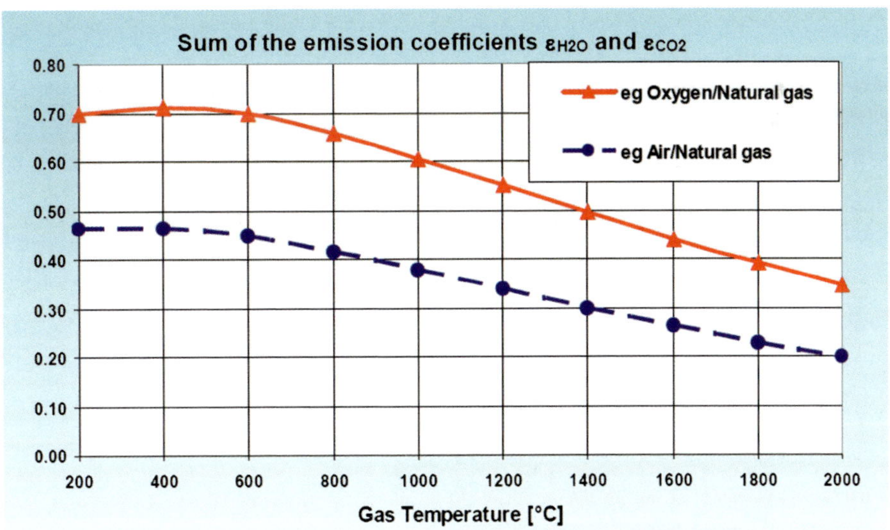

Fig. 3.55: Calculated emission coefficients of products of combustion from natural gas/oxygen and natural gas/air burners. The higher emission coefficient indicates higher heat transfer through gas radiation (Source: Linde AG)

3.3.6.2 Heat balance

The heat provided by the fuel can be regarded as the offered energy to the process. The remaining or available heat to melt or process the goods can be calculated as follows:

$$Q_{FH} - Q_{FGH} = Q_{AH}$$

Where: Q_{FH} = Fuel heat, i.e. calorific value, pre-heated air

Q_{FGH} = Flue gas heat

Q_{AH} = Available heat

The fuel heat can easily be determined by the lower calorific value of the fuel. The flue gas heat depends on the following:

- Flue gas volume/mass flow
- Flue gas composition
- Flue gas process exit temperature.

The available heat Q_{AH} is the heat which can be used in the furnace.

$$Q_{AH} = Q_{WL} + Q_{OL} + Q_G$$

Where: Q_{AH} = Available heat

Q_{WL} = Heat wall losses

Q_{OL} = Heat other losses

Q_G = Heat to process goods

The available heat can be split into wall losses and other losses such as leakages, door open times, process steps etc. and the heat which heats the goods (**Fig. 3.56**).

When assuming that the wall losses and the other losses remain nearly the same for the same furnace temperature and conditions, then all the reduced flue gas losses will be transferred into the heating of the furnace goods.

3.3.4.3 CFD modelling

CFD (computer fluid dynamics) modelling is used to analyze problems involving fluid flows. High speed computers are used to simulate the interactions of gases and liquids with the surface bas-

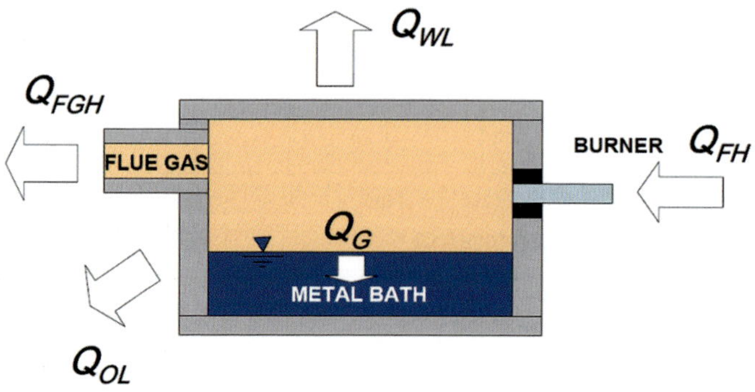

Fig. 3.56: Heats of melting furnace (Source: Linde AG)

ing on small surface elements (**Fig. 3.57**). The efforts required for such analysis are far too intense to use for the standard furnace design. For investigating the pros and cons of a new technology the CFD modelling is a very excellent tool. As an example, a brief description of the analysis of oxy-fuel firing is described.

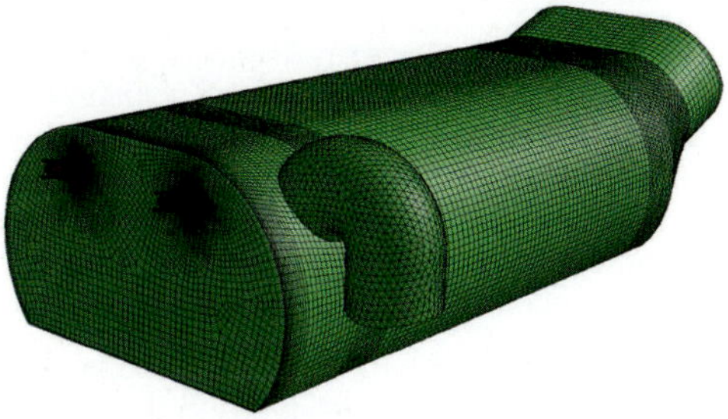

Fig. 3.57: Example of mesh size and design (Source: Linde AG)

A melting operation consists of various specific elements that need to be transferred into the computer model. However, not all can be transferred and not all can be modelled. Hence, the model will only be a model and not real. Aspects, such as charge material quality, quantity, composition and mixtures, are very difficult to describe in a model. Charging material storage and altering process conditions as well as the melting itself with all changing physical properties are hard to specify and model. The combustion space of a furnace can be modelled with a justifiable effort. The system of the combustion space then needs to be defined and described well enough to come to reasonable results that can reflect reality. The model will preferably describe steady state conditions, i.e. altering firing rates and flame shapes cannot be characterized in one single step. The model needs to be connected to reality and will need to be verified with the current operation.

The melting or heating operation (**Fig. 3.58**) to be modelled typically consists of a specific furnace design (rotary furnace, reverberatory furnace, tower furnace, shaft furnace and many more). This furnace design, together with refractory material, flue openings, burner positions and designs, bath level and other protruding elements, define the combustion space. Designs, thicknesses, and physical properties of the materials then are put together to the model. The mesh size of the model describes how detailed the combustion process will be described in a specific location of the model. The mesh size can vary across the furnace. Typically, it needs to be very small (detailed) where rapid changes in either design or chemistry are expected. The mesh is refined near the region of the burner to capture the gradients effectively. Homogenizing with slow dynamics areas will have wider mesh sizes, when one expects reduced activity. A typical aluminium reverberatory furnace model with a capacity of 30 t and a footprint of 50 m^2 will have 500,000 knots with an average distance of 0.15 m. Where each knot represents that all equilibrium equations (heat, energy

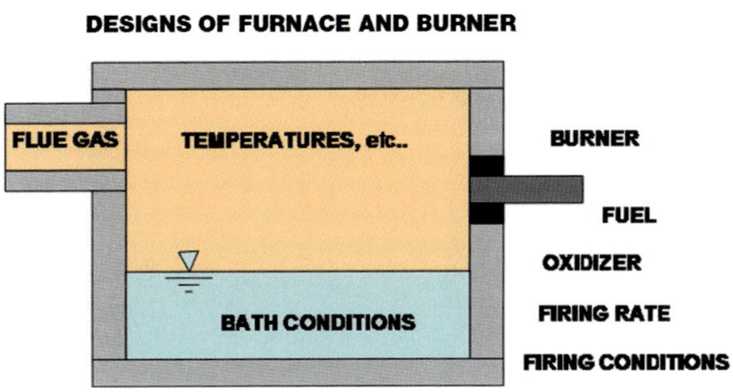

Fig. 3.58: Basic elements of aluminium melting process (Source: Linde AG)

and mass) will be solved. The model works its way through the furnace system by moving from knot to knot (Fig. 3.57). It can take days or weeks to simulate one single steady state operating point in a way that the results make sense and are good enough for verification. Recent advancements have enabled to solve the governing equations in parallel using multiple CPUs which reduce the computing time significantly.

Typical combustion systems involve high speed flows. Hence, choice of proper turbulence models is very important to predict the solution accurately. In most of the cases, reaction is mixing-controlled. Hence, turbulence chemistry interaction needs to be resolved correctly to accurately predict the flame shape and temperature distribution. Furthermore, since high temperatures prevail inside the furnace, radiation heat transfer plays a significant role in transferring energy from the combustion space to the metal bath. Hence, choice of the radiation model will also play a vital role in the accuracy of the solution.

The combustion system is a very complex system of designs, kinetics and chemical reactions. When modelling, the modeller has to decide which condition to model and why. Very often typical and characteristic operating conditions are being modelled. For example, the high temperature intervals if looking at specific heat transfer and temperature profiles. Flame shapes and sizes for refractory material investigations. NO_x and CO profiles for gaseous emissions from combustion processes.

Traditional cost-intensive CFD modelling has been used for major furnace projects, where the modelling cost was only a fraction of the capital investment. The simulation was used to make informed decisions about heat transfer, temperature profiles, emissions, burner and flue locations and efficiencies/cost. Large capacity steelmaking plants, power plants and large scale glass plants have been using CFD modelling for a long time and have good experience in getting useful results for decision-making processes. For aluminium and metals melting and recycling there are only few examples how and where CFD is being used. Linde Gas has put effort in understanding and detailing combustion processes with CFD for melting of metals such as aluminium and copper.

By simulating combustion processes, the effects of different flame shapes can be analyzed and compared. Conventional oxy-fuel is often regarded as pipe in pipe and round flame designs (**Fig. 3.59, 3.60**).

By comparing such different flame specific characteristics in one single furnace design, there will be differences in heat transferred into the metal bath area and towards the walls. When changing burners and/or burner locations and comparing the effects on the metal bath and the walls, this will lead to an optimization process.

Air-fuel operated combustion processes for melting aluminium has benefits and disadvantages that are listed below:

Benefits:

- Low flame temperature
- No oxygen cost
- Good convective heat transfer
- Slow reactivity

Disadvantages:

- High gas volumes
- High energy demand
- Low efficiency
- Noise emissions
- Dust emissions

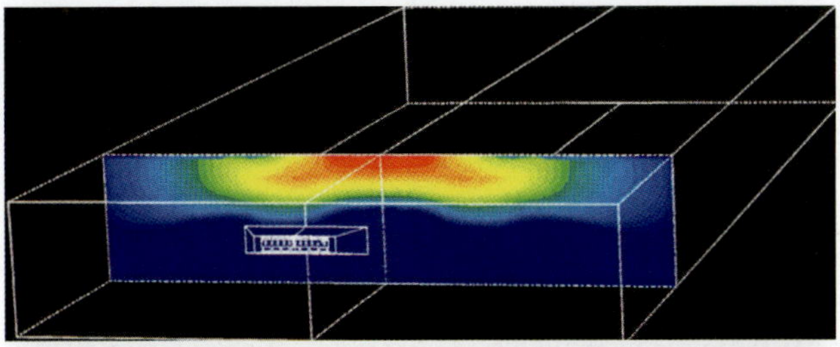

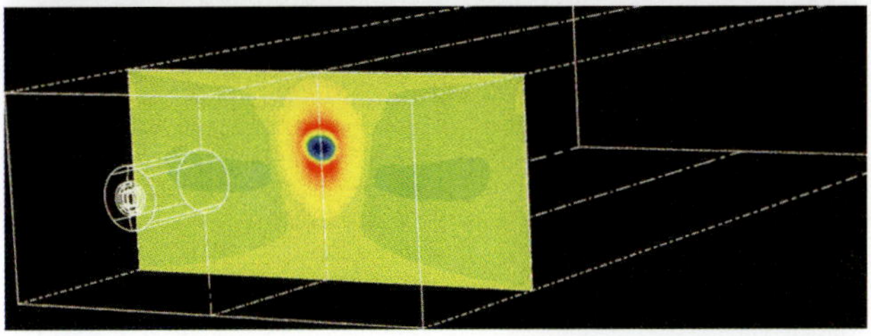

Fig. 3.59: Flat jet oxy-fuel burner flame and combustion area temperature profiles (Source: Linde AG)

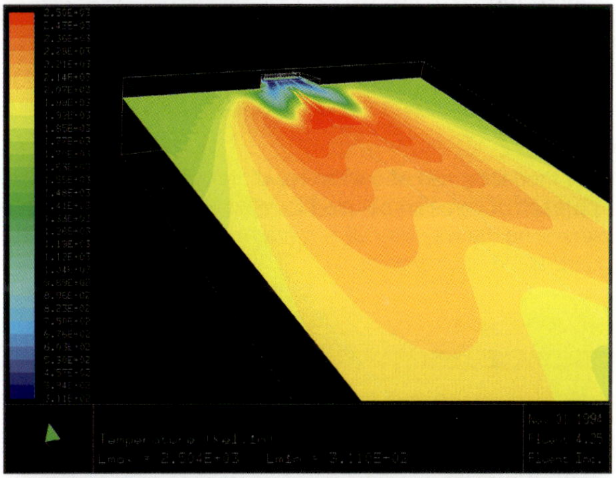

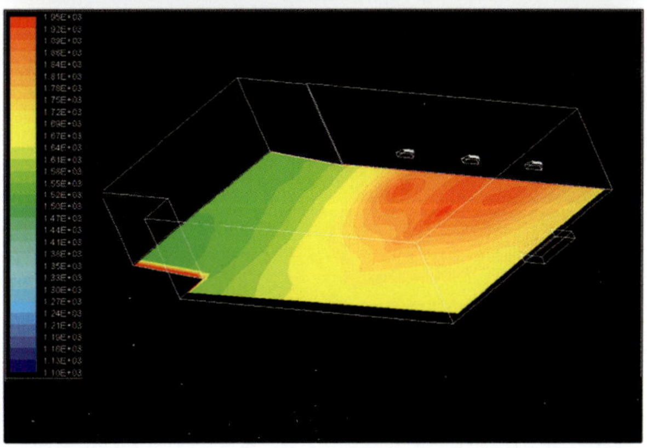

Fig. 3.60: Example of temperature profile above metal bath (Source: Linde AG)

Linde Gas has developed a combustion technology that combines the advantages of air-fuel combustion by avoiding the disadvantages at the same time. This technology was developed from the need of the aluminium industry to avoid local overheating and, hence, reduce oxidation of the metal. The Linde response is low temperature oxy-fuel combustion technology. The low temperature oxy-fuel combustion process combines the benefits of air-fuel and oxy-fuel combustion processes. This means the low flame temperature and high convective heat combined with high energy efficiency from oxy-fuel. CFD modelling is used to describe and evaluate the changes that oxy-fuel would bring in such a situation. The experience from converting to oxy-fuel at a cast house operated by Hydro Aluminium in Norway has been described.[3)] There an oxy-fuel burner has been used to compensate hot pot room metal with ambient temperature solid scrap. This is an example of how oxy-fuel lifts existing limits and borders to the next level.

These days the analysis of combustion processes, especially in the aluminium sector, focus on highly efficient heat transfer without local overheating. Comparing various oxy-fuel cases leads to the following conclusions:

- oxy-fuel and oxy-fuel can be different
- flame shape and volume are important
- flame temperatures matter
- furnace gas recirculation has an impact on heat transfer

When evaluating all these different parameters oxy-fuel can be optimized versus oxy-fuel. The next exciting question is: How does optimized oxy-fuel compare to regenerative pre-heated air fuel operation? Here Linde has done extensive research to better understand the specifics and details that are important to keep all benefits from air-fuel firing and all benefits from oxy-fuel firing.

CFD modelling has shown to have many benefits and it avoids the often applied “trial and error” approach. It can also be used to maximize oxy-fuel benefits and minimize emissions. Linde Gas is pioneering the way into CFD modelling for nonferrous metals melting in combination with oxy-fuel combustion technologies.

3) Niehoff, T. (2010)

3.4 Energy losses

Christoph Schmitz

As in any technical process, furnaces cannot operate without losses. The difficulty with our technologies is that we have different conditions in every stage of the furnace cycle. This does have a significant impact on the heat transfer mechanism with all the resulting data for temperatures and heat quantities. Although we are able to study the conditions in a furnace at different stages, this will be meaningful for introducing a new technology – whatever that means – for the general application. It is too time-consuming to try to look at all details. These will change whenever different types of scrap have to be processed.

It is of limited use if experts and suppliers of equipment provide data and results obtained for one specific condition of the furnace process and may be also related to a very specific type of material. Therefore, we consider average values with a good estimate of data and conditions for the entire process.

And still, we should never forget that a complete modelling of a melting process is not yet possible. Future studies may permit to go into more detail which also could be helpful to improve the efficiency of our furnaces as it has been done in the past.

3.4.1 Metal discharge

During the furnace cycle the metal has been melted and superheated to the temperature as desired by production for further processing. The energy input has resulted in the increase of enthalpy of the metal. Some contaminations, such as oil, plastics or binders of paint, have contributed to the energy input. Others require additional energy. These are metals not melted during the process or metal oxides. There are also melting additives required for the processing which have to be heated to the process temperatures. This is mainly the salt mixture used in the rotary drum furnace. All components not dissolved in the metal have to be removed from the furnace as slag. Their heat has to be considered in the energy balance of the furnace as lost when related to the specific energy requirements for processing the metal. Usually the energy requirement of a furnace is indicated as kWh/t material charged. But this includes everything that goes into the furnace. If the recovery of the scrap charged is 70 %, the specific energy related to the metal will be 30 % higher. Furthermore, the salt quantity has to be considered. Using for instance 200 kg salt per ton of aluminium, the specific energy will increase by another 20 %.

When evaluating the economics of a plant, specific values are related to the net metal output of the operation. This certainly refers to the specific energy consumption. The calculation related to the net metal output must therefore read:

$$q_{Al} = p_C \frac{G_{scrap} + G_{salt}}{G_{Al}} \tag{3.95}$$

with q_{Al} = specific energy consumption in kWh per kg of net aluminium production, G_{scrap} the weight of material charged in kg, G_{salt} the salt quantity in kg and G_{Al} the net aluminium quantity recovered.

Since $G_{Al} = \eta_r \cdot G_{scrap}$, the equation reads

$$q_{Al} = \frac{p_C}{\eta_r} \cdot \left(\frac{G_{salt}}{G_{scrap}} \right) \tag{3.96}$$

In this equation η_r is the well-known recovery of the scrap.

What about the dross skimmed off from a furnace?

Dross, a mixture of aluminium oxide and metallic aluminium, is obtained when an oxide layer on a metal bath is removed; i. e. the bath is skimmed. The aluminium oxide is formed during the melting and heating process by reaction of aluminium with the oxygen in the combustion products. This is an exothermic reaction, meaning in our case that aluminium is burned and accompanied by generation of heat. Although the quantity of aluminium burned is to be regarded as loss in the entire production process line, the exothermic heat contributes to the overall energy balance as positive factor. However, aluminium is a very expensive fuel and thus the reaction with oxygen should be suppressed whenever possible. It is inevitable that metallic aluminium is trapped in the oxide when the bath is skimmed. The enthalpy of that metal is definitely a loss of energy. But considering the comparatively small quantity, we can fairly assume that the heat generated by the oxidation of aluminium compensates for this loss.

There is a permanent loss of aluminium during the different production stages from scrap to final product. And there are sometimes no-go charges. This metallic aluminium is not lost to the process but it has to be remelted again, requiring energy. Looking at the overall economy, this is definitely a loss of energy that must be considered in the total energy balance of a production process.

Upgrading the energy efficiency of a furnace also includes the improvement of heat transfer into the liquid metal bath. As we have discussed in one of the previous chapters, heat transfer to the bath surface is limited by the stable stratification of the liquid metal bath. The bottom layer will have a lower temperature than the top layer and there is no possibility to change this situation by natural draft. If the metal bath is circulated by a pumping system, the stable stratigraphy is disturbed and the metal from the bottom can surface and thus be heated up. This helps to reduce heating time and overheating of the surface, resulting in less energy input, i. e. better energy efficiency.

3.4.2 Stack losses

It is the principle of a furnace process to convert chemical energy of fuel into sensible heat of combustion products which are then used to increase the enthalpy of the furnace batch, i. e. melt the metal. As we know, the temperature of the combustion products decreases due to the process of heat transfer. According to the second law of heat processing, this temperature cannot drop below the temperature of the batch. The flue gas leaving the furnace must have at least the metal temperature. In reality this would be true for a steady state condition when the temperature of the combustion products has been lowered gradually before and no losses occur in the metal. In reality, there is always a difference between the temperatures of metal and flue gas. The temperature increases as soon as no or only little energy can be transferred due to the conditions of the batch. With a liquid bath surface, heat transfer is slow and the temperature difference between metal and gas is high. Solid material with a larger surface area is able to absorb much energy and, consequently, the temperature difference between batch and combustion products is small. If we assume that the metal has received the energy required to reach a metal temperature of 720 °C, the flue gas temperature will be in the range of 1100 °C. Then the energy leaving the furnace is quite substantial:

$$q_{flue} = c_{pf} \cdot \Delta T \tag{3.97}$$

$$q_{flue} = 1.35 \cdot (1100\ 273) = 1853 \text{ kJ/m}^3 \text{ or } 0.515 \text{ kWh/m}^3$$

From experience we know that the energy required to melt and superheat 1 ton of aluminium is 800 kW in a well-designed furnace without heat recovery. Since the specific heat of gas is 10 kWh/m³, a gas volume of 800/10 = 80 m³ per ton of aluminium is required. This will produce

12 x 80 = 960 m^3 flue gas (based on the approximation of 12 m^3 flue gas per 1 m^3 of natural gas). With the value of 0.515 kWh/m^3 of flue gas we obtain

$$q_{flue} = 960 \cdot 0.515 = 494 \text{ kWh/t of aluminium.}$$

With the theoretical energy requirement of 316 kWh/t for melting and superheating one t of aluminium, we summarize that the energy required would already exceed the figure of 800 kWh/t: 316 + 494 = 810 kWh/t. The reason is that we have assumed that the flue gas temperature is 1100 °C. But this is true for a short period towards the end of the melting cycle only. The average temperature of the flue gas will be in the range of 900 °C.

We calculate again:

$$q_{flue} = 1{,}35 \cdot (1100 + 273) = 1583 \text{ kJ/m}^3 \text{ or } 440 \text{ kWh/ t of aluminium}$$

and

$$960 \cdot 440 = 422 \text{ kWh/t}$$

We can now summarize

theoretical energy	316 kWh/t	equal to	40 %
flue gas losses	422 kWh/t	equal to	53 %
wall and door losses	62 kWh/t	equal to	7 %
total	**800 kWh/t**	**equal to**	**100 %**

The figures show that the flue gas losses are higher as the net energy required. But the calculation gives a more realistic figure for the wall losses as well. The total efficiency of the furnace without recovery of energy is as low as 26 %. Considering the high value for the stack losses, it appears to be worthwhile to look into methods to reduce these losses and to recycle the energy to the process.

In some cases it may be economical to use the energy of the stack losses for other applications. One possibility is to produce hot water for the central heating system for offices and other buildings. There is always a problem involved when trying to use process heat for other purposes than recycling it to the process. The sequence of the furnace does usually not meet the cycles for the exterior application. In the case of the central heating system, a careful balance of the entire plant system is required. For instance, in the summertime much less energy is required for heating the buildings than in wintertime. The furnace will also supply, in fast changing sequences, more or less heat into the system. By applying furnace operation measures, a certain control is possible by changing the cooling air requirements for the flue gases. If a buffer system, such as warm water balancing in the central heating system, the application of a heat exchanger provides a good means of energy savings for the entire plant.

3.4.2.1 Recuperators

In order to recycle energy from the flue gas to the furnace process, the combustion air for the burners will be pre-heated. This measure increases the temperature of the flame and, as consequence, the temperature of the combustion products. The highest temperature, which is theoretically possible, cannot exceed the temperature of the flue gases. But there are other restrictions as well. At high air temperatures the generation of the poisonous NO_x will exceed the permissible values as defined by the environmental regulations. Therefore, the temperature of the combustion air is limited to 450 °C.

Assuming a c_{pair} of 1.12 kJ/m^3 and an ambient temperature of 20 °C, the maximum of energy that can be transferred to the combustion air following the above limitations will be

$$q_{air} = c_{pair} \cdot \Delta\vartheta \tag{3.98}$$

$$q_{air} = 1.34 \cdot (450 - 20) = 482 \text{ W/m}^3 \text{ or } 0.13 \text{ kWh/m}^3$$

(Actually we should use the absolute temperature T. However, the difference will be the same. Thus, we can calculate in °C)

On the basis of the specific energy consumption of 800 kWh/t_{al}, the natural gas volume of 80 m^3 will be required per kg of aluminium, as we calculated before. The required combustion air quantity will be v_{air} = 80 x 11 = 880 m^3/kg of aluminium. This again is based on the approximation developed in one of the previous chapters.

We now are able to calculate the energy recycled by combustion air at

$$q_{air} = 880 \times 0.13 = 114 \text{ kWh/t of aluminium.}$$

The stack losses, as calculated above, now result in the difference between the energy transported to the stack by the combustion products, reduced by the energy recycled by means of the combustion air back to the furnace. Our summary now reads

theoretical energy		316 kWh/t	equal to	46 %
flue gas losses	(420 – 114)	306 kWh/t	equal to	45 %
wall losses		62 kWh/t	equal to	9 %
total		**684 kWh/t**	**equal to**	**100 %**

We notice that the furnace efficiency has increased from 40 to 46 %. This is quite a success. The fraction of the wall losses has increased to 13 % now, a factor that is worthwhile to look at. The value obtained now comes close to that of a tiltable rotary drum furnace with its excellent heat transfer mechanisms.

Recycling of energy results in higher combustion temperatures if the conditions established before remain unchanged. Since the resulting temperature in the furnace would be higher than in a furnace without heat recovery, conditions for the heat transfer would change with the resulting impact on the refractory material and the heat losses through the walls. Furthermore, there are higher flue gas temperatures. All this results in a negative effect on the furnace efficiency and thus decreases the advantage of the heat recovery to a certain extent. In a well-designed furnace, this will not be the case. Temperatures are already set to the maximum possible level. Instead of increasing the temperature in the furnace, the input of fuel will be reduced. This actual benefit can be measured directly.

The recovery system is designed for the most efficient heat transfer with the lower value of the flue gas temperature. The peak temperature must not be considered due to the environmental restrictions. If a lower temperature of the flue gas is considered for the size of a recuperator, the resulting air temperature at the peak flue gas temperature will be far too high. In such a case the flue gas must be cooled by fresh air which may be convenient in any case to adjust the temperature to the value required for a waste gas scrubbing system.

We should note that we have calculated on the basis of average values. This means that the recovery could be increased by setting the combustion air temperature to higher values.

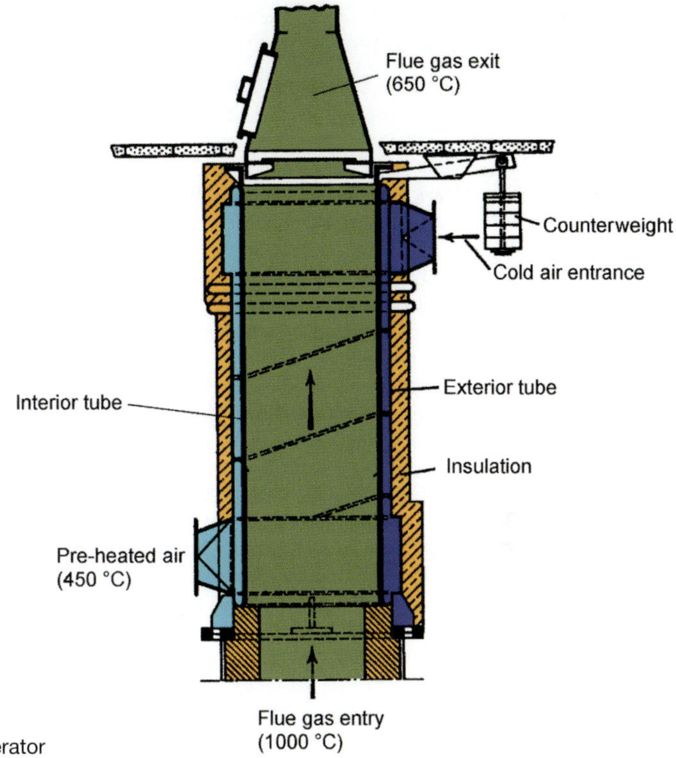

Fig. 3.61: Principle of a recuperator

Here we have the problem of NO_x generation. The value of 214 kWh/t of aluminium, that can be recovered on the basis of air pre-heating, is the upper limit if the environmental impact gets due consideration. The environmental regulations are based on values measured over a certain period of time. Considering the entire furnace cycle, including downtime for charging, holding and the like, it might be possible to exceed the combustion air temperature for some time. The whole complex requires a careful analysis of all plant conditions to arrive at the optimal balance between energy consumption and the generation of harmful gases to obtain the highest possible level of environmental protection.

In order to heat the combustion air, both gases pass each other separated by a wall. The basic concept of a recuperator comprises an inner tube that carries the hot flue gas, surrounded by an outer tube for the air to be preheated. The outside of the system is insulated by refractory material (**Fig. 3.61**). The gap between the outer tube and the inner tube is small in order to achieve high velocity of the air for improving the convective heat transfer.

Recuperators can be constructed out of refractory material or steel. Refractory materials are preferred for very high temperatures but the units are heavy and require large space for installation. The aluminium industry prefers steel recuperators. The components coming in contact with the hot and sometimes aggressive combustion products are generally made of heat-resistant stainless steel. In some cases cast iron elements are used. The steel recuperators have the advantage of very good heat transfer combined with acceptable space requirements. One further advantage is the low heat-storing capacity which is preferred for fast changing conditions within the furnace cycle.

Hot combustion products and air pass the recuperator as counter-current, parallel flow or cross-flow.

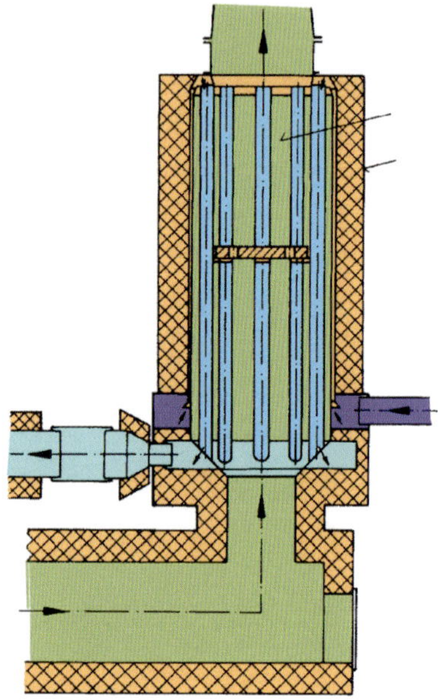

Fig. 3.62: Radiative recuperator with center cage

In the basic concept of the recuperator, the heat exchange from the inner tube to the outer tube will be by radiation. This type of recuperator is therefore called radiation-type recuperator. This expression may be misleading to a certain extent. The transfer of energy to the combustion air can only be by convection since radiative energy passes through clean air and is not absorbed by this medium. Consequently, the heat is radiated to the outer tube and the combustion air passing this area will receive energy by convective heat transfer as it does from the inner tube. Radiation is the most efficient means of heat transfer in upper temperature regions. For lower flue gas temperatures, the area for convective heat transfer must be increased. Better efficiency can be obtained if the combustion air passes through tubes arranged around the center as cage of tubes (**Fig. 3.62**). If the furnace operates mainly at lower flue gas temperature, the air is directed through a bundle of tubes (**Fig. 3.63**). For the heat exchange it is of no importance whether the combustion air or the flue gases pass through the inner tubes of this heat exchanger. Since flue gas carries some dust and also gaseous contaminations, scaling and dust layers must be expected at contact surfaces. It is easier to clean the outside of the tube bundle then and, consequently, only clean air should pass through the tubes.

The radiant-type recuperator is of uncomplicated design, easy to clean and therefore very reliable. But it requires space for installation. **Fig. 3.64** shows the installation of recuperators. They are arranged in the flue gas duct next to the furnace in order to have a short distance for the insulated hot air pipes.

Fig. 3.63: Installation of radiative recuperatos in a remelting plant

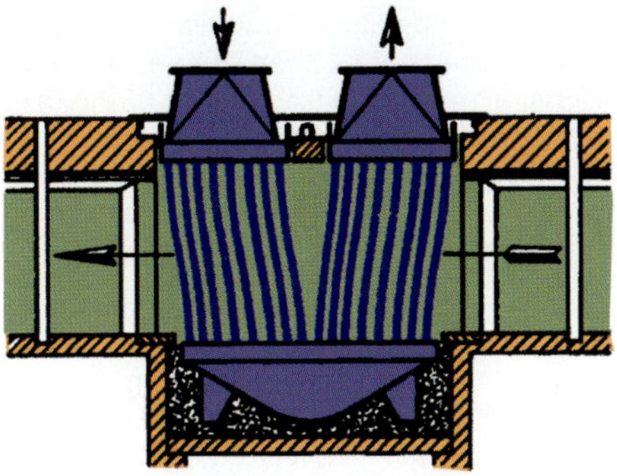

Fig. 3.64: Tube bundle recuperators installed in the ducting system

Fig. 3.65: Tube bundle recuperator (Source: Peiler-Montanwärme)

Due to their small size, tube bundle recuperators (**Fig. 3.65**) can be installed in the flue gas duct in horizontal, vertical or inclined position. These units as well should be installed as close as possible to the furnace to obtain short distances for the hot air pipes.

The service life of a recuperator depends on the flue gas temperature and the presence of aggressive components. Treatment of metal generally takes place with burners switched off. No flue gas is generated in such case and the recuperators are not effective. For the escape of the aggressive gases generated by the treatment agents during this process, a by-pass to the recuperator is a simple technical solution. The by-pass is connected to the pipelines for the central waste gas treatment system. Another possibility would be to dilute the waste gases by introducing fresh air to a level where the concentration of the aggressive components is not harmful to the equipment anymore. From an environmental point of view, this method is not advisable.

The impact of temperature can be reduced by operating the recuperator in the parallel flow mode (**Fig. 3.66, 3.67**). The cold air is introduced at the point of the highest flue gas temperature, thus

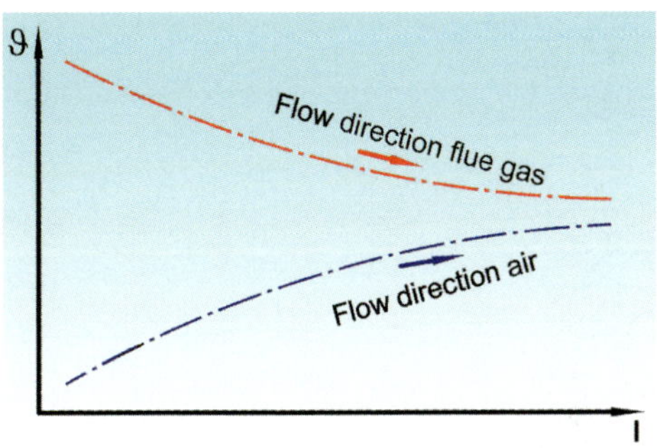

Fig. 3.66: Principle temperatures at different flow condition, counter-current flow

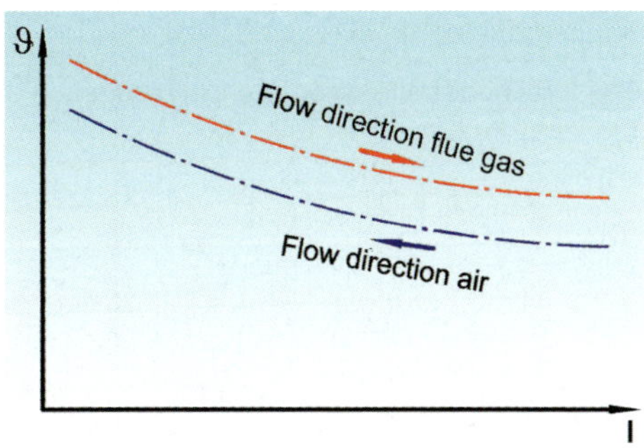

Fig. 3.67: Principle temperatures at different flow condition, parallel flow

cooling the steel components efficiently and keeping the wall temperature low. When cooling the flue gases is required, the necessary cooling air can be introduced at that point as well. However, one should carefully look into this method since additional air may increase the size of the unit. The counter-current flow of flue gases and air in the recuperator is the most efficient mode when energy exchange is concerned. The air is introduced at the point of the lowest flue gas temperature, i. e. at the flue gas exit and flows counter-currently to the flue gases. This mode ensures and maintains a constant temperature difference between flue gas and air.

Parallel flow mode, introduction of fresh air and a by-pass for harmful gases protect the materials, thus increasing their service life.

In the radiative-type recuperator, the heat passed to the combustion air comprises a radiative proportion and a convective proportion. The air receives heat from the surface of the inner tube and from the outer tube as well. The outer tube is heated by means of radiation from the inner tube.

The convective heat Q_C is calculated by the equation

$$Q_c = A \cdot \alpha_w \cdot \left(\vartheta_w - \vartheta_{air}\right) \tag{3.99}$$

with A = surface area of the tubes, α_w = heat transfer coefficient, ϑ_w = wall temperature and ϑ_{air} = air temperature. Since both tubes are made of steel, aw will be identical for both walls and will be in the range of 25 – 35 W/m^2 K.

Energy from the flue gases to the walls of the inner tubes comprises radiation from the flue gases and of convection. Taking into account that the combustion air "cools down "the inner tube, we can fairly assume that the steel temperature will be at least 300 to 400 °C lower than the flue gas temperature. This depends, however, on the flue gas temperature. The temperature difference between the inner temperature of the tube and the outside temperature is small due to the good heat conduction of steel and the thin wall. The outside surface of the inner tube radiates heat to the outer tube according to the equation

$$Q_r = A \cdot C \cdot \varepsilon \cdot \left[\left(\frac{T_{iw}}{100}\right)^4 - \left(\frac{T_{ow}}{100}\right)^4\right] \tag{3.100}$$

We already know the equation rom previous chapters: A = surface area (since the distance between outer tube and inner tube is small, we can assume that the surface areas are almost identical), C = radiation coefficient of the black body = 5.67 W/m^{2}K^4. ε the emission coefficient for steel with oxide skin = 0.8 (this is similar to the coefficient for refractory material) and T the wall temperatures in K. This will result in a temperature of the outer tube that is lower than the temperature of the inner tube.

The total heat transferred to the combustion air comprises the proportions of the convection from the inner tube and the outer tube:

$$Q = Q_o + Q_i \tag{3.101}$$

$$Q = A \cdot \alpha \cdot \left(\vartheta_{wo} - \vartheta_{air}\right) + A \cdot \alpha \cdot \left(\vartheta_{wi} - \vartheta_{air}\right) \tag{3.102}$$

A and a are identical from which follows with ϑ_{wo} and ϑ_{wi} the temperatures of outside and inside tube

$$Q = A \cdot \alpha \cdot \left(\vartheta_{wo} + \vartheta_{wi} - 2\vartheta_{air}\right) \tag{3.103}$$

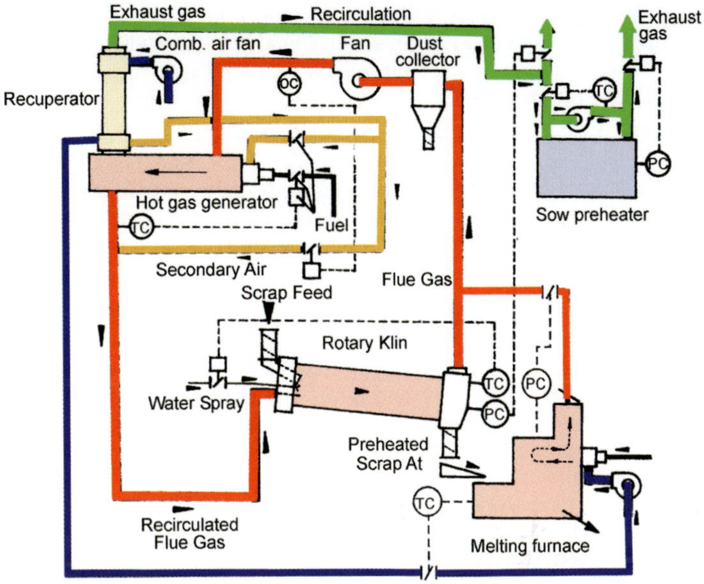

Fig. 3.68: Heat recovery system for bulk material

The calculation of the relevant dimension can be executed by a computer program which is available from the suppliers of recuperator equipment and manually by iteration as we have done in case of the refractory lining. Since the parameters depend on each other, certain values are estimated and the calculation with these values is made until the calculated values match the estimated data with sufficiently low difference. The calculations for a tube bundle recuperator are based on heat transfer by convection. Since they are used in the lower temperature range, the radiative portion can be neglected. Using the equations for the convective heat transfer, the total surface of the tubes, as well as the mode of flow in the unit, need to be considered. The calculation method is either a computer model or iteration.

In order to cover the entire temperature range a combination of both the radiative recuperator as well as the convective recuperator could be arranged in line to each other or as combined unit. What to do is in the end a question of economics.

Although a standard recuperation system for one furnace is simple and easy to arrange, a complete arrangement for heat recuperation may become quite complex if combined with other elements. **Fig. 3.68** shows the principle flow sheet of such an arrangement. The unit is designed for the installation of a twin chamber furnace. It includes the entire treatment of flue gases, including their recycling, and shows the optimal solution that cannot be achieved with other firing systems. The schematic drawing also gives the different energy quantities in kJ/kg of aluminium. We should note that the liquid aluminium discharged will have a temperature of just about melting point which was set by production. The system includes scrap pre-heating and utilization of the heat of combustible contaminations. The heat is used to the extent of the ability of heat absorption by the liquid metal bath in the furnace. An incinerator will burn the organics removed from the surface of the material. Part of the hot combustion products is recycled directly to the furnace. The remaining combustion products are passed to a recuperator to heat up the combustion air for the burners of the hot air chamber of the furnace and the incinerator. The flue gas travels further down the line to a quenching chamber. By water injection the flue gas is cooled suddenly to suppress the re-formation of dioxins cracked in the hot combustion products. The final stage is a filtration plant with calcium hydrate injection for final cleaning and neutralization of acidic contents in the flue gas.

3.4.2.2 Regenerators

Although based on different principles, regenerators are also used for recovering heat from the flue gases leaving the system through the stack. Regenerating is a very old concept that was used in the early days of the steel industry. But with ever increasing cost for energy, systems have been developed to use this technology for furnaces in nonferrous metallurgy as well. The first unit attached to a burner appears in 1983 for a small glass furnace and its use has made remarkable improvements and has been successful in other industries as well.

In a regenerator, flue gas and combustion air alternately flow through a chamber filled with a heat storing medium, charging heat with waste gas flow and reclaiming it with air flow (**Fig. 3.69**). Regenerators, if constructed out of materials which will take and hold the full flue gas temperature from the fired chamber and resist any corrosive elements contained therein, operate very efficiently and reliably. A ceramic heat storing medium can be used in form of small uniformly shaped pieces or balls. Each piece is solid and supported by those adjoining it, thus removing the potential for thermal stress from it. In some regenerator beds, specially shaped honeycomb inserts are used as heat storage medium. These inserts are more exposed to thermal stress problems than the beds packed with small pieces. Generally, suppliers of regenerative burners claim that the heat transfer performance of the regenerator is significantly better than that of a recuperator. This may be true if one considers that the specific surface area of the regenerator bed is much bigger than in a recuperator. It cannot be seen that the heat transfer coefficient is also increased. As outlined in another chapter, the flow of gas and the resulting heat transfer are very complex and

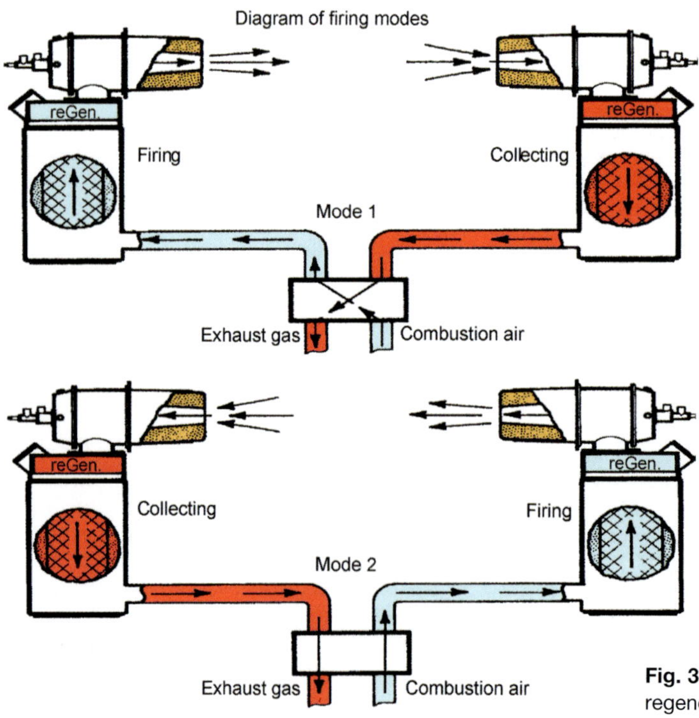

Fig. 3.69: Principle operation of a regenerative burner system

cannot be judged easily. The main difference is certainly the higher application (service) temperature. It is claimed that the air is preheated to a value of close to 1,000 °C or even higher. This will inevitably result in an extremely high NO_x generation. The first step was to develop a waste gas recycling system (EGR) which was able to reduce NO_x generation, of course to the disadvantage of energy efficiency. Later development work was done to obtain a low NO_x level by introducing a low NO_x burner which certainly can be used as hot air burner as well, permitting perhaps a higher pre-heat temperature for a recuperator. Data from operating plants, based on the operation of one individual furnace, have apparently not been published. Anyhow, the content of NO_x in the waste gas must be monitored for the entire plant over a certain period of time to arrive at reliable data for plant operation. Snapshots are not sufficient to define criteria.

The regenerative burner system consists of a high temperature burner coupled to a regenerator bed. The burner serves double duty, acting also as the exhaust port for the fired chamber. One regenerative burner system comprises two burners and two regenerator beds connected via a reversing valve and the required reversing logic (Fig. 3.56). While one of the burners fires by using cold air fed to the base of its regenerator, flue gas is drawn through the other burner and through its associated regenerator bed to pre-heat the packing and then discharged to the atmosphere. When the regenerator is heated and sufficiently charged, the reversing system operates and cold air flows to the newly heated regenerator and is pre-heated while the previously cooled regenerator is now reheated by the flue gas from the other burner now firing. The firing period and thus the changeover will have to be in short intervals in the range of 20 seconds. This depends on the ability of the regenerator bed to absorb heat and pass it again to the air being preheated in the reverse mode.

For more details on regenerative burner systems please refer to section 3.3.3.2

We should now summarize and compare the methods of heat recovery from waste gas leaving the furnace. For all systems extracting heat from the flue gases, the efficiency of the system depends very much on the flue gas temperature. In the lower range of a furnace cycle, the efficiency of the system is comparatively low. This is different in the upper ranges of the temperature which will occur towards the end of the melting cycle. It applies very much for the twin chamber furnace where most of the heat is introduced via the liquid metal bath in the heating chamber. The increase of efficiency is due to the higher combustion temperature which in turn depends on the temperature of the pre-heated flue gas. This allows a decrease of gas input to the furnace. Higher combustion temperature will result in higher NO_x generation. Therefore, there will be a limitation to the combustion air temperature. Generally, the regenerator is able to produce higher temperatures due to the ceramic components of the generator material. The critical section will be components of the burner unit. With the recuperator the metallic components restrict the air temperature obtained from the unit. Since it is necessary to stay within the ranges as defined by the environmental regulations, there will be no particular problem.

Regenerator burners are supplied as compact units attached to the furnace. They do require some space around the burner unit and also for piping and valves. Furthermore, a minimum of two units is required. These space requirements have to be considered when designing the furnace. Easy access to the generator bed for removal and cleaning the generator must be ensured. Recuperators are arranged next to the furnace in a way to allow access for cleaning as well. Apart from the large insulated pipes for the hot combustion air, no particular space is required at the furnace.

All systems used for heat recovery are sensitive to dust and aggressive components of the flue gas. These are usually produced by treatment salts or gases in the furnace. In case of the regenerator, the regenerator bed will clog immediately and in the case of the recuperator the steel components will wear very fast. Special measures are required to by-pass the units during the treatment phase and shield the burner nozzle. All systems fail in dusty furnace atmosphere. They can only be used for clean metal processed in melting furnaces of the downstream industry or for holding furnaces. But also in these furnaces some dust is generated. Thus, frequent cleaning of the units is essential. The cleaning periods depend on treatment and material processed.

The decision to use waste heat recovery for a specific furnace has to include the analysis of the entire plant operation. This must consider environmental impacts, operating and maintenance cost as well as investment. This all is compared with the energy savings.

One important factor that cannot – yet – be expressed in economical figures is common to all waste heat recovery systems. The quantity of energy required is reduced and so is the amount of carbon oxide released to the environment. This advantage must not be lost by increasing the quantity of poisonous gases such as NO_x.

3.4.2.3 Incinerators

We know meanwhile that scrap material charged to the melting furnace usually contains a certain quantity of organics. These have to be removed from the metal under the condition that no or at least very little oxidation occurs at the surface of the individual pieces of metal. This request can only be fulfilled if the temperature on all material surfaces is below 400 °C and no oxygen is present for reaction with aluminium. Consequently, the flue gas leaving the furnace or the furnace section carries a substantial amount of not reacted, i.e. not burned combustibles. These cannot be released into the open in addition to the fact that their chemically bonded energy could contribute to the heat economy. If these combustibles cannot be burned directly in the furnace, as is the case in a tiltable rotary drum furnace, a separate incinerator or afterburner unit (**Fig. 3.70**) is required. This comprises a combustion chamber consisting of a refractory lined housing, equipped with a burner at one side and the flue gas exit at the other side. The flue gases enter from the furnace and mix with the hot combustion products of the incinerator burner to raise their temperature above the igniting temperature of approximately 900 °C. Combustion products and flue gas from the furnace must be mixed thoroughly so that no unburned components leave the combustion chamber. The firing side of the chamber is designed as separate quarrel (square block) with small throat which increases the velocity of the combustion products. The flue gases from the furnace are injected by one or more entry points arranged tangentially at the circumference of the housing. This creates extensive turbulence and good mixing conditions. The incinerator is sometimes designed as rectangular housing with cross baffles to obtain the good mixing behavior. To make sure that all combustibles are burned, the environmental regulations require a residual time of the flue gases of at least 3.5 seconds. This requires quite large sizes for the incinerator. The time required can be obtained either by reducing the velocity of the flue gases by having a large cross-section or by a long distance to travel at higher velocity. Mixing is efficient at high velocities. Thus, additional flow obstruction is not required. At low velocities some baffles are helpful.

Incineration is not always an energy-saving method. The flue gases of the furnace may have a temperature that is below the igniting point. Before afterburning can come into affect, the temperature of the flue gas must be lifted to at least 900 °C. This requires additional energy input by the incinerator burner. The additional energy can in most cases, together with the energy obtained from burning the combustibles of the flue gas, be recycled to the furnace. Part of it has to be taken out of this closed loop since all combustion products introduced have to leave the system

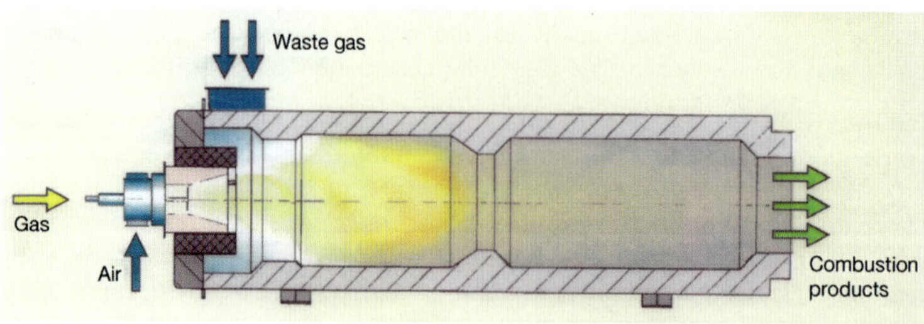

Fig. 3.70: Incinerator

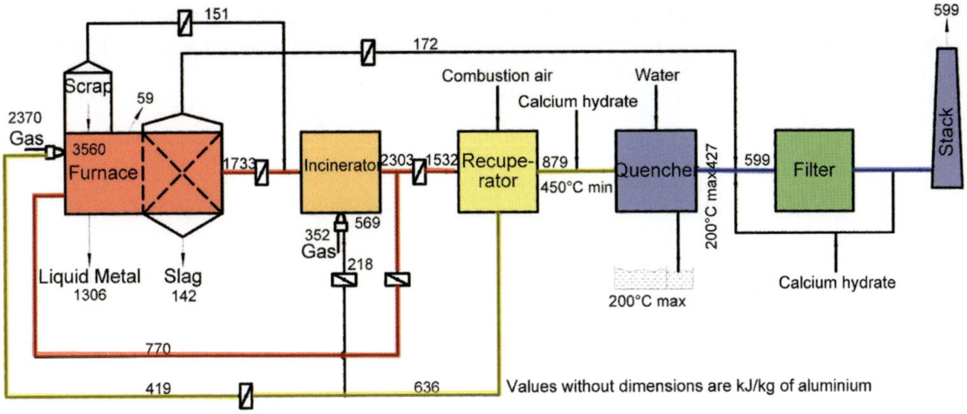

Fig. 3.71: Flow sheet of a heat recovery system for a twin chamber furnace

eventually. A recuperator installed downstream of the incinerator will be able to recover some of the energy. Please refer to **Fig. 3.71**, showing the principle flow sheet for such an installation. It may be possible that the content of combustibles is comparatively low and not much energy can be obtained. It may be still be necessary to operate an incinerator since all carbon and volatiles must be removed from the flue gas to fulfil the environmental regulations.

The burner input to the incinerator is needed only as long as organics are present. During this period combustion air, as required to provide the necessary oxygen, is fed to the incinerator together with the waste gas and both components are thoroughly mixed in the unit. To do so the combustion air blower for the incinerator burner will supply the necessary quantity of air. As the content of combustibles decreases, the content of oxygen in the final combustion products will increase. Measured by a lambda probe; air and energy input will subsequently decrease automatically down to a level where no energy input is required anymore since all organics are removed from the scrap.

If flue gas is recycled directly from the incinerator to the furnace, it is advisable to install a hot gas cyclone for removing dust from the flue gas. The dust load of the waste gas depends very much on the scrap charged to the furnace. Many of the scrap materials are, for instance, painted. This paint consists of metal oxides providing the color and organic binder that will evaporate by the pyrolytic process in the furnace. But even with clean metal, a certain amount of dust is generated because of the unavoidable oxidation of aluminum.

We will look at an example of an incinerator system.

The material charged is contaminated by oil and paint. The contents of the organic contamination (oil and binder of paint) will be 5 %. The melting rate shall be 1 t/h. At the first stage the furnace shall be loaded with 2 tons of scrap.

The quantity of combustibles is

$$G_C = 0.05 \text{ x } 2{,}000 = 100 \text{ kg}$$

With lower heating value of 41,800 kJ/kg, the chemically bound energy Q_C of the combustibles is

$$G_C = 41{,}800 \text{ x } 100 = 4{,}180{,}000 \text{ kJ or } 1{,}161 \text{ kWh}$$

This is more than required for a melting rate of 1 t/h, which is 600 kWh/t for a furnace with heat recovery for melting and superheating. On the basis of a lower heating value for gas of 10 kWh/m³ for natural gas, we obtain the required gas quantity V_G

$$V_g = \frac{600}{10} = 69\ m^3/h$$

With our approximation (12 m³ of combustion products per m³ of gas) we calculate the volume of flue gas to be

$$V_{flue} = 12 \cdot 60 = 720\ m^3/h$$

The flue gas temperature entering the incinerator shall be 700 °C. We have to lift this temperature to the igniting level, plus some safety margin, to 900 °C. The energy Q_B to do this will be

$$Q_B = V_{flue} \cdot c_{p\,flue} \cdot \left(\vartheta_{ign} - \vartheta_{flue}\right) \tag{3.104}$$

$$Q_B = 720 \times 1.38 \cdot (900 - 700) = 60{,}720\ kJ/h \text{ or } 16.9\ kW$$

The burner required to raise the flue gas temperature to this level will require

$$\frac{16.9}{10} = 1.69\ m^3/h \text{ of natural gas}$$

resulting in a flue gas volume of

$$12 \times 1.69 = 202\ m^3/h$$

As we have calculated before, the energy of the 100 kg of combustibles is much more than required for the melting rate of 1 t/h and also much more than required to raise the flue gas temperature. After ignition, the temperature of the combustion products will increase to approximately 1,200 °C. Thus, the burner of the incinerator could be switched off after ignition as it may be the case with the main burners in the furnace. For safety reasons, it is usually kept in operation at low fire. Furthermore, the oil concentration in the flue gas fluctuates and sometimes it may not be sufficient to maintain a steady state operation.

For the combustion of the oil content, air must be injected to provide the necessary oxygen. Using our approximation of 11 m³ of air per kg of oil we arrive at

$$V_{air} = 11 \cdot 100 = 1100\ m^3/h$$

Now we summarize the total exhaust gas quaNow we summarize the total exhaust gas quantities leaving the incinerator.

flue gas	720 m³/h
pilot burner	202 m³/h
oil combustion	1100 m³/h
total	**2022 m³/h**

From this we can only recycle 1,300 m³/h to the furnace because 720 m³/h are newly injected by the furnace burners. Actually we would be able to run the furnace with the energy leaving the incinerator without additional burner input only on the basis of the content of combustibles in the scrap. The temperature of combustion products leaving the incinerator will be approximately 1,200 °C.

For evaluating the above figures, we have assumed that the decoating, i. e. removal of combustibles, takes one hour and that the values are based on averages for this period. However, the time for this process will last 10 to 20 minutes only. During this time, the furnace burners may operate at their peak capacity which is generally 30 to 50 % higher than the average required for melting the aluminium and superheating the aluminium as well as for compensating the wall losses. For comparison we will convert the values to rates per minute. The burner input will be based on an excess input of 50 % and then needs to be divided by 60 and likewise the input of the pilot burner. The decoating lasts only 20 minutes. Thus, the relevant values are to be divided by 20.

The above figures would now read

flue gas	18.0 m^3/min
pilot burner	3.4 m^3/min
oil combustion	55.0 m^3/min
total	**76.4 m^3/min**

From this we can recycle 58.4 m^3/min to the furnace; 18 m^3/min are newly injected by the burners. The heat transfer in the furnace must be considered for final evaluation of the energy balance. It may very well be that the heat generated by the combustibles in the scrap is more than the energy that can be absorbed by the metal. The contamination may also be less than 5 % and then the conditions change altogether. The critical question is now: How much energy can be recycled anyhow? Definitely, all quantities of generated flue gas must leave the system. Consequently, finally all flue gases leaving the incinerator have to go to the stack eventually. We are able, however, to decrease the input of the furnace burners and supply the same quantities of combustion products normally generated by these burners to the furnace and keep them in a closed circulating loop. We could even have more flue gas in such a closed loop, provided the batch is able to absorb the equivalent energy.

The flue gases leaving the incinerator are free of unburned organics. They could be charged to the furnace and released from the system by a flue duct in the furnace controlled by a damper to branch off the quantity required for decoating the scrap. This would reduce the load in the closed loop of the furnace system.

This method will result in higher temperature of the flue gas. As in any heat recovery system, the heat transfer is improved but the limits are set by the refractory lining in the furnace and ducting. Anyhow, since the energy released from the contaminations is usually not considered in the energy balance (and is free of charge), recycling of these gases contribute to the improvement of the fuel economy of the furnace.

For sizing the incinerator, all volumes are at mean conditions: for the actual flow rates we have to relate the volumes to actual conditions.

$$V_{act} = V_m \cdot \frac{T_{act}}{T_m} \tag{3.105}$$

With V_{act} = volume at actual condition and V_m = volume at mean conditions (ϑ =20°C), T_{act} = actual temperature in K and T_m = mean temperature (273 K). We have calculated before a flue gas volume of 70.4 m^3/min at mean conditions. This will now be

$$V_{act} = 70.4 \cdot \frac{1200 + 273}{273} = 380 \ m_a^3/min \text{ equivalent to } 6.33 \ m^3/sec.$$

The combustion chamber shall have a diameter of 1.5 m with a cross-section of 1.8 m^2. From this, the actual gas velocity is $\frac{6.33}{1.8} = 3.51 \, m/s$.

In order to fulfil the requirement for a residual time of 3 seconds, the combustion chamber will require a length of 3.51 x 3 ≈ 11 m. This is too much and we must increase the diameter to 2.5 m with a cross-section of 4.9 m and now get a gas velocity of $\frac{6.33}{4.9} \approx 1.3$ m/s with the resulting length of 3 x 1.3 = 3.9 m.

Finally we arrive at reasonable dimensions.

The installation of an incinerator requires some space and capital investment. It comprises a steel housing that is lines with refractory materials and equipped with a burner system as well as the related control equipment. If the furnace design permits, the flue gases can directly be recycled to the hot metal chamber of a twin chamber furnace. In this case the furnace chamber works as incinerator. This is particularly efficient if using oxygen.

3.4.2.4 Charge pre-heating

In the above section we referred to a furnace that was charged with batches of a defined size. The combustible contaminations where released in a short period of time with the effect that the total quantity of energy available during this time could not be fully utilized in the furnace. A more balanced energy supply is possible if the materiel is preheated prior to charging. In a twin chamber furnace this is done in a separate chamber. Apart from the advantages in regard to de-coating and metal recovery, there are more benefits relating to energy efficiency. Placing the material on the preheating bridge within a reverberatory furnace offers certain advantages but does not improve the heat economy remarkably. The extension of the preheating bridge in the furnace as dry hearth also improves the heat economy since the hot waste gases have to pass the bulk of scrap on their way out, thus transferring their heat to the scrap. This is also the case in a twin chamber furnace. The dry hearth can be designed as separate unit, arranged in front of the furnace (**Fig. 3.72**), whereby the material is fed via a conveyor system into the furnace.

Only preheating outside of the furnace, using the waste gases, results in a notable increase of the heat efficiency of the system. For block material, preheating can be realized in a separate chamber through which the flue gases of one or more furnaces pass on their way to the stack. The idea sounds simple but the realization is more difficult. When transferring the block from the preheating chamber to the furnace, time is of essence. Charging of the furnace and discharging of the preheating chamber must be as fast as possible. Using a forklift truck, the blocks can be placed into the furnace quite efficiently. It is of very limited use to have a small quantity of blocks preheated. Thus, a larger size of the preheating unit is essential. This must also allow handling of boxes filled with small metal pieces or stacks of ingots. Therefore, the preheating chamber requires some special loading and reclaiming facility. A forklift truck can now pick up the material and transport it to the furnace. A definite advantage in using such a system is the possibility to preheat sow ingots. These are often stored in the open or transported over a long distance. During transportation, water is collected at the surface and in small cracks, fissures and shrinkage cavities. If the sows are charged into the furnace directly, the trapped water may cause an explosion and endanger operators and equipment. In general, however, it appears that separate heating chambers cannot be justified economically if only considering heat recovery and today's cost of energy. This may change in the future since steady increase of energy cost seems to be

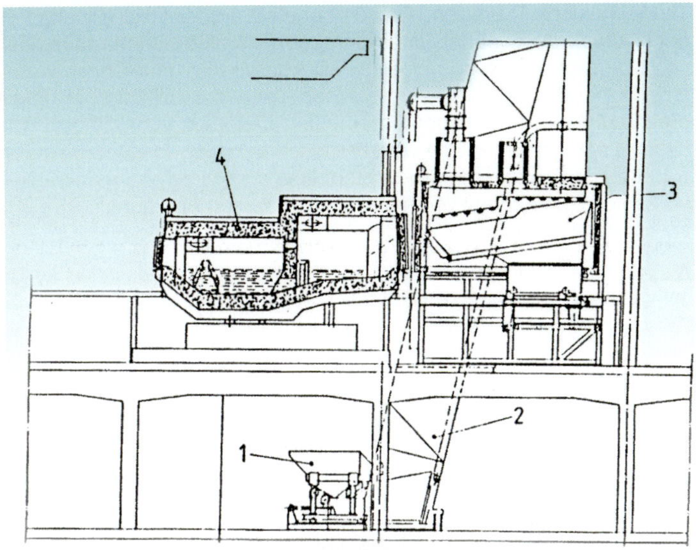

1 - Feed line
2 - Skip hoist
3 - Pre-heat chamber
4 - Melting furnace

Fig. 3.72: Preheating chamber for scrap (Source: Maerz Gautschi)

inevitable. Preheating of smaller pieces supplied as bulk is easier to handle. Used beverage containers (UBC) as well as swarf and chips from machining mostly undergo pre-treatment to remove oil or paint. This is done in special plants located in an area not necessarily close to the melting plant. Energy is required for this processing and the material needs to be cooled for handling. If located next to a furnace, waste gases can be used as energy source and the heated material is passed to the furnace without cooling. This is already a two-step improvement: Heat generated from the combustibles in the scrap can be utilized in the furnace and the material has reached a certain temperature level prior to charging. Such a system is depicted in Fig. 3.68. The flue gases leaving the furnace are directed to a hot gas generator (incinerator) for burning the combustibles. Hot gases leaving are then passed to a recuperator which provides preheated combustion air to the furnace burners as well as to the de-coating and preheating unit. Normally this unit comprises a rotary drum but could also be a special linear conveyor or a fluid bed system as described in chapter 2. Except swarf and UBC, the drum de-coater can also process small metal pieces such as shredder. Material passes the drum in parallel flow to the combustion products. Additional combustion is provided by the organics contained in the scrap. While flue gas leaving the drum at the material exit is directed to the incinerator, the treated material reaches the furnace via a chute. The furnace requires a suitable system for submerging the small material particles in the liquid metal bath. This has been described in previous chapters in detail. Some or all of the waste gases leaving the preheating drum may be directed to the furnace for better heat exchange. The variations have to be evaluated considering plant concept and material basis.

Such combined systems offer advantages because of the possibility to continuously feed material to the furnace, thus providing also a more or less constant quantity of combustible contaminations. The constant feed ensures much better utilization of the energy obtained by burning the organics. Metal can be tapped frequently as soon as the furnace is filled with a sufficient quantity of liquid metal, while melting commences without interruption. As you may already assume, there are limitations. For such continuous operation, alloy composition of both product and feed material must be identical. Minor adjustments to the analysis can be made in a downstream casting furnace.

The technology is efficient if tailor-made for a specific application. In a normal operation different kinds of raw materials are processed. Therefore, the furnace needs to be designed to permit charging of a wide range of materials with all their differences in composition. Machined chips, shredder and small pieces of scrap are usually a collection of different alloys which means the organization of a continuous production becomes very difficult. The melting furnace could be used as casting furnace as well. But since furnaces, in a setup as described, must necessarily be a stationary furnace, the problems of flow control as well as temperature management will come up as described in one of the previous chapters. Additionally, the process can be extended by a sow preheating station.

A system for preheating material must be economically justified by the energy savings only. Since investment cost is high, maintenance is substantial and operation requires skill. Such a justification might be achievable in very specific cases only.

3.4.2.5 Tower furnaces

In this type of furnace the scrap is fed to a vertical shaft. (Please refer also to section 2.1).The waste gas must pass the entire bulk of material on its way to the stack – the flue gas is cooled, the material is heated. The heat exchange is very intensive since the gas flows around every individual particle. Aluminium in the lower part of the furnace is melted and the liquid metal flows down into the collecting chamber of the furnace. The heat recovery is very efficient and the system is successfully applied for small melting rates and for melting of ingots and risers as well as no-go castings as typically required in die-casting plants. Although the principle should be useful if small metal pieces are to be melted, practise has shown a very high metal loss in this case. The reason could be that metal starts melting in the upper region of the tower already, solidifying when coming in contact with colder metal on its way down and being melted again later. The resulting high melt loss is the reason that this type of furnace is not used in the secondary aluminium industry, although very heat-efficient.

An interesting solution involving a tower furnace for avoiding the problem of pre-melting to a certain extent is shown in **Fig. 3.73**. The tower is separated into different preheating sections arranged on top of each other which drop the material from section to section via bottom flaps. The flaps must be designed to resist the temperature, let the flue gas pass and still hold small pieces of metal. As soon as the material on the lowest level has reached the required temperature – higher for larger pieces such as ingots and lower for smaller particles such as machined chips to avoid extensive oxidation – the bottom flap opens and the metal drops into the liquid metal section

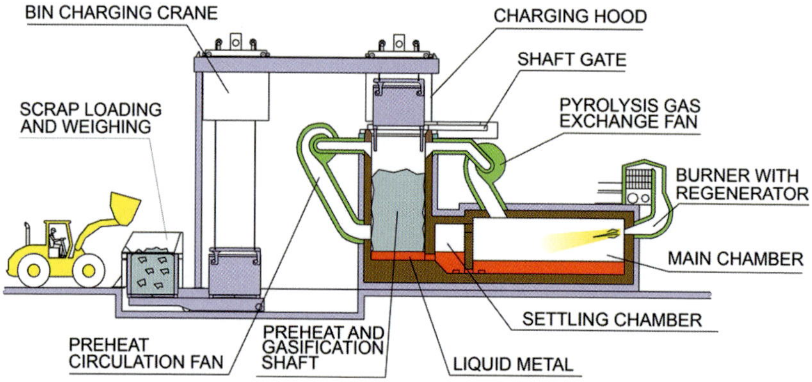

Fig. 3.73: Tower furnace with different pre-heating stages

of the furnace. Now the other levels are charged to the level below step by step until the upper level is empty and can be charged with fresh material. The temperature of each preheating level can be set individually. This permits processing of different kinds of scrap and the heat efficiency is good. The top layer is charged by means of a skip hoist and no further handling of hot scrap is required outside the furnace, as may be the case with other technologies.

The Hertwich system is such a special design of a tower furnace. One charging box is arranged above the melting chamber of a twin chamber furnace. Small and oil-bearing material can be charged into this box and released via a bottom flap into the melting chamber. Of course this furnace must be equipped with a pumping system for metal circulation that should be standard in a twin chamber furnace anyhow. The technology is comparatively flexible in regard to the kind of scrap, provided the contaminations of the scrap consist of organics with small quantities of metallic and non-metallic additions.

The stack losses represent the highest proportion of losses in an aluminium recycling furnace. There are different methods for recovering heat otherwise lost to the atmosphere. The most efficient processes to recover this energy are recuperators, regenerators and charge preheating. The decision depends very much on raw material and product as well as on plant setup as well as on which technology is the most promising and economically sound. Some technologies are sophisticated and require high capital investment. A balance has to be found between operation cost, i. e. price of fuel, operation and maintenance cost, ecology and economy. The decision, which and to what extent heat recovery systems should be installed, is the result of a careful analysis and may lead to different results for different projects.

3.4.3 Wall losses

Inside temperatures of a furnace are defined by the furnace process but outside shell temperatures must have a value that is not much above the ambient temperature. This would be the ideal case that cannot be achieved in practise. But the higher the temperature difference, the higher the losses through the shell, i.e. the furnace wall. An acceptable compromise has to be found. There must be no danger for operating and maintenance staff when touching the furnace shell. A temperature in the range of 100 °C is commonly accepted in aluminium recycling plants. As we have seen in previous chapters, the sizing of the refractory lining and the selection of lining materials require balancing of the parameters to arrive at optimal results. As we have defined, the 658 °C isotherm should preferably be within the hot face layer. This is to avoid maintenance problems but the temperature of the outside shell should be as low as possible to reduce heat losses. Well, this could be achieved by a very thick refractory layer, a measure that is prohibited by economics.

One other factor is usually underestimated. It is common practise that furnaces are painted with silver color. The walls do radiate heat to the environment. This is expressed by the equation which is meanwhile very familiar to us:

$$Q_{rad} = A \cdot \varepsilon \cdot C \cdot \left[\left(\frac{T_W}{100} \right)^4 - \left(\frac{T_A}{100} \right)^4 \right] \tag{3.106}$$

with A = surface area, ε = emission factor, C = emission factor of the black body, T_W and T_A = temperatures of wall and surrounding. The emission factor ε for a silver painted surface is 0.3, for a grey wall the factor is 0.8. As we see immediately, the wall loss caused by radiation of a silver painted wall is more than double that of a grey surface. The consequence should be: no silver paint for furnace walls anymore, if an energy-conscious philosophy is applied.

The heat losses through the shell will be different for specific areas and changing furnace sequence. In the bath area the inside temperature is lower than for the upper part of the furnace but the heat transfer increases as soon as the metal is tapped since the bath area is now exposed to the much higher temperature of the upper furnace. Since the refractory lining has to be designed to resist the particular conditions of the metal bath, the temperature of the shell should theoretically increase but, due to the fly wheel effect, no temperature increase can be measured at the shell, at least while the empty furnace is not kept at peak temperature for hours which is very unlikely. Similar applies to the upper furnace. The hot face in this area is exposed to the high temperature and occasional metal splashes. Therefore, the rear refractory lining is designed for optimal heat insulation. Temperature fluctuations inside the furnace can also not be noted at the outside shell of the upper furnace area. During heating up of a cold furnace, which could occur after a complete relining or after a lengthy maintenance period, the shell temperature will gradually increase until a steady state is reached. Different areas might have slightly different temperatures. But sometimes high temperatures are noted at a certain location on the steel shell. These so-called hot spots are an alarming indication. They only occur if there is some trouble in this area of the furnace. It could be that metal has penetrated the refractory lining or that the refractory lining is damaged due to a mechanical impact. Sometimes the hot spot is large and shows slow increase of temperature. This indicates wear of the refractory lining. Monitoring the shell temperature and comparing the new data with the previous records give a good indication of the shell losses and the condition of the refractory lining. This should be done at regular intervals and should be recorded within the overall energy balance of the furnace. Temperature readings can be done with a touch thermocouple or with a heat picture camera. The camera gives a very comprehensive view of the various temperatures and the identification and location of hot spots to initiate or decide about immediate action or planned maintenance.

3.4.4 Door losses

Door losses are defined as losses through doors and other openings of the furnace. They comprise losses caused by hot gas flowing out by furnace leakage. This is mainly the large door when opened for charging, skimming and tending. Other leaks are tapping spout and charging spout for liquid metal as well as additional smaller doors which may be required for operation. In the sophisticated design of furnaces of the newer generation, lids for these openings seal the furnace even when operated under overpressure. Very much care has been taken for the design of a large main door, as described in section 1.

During operation a small overpressure is maintained in the furnace to avoid ingress of cold air. This adds to the heat economy but may have a negative effect on door losses. If leakages occur in case one of the sealings fail, hot gas will leave the furnace and its energy must be regarded as lost. Also when opening the covers, some gas may leave the furnace. Therefore, opening of any lid or door should be restricted to a minimum. For a short period of time the furnace overpressure can be switched to negative pressure to protect the operators working in this area. This is not possible during casting or liquid metal feed. Door losses will be at their maximum as soon as the large furnace door is opened for furnace charging and tending. Additionally to the hot gas leaving the furnace, radiation from the interior will cause loss of energy. During this period the burners are switched off but the temperature difference between furnace interior and furnace surroundings causes a natural circulation of air between inside and the furnace surrounding. Furthermore, in order to maintain an acceptable working atmosphere, the hot air or even the hot gases occurring during charging and metal treatment are removed from the front of the furnace by a large fume-collecting hood arranged above the door. Protection against radiation of heat can only be realized by proper clothing and face protection of the operators. In order to reduce the door losses, all openings must be closed immediately as soon as there is no work undertaken requiring an open door.

A proper measurement of the door losses is very difficult. Their value is small compared to the other losses and enters the heat balance with a factor of 2-3 % only. In the heat balance, door losses will be a remaining factor obtained by calculation. It will include all inaccuracies of the other measurements and calculations are also included in the values. Regarding a calculation, different parameters, such as gas composition, gas quantity leaving the furnace as well as the accurate temperature, have to be known. Therefore, it is quite clear that the calculation also bears some inaccuracies.

Basis for a calculation of the quantity of the gas leaving the furnace is the outflow equation of hydrodynamics

$$V = \mu \cdot A \cdot \sqrt{\frac{2g \cdot p}{\rho}} \tag{3.107}$$

with V = gas quantity, μ = flow factor (0.6-0.7), A = outflow cross section, ρ = specific density of the gas and p the pressure in the furnace. Under the condition that the density of the flue gas is in the range of 1.3 kg/m^3 in standard conditions and the furnace temperature is 1,200 °C, the gas quantity can be calculated as

$V = 1740 \cdot A \cdot \sqrt{p}$ expressed as mean m^3/h

The pressure should be set to a zero value when working around the furnace. However, the draft changes with the height of the furnace. The furnace pressure is measured at the furnace roof. Thus, adjusting the pressure to the working level requires some correction. Looking at a furnace temperature of 1,200 °C, the draft in the furnace will change with every meter of elevation by 9.6 Pa.

Example:

The over-pressure of a furnace is set to 50 Pa. The width of the tapping spout is 0.2 m, the height above the metal level is 0.05 m. The burner rate of the furnace is 3 MW, equivalent to 300 m^3 natural gas with a lower heating value of 10 kWh/m^3 generating a flue gas volume of 12 x 300 = 3,600 m^3/h at mean condition. From this the gas velocity can be calculated

$V = 1740 \cdot 0.05 \cdot 0.2 \cdot \sqrt{50} = 123\ m^3/h$ at mean conditions.

This would be, if the furnace is working at maximum rate,

$$\frac{123}{3600} \cdot 100 = 3.4\ \%$$

A value that is quite high. Since over-pressure is established in the roof, the actual pressure at the bath level is lower. At a height of 2.5 m above the bath level, the pressure will be

$$50 - 2.5 \times 9.6 = 26\ Pa$$

From this the resulting flow rate out of the tapping spout will now be

$$V = 1740 \cdot 0.05 \cdot 0.2 \cdot \sqrt{26} = 88.7\ m_n^3/h$$

which is only 2.4 % of the total gas flow.

The example shows that the loss occurring by gases escaping through openings and leaks of the furnace should not be underestimated. Also the interior furnace pressure should not be set at a value that is much higher than required. At metal bath level, a pressure of 1 Pa should be sufficient. The resulting pressure-setting would be in such a case

$$1 + 2.5 \cdot 9.6 = 25 \text{ Pa}$$

with a resulting percentage of escaping flue gas of 0.5 %.

The example also shows again that settings and definition of parameters in a furnace should be made with care. If the furnace is operated at a pressure below the atmospheric pressure, door losses occur. Cold air flows into the furnace and is heated up. This energy is to be regarded as lost as well. For the calculation, equation (3.107) can be used as well.

The major loss will be by opening the large furnace door. The gas flow out of the furnace will be ruled by the natural draft of air generated by the heat exchange between furnace temperature and ambient air. The largest quantity will be due to radiation from the furnace to the outside.

Example:

The above furnace shall have a door of 5 m length and 2.5 m in height. The emission factor of the refractory lining shall be 0.8. Interior furnace temperature shall be 1,200 °C or 1473 K, outside temperature shall be 20 °C or 293 K.

The heat radiating through the open door is calculated using the equation

$$Q_D = A \cdot \varepsilon \cdot C \cdot \left[\left(\frac{T_F}{100} \right)^4 - \left(\frac{T_o}{100} \right)^4 \right]$$

$$Q_D = 5 \cdot 2.5 \cdot 0.8 \cdot 5.67 \cdot \left[\left(\frac{1473}{100} \right)^4 - \left(\frac{293}{100} \right)^4 \right] = 2{,}655{,}700 \text{ kJ or } 738 \text{ kW}$$

This would be the energy the furnace would lose within one hour by radiation if the door is kept open. Certainly the burners would be switched off and the heat must be provided by the furnace. Considering an empty furnace, the energy must be provided by the refractory lining. Having the door open for a long period of time would result in decreasing the interior temperature down to approximately 400 °C. But even with an average temperature of (1473 + 673) ∏ 2 = 1073 K, the heat loss will be 207 kW.

The open door blade will also radiate heat to the equipment installed at the roof. For protection, a heat shield is arranged opposite to the door blade in the upper position. This shield is lined with refractory materials which also reduce the heat losses of the refractory lining. In a closed position the door losses due to heat conduction are part of the wall losses.

In former days, the door frames were water-cooled to avoid distortion with the resulting gaps, i. e. door losses. The heat removed by the cooling water contributed as well to the door losses.

Much care has been taken by the furnace designers to reduce door losses. These efforts should not be diminished by poor operating practises. Sealing of doors, taphole and charging well should be monitored frequently and leakages should be stopped as soon as possible. The door must always be closed and only opened for the necessary operation which should be performed as fast as possible. Operators should prepare charges prior to opening of the door to reduce charging time, i.e. time with the door open.

Although door losses represent only a fraction of the total energy losses, all efforts to save energy are finally part of the total energy balance of a plant.

3.5 Furnace design

Christoph Schmitz

3.5.1 Mechanical structure, general

The structure must be designed to resist all mechanical forces due to furnace movement and thermal stress without deformation. There are different forces acting on the steel structure. First, we have the weight of the metal charge. Then there is a very heavy load which is the weight of the refractory lining and, last but not least, the steel shell must also carry its own weight and, on top of that, all attachments such as burners and burner control, combustion air blowers, pressure control dampers. Finally, there are dynamic forces caused by impacts from metal charging and furnace movement. In case of the rotary drum furnaces, the movements of materials sometimes cause very severe dynamic forces. Some of these forces require due consideration. In a stress analysis, some forces are difficult to obtain and require good judgement. Some do not need consideration since their values are negligible in comparison to others. Usually, even the weight of the batch is small compared to the weight of refractory material and shell. A reverberatory furnace, having 15 tons bath capacity, requires approximately 80 tons of refractory material; the weight of the steel shell is roughly 30 tons, adding up to 110 tons total load. Comparing these loads, we notice that the weight of the metal amounts to only 14 % of the dead load or 12 % of the total load. Looking at a stress analysis, also the weight of attachments is generally already included in the safety margin of the calculation. In some cases the specific area of the furnace must be looked at. This may be the case, for instance, if heavy regenerative burners with their regenerator bed are to be installed.

Prior to undertaking a stress analysis, the furnace designer must make sure that all forces are taken care off and that their impact results only in a minimum of deformation of the structure. Weak furnace structure will inevitably cause numerous interruptions of production, extensive maintenance work and finally total or partial destruction of the furnace. And there is another fact to be considered. It is very difficult to design and to build any equipment in a casthouse to be "operator-proof". The target is to get optimal production, very often at the limit of the capabilities of the equipment. Consequently, the operators often do not operate their equipment very carefully to the regret of the designer.

Due to the importance for production, hearth furnaces and rotary drum furnaces are discussed in more detail. For other furnaces, such as induction furnaces, we will have a short overview of design principles only.

3.5.1.1 Hearth furnaces

Furnace shell

Stiffness of the furnace is a very important criterion for the design of hearth furnaces. Elastic movement of the shell would cause damage to the much less flexible refractory lining. Already little stress applied to the refractory material can result in remarkable reduction of service life.

The stationary furnace (**Fig. 3.74**) generally comprises a rectangular, open box with one side, i. e. the door side, not existing. The shell consists of a steel plate of adequate thickness that rests on heavy steel members across the complete furnace width which, in turn, sits on concrete slabs. The individual base members are reinforced by steel shapes. Up-going structures should also be built sturdily where possible. Basically, the walls consist of steel plate reinforced by structural members. Where openings are required, i. e. burner ports or casting spout area, the shell is locally reinforced with structural steel members or heavy steel plate.

For tilting furnaces, (**Fig. 3.75**) the design becomes more difficult. Tilting forces have to be taken into account when doing the stress analysis. If the furnace could be regarded as closed box, this

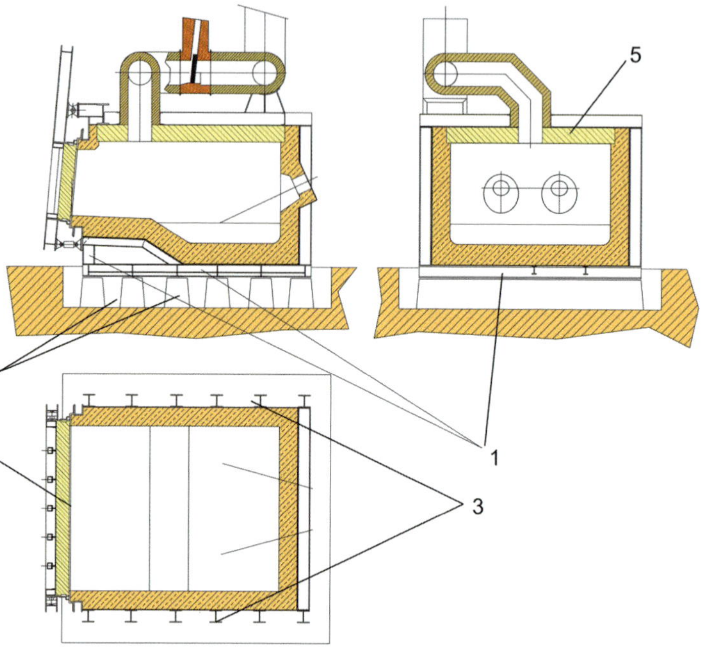

1 - supporting structure, 2 - concrete sleepers, 3 - wall reinforcement, 4 - front door, 5 - roof elements

Fig. 3.74: Stationary reverberatory furnace

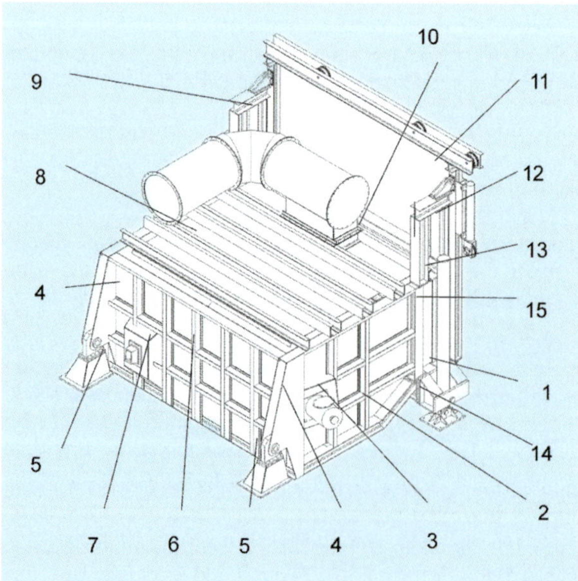

1 - cross beam,
2 - supporting girder for sidewalls,
3 - burner,
4 - tilting stiffeners,
5 - tilting brackets,
6 - back wall with reinforcement,
7 - burner and burner box,
8 - roof girders,
9 - door support,
10 - cross beam above door,
11 - heat shield,
12 - door lifting mechanism,
13 - door guide rails,
14 - lower shell support,
15 - door lifting cylinder

Fig. 3.75: Principle design of a tiltable reverberatory furnace

Fig. 3.76: Furnace shell (Source: Alcutec)

task would be easier. However, one complete furnace wall is open because of the large furnace door and also the roof may not be able to carry much load because of the panel design. Although the structure can be designed to accept tilting forces very easily, it appears to be more difficult to handle the torsion imposed due to the load applied by one cylinder only. Therefore, the shell must be designed in a way that the bottom of the steel structure will carry most of the load. This requires large structural beams with adequate reinforcement. One large size beam is located underneath the door opening. Depending on the size of the furnace, it is made of one or two heavy structural members. The tilting cylinders are connected to this structure at both sides of the door opening by means of pivot bearings. Starting from the beam, the furnace bottom is structured. As in the case of the stationary furnace, the bottom is made of heavy steel plates with a thickness of 12 to 20 mm. This plate is necessary to provide a good and sturdy base plate for the refractory lining. Rectangular to the door, parallel arranged beam steel members form a supporting frame for the furnace bottom and the furnace walls. They are connected by a cross-beam at the end, opposite to the door. The frame structure is stiffened additionally by structural shapes arranged parallel to the large door beam (**Fig. 3.76**). The furnace bottom is bent upwards near the door in order to provide the basis for the charging ramp and the slope in the furnace for de-drossing. Starting from the bottom structure, the up-going walls are built up. They also comprise a steel plate, reinforced by vertical and horizontal steel members. At the tapping spout section, a box-type structure is built up to provide the required distance of the tapping spout to the tilting axis. Around the burner area the shell is reinforced locally as well. The roof is not closed yet. It will be closed by pre-fabricated slabs which comprise a steel plate structure reinforced by structural steel members. They will be installed after the refractory lining and connected to the sidewalls by bolts.

Above the door opening a cross-beam will be permanently connected to the sidewalls. This beam takes the forces from the sidewall above the large door opening. It will also carry the door mechanism with the door and support the flue gas discharge.

Generally, the tilting supports are arranged opposite the furnace door. In order to obtain a working level for the transfer launders, the tilting axis is in the lower part of the furnace. This also determines the surface level of the metal in the furnace. The reaction forces of the full furnace load appear as vertical load as well as horizontal force, depending on the tilting angle. Both loads result

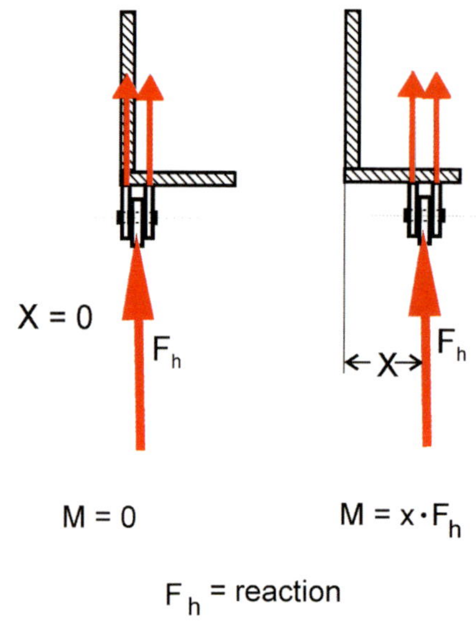

Fig. 3.77: Horizontal forces

1 - furnace shell, 2 - swivel bearing, 3 - furnace door, 4 - tilting brackets, 5 - sidewall, 6 - swivel cylinder for door opening

Fig. 3.78: Tilting brackets (Source: Alcutec)

in momentum acting at the front wall (**Fig. 3.77**). We do have two supports for the tilting forces, one at each side of the furnace. In order to distribute the momentum over the entire sidewall, the supports should be welded directly in line with these walls ($x = 0$, $M = 0$) or be located at least as close as possible to the one or the other side which will result in a momentum $M = x \cdot F_h$ acting on the front wall. Sometimes it is unavoidable to locate the swivelling points in some distance from the sidewall due to the door arrangement or the size of the furnace. But in general, the designers aim to have the horizontal forces act directly on the sidewalls. In any case, they have to be distributed over the full height of the furnace. This is obtained by full length brackets at both sides of the respective swivelling point (**Fig. 3.78**) which is reinforced by stiffeners.

Static determination and stress analysis

Based on experience, most hearth furnaces are not determined statically. This calculation combined with the stress analysis requires much time, although some computer-aided procedures can be applied. Many designers know from experience which structural shapes they have to place in which areas of the furnace shell. But for new furnace designs, either deviating from the standard concept or unusual sizes, a statical determination and stress analysis are very helpful to not "over-design" the equipment. Standard computer models do not work entirely for furnaces. The relevant calculations are a combination of computer-aided procedures and individual analysis performed manually. Generally, the base is a computer program developed for steel structures. However, these models to not fully match the requirements of the furnace designer. Looking at the loads of a furnace, to some extent the calculation procedures, as applied for pressure vessels, may be accurate. The stress generated is mainly caused by forces acting from the inside of the

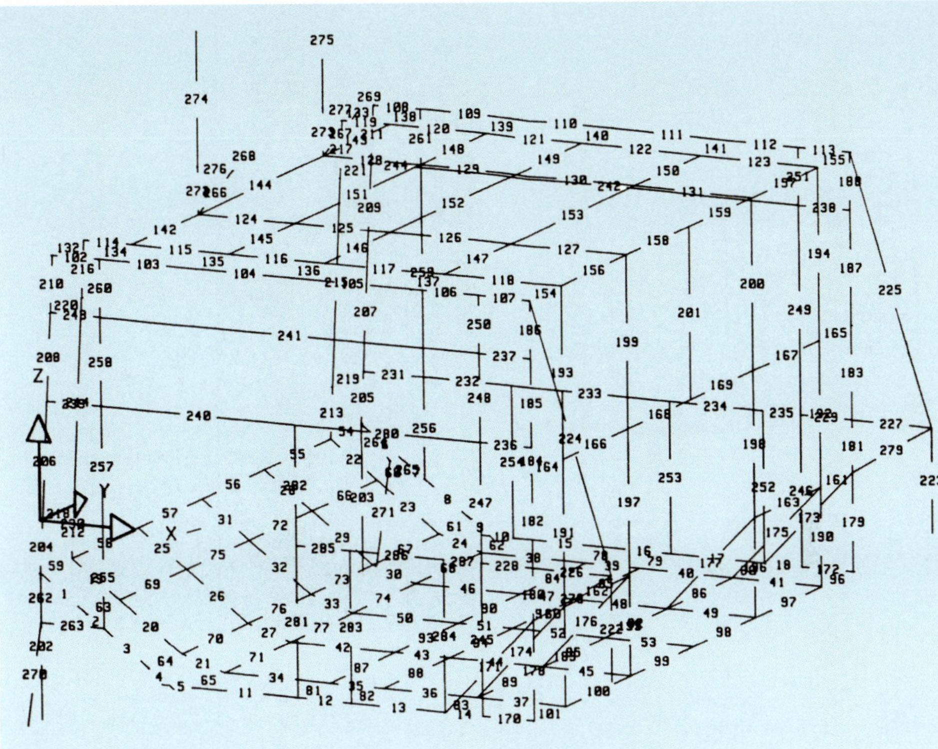

Fig. 3.79: Structural grid for the steel structure of a tiltable reverberatory furnace

unit. Loads can be regarded as internal pressure, but even not fully, since these forces change due to the movement of the tilting furnace and the reaction forces to support the furnace acting from the outside.

The procedure, as described, is used for the calculation of a tiltable reverberatory furnace having a bath capacity of 30 tons. It certainly covers general procedures since most furnaces show very similar design features.

One of the very first steps is to determine the grid of the structural members as designed (**Fig. 3.79**) with the determination of individual members and numbers for the junction points of members. This is required for the identification during calculation. Numbering has to be done for all sections of the furnace, i. e. walls, roof and bottom. The computer calculation is executed with a method for finite elements. The determination of the input is manual work. Different load cases have to be investigated: Furnace in level position, without and with metal filling, furnace tilted by both cylinders, furnace tilted with one cylinder only. Internal forces by expansion of the refractory material and the influence of heat are not considered. Their influence is in the limits of accuracy of the design load estimates and safety margins applied. Having completed the definition of all data and cases and after feeding these to the computer, the program can print out an overview (**Fig. 3.80**) and a listing of the coordinates, loads on every element and the acting forces. Next manual input is required again since the various cross-sections are not standard structural members but a combination of different steel shapes and steel plate. For dimensioning the different steel members, the moment of inertia is required. Finally, the size

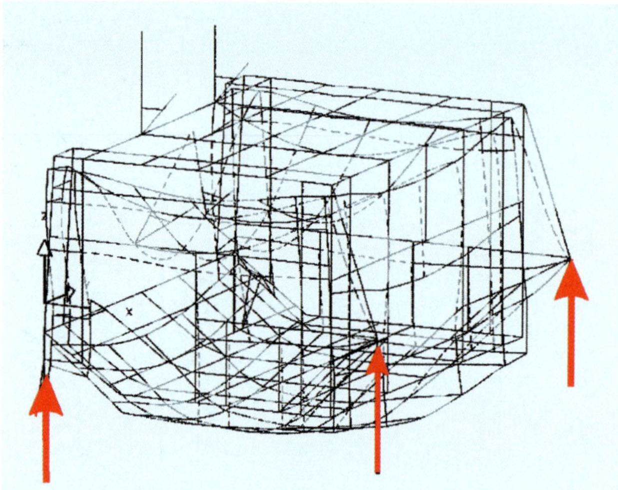

Fig. 3.80: Deformation study of a furnace shell

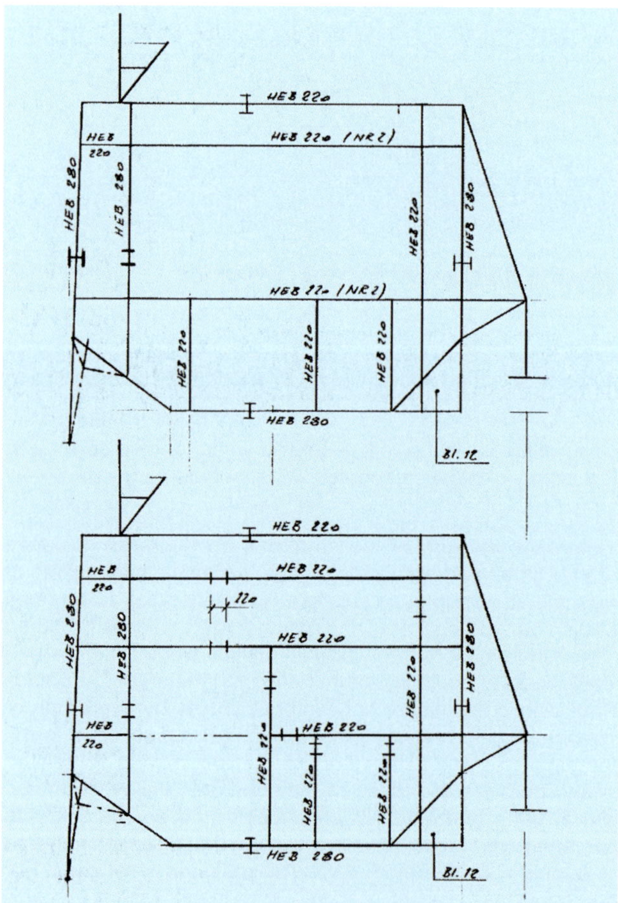

Fig. 3.81: Definition of structural members

of the structural members can be determined (**Fig. 3.81**) and the stress analysis can be done. Furthermore, the vertical and external loads, which are essential for the design of the foundations, are obtained.

The statical analysis, the determination and the stress analysis are very complex tasks. The number of pages amounts easily to more than 100. It is certainly not required to go through this every time a furnace is designed. However, one very interesting and valuable result is a computerized analysis of the deformation of a furnace. To do so, it will be assumed that only one tilting cylinder will take the full load of the furnace (Fig. 3.80). The diagram shows the local deformation while the table generated indicates specific areas. Most of the deflection is noticed in the upper girder of the sidewall where the single cylinder would be acting. This could be expected. The deflection is approximately 10 mm in that very unlikely case. But considering the area, this would not be critical with regard to the refractory lining. To have such a case is not possible. The cylinders are designed to take 50 % of the load each with some safety margin added. Thus, one cylinder would not be able to lift the furnace. Furthermore, cylinders are operated by one valve. One of the cylinders needs to be blocked completely, if this could happen. But still some additional load would result in some more deformation and stress, if it could happen.

Tilting supports

The tilting furnace is supported at four points: the two hydraulic cylinders for tilting action and the swivelling supports. The tilting shafts are supported by maintenance-free bearings which are made of special sinter bronze (**Fig. 3.82**). For lubrication these bushings contain pockets filled with a mixture of graphite and molybdenum oxide. No lubrication is required for the service life of the bushings. The furnace is supported by sturdy supports which are bolted to the foundation.

Fig. 3.82: Swivel bearing

The tilting cylinders are supported by heavy steel plates resting on the concrete foundation. The arrangement is “upside-down”, meaning that the cylinder eye is connected to this foot plate. The cylinder body is fastened on the steel structure of the furnace by means of a cardanic mounting. Due to this arrangement, no pits are required for the hydraulic cylinders.

Flue gas discharge

The best position for the flue gas discharge is above the furnace door. The pre-heating ramp is located there. Having the flue gases pass this area, a good heat transfer is assured for heating the material. Furthermore, the flue gases are not so hot in this area since they have “lost” most of their energy in the furnace interior. Thus, the impact of heat on the furnace door is lower than in other areas. The flue duct may also be located differently and still enable good heat distribution in the furnace. The high velocity burners create a substantial turbulence in the furnace with the result of heat transfer no matter where the main flow direction is. Thus, the designer will have the freedom to deviate from the ideal position, if plant layout or other circumstances require.

The furnace discussed shell is a typical design. As it was in the past, furnace designs have many features in common, no matter who the designer and the manufacturer may be. Consequently, the features outlined are typical for reverberatory furnaces used in industry today.

Furnace door

Similar applies to the large main door which is as large as one long furnace wall. It is exposed to the heat inside the furnace and the cold ambient air. The structure must be light enough to enable easy movement for opening and closing. At the same time, it must allow proper sealing at the door frame in order to maintain the positive pressure inside the furnace. This requires a design which avoids or compensates distortion without using a heavy structure. The door frame installed on the furnace structure is one of the most critical items. Some conventional systems comprise a hollow profile with interior water circulation. The door frame is cooled by this with the result that deformation due to heat could be kept within reasonable limits. For this advantage an energy loss of 6 to 8 % has to be accepted. There is also a potential danger of water coming in contact with liquid metal which may cause an explosion if one of the fittings fails or a frame profile is damaged. Many accidents have occurred in the past. The better design of the "dry door" comprises a sturdy frame made of individual sections of heat-resistant castings. The frame is embedded in the refractory material to a high degree so that the area for heat transfer is very small. This frame is absolutely distortion-free. Heat losses could also be reduced to 2 to 3 %. In case of damage only one or two sections of the door frame will be replaced, an action which does not require long downtime and repair cost. **Fig. 3.83** shows the door frame made of individually cast iron elements.

Fig. 3.83: Steel shell with door frame (Source: LOI Thermprocess)

The door is designed as sturdy frame which carries the door seal system (**Fig. 3.84**). The door blade lined with refractory materials comprises individual panels fixed to the outer frame to allow some free movement (**Fig. 3.85**). This ensures that forces generated, due to heat expansion, are compensated and not transferred to the frame. For the door sealing, a heat-resistant and asbestos-free mineral wool rope is used. The rope holders are made of individual sections and allow adjustment of the sealing in different areas as it may be necessary. For closing, hydraulic cylinders will press the entire door against the door frame which ensures a tight sealing. Some furnace designers apply the sealing force by moving the door guide rollers over a curve in the lower part of the door. This concept is very simple and does not require hydraulic cylinders but should not be accepted by the casthouse operators since a certain wear of the sealing rope cannot be avoided, resulting in a damaged sealing after a short period of time. For proper application of the sealing force, the guide profiles for the door rollers are designed as swivel arms which allow the complete door structure to be moved away from the door frame for lifting, thus clearing the sealing rope completely or to allow powerful pressing of it against the door frame (**Fig. 3.86**). Hydraulic plunger cylinders have proven to be most reliable and efficient lifting gear. Speed and door position can be adjusted at any time and the hydraulic forces do not require complicated designs for counterweights. The described lifting system and door sealing also permit to design a two-part door with-

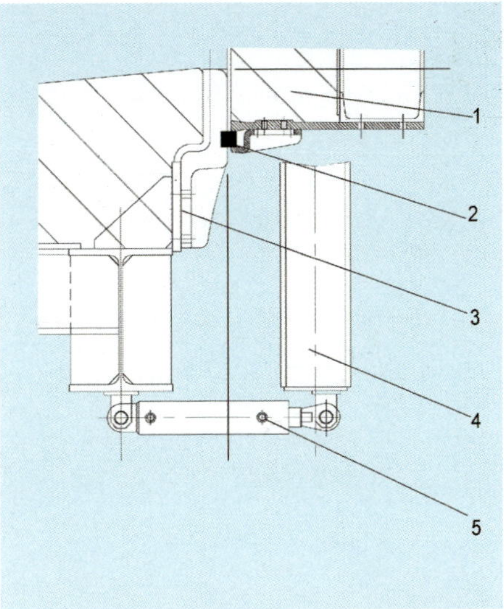

1 - door blade, 2 - ceramic fiber sealing rope with brackets, 3 - cast iron door frame element, 4 - door guide rails, 5 - door swivel cylinder

Fig. 3.84: Door sealing

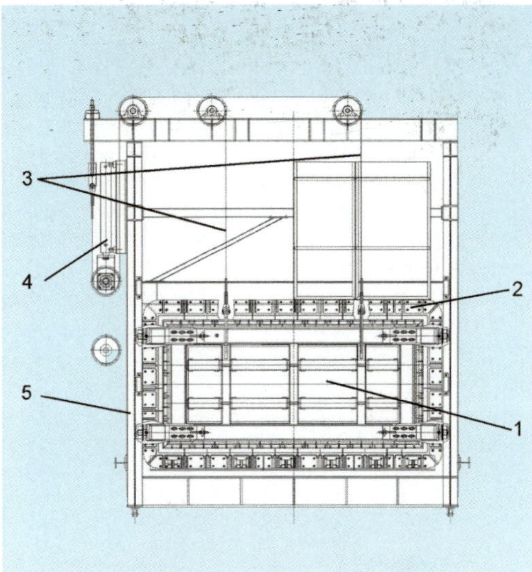

1 - door blade, 2 - cast iron door frame, 3 - lifting steel ropes, 4 - lifting mechanism, 5 - guide rails

Fig. 3.85: Door design (Source: Alcutec)

out center column. Each part of the door can be opened individually. This may be required if a large furnace requires an extended door size. In this case it may not be convenient to open the large door for operations such as taking a sample.

Pouring spout

A suitably sized pouring spout allows discharging the metal into the launder system to the casting machine or into a crucible. In a stationary furnace, the pouring system comprises tapholes, arranged at the lowest

1 - door structure,
2 - sealing rope holders,
3 - fastening brackets,
4 - lifting steel rope,
5 - door guide rail,
6 - cylinder pulley,
7 - door-lifting cylinder

Fig. 3.86: Door-lifting mechanism

Fig. 3.87: Double taphole
(Source: LOI Thermprocess)

Fig. 3.88: Tapping spout in a tiltable furnace

point of the furnace (**Fig. 3.87**). Since these are critical components, usually two tap points are installed. If arranged at different levels, they can be used to maintain a metal heel in the furnace. The tapholes are closed with stoppers or plugs that have a covering made of refractory material. The tapping system used in tiltable furnaces is different. It is arranged at bath level. Thus, metal will flow only when the furnace is tilted. For discharging into a launder, the pouring spout allows metal transfer without cascading of metal. It comprises a section bolted to the furnace shell and a section attached to the launder. They are connected by means of a swivelling bearing outside the metal trough. Both parts are sealed against each other thus avoiding metal leakage (**Fig. 3.88**, see also Fig. 2.15). Since the swivelling axis of the furnace is located in some distance to the furnace shell for design reasons, the swivelling point of the pouring system must be there as well. Therefore, a pouring box is integrated in the furnace shell structure. The spout is equipped with a hydraulically or manually operated lid. This is important for maintaining the positive pressure in the furnace.

Liquid metal charging

Liquid metal may be charged through the large furnace door or, which is the better solution, through a charging well (Fig. 2.6). The cross-section of this well is just large enough for the quantity of metal to be charged. Penetration of too much air, which would result in increased dross formation, is reduced. The metal flows into the bath underneath the metal surface once a certain metal heel is reached. Turbulence cannot be avoided totally but some of it may be desired to remove some trace elements such as sodium or magnesium during charging. The metal charging well can be closed by a hydraulically operated lid and will have a sealing similar to that of the large door.

Liquid metal could be charged to the furnace by means of a transfer launder. In this case, the metal charging well is much smaller than required for pouring metal from a crucible.

A special charging well can be avoided if liquid metal is poured into the furnace via the liquid metal discharge spout. This method should be considered when designing a new plant or an extension to an existing plant. Charging metal by this method enables metal transfer without cascading. At the beginning of the transfer, the furnace is in the upper tilting position and is then moved back as it is filled by liquid metal. The procedure can be controlled automatically by a level control system in the transfer launder.

In order to maintain the very low positive pressure inside the furnace, all openings in the furnace wall must be closed tightly if not required for operation. We have stressed that for the furnace door or doors, the pouring spout and the liquid metal charging well.

Thermocouple port

Another opening is required for the thermocouple for measuring the liquid metal temperature. In a melting furnace this thermocouple may be damaged or even destroyed when charging solid metal. It is, therefore, lifted by a hydraulic cylinder out of the bath range during this procedure (**Fig. 3.89**). As soon as liquid metal is present in sufficient quantity to form a bath, the thermocouple is moved downwards into the furnace. A small opening in the furnace shell is required to move the thermocouple enabling combustion products to escape at positive pressure in the furnace. Although this quantity is too small to upset the furnace pressure, hot flue gases in the area are unpleasant and may also be harmful to limit switches or other elements located there. The small opening can be closed very efficiently by a small air curtain. A hollow steel ring with small holes drilled into it on the inside will provide the required sealing air.

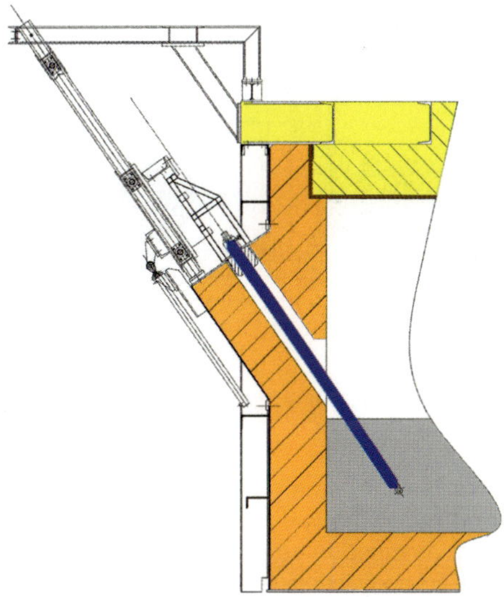

Fig. 3.89: Thermocouple arrangement

Dross removal

A large shield is placed underneath the large door in order to avoid that hot dross may fall into the foundation. The shield is sloped towards dross pans placed next to it for collecting the skim. The steel plate construction is adequately reinforced to resist the impacts of the de-drossing equipment but also to resist the deformation caused by the hot dross.

3.5.1.2 Twin chamber furnaces

Separating wall

All hearth furnaces designed on the basis of a rectangular box require identical or at least similar concepts for the design of shell, doors and other attachments. There are certainly some particular features to be thought about in the design. A specific item is the separation wall between two furnace chambers (**Fig. 3.90**). Heat is transferred to that wall from both chambers and the steel wall in between is not able to pass this heat to the surrounding area. The result would be that there might be no wall losses but steel – and refractory material – would heat up to the temperature of the furnace interior. Since this is certainly not of advantage, some measures have to be taken in order to

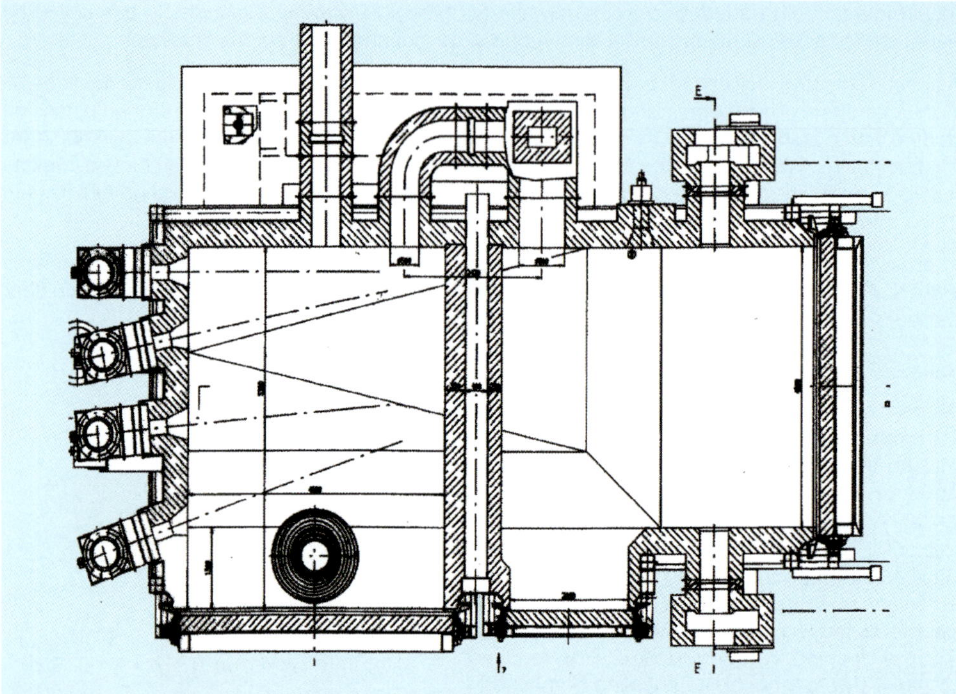

Fig. 3.90: Separating wall in a twin chamber furnace

avoid this uncomfortable situation. This is realized by having two different steel walls in this area arranged in some distance to each other. Due to convection, a good circulation of air is obtained since the heated air will flow upward, thus drawing in cold air from the surrounding area. If this cooling effect is not sufficient, cooling air can be injected by forced draft generated by a separate blower.

Hot air blowers

One characteristic feature of a twin chamber furnace is the recirculation of hot combustion products from the melting chamber to the hot air chamber (**Fig. 3.91**). This requires hot gas circulating fans operating at a gas temperature up to 600 °C. The housing, including the spiral casing, can be lined with refractory materials to resist the temperatures. This is not possible with the impeller. The design must also compensate for some wear by dust. Materials having high temperature resistance and good resistance to

Fig. 3.91: Hot air blower for a twin chamber furnace (Source: LOI Thermprocess)

wear have to be selected for this critical component of the blower. Considering the operating temperature range, this is not an easy task. Impellers with internal cooling have been used with some success requiring additional components to supply the cooling air to the rotating impeller. But still, the impeller will be a wear part with limited service life.

Doors, flue gas ducting, pouring spout, thermocouple installation and the other components are identical to those of a reverberatory furnace. This applies also to tower furnaces and dry-hearth furnaces which are developed from the basic concepts of the hearth furnace. Their particular features differ from one another by the principal arrangements and production methods rather than by the principle concept of the furnace design.

For the arrangement of charging equipment, different concepts are used. This will be covered later on.

3.5.1.3 Oval furnace

An entirely different design principle is used for barrel-type furnaces and oval furnaces. Since the shell is formed like a vessel, it is self-supporting and very little reinforcement is required as long as sufficient wall thickness is selected. For distributing the external forces over the circumference of the shell, girders made of steel plate are welded to the furnace body (**Fig. 3.92**). The cardanic mounting of the hydraulic cylinders are fitted and opposite the pivot bearings.

The static determination of the furnace differs from the one of the hearth furnace. It is also based on the method of the finite elements. Instead of a grid of steel members, areas on the furnace shell are defined (**Fig. 3.93**). Each of the two front ends is usually equipped with a door. This eases skimming since the furnace is usually quite long in order to have an acceptable bath capacity. The principle design is the same as for hearth furnaces. As in the reverberatory furnaces, the door frame is made of cast iron elements. However, due to the shape of the furnace, the doors are comparatively small. This arrangement provides space for the burners which are arranged left and right of the door (**Fig. 3.94**). The burners are equipped with a protection device which covers them when the aluminium in the furnace is treated with aggressive chemicals such as chlorine.

Fig. 3.92: Fastening of the tilting cylinder in an oval-type furnace
(Source: LOI Thermprocess)

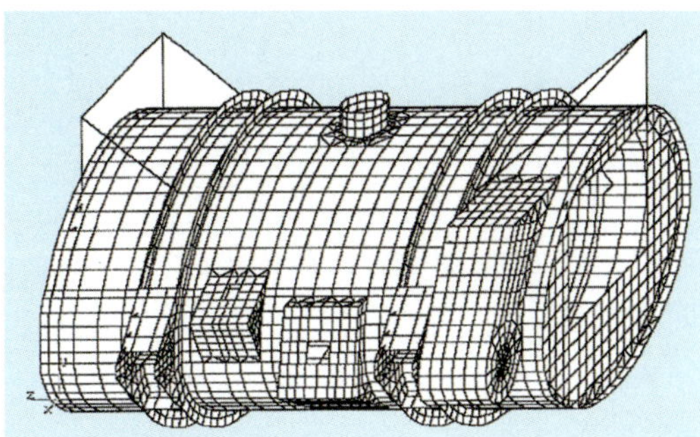

Fig. 3.93: Static system for the calculation of an oval-type furnace

Fig. 3.94: Burner arrangement in an oval-type furnace
(Source: LOI Thermprocess)

Fig. 3.95: Stationary rotary drum furnace

The tapping spout is arranged at bath level in the tilting axis of the furnace. Here again, metal can be tapped only when the furnace is tilted. The design of the spout is also similar to the one used for the reverberatory furnace. Pivot bearings and tilting cylinder mounting are similar as well.

3.5.1.4 Rotary drum furnace

Rotary drum furnaces are designed as drum with dished ends. A central opening at one end permits charging and the installation of the burner while the flue gases are discharged through a central opening at the other end. The drum is supported by two rings fastened to the drum. These rings are responsible for the smooth rotation of the entire drum. They are machined to be able to obtain the required concentric running. The drum is not machined and, therefore, some tolerances regarding the concentricity must be expected. The rings are fixed to the drum by means of a number of wedges arranged at the circumference of the drum (**Fig. 3.95**). This system also allows aligning the rings with the drum in order to obtain the concentric run. A high-tech system is shown in **Fig. 3.96** and explained in more detail with the example of the design of a tiltable rotary

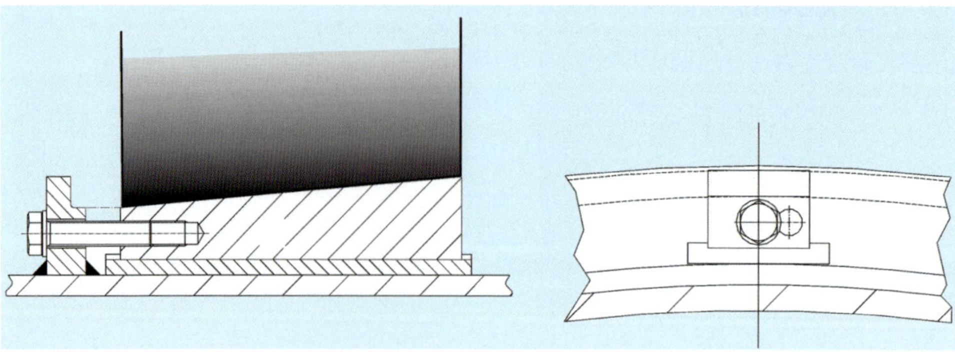

Fig. 3.96: Fastening wedges for supporting rings

Fig. 3.97: Roller station for a tiltable rotary drum furnace (Source: Alcutec)

drum furnace. The entire furnace is supported by four roller stations. These are fitted to a base frame, acting as supporting structure. The pressure between ring and roller surface is calculated for Hertzian stress. The rollers not only have to take the full load of the furnace comprising drum, refractory lining and batch, but dynamic forces originating from the unbalanced movement of the batch in the furnace. These forces are taken into account by multiplying the static forces with an impact coefficient. Horizontal forces also have to be considered. First there is the heat expansion resulting in a sliding of the drum in horizontal direction. Thus, the furnace cannot be fixed in this direction. To avoid uncontrolled movement, a supporting roller system is fitted to the base frame. It comprises two rollers at both sides of the front ring. In order to allow heat expansion, the rollers are not fitted tightly. They are set with some space, allowing the furnace to travel a short distance back and forward, if the furnace is aligned properly between the rollers. This arrangement is not possible in a tiltable furnace. There is a substantial horizontal force in any tilted position. Therefore, a fixed bearing is installed at the bottom end of the furnace. The rollers do not only have to take care of the vertical forces but also of the horizontal forces due to heat expansion. Therefore, also horizontally acting bearings are part of the roller station (**Fig. 3.97**).

Fig. 3.98: Burner and charging door of a rotary drum furnace

The burner is integrated in the charging door (**Fig. 3.98**). It has to be moved for furnace charging. The structure to do this may be a swivel arm or a sliding system for horizontal movement. This is required if the burner moves for a short distance horizontally into the charging opening in order to obtain a good sealing of the furnace. The flue gas discharge is arranged at the opposite end and is sealed by a labyrinth system with sealing air.

The definition of the required power for the drum drive is based more or less on experience. Principally, the calculation methods applied for ball mills could be useful. However, since comminution is involved in case of a ball mill, the calculation is based on specific power consumption and is, therefore, not of much help for the rotary drum furnace. To be on the safe side, furnace designers tend to use the assumption that the entire batch sticks to the wall thus creating a torque which has to be compensated for by the drum drive. This is not very realistic. It can only happen if the batch is frozen. If such an incident happens, the furnace will not be turned in order to avoid a dangerous situation that may even lead to substantial damage to the furnace lining. But in any case, calculation on the basis of this assumption results in safety margins that take care of acceleration of the drum rotating speed.

The moment of turning is calculated by using the equation

$$M_d = G \cdot g \cdot 0.5 \cdot d_i \quad (3.108)$$

with M_d = turning moment in Nm, G = weight of the batch in kg, g = gravity acceleration and d_i = inside diameter of the drum. To obtain requirements for drive and drive motor, the ratio of speeds in the drive needs to be taken into account.

The required power can be obtained by

$$P = M_d \cdot 2 \cdot \pi \cdot n \cdot W \quad (3.109)$$

with P = required power in kW, n = rpm, W = conversion factor 60 x 10^3 kW/Nm min^{-1}.

This is the equation. Considering, however, that no operator will operate the drum at full speed, if there is a frozen lump of material adhering to the wall, a speed of only 2 min^{-1} will be assumed.

Example:

A rotary drum furnace with a bath capacity of 25 tons has an inside diameter of 2.1 m. How much power is required for the drive?

$$M_d = 25{,}000 \cdot 0.810 \cdot 0.5 \cdot 2.1 = 257{,}513 \text{ Nm}$$

The power required amounts to

$$P = 257{,}513 \cdot 2 \cdot \pi \times 2 \cdot 60 \cdot 10^{-3} = 54 \text{ kW}$$

Considering an efficiency of h = 0.8, the drive motor must have a rating of 67.5 kW.

A very common solution for the drive arrangement of a rotary drum furnace is to have it working on supporting rollers. This can be realized by means of a gear motor with frequency control or speed-controlled hydraulic drive (**Fig. 3.99**). The question is, however, if this system is sufficient to transfer the necessary torque.

We look at our example. The furnace is supported by the roller station arranged at an angle of 60 degrees. The load proportion from the total weight per roller will be

$$F = G \cdot g \cdot \cos \alpha = 25{,}000 \cdot 9.81 \cdot 0.866 = 212{,}400 \text{ N}$$

With a friction factor m = 0.16, an outside ring diameter of 2.7 m, the torque that can be transmitted will be 212,400 x 2.7 x 0.16 = 91,760 Nm per roller station, amounting to 183,540 Nm in total.

Fig. 3.99: Hydraulic drive unit for a rotary drum furnace

This is only 83 % of the value that was calculated before and may explain why such type of drive sometimes slips in extreme conditions.

Fig. 3.100 shows a modern tiltable rotary drum furnace. We will use this furnace to explain some typical design features. It is built on a heavy steel base frame carrying the sturdy tilting frame via solid pivot bearings that are set in motion by two hydraulic rams and the drive unit. The tilting frame carries the drum-supporting roller stations and the axial bearing system. These bearing

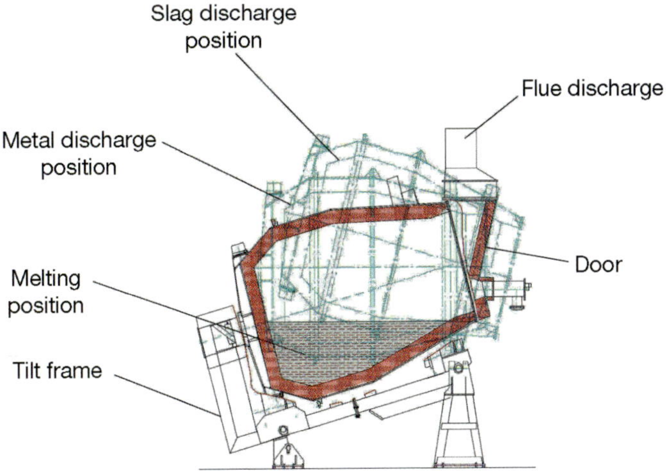

Fig. 3.100: Tiltable rotary drum furnace (Source: Alcutec)

systems support the furnace drum lined with refractory materials. The charging and tapping opening of the drum is closed by a sealed door that can be swivelled on a separate column. The door carries the burner. The upper section of the opening is closed by a fume hood that is also sealed against the furnace drum. For an optional design, the burners are mounted to the fume-collecting hood.

The original concept of the tiltable rotary drum furnace is based on a cylindrical shape of the furnace drum. For smaller furnace capacity, this design is very convenient since it offers the maximum size of the charging opening. As capacity increases, this shape becomes critical. Increasing the size of the furnace opening does not result in faster charging and discharging anymore and energy losses through the large diameter get unacceptably high.

Increasing the capacity, by extending the drum length, is also not possible. There is a limitation arising from the burner momentum and a long furnace will generate a cold rear wall. Thus, the length/diameter ratio should be close to 1 to 3. There is a manufacturing problem as well if the furnace diameter gets too spacious. Large machine tools to turn supporting rings and drum are not readily available any longer. Consequently, some solution has to be found. One possibility is to shape front and back of the drum as a steep cone. As explained in a previous chapter, this creates some problems. It would help in regard to charging opening and energy loss but it inevitably leads to an excessively large tilting angle that may result in a dangerous situation at the end of the slag discharging operation due to uncontrolled sliding of hot slag. The better solution is to shape a very slim cone at the front end of the furnace and to shape the furnace like a Bessemer Steel Reactor. The tilting angle is now within acceptable limits with no danger of slag sliding. There is one more advantage of this shape. Most of the batch is located in the back of the furnace. The firm section of the burner jet passes as a very stable flow through the comparatively small area in the front part of the drum. Furthermore, the reduction of pressure helps the burner jet to extend far into the furnace thus maintaining a hot rear wall. On the other hand, the reflected gases will help to stabilize the burner jet after leaving the burner brick and direct the flue gas to the exhaust hood.

Furnace drum

Considering the required high rotating speed of the drum, the accuracy of the supporting ring arrangement needs to be quite precise and the manufacturing quality of the drum as well. It is very comfortable, but also very expensive, to have the fit for the rings machined and the fastening arrangement as well. Fig. 3.100 shows the clamping arrangement for the supporting rings. It comprises a number of cones sliding in a machined bed and will be fastened by bolts. To obtain an accurate fit, the top of the cones is also machined when preliminary fixed in the sliding shoes arranged on the circumference of the drum. Together with the tapered inner diameter of the supporting rings, this method ensures accurate rotation of the drum. For large furnace sizes, the drum is manufactured in two parts for transport reasons. Considering the precise machining of the ring system, the fit for the drum parts is also machined. After testing the concentric run, the two parts of the drum are permanently fixed by welding.

The drum also carries the shaft for the axial bearing and the sealing ring for door and exhaust hood.

Bearing system

During rotation of the furnace drum at fairly high speed, unbalanced forces are generated by the movement of material in the furnace and perhaps also by large pieces charged or material adhering to the drum wall. Therefore, the design of the supporting bearings housed in the roller stations needs careful attention (**Fig. 3.101** and Fig. 3.97).

There are two different kinds of forces to be considered: the load of the batch and, even more, the weight of drum and refractory lining, as well as the dynamic unbalanced forces which generate

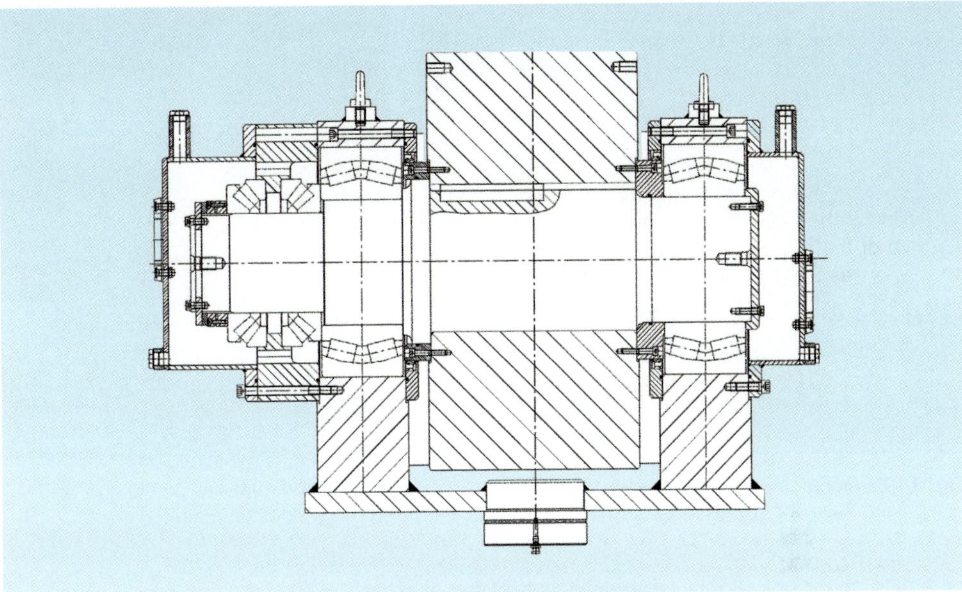

Fig. 3.101: Roller station for a rotary drum furnace (Source: Alcutec)

radial loads on the bearings. These are fully taken in by the supporting rings at two supporting points with two roller stations each (Fig. 3.101). This ensures clearly defined bearing of the radial forces. There are also some axial forces that have to be compensated by the axial roller supports. Due to heat expansion and friction at the rollers, axial forces are generated. This is essential and must be considered in the design by arranging axial bearings, too.

At the rear wall of the drum a central bearing system holds up the axial forces caused by the axial loads dsue to the tilting angle of the furnace. This bearing is designed as pendulum support, a kind of universal joint that permits free movement to the Z and Y axis but is firm in direction of the X axis, i. e. the axial direction of the furnace drum. Additionally, the bearings are able to compensate for misalignment of the furnace axis.

Many tiltable rotary drum furnaces of today's design are equipped with one supporting ring with two roller stations and one central bearing to take in radial and axial forces. The theory behind this is that a statically defined system can be achieved. This assumption appears to be incorrect. Within the roller stations at least two bearings are required for supporting the shaft. Pivot bearings within this system are only able to compensate the internal misalignment. To be able to adjust to misalignments of furnace axis and support, these bearings need to pivot with the complete roller stations. For obvious reasons this is very difficult to achieve. The so-called edge running of bearing and rollers is inevitable.

Tilting frame

It is not so difficult to support a stationary rotary drum furnace to take the forces generated by the furnace load and during operation. The roller stations are fixed to sturdy concrete slabs and fixed thoroughly. This is more difficult with a tiltable rotary drum furnace. All forces need to be carried by a moving structure. Therefore, this structure must be designed to take all loads during varying

Fig. 3.102: Frame of a tiltable rotary drum furnace

Fig. 3.103: Front view of a tiltable rotary drum furnace (Source: Alcutec)

operating conditions. Additionally, deflection of the members of the structure and, last but not least, vibrations have to be kept to an absolute minimum.

Fig. 3.102 shows a structure made of heavy hollow shapes that can not only fulfil the requirements, as mentioned above, but also minimizes collection of dust on the structure. The pivot points should be of oversized design in order not to weaken the heavy structure.

The two hydraulic tilting cylinders are connected to mounting plates to take the tilting frame by means of universal joints. The roller stations are bolted to the tilting frame. After they are properly aligned with the furnace drum, e.g. the supporting rings, they are fixed by pins and high tensile bolts. Depending on the size of the furnace, it may be required to have the tilting frame made in two parts in order to simplify the transport to site. In such a case, the two parts will receive fits for precise joining during erection.

Tilting supports

The tilting frame rests in two tilting columns designed as welded steel structure (**Fig. 3.103**). The tilting axis is at level with the lowest point of the charging opening. This reduces the height of the cascade of liquid metal. The pivot points are designed with permanently lubricated bearings as in the case for the tiltable reverberatory furnace.

Base frame

The tilting frame is supported by a sturdy base frame that is set on the heavy concrete foundations. The connections for the pivot point are precisely machined but they still allow for adjustments. The supports for the hydraulic rams are also supported on this base frame. The concept of the sturdy base frame ensures solid fixing of the complete furnace to the concrete foundations and thus vibration-free operation of the furnace.

Drive unit

The drive unit must ensure that the drum rotates at any set speed during all operating conditions. Sometimes scaling at the drum walls is unavoidable and big lumps of material adhere to the wall. It may also be the case that big and heavy pieces of material have been charged producing an unbalanced force. In these conditions the drive unit must provide enough torque for rotation. The method of combining a frequency-controlled electric motor with a gearbox at fixed ratio and a chain around the furnace drum (**Fig. 3.104**) may appear somewhat old-fashioned but it provides reliable operation at any rotating speed. The heavy chain elements are precisely machined as well as their fit to the drum so that smooth movement is ensured. It is also possible to exchange individual elements without problem and avoid a long downtime in case of trouble. There are certainly other methods of supplying the torque to the furnace drum. Friction drives via the supporting rollers with gear motors or hydraulic motors are also common. As we have noticed in the example, the drive via the roller stations may not be very reliable. Central drive units are also in use in the rotary drum furnace. They are connected to the central shaft near the central bearing. A very common design is a speed-controlled hydraulic drive that generates a large torque at low speed. As long as the requirements are fulfilled, the selection of the drive system depends on the personal choice of the plant manager.

Furnace door and fume collection

The door is one on of most important topics of any design. Proper implementation reduces heat losses and improves the environmental conditions around the operating plant. Consequently, the door must be properly sealed and easy to operate. **Fig. 3.105** shows a solution to fulfill these requirements.

The door covers the lower half of the drum opening. It is sealed against the rotating drum by a dynamic system with air lock, similar to that of the stationary rotary drum furnace. Air is provided through the swivel arm of the door. In the precisely machined drum section and the counterpart of the door a small gap can be maintained and, consequently, no gas can escape even at a small positive furnace pressure. A small gap is also maintained between exhaust hood and door. The door opens by means of a hydraulically-operated swivel arm that is supported by a rotation tower, equipped with special antifriction bearings for smooth movement. Generally, the door also carries the burner equipment.

The exhaust hood is arranged in the upper half of the furnace opening. It is sealed against the rotating drum by using the identical system as for the door. With proper sizing of the hood, the flue gases escaping from the drum, whenever the door is opened, are taken efficiently into the waste gas system. The gases now flow through a ducting lined with refractory materials to the filter plant.

Fig. 3.104: Drive unit for a rotary drum furnace (Source: Alcutec)

Fig. 3.105: Installation of a tiltable rotary drum furnace (Source: Alcutec)

To connect the tilting furnace to the stationary waste gas system, a rotary joint is arranged in the tilting axis of the furnace.

Proper door and hood design are a given requirement for the environmental-friendly operation of the furnace. The door is closed during de-coating, preheating and melting. The door is open for metal tapping and slag removal as well as for charging. Some dust may arise during the actual movement of material into the furnace during charging. To avoid that this dust may escape into the furnace environment, the charging machine docks tightly to the furnace hood. During the short time, while the charging machine is advancing or retracting, fresh air is taken in by the exhaust hood and, of course, a small amount of furnace gases. To compensate this, the burner can also be arranged in the fume-collecting hood. This permits operation of the burner at low fire when the door is open. Organics escaping from the furnace during charging are burned – at least partly – on their way to the scrubbing system.

As soon as material is charged, some de-coating already takes place and the furnace is hot. The fumes arising from this are taken out efficiently by the exhaust hood (**Fig. 3.106**). After charging, the furnace tilts back and decoating takes place in the closed furnace drum. During this operation phase excess air or oxygen is provided as required for burning the pyrolyzed organics. Consequently, no additional after-burner chamber is required in the flue gas system.

Fig. 3.106: Gas collection hood on a tiltable rotary drum furnace (Source: Alcutec)

The material feed-side of the conventional stationary rotary drum furnace is not easy to keep free of smoke and dust. In order to collect these pollutants the area in front of the furnace is completely housed inside a large chamber made of steel plate. The access is provided by large doors mechanically operated or by plastic swing doors. The entire setup does not even ease operation, requiring a substantial amount of air for efficient operation. This air volume requires a large filtration plant.

Contrary to the large housing of the open furnace, the efficiency of the flue gas exhaust does not take in a large volume of fresh air. Therefore, the flue gas quantity is reduced. Apart from the reduced volume, that may even be smaller when using oxygen, the separation efficiency of the filter plant is much better due to the higher dust load of the flue gas.

Metal tapping and slag discharge

Metal is tapped from the stationary rotary drum furnace via a taphole located in the flue gas wall at the circumference of the drum. It is closed by a stopper identical to the one used in a stationary reverberatory furnace. For tapping, the drum is rotated into the lowest position of the taphole.

Fig. 3.107: Taphole for slag discharge in a stationary rotary drum furnace

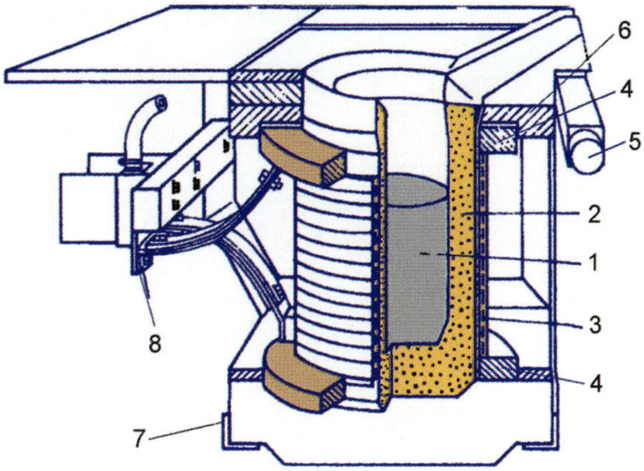

1 - metal batch, 2 - refractory crucible, 3 - induction coil,
4 - top and bottom supporting rings, 5 - swivel point,
6 - top holding structure, 7 - steel structure,
8 - cable and hose pipe bracket

Fig. 3.108: Crucible induction furnace

After metal tapping, slag is discharged. This is done by opening a large taphole in the furnace drum. After the furnace is emptied, this hole is closed by means of a clay plug that is securely tightened by means of a lid (**Fig. 3.107**).

The tiltable rotary drum furnace does not have special facilities for metal tapping and slag discharge. These operations are performed through the large furnace door by simply tilting the furnace.

The slag is discharged into slag containers arranged underneath the drum or, as in the case of the tiltable rotary drum furnace, next to the furnace door. The individual slag containers are placed on trolleys and pulled by a chain drive or a pulling trolley. The slag train is arranged in a way that the filled slag containers can be removed and empty containers replaced easily.

3.5.1.5 Induction furnace

Crucible induction furnace

The crucible induction furnace consists of the refractory crucible, the induction coil, surrounding the crucible, and the supporting steel structure (**Fig. 3.108**). The crucible is made of monolithic refractory material. It is the wear part of the furnace and must be changed at frequent intervals. Most plants keep one or two crucibles in stock for fast exchange of the worn out part. The coil is held by ceramic supporting rings at the bottom and the top. Additional yokes may be installed to concentrate the magnetic field. The induction coil windings are made of rectangular copper tubes. This is required for the cooling water flow through the coil. Crucible and induction system are supported by a welded steel structure. The structure accommodates the covering of the crucible surroundings and the operator's platform. The furnace is tilted by means of a hydraulic cylinder. The tilting axis is level with the top of the crucible and is arranged at one side of the steel structure. The pivot axles are supported by sturdy tilting supports bolted to the concrete foundation. A bracket for holding the cables for the power supply and the pipes for cooling water is attached

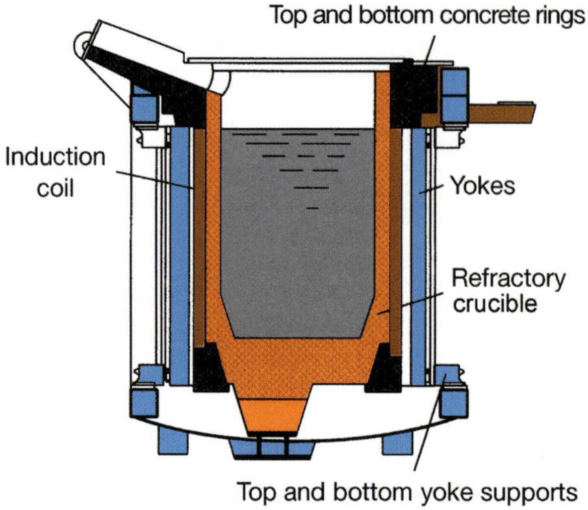

Fig. 3.109: Pouring spout arrangement in a crucible induction furnace

to the steel structure opposite of the tilting axis. The crucible may be covered with a lid that can either be tilted or swivelled for charging. Material to be melted can be charged directly from a chip conveying system into the furnace. In such a case the lid must swivel to one side. Another possibility is the use of a charging car with the lid tilted upwards. Liquid metal is discharged by tilting the furnace. For this reason the crucible may have a collar with a pouring spout (**Fig. 3.109**). The waste gas discharge can be connected to this collar as well. It is connected to the plant's waste gas scrubbing system via a rotary joint in the tilting axis.

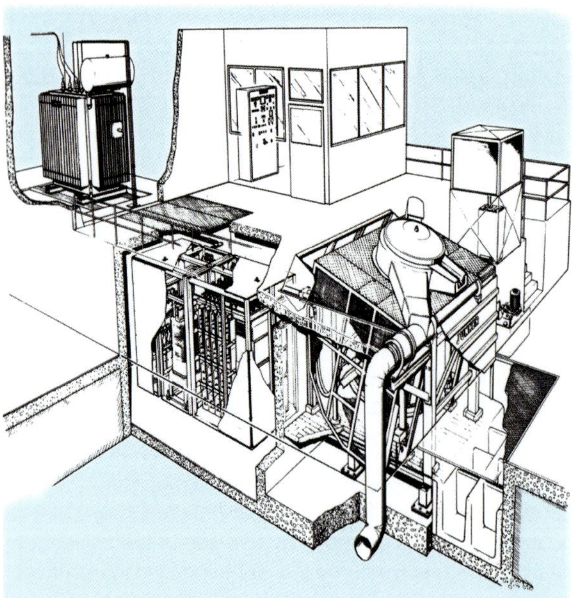

Fig. 3.110: Induction furnace plant (Source: Otto Junker)

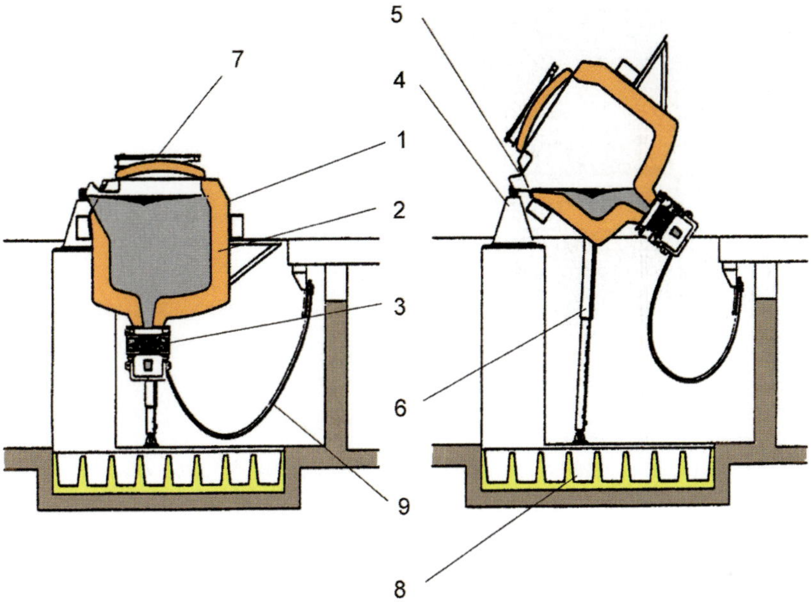

1 - furnace shell, 2 - refractory lining, 3 - inductor, 4 - swivel point, 5 - pouring spout, 6 - tilting cylinder, 7 - furnace lid, 8 - refractory collecting boxes (required if the sealing of the inductor fails), 9 - cable and hose pipes

Fig 3.111: Channel induction furnace

A crucible induction furnace requires some space for its electrical system. Thus, the plant layout may become a bit complex, particularly with a large furnace (Fig. 3.109).

There is a habit in aluminium recycling plants. When swarf needs to be processed – and this is generally the reason for the installation of a crucible induction furnace – frequent charging at short intervals is required. Thus, the operators leave the lid open and just continue to overfill the crucible. For such cases, a fume-collecting hood should be installed above the furnace.

Channel induction furnaces

The channel induction furnaces used in the aluminium industry are generally designed as round steel vessel lined with refractory materials (**Fig. 3.110** and **3.111**). The sturdy tilting brackets are adequately reinforced and welded to this self-supporting shell. Similar to hearth furnaces, the tilting supports are bolted to the concrete floor. The pivot bearings are of a maintenance-free sinter bronze type. Depending on the size of the furnace, one or two hydraulic cylinders tilt the furnace. The tilting axis is level with the top edge of the furnace shell. A special spout, which can be sealed as the lid is closed, is arranged in the tilting axis and permits discharge of metal into a launder system without cascading the metal. The tapping spout design is identical to that described for the tiltable reverberatory furnace.

The top of the furnace is sometimes shaped as cone to reduce the diameter of the lid, which is lined with lightweight refractory mixes and operated by means of a hydraulic cylinder. The swivelling support is also bolted to the concrete floor next to the furnace shell.

One or more inductors are arranged at the bottom of the furnace shell. These units do have a limited service life and must be replaced at certain intervals, usually 6 to 10 months. Therefore,

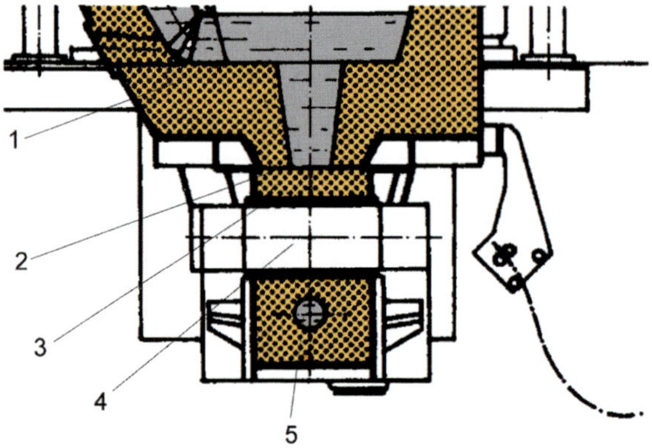

1 - furnace body,
2 - sealing,
3 - induction coil,
4 - iron core,
5 - metal channel

Fig. 3.112: Flanged Inductor in a channel induction furnace

they are bolted to the shell to allow fast and easy exchange (**Fig. 3.112**). Inductors and integrated channel are generally rammed with refractory mixes. A good sealing with refractory fiber material is required between furnace bottom and inductor flanges in order to avoid leakage of metal at the connecting areas.

3.5.1.6 Hydraulic equipment

Many of the different motions in furnaces are activated by hydraulic cylinders or hydraulic motors. In earlier days, these operations were performed with geared motors acting with chain hoists, chains, spindles and the like. The reason was that hydraulic equipment was regarded as not safe and their controls and accuracy were not to the satisfaction of operators and maintenance personnel. These handicaps could be overcome with the development of components and the related electrical control. The use of synthetic hydraulic liquids did improve the safety as well.

There are different circuits in a furnace in the secondary aluminium industry.

Furnace tilting

The most useful type of hydraulic cylinders is the single-acting plunger cylinder (**Fig. 3.113**). Only one connection of hydraulic pipes is required. The return into the starting position is due to the weight of the furnace. Thus, no additional pressure is required. If double-acting cylinders are used, an additional pipe is required to fill the other chamber of the cylinder in order to avoid corrosion. The heavy cylinders are equipped with a pipe breakage valve that closes automatically in case of a sudden pressure drop which may occur if a fitting, hose pipe or a pipe fails. In normal operation the valve is open. There are two steps of the tilting movement. Fast tilting is required to move the furnace into casting position. As soon as this is reached, a different valve is initiated for precise

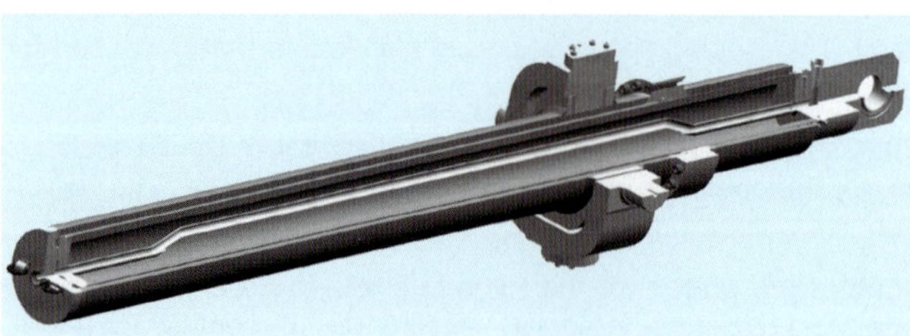

Fig. 3.113: Plunger cylinder (Source: Hoven)

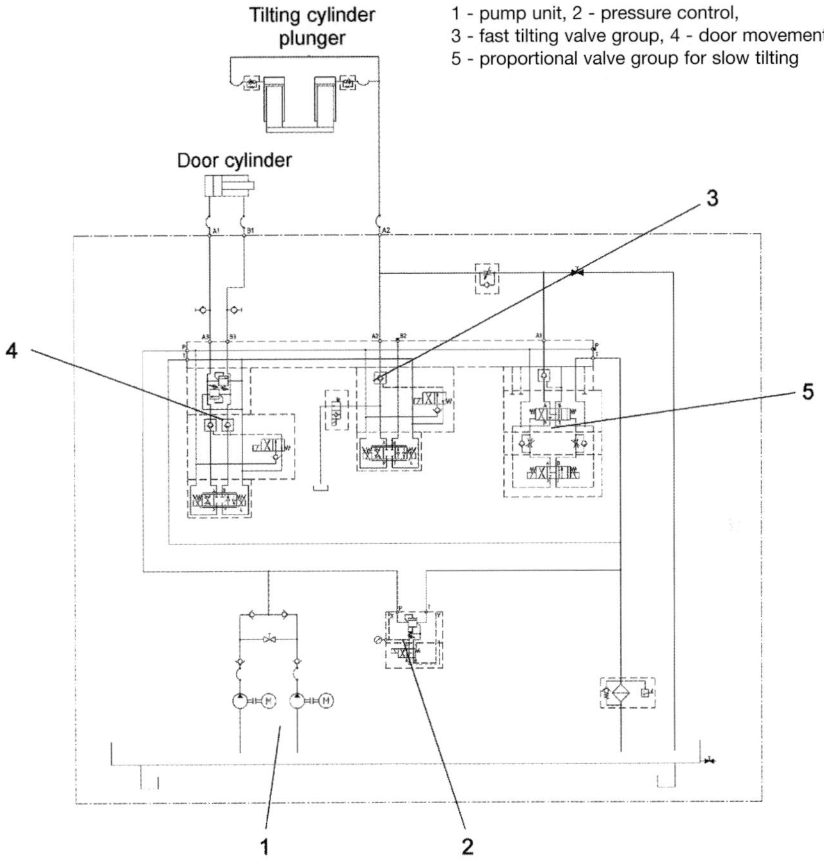

1 - pump unit, 2 - pressure control,
3 - fast tilting valve group, 4 - door movement,
5 - proportional valve group for slow tilting

Fig. 3.114: Hydraulic flow diagram

movement. This is required for casting control via level indicators, i. e. a laser level measuring system. The speed can also be controlled by a pump having a variable flow rate. This type of control is energy-efficient but slow. It is not suitable for precise flow control. As soon as casting or pouring is terminated, fast movement of the back tilting is initiated.

The flow rate of the hydraulic pumps is calculated for the maximum velocity. Fine-tuning of the motions in all directions is done by flow control valves which act in one direction and allow full throughput in the other direction. In case of a power failure, the furnace must tilt back automatically in fast motion as set for normal operation. For this action a spring-loaded valve is built in that shifts into the return flow position as soon as there is no electric power on the magnet for the other position. **Fig. 3.114** shows an example of a hydraulic diagram with tilting and swivel door circuit.

Door operation, hearth furnace

The door operation consists of two steps. First the door is tilted away from the closed position by single-acting cylinders. In the end position, the lifting cylinder is initiated. Here again, a plunger cylinder is used since the door is able to return to its starting position due to the weight of the door. In order to avoid a pressure shock, a proportional valve is used which closes via an electronically controlled ramp to slow down the motion in the end position or when stopped manually. As soon as the lifting cylinder reaches the lowest position, the door tilting cylinders press the door against the sealing. A pressure reservoir ensures that a positive pressure is maintained even if the hydraulic pumps are switched off. If pressure falls below a set value, a pressure switch will start the relevant pump again.

Door operation, rotary drum furnace

This door is not lifted but swivelled into the open position since the door is quite heavy with the related large inertia. This is not that critical for the acceleration of the motion which is limited by hydraulic pressure and pump capacity. It is different if stopping the unit because the moving mass of the door must slow down before the end position is reached. A so-called ramp, programed in the electric control unit (PLC), controls the slow closing of the hydraulic proportional valve. The signal for operating the servo-coil of the magnetic valve changes slowly to the value of current or voltage for the end position. The control is also active when the door will be closed; however, in the reversed direction. Both directions of the door movement require power; therefore, double-acting cylinders are required.

Auxiliary functions

Different functions do not require precise control. These are operation of charging and pouring lid as well as thermocouple motion. The velocity of these movements is controlled by flow valves for both directions of the double-acting cylinder.

Hydraulic unit

The hydraulic unit comprises the pump section with reservoir for the hydraulic fluid and the valve rack or valve block with all control valves.

The reservoir comprises a steel tank with oil-resistant piping or stainless steel which is the better solution; however, more expensive. A separating wall separates the tank into chambers for returning oil and for oil-intake by the pumps. The lowest edge of the connecting opening in the wall, separating the tank sections, should be at some distance from the tank bottom to hold back dirt returned with the oil. The suction pipe for the pump should also start at some height from the bottom. In order to collect leaking oil, it is advisable to have a collecting trough underneath the reservoir. The bottom of this trough should be sloped towards a drain plug. The reservoir is equipped with optical and electrical oil level indicators as well as temperature gauges, electrical and optical as well. Filters arranged in the return lines are equipped with differential pressure switches to indicate choking of the filters by dirt. It is good to have a double filter system, allowing cleaning without disturbance of production.

We notice that a unit like a hydraulic tank, appearing very simple at the first glance, requires some considerations to ensure safe operation.

A very efficient system to maintain a clean hydraulic fluid is a separate filter circuit with low-pressure pump with small flow rate and low pressure that will always generate a flow of oil through an inline filter. At low outside temperature the oil may be too stiff for trouble-free operation. Contrary to that, the oil could become too hot in warm climate. Therefore, the filter circuit should also incorporate oil cooler and inline heater to be switched on as required.

The main pumps are switched off via a timer after no operation is initiated for a certain period of time. They will automatically be started by the electric control system if action, such as door opening or furnace tilting, is initiated. The pump unit is equipped with pressure relief valves, check valves and pressure witches. A sufficient quantity of pressure indicators is also very important. These are indispensable for commissioning of the plant. At least quick connecting points must be installed for temporary connection of pressure gauges.

Fig. 3.115: Hydraulic room

The required valves for the different circuits are mounted either on the tank or on a separate valve rack. Any arrangement must permit easy access to all elements for setting or replacement. If installed on the tank directly, the valves can be mounted on machined blocks. The interconnecting pipes are replaced by bore holes drilled in such blocks. This reduces leakage but may hinder the required access. The better solution for maintenance is the valve rack (**Fig. 3.115**). The valves are mounted on individual base plates which are connected with pipes from the back. This provides easy access to all valves and excellent overview over all components installed.

Additionally to the control valves, hand-operated ball valves can ease maintenance. It may, for instance, be required to lower a furnace manually. For safety reasons the levers of such ball valves must be equipped with limit switches. Additionally, the handles of the valves must be removed if no maintenance work is performed.

As general rule, one individual hydraulic unit must be used for every operating unit, i.e. each furnace, casting machine or auxiliary equipment. Similar to the electrical cabinets, the hydraulic units should be installed in a separate room. Since pipelines are filled with hydraulic fluid, all actions are transferred immediately. Temperature is, except for extremely cold conditions, not critical since cylinders are not sensitive and returning fluid passes through the tank before it gets to the sensitive valves.

Hydraulic cylinders

The hydraulic cylinders used for the equipment in the aluminium melting plant are of standard design whereby the heavy series is usually applied. The cylinders for tilting and door lifting deviate from the standard design. For these purposes, plunger-type cylinders seem to be the most convenient type (Fig. 3.115) of equipment. Since the weight of the furnace or the door acts in a direction, only lifting is required. Therefore, the single-action cylinder is very useful. It has no piston but a very strong rod that moves as soon as hydraulic fluid is pressed into the cylinders. The rod diameter is close to that of the cylinder permitting good guidance, too. The rod has a swivelling pivot bearing which, in combination with the cardanic mounting of the cylinder, permits compensation of angular misalignments. The cylinders used for door lifting usually do not require this freedom of movement. They have a fixed mounting to be able to carry the rope rollers for the door lifting.

Pressure range

High pressure in a hydraulic circuit results in small-size components but also more difficult sealing of fittings and problems with hose pipes. Low pressure requires larger equipment sizes which may be difficult to position in the structure.

A good compromise is a pressure range of 120 to 175 bar. For this range a large selection of standardized off-shelf components are available.

Hydraulic fluid

The use of standard mineral oil for hydraulic fluid is not really critical. But sometimes a fitting may blow and oil may come in contact with liquid metal and ignite. Although such incidents do not happen very often, the faintest chance for an accident should be avoided. It is, therefore, recommended to use synthetic fluids such as Quintolubric. This kind of fluid is hardly inflammable and, therefore, safe for operation in the surroundings of furnaces. It requires, however, special sealings in cylinders and valves. This must be considered when specifying hydraulic equipment.

Fittings

Fittings must be selected to ensure that no leakage can occur. Fittings with O-rings are very safe but require careful installation. However, this is required in any case. Experience and know-how is required for the installation of pipework on site. There is, for instance, the common mistake to have pipe bends not fixed properly. Pressure shocks act very strongly if the flow direction changes. Therefore, a pipe bracket is indispensable on both sides of the bend.

Before start-up, all pipes must be rinsed by an external system before the hydraulic unit is started. After this, the unit is started carefully and speeds and pressures are set as required. During this procedure the oil filling must be checked since all pipes and cylinders will fill with fluid now. Returning hydraulic fluid may also cause overflow of the tank if the additional filling has not been performed at the correct position of the elements of the hydraulic circuit.

Considering the typical aspects of the hydraulic circuits used in aluminium plants, a reliable and easy to operate system is available that operates trouble-free over a long period of time.

3.5.1.7 Auxiliary equipment

Charging machines

The furnace equipment is designed for effective operation. Charging and furnace tending must match the high performance operation. Manual charging or charging with forklift and front-end loader requires too much time and results in extensive non-productive ancillary times. Manual dedrossing is a hard and time-consuming task. Thus, proper equipment will help to improve furnace performance.

Charging systems

A very simple and effective charging system comprises a steel table placed in front of the furnace door. Material such as stacks of ingots or similar are placed on this table. After opening the door, the material stack is pushed onto the pre-heating bridge or even pushed further into the furnace with the aid of a forklift. After charging, the door is closed and a new batch is placed on the table. After charging work is completed, the table is removed and placed in the vicinity of the furnace. The front of the furnace is now clear again for other operations.

Mold-type charging machine

The most effective charging machine for reverberatory furnaces is the mold-type charging machine (**Fig. 3.116**) which is usually designed to a load capacity of 10 tons and comprises a trough with open front and top. The material is loaded from the top by means of a front-end loader or via containers with the prepared batches and forklift with turning device. A pusher consisting of a rigidly built shield and a hydraulic pushing cylinder pushes the complete batch into the furnace. To do this the filled charging machine is placed in front of the furnace. After opening the door, the machine moves forward and partly into the furnace and the pusher charges the material.

Fig. 3.116:
Mold-type charging machine

After charging, pusher and mould move backwards and the furnace door is closed. The charging machine now moves into the loading position where the next batch is placed into the mold.

For these operations the machine is mounted on a cross-carriage moving on rails in the concrete floor. The cross-travel drive consists of heavy crane travel gears. The cross-travel carriage is designed as sturdy steel structure made of structural steel members and steel plate. The carriage is supported by rimmed wheels and fit to one shaft with sturdy bearings per pair of wheels. The forward carriage mounted to the cross-travel carriage requires a very sturdy design. It consists of a heavy steel structure carried on roller guides moving or rather travelling over stiff guide beams for charging trolley and mold carriage. The entire mold unit is pushed forward by means of hydraulic cylinders.

One charging machine can serve different furnaces if they are arranged in one front line. Instead of a mold permanently fixed to the machine, different molds can be filled for a batch preparation and stored in the reach of the charging machine which picks up the new mold by means of a clamping device. The charging machine can be parked out of the reach of any other activities.

To have more freedom in movement, the charging machine can be designed as a mobile unit (**Fig. 3.117**). All movements, including drive, have infinitely adjustable speed to allow precise movement and accurate positioning. The molds or charging boxes are carried by forks similar to a forklift. Via a parallelogram gear the box can be moved up and down as well as tilted. A pusher,

Fig. 3.117: Self-propelled charging machine (Source: Glama)

Fig. 3.118: Vibrating trough charging machine (Source: Alcutec)

integrated in the mold carrier, moves the back wall of the mold forward for charging the furnace. Since the machine can pick up boxes from any storage position, operators are able to prepare the batches as required. The machine will pick them up in the defined sequence and charge the furnace as required.

Vibrating trough charging machine

Box-type charging machines are ideally suited for use in reverberatory furnaces. They are not so useful for rotary drum furnaces. In this case a charging machine having a vibrating trough works more efficiently (**Fig. 3.118**). The trough is mounted on a sturdy undercarriage resting on heavy springs. Two unbalanced electric motors are attached to either side of the vibrating trough generating a directed vibration at a forward directed angle. This movement throws pieces located on the trough, as well as parts lying on the layer of pieces, in the direction of the vibration thus causing a forward movement of the whole material layer on the trough. The unbalanced weights can be calibrated to the desired conveying amplitude, i. e the conveying speed. As soon as the electrical energy is switched off, the motors will slow down. Doing so, they will pass the critical frequency causing a resonance amplitude of the oscillating system that may cause damage to the machine. In order to avoid this unwanted situation, the motors are equipped with a brake that is initiated as soon as the power to the motors is switched off.

A loading hopper is arranged overhead the vibrating trough. The hopper is filled with material which is then reclaimed by the vibrating trough. The hopper is also supported on the sturdy undercarriage. This comprises a steel structure that is moved back and forth on rails in front of the furnace. This movement is required to get the vibrating trough out of the furnace and to allow movement of the furnace door with integrated burner unit. As in the case of the mold-type charging machine, the travel drive consists of heavy crane travel gears. The undercarriage is designed as sturdy steel structure made of structural steel members and steel plate. The carriage is supported by rimmed wheels, fitted to one shaft with sturdy bearings per pair of wheels.

As in the case of the mold-type charging machine, the complete machine can be mounted on a travelling carriage. This may be required if not enough space is available for a loading position or if more than one furnace is to be served by one machine. The charging machine can be interlocked with the door operations of the rotary drum furnace.

Skip hoist charging machine

Particularly in the case of tower furnaces or twin chamber furnace with a top loading system, neither a mould charging machine nor a vibrating trough charging machine is suitable for the task. Material has to be charged to the furnace at an elevation that does not permit floor operation of the standard charging unit. The appropriate solution is a skip hoist that can be filled on the floor level. It comprises a charging bucket that is pushed upwards by means of a chain drive or a hydraulic cylinder. The bucket is guided by rails made of structural steel members leading upwards on top of the furnace to the charging opening. The guide rails are bent on top of the travel so that the front end of the bucket tilts and empties its content into the furnace. The batch is charged directly into the upper end of the tower or into a basket. Here the material can either be pre-heated or dried or both can take place. The basket can be designed as halve round shape (**Fig. 3.119**). A very interesting design was developed by Hertwich Engineering. It comprises a surge hopper with bottom flaps. As it is the concept of the tower furnace, the material now comes into contact with the hot products of combustion and will be pre-heated. A portion of the energy leaving the furnace normally through the stack is thus utilised. Oil or other organics, which may be contained in the scrap as unwanted contaminations, are also removed and their energy can be recycled thus contributing to the energy required for melting. If the excess air content is set to a value just sufficient to burn the organic components just removed from the scrap, metal loss is kept low as well. This has been described before.

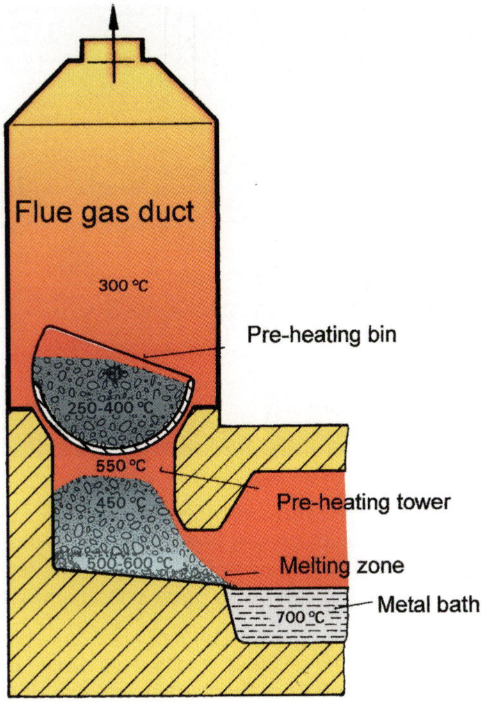

Fig. 3.119: Charging system for a tower furnace

Furnace tending machines

The furnaces used in the past have been comparatively small units of 10 tonnes capacity or even less. Operations in the furnace, such as stirring and skimming, could be done manually with not too much effort. As the furnace sizes were growing bigger and bigger, furnace tending became a power-intensive task for the operators. The first step was to replace the manual tools consisting of a steel rod with a steel plate welded across the front end by a skimmer that could be attached to a forklift. These tools are still in use. They comprise a rectangular tube with the skimming blade welded across in front of the tube. The opposite end is designed as supporting structure with shoes for the forks of a forklift truck. In order not to lose the tool when cleaning furnace walls and bottom, the holding structure carries holding clamps. The tool is still difficult to handle and it requires some practise of the operators to use it efficiently. When skimming the furnace bath, the blade takes a large quantity of liquid metal into the skim. Due to its rough structure, the skimming is not complete or the skimming blade submerges too deep into the bath; additional metal loss is the result.

Therefore, mechanised skimming machines have been designed. The skimming blade is attached to a telescopic arm which can be operated sensitively since the movements are controlled by hydraulic servo motors. Contrary to the fixed design of the skimming tool for forklift trucks, the

Fig. 3.120: Self-propelled furnace tending machine (Source: Glama)

telescopic arm of the tending machine permits the operator to get closer to the point of action and operate the machine very precisely with the result that less liquid metal is trapped in the skim. For cleaning the sidewalls the tool can be swivelled. Stirring is also easy but there are more efficient systems available, such as pump units or the electro-magnetic stirrer.

The tending machine can be mounted on rails in the concrete floor with cross-travel carriage or as self-propelled unit (**Fig. 3.120**). This design gives more freedom in operation and is very useful if the furnaces are not lined up but installed at different locations.

Considering these facts, a tending machine is very useful if more than one reverberatory furnace is operated. Having only one furnace, the investment seems not to be justified.

Metal transfer launders

If the plant is ideally designed, launders transfer liquid metal from the melting furnaces to the casting furnaces and to the casting machines. These launders can be sloped or operate level. Sloped launders can transfer metal in one direction only, while level launders allow transferring metal also in the backwards direction. This may be required when the tilting spout for casting is used for charging liquid metal as well. This is a very convenient method when arranging furnace and casting equipment. It furthermore simplifies the sealing of the furnace since there is one less opening in the walls. The level launders also reduce turbulence in the metal flow thus resulting in less oxide inclusions in the metal. The cross-section of the launders should be large enough to permit fast metal transfer without turbulence but not too large to have the metal staying in the launder too long. It is quite obvious that the launders are not the best selection for dosing the metal flow to equipment, for instance the casting machine. For transferring metal, a large launder shortens the ancillary time of the furnace since the transfer time is much shorter than the one of a launder with small cross-section.

Transport crucibles

If metal cannot be transferred by launders, transport crucibles will be required. This may be the case when the liquid metal is required in distance plant sections or if the layout of the production plant does not permit the use of launders. This could also be a newly designed plant with a strict separation of melting plant and casting plant. Generally, in a primary aluminium smelter, transport of metal from the potlines to the casthouse is performed by means of crucibles. Crucible transport is also required if a customer wants the supply of liquid aluminium.

Within the production plant the crucibles are transported by crane or transport vehicles. The furnaces receiving the liquid metal will have to be equipped with a suitable port for liquid metal charging. Only if an existing furnace plant shall be served by crucible, but not having the required means of liquid metal feed, special devices are necessary for charging. This could be a tilting device with feeding launder mounted to a transport vehicle. However, this handling is complicated and should be avoided wherever possible.

The crucibles for transport of liquid metal inside plant facilities are designed as steel bucket lines with refractory materials. It is equipped with a yoke for crane handling. The suspension points are arranged slightly above the point of gravity. This ensures safe transport but easy tilting of the crucible for pouring. A gear, either hand-operated by means of a wheel or a mechanical gear, is arranged at the point of gravity to perform tilting. Another tilting method is by means of a crane jack ring welded to the bottom of the crucible. An auxiliary crane hoist will hook in there and lift the crucible for tilting.

The refractory lining of the crucible can be built as multi-layer system with bricks or monolithics. The interior of the crucible must be kept clean and possible scaling must be removed prior to using it. In order to avoid formation of corundum, the crucible must be pre-heated. This is done at a special pre-heating station comprising a swivel lid with integrated burner. The location of this pre-heat station or stations is close to a wall of the production building so that there is no impairment of operation in the building. If there is crane handling, the station must be in the vicinity of the crane hook.

When selecting the crane for crucible transport, the permissible load must be large enough to lift the metal and the crucible with lining and yoke. In general, the total weight is as much as twice the weight of metal, i. e. 10 tons for 5 tons of metal or 20 tons for a 10 t crucible.

Transport crucibles for road transport (**Fig. 3.121**) require a different design. The crucible structure including refractory lining must be light in order to reduce the load for the truck with the metal

Fig. 3.121: Crucible for road transport of liquid metal (Source: LOI Thermprocess)

content as much as possible. The heat insulation capability requires extreme values since the transport may take several hours and the metal temperature must not drop too much. Thus, an optimal design has been found. The refractory system comprises a refractory material which also resists the direct contact of the metal. The multi-layer system is made up of a thin front layer and a rear layer with very good heat-insulation characteristics. This is accommodated in a steel shell, designed as closed vessel. Prior to filling, the transport crucible is pre-heated through the filling and pouring spout at a special heating station within the plant. No further heating is possible and required until the crucible reaches its destination and is emptied and returned to the production plant. The pouring spout is closed by a sealed lid during transport. Transport shoes at the bottom of the crucible are used for the forks of a lift truck for loading the crucibles on the truck. Generally, three crucibles make up a truck load.

The refractory lining has to be repaired at frequent intervals. Considering the small capacity of the crucibles of 1 or 1.5 t and their permanent use, a special refractory material repair shop is to be operated in every production facility supplying liquid metal to customers.

3.5.2 Refractory lining

One of the most important components of a furnace is the refractory lining. No furnace can be operated without it. The service life and efficiency of any technical furnace depend largely on the suitable choice, the quality and – this is very important – the correct installation of the refractory material. The furnace designer, furnace builder, refractory manufacturer and the engineer responsible for operation are well aware of the relationships between these factors. It is indispensable to have a basic understanding of the particular characteristics of refractory materials to be able to select suitable refractory material grades for the specific purposes.

3.5.2.1 Requirements of refractory linings

Temperature resistance

The refractory lining must resist the temperature of the furnace interior and the material charged to the furnace. There are always fluctuations of the temperature in the furnace during an operation cycle. Shortly before the metal is ready to be tapped, the temperatures in the different sections have reached their highest level. Metal has obtained the temperature set by the operator. This is generally in the range of 700 to 800 °C. The upper part of the furnace must now have a temperature level to be able to transfer sufficient quantity of heat to the metal bath. This temperature should be as high as possible, since this heat transfer will be by radiation only. Here the first compromise is required since the limiting factor is the selected refractory material itself and its service temperature needs to be balanced with the other characteristics required. During the furnace cycle the temperature profile changes. As the metal is tapped, the bath area is exposed to the much higher temperature of the entire furnace. This needs careful consideration since the contact with metal in the bath area requires specific characteristics which may be somewhat in contrast to the resistance to high temperature. Finally, the furnace is empty and is ready to receive a new charge. This requires opening of the generally large furnace door with the resulting heat loss; the furnace gets cold.

A very specific change in temperature occurs in the rotary drum furnace. During every full rotation of the furnace drum, the refractory lining is covered partly with material and then exposed to the hot combustion products. This is combined with the mechanical impact of the moving material and the different stages of a complete furnace cycle. The designer of the refractory lining must certainly consider these aspects when selecting the refractory material.

In many furnaces regenerative burners are installed. Due to the heat recovery, these burners will have a very hot flame that could impinge on a comparatively small area of the furnace wall or the roof. As the cycle of the burners continues, flame temperature gets lower down to a minimum and then the cycle starts in another area. The result is changing flame temperature in different areas of the furnace lining.

The changing temperature levels, which will certainly occur during every furnace cycle, need due consideration for the design of the refractory lining.

Resistance to chemical attack

Aluminium tends to penetrate into the crystal structure of the ceramic material. A typical important property is the slag resistance. It characterizes the resistance of refractory material to chemical attack of any kind, including that of gases and vapor. The destructive agents coming in contact with the refractory material are of different composition and destructive processes are correspondingly numerous. In hearth furnaces there is the attack of the aluminium and its alloying elements as well as the agents used for metal treatment. In case of the rotary drum furnace, the impact of the salt with its components must be considered. Usually the salt mixture contains a certain percentage of aluminium fluorides or even cryolite which is used for the digestion of aluminium ore. A very critical component is the content of silicon in certain aluminium alloys. This element reacts with the silicon oxide of the refractory material.

Choosing the right refractories for the lining of an aluminium holding furnace or melting furnace can be difficult, especially considering the aspect of corrosion resistance. The problem is complex and involves chemical reactions as well as physical phenomena. Since the corrosion conditions differ from one area to another in these furnaces, different refractory materials are required to resist corrosion by the various mechanisms. A critical area in a heath furnace is the so-called "belly band". This is the zone between the metal bath and the melt-free section of the furnace. Refractory material having a good resistance to corrosion below the metal line may not necessarily be resistant at the belly band. In this area, corundum forms at the surface of the molten metal where plenty of oxygen is available from the furnace atmosphere. This corundum is a composite material containing molten metal channels. As it grows, it may come into contact with the furnace wall above the molten metal level. In this way, molten metal is brought into contact with the refractory wall above the metal line where the temperature is much higher than below. The high temperature in combination with the action of the corundum growth creates favorable conditions for metal penetration (**Fig. 3.122**).

Almost all of the alloying components of aluminium react with refractory material resulting in additional wear. But this has to be accepted to a certain extent. Selecting the specific refractory

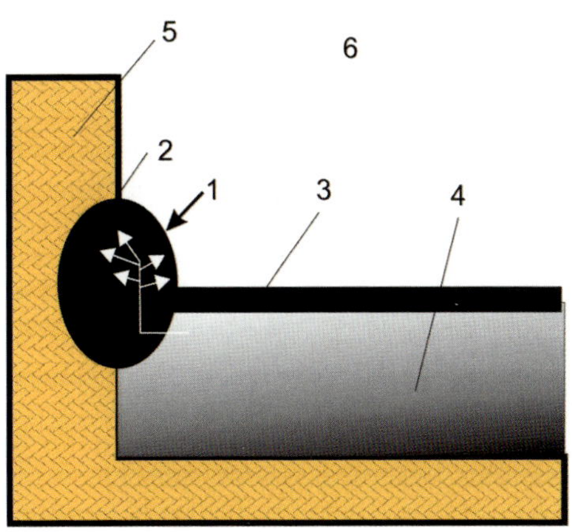

1 - oxygen,
2 - corundum growth,
3 - skin,
4 - aluminium 720 °C,
5 - refractory material,
6 - furnace temperature 1,000 °C

Fig. 3.122: Corundum growth in the belly band of a furnace

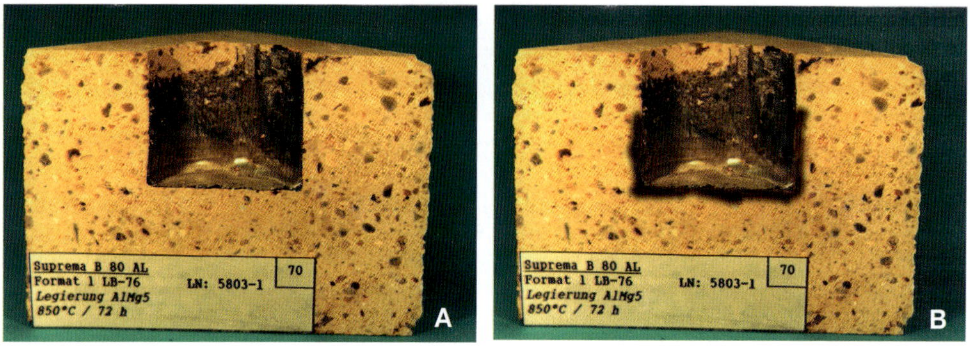

Fig. 3.123: Penetration test for a refractory lining

material grade, the effect of chemical reaction is very limited. In most of the reverberatory furnaces refractory material service life is five years or even more, provided wear of the refractory material is monitored carefully and minor repairs will be made frequently.

The refractory lining installed in rotary drum furnaces has to face the attack of NaCl and KCl since these are used as flux for processing of scrap. The additional CaF_2 used in such flux is even worse since it dissolves the aluminium oxide forming part of the refractory material. To compensate for this, the alumina content of the refractory material is reduced.

The slag resistance is determined by the so-called crucible method. A crucible is cut from a brick to be tested and is filled with aluminium. This crucible is held for a specified period of time at a temperature that is 100 °C higher than the envisaged metal temperature. After cooling, the crucible is cut open and the dissolution of the brick substance and the impregnation are compared. **Fig. 3.123** shows a sample of such a test which was successful in case B. If aluminium would have penetrated into the refractory material, material in the vicinity of the crucible shows dark color indicating the penetration zones as seen in case A.

Another testing method is the so-called finger probe, which is used to determine the corrosion resistance. Small samples (fingers) are tested at a defined temperature over the defined period of time. The result is evaluated by visual inspection.

Resistance to mechanical impact

Usually operators do not treat their furnaces nicely and carefully. They are always under pressure to achieve the maximum production rate. For charging the furnace with scrap, which very often comprises large aluminium pieces, heavy equipment is used. The scrap is dumped into the furnace resulting in a severe impact on the refractory lining. When cleaning the furnace walls, scaling from previous production has to be removed. Even if the furnace is still hot, intensive scraping is required. Some damage to the refractory lining can happen easily. Another action that results in excessive impact in the door area of a furnace is de-drossing. Here again, mostly heavy equipment is used which may cause damage in the door area. These areas of the furnace do require some maintenance at frequent intervals to avoid problems with increasing damage.

The mechanical impact in rotary drum furnaces is even more severe. It is characteristic for this process that the material enters into a tumbling movement. This leads to a grinding effect on the furnace wall causing substantial wear of the refractory lining with the result that this lining must be replaced every two to three years. Frequently heavy pieces of scrap are charged to the furnace which are lifted due to the rotation of the drum and, having arrived at a certain level, fall back to the lower section of the drum hitting the lining intensely. This is definitely something the design engineer is not very fond of since it reduces the service life of the refractory material substantially.

Good heat insulation

The energy entering the furnace should be used for the process. Losses must be as low as possible. Furthermore, it is certainly understood that the steel parts of a furnace can not be operated if red hot. Hence, the heat insulation properties of the refractory material are an indispensable criterion. The heat conductivity lowers with the density while the mechanical strength of the refractory material depends on a dense refractory material. The heat resistance is also lower with material having better insulating characteristics. Considering these features, the refractory lining of a furnace comprises a multi-layer system built of materials having very specific characteristics.

Later on we will come back to the design of the refractory lining once again.

3.5.2.2 Groups of refractory materials and their raw material

The refractory material used in aluminium technology is based on silica and alumina. The oxides SiO_2 and Al_2O_3 form the basis of the most important group of refractories. It was recognized several centuries ago that materials made from these oxides are fire-resistant and this knowledge permitted early melting of metals.

Silica material

This material mainly consists of silicon dioxide (SiO_2) which is generally described as silica. This occurs in a variety of crystalline modifications and its super-cooled melt is quartz glass. Silica bricks are used as refractory lining in glass melting furnaces, coke ovens and gas furnaces, hot blast stoves and in the roof of electric arc furnaces.

Fireclay

Whereas silica material largely consists of a single oxide component, fireclay has two main components, i. e. 10 to 45 % Al_2O_3 and 5 to 80 % SiO_2. Fireclay products are manufactured out of unfired refractory plastic bond clay (chamotte). The variety of clays and manufacturing techniques allow the production of numerous types of bricks appropriate for particular applications. The usefulness of fireclay bricks is due to the presence of the mineral mullite which forms during firing and which is characterized by high refractoriness and low thermal expansion. At higher temperature, a glass phase is generated. The softening behavior of fireclay bricks is determined by the amount and composition of the glassy phase. Due to the alkali content and the presence of other impurities, this phase starts to soften at 1,000 °C and it gives a high softening interval to fireclay bricks which is of great importance for their application.

Fireclay bricks are used in furnace construction, blast furnaces and hot blast stoves, steel foundries, coke ovens and gas plants, glass industry and the cement industry. In the aluminium industry a large quantity of fireclay bricks are used for the construction of baking furnaces for anodes.

High-alumina material

Increased thermal and chemical stress on refractory linings require materials of higher quality for many industrial furnaces. The chemical composition of fireclay products is limited by the use of clay as raw material base. Therefore, it is necessary to utilize high-alumina material (Al_2O_3 >45 %) with a different raw material base.

As the two-component system silicon-alumina (**Fig. 3.124**) shows, refractoriness increases with growing alumina content. At a composition of 94.5 % SiO_2 and 5.5 % Al_2O_3, the eutectic is at 1,595 °C. As the proportion of alumina is increased, the melting point of the mix rises to the melting point of pure corundum which is at 2,050 °C.

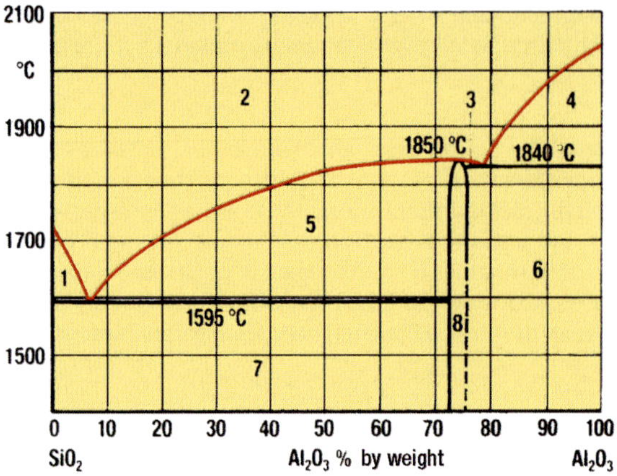

Fig. 3.124: Phase diagram SiO_2 – Al_2O_3

The only stable compound in the system SiO_2-Al_2O_3 is mullite which has a defective space lattice and decomposes to corundum and a liquid phase (incongruent melting) at about 1,840 °C. The chemical composition of mullite is variable. At extremely high temperatures, e. g. during melting, a modification of about 78 % Al_2O_3 ($2Al_2O_3 \cdot SiO_2$) is formed and at lower temperatures with about 72 % Al_2O_3 ($3\ Al_2O_3 \cdot 2\ SiO_2$). Up to 8 % Fe_2O_3 and 2 % TiO_2 can be dissolved within the mullite lattice.

3.5.2.3 Raw materials

For the production of high-alumina refractories, natural raw materials as well as synthetic materials are used (**Table 3.18**).

Table 3.18: Natural and synthetic materials for alumina refractories

Natural raw materials	Synthetic raw materials
kyanite	sintered mullite
sillimanite	fused mullite
andalusite	alumina
bauxite	sintered corundum
shale clay	fused corundum

The minerals kyanite, sillimanite and andalusite have the same chemical composition according to the general formula $Al_2O_3 \cdot SiO_2$. Their crystal structure differs widely according to formation and origin and so do their properties, e. g. density. During firing these minerals change into mullite and a liquid phase. This transformation occurs at different temperatures. The transformation of kyanite and andalusite starts at about 1,300 °C, whereas sillimanite starts to form mullite at appreciable quantities only above 1,500 °C. Depending on the starting material, this mullite formation results in varying expansion during firing. Due to its high density of 3.5 to 3.6 g/cm^3, kyanite increases in volume by about 15 % during firing and must, therefore, be calcined before use. The expansion of the volume of sillimanite is only 5 to 8 %.

The raw material bauxite is used for the production of primary aluminium. Only bauxite with a high Al_2O_3 content and low iron content can be used for the production of refractories.

The increasing demand for refractories made it necessary to become independent of the property fluctuations in the natural raw material deposits. Therefore, more and more synthetic materials in the sintered or fused form are in use. Sintered corundum is obtained by heating refined alumina to a temperature just below melting point in rotary kilns. Fused corundum is preferably melted from bauxite in electric arc furnaces.

High-alumina refractories are used for furnaces in the aluminium industry. In other metallurgical industries different refractory materials are applied. This is due to the different requirements in these technologies. One large group are the magnesia bricks which are used, for instance, in the copper industry. Basic bricks containing chrome are used in the steel industry. The glass industry prefers silica bricks for the lining of their furnaces.

Parts made of silicon carbide are also used in the aluminium industry, e.g. in crucible furnaces. The raw material for silicon carbide is obtained from quartz and carbon in an electric process in reducing atmosphere. SiC is used if good heat conductivity is required.

3.5.2.4 Types of refractory material and manufacture

Two major groups of refractory materials are available: bricks and unshaped materials (monolithics). The basic compositions and the material mixes are identical for both types of refractories. With regard to chemical composition and possible stress, the same selection criteria apply for bricks and unshaped material. Also manufacturing is identical up to the mixing of the material.

Unshaped materials (monolithics) are ready for use after mixing while bricks and special shapes undergo shaping, followed by firing to arrive at the final product.

Various methods are employed in the refractory industry for the manufacture of refractory materials. The simplest procedure to manufacture bricks is to saw shapes from natural or artificial raw materials. Another method is the casting of melts of specific compositions in molds in order for the melt to solidify into blocks or bricks. Materials manufactured in this manner are called fused cast products. During recent decades, methods formerly used in the ceramic industry have been used increasingly in the production of high-duty refractory materials.

Today the so-called heavy clay ceramic method is preferred for the manufacture of refractory materials. The production step comprises comminution by crushing and grinding, classification and mixing the graded material to obtain a correct mix. During this step all selected raw materials are mixed as required in the grain size distribution as required for the mix. The selected binder is added during mixing. After fine crushing and wet grinding - as manufacturing stage materials - the process is completed for unshaped refractory material but continues for bricks with shaping by means of casting, extrusion or isostatic pressing.

Bricks are manufactured as standard shapes or as shapes shaped for special applications. Standard bricks offer the consumer many advantages. Pressed bricks are of high quality and uniformity. They can be supplied quickly at lower cost than special shapes.

For the production of bricks the mix obtained after mixing all the ingredients is sent on to the pressing or shaping plant. The classic process of hand-shaping is now replaced by heavy presses. For shaping mostly high-duty hydraulic presses are used which can shape bricks with very narrow tolerances. Each individual brick is shaped to dimensions required whereby the shrinkage occurring at later production stages has to be considered.

After shaping, the bricks are generally subjected to a drying process in order to remove water or other liquid that was added for the shaping process. Only completely dry shapes can be fired without the danger of cracking. Drying is mostly accompanied by shrinkage that is proportional to the amount of liquid which evaporates. This has to be considered when building the molds for shaping.

According to the type of brick, the dried shapes are fired at temperatures in excess of 1,000 °C for about three days. During this process all volatile components, such as residual water, water of hydration, CO_2, SO_2 etc., are evacuated. In addition, the texture of the brick is formed as result of reactions, re-crystallization, transformation or the formation of a liquid phase. As these processes are connected to contractions or expansion, the dimensions of the refractory brick often change during firing and this requires extensive experience when designing and manufacturing the molds for shaping. The result is a brick with defined characteristics and accurate dimensions.

This is the advantage of using bricks for lining the furnaces. The refractory lining made of bricks has homogeneous characteristics throughout the specific furnace area, thus becoming a high quality lining. Behavior during operation and maintenance periods becomes predictable. Also curing after installation of the lining is less critical than curing of a monolithic lining. This curing is still necessary since a bond between the various bricks must be established. To obtain this bond, a mortar or mastic corresponding to the material of the bricks is used. However, the bricks must be laid carefully and the bricklayers need to have the necessary skill for this type of work.

The result is a very reliable refractory lining with good service life and characteristics fulfilling the requirements of the specific furnace.

The link between unshaped (monolithic) products and bricks are the so-called prefabricated parts. These are manufactured from unshaped materials using the typical processing techniques, e. g. ramming or vibrating. In many cases the parts are dried and tempered at the refractory material manufacturing plant. In general, prefabricated parts are produced in large sizes in order to enable a lining with few joints.

Depending on the application, there are three major classes of unshaped products: Batches for monolithic structures and repairs, generally described as "mixes", material for laying and joining and materials for coating and surface protection.

The mixes are used to construct a monolithic lining. They are classified according to the water content during application and the type of setting.

Ramming mixes and so-called dry ramming mixes are mixes of friable consistency before installation which contain bonding agents. They harden under the influence of heat above room temperature. They are rammed by hand or with the aid of mechanical equipment. The mixes are used dry, as delivered, or after addition of water or other liquids. Dry ramming mixes contain particularly low melting point additives.

Plastic mixes are available as friable or block mixes ready for use, consisting of granular refractory material and plastic bonding clays. They can also contain chemical bonding agents. The efforts involved for installation and shaping are considerably less than that required for ramming mixes. Plastic block mixes have the advantage of not requiring molds.

Refractory castables are batches of granular refractory charges and bonding agents which, after water or another mixing liquid has been added, solidify and harden without a special drying procedure at room temperature. Standard castables can be utilized up to a temperature of 350 °C. A typical application in an aluminium melting plant is the lining of waste gas pipelines. For temperatures exceeding this value, refractory castables have to be used. These castables are installed by pouring, followed by compaction by way of vibrating.

Gunning and slinging mixes are named according to the method of installation. The gunning is executed by way of shot blasting the mix by means of water and compressed air. This is a very convenient method for refractory material repair.

The durability and the operational safety of the refractory brickwork depends not only on the proper selection of the refractory bricks but also on the correct laying and joints filled with fine grain material such as mortar. The characteristics and composition of the refractory mortar must correspond to that of the bricks. The consistency of these materials is an important property. Laying and joint materials (refractory mortars, mastics, adhesives) should have good water retention properties to avoid drying out of the mortar due to penetration of water into the refractory bricks. Correctly mixed with water and properly used, they ensure tight brickwork structures and considerable increase in strength and no premature wear of joints. Generally, monolithic linings require sealing of the surface against the various attacks on the part of chemicals or gases. This protection coating material has a lower viscosity as the joint material. They smoothen the surface and close the small fissure generated during drying and curing.

We have discussed the different refractory materials but still do not know why the collections of grains do not disintegrate. The reason is that there are different types of bonds which create a solid structure of the refractory lining having good strength as required.

The ceramic bond requires a mix that is generally composed of a refractory charge with clay as the carrier of the bond. The ceramic bond is obtained by sintering at high temperature. This is preceded by the volatilization of residual water and other minor additives. Sintering, in the case of high-alumina material occurring as dry sintering, starts at a temperature of 100 °C for high-alumina material. It is stimulated by the effort of the components to reduce surface energy. This takes place by self-diffusion via vacancies in the crystal lattice and by the growth of energetically favored nuclei. The infiltration through pores of the material forms a sturdy structure of the entire material and establishes links between the different grains.

Additions of organic binders, such as sulfite lye but also tar, pitch etc., can improve bonding in the low temperature range up to 300 °C by providing a chemical bond. The water-soluble additives decompose during drying thus causing adhesion and during heating in oxygen-rich atmosphere they burn up. With further increase of temperature, a ceramic bond occurs. Inorganic binders, such as aluminium phosphates and phosphoric acid, offer high strength at quite low temperatures and can, therefore, be used to bridge the range between organic setting and ceramic sintering. Additions of phosphoric acid or aluminium phosphate to mixes in the system SiO_2-Al_2O_3 offer an effective setting from 300 to 500 °C which cannot be dissolved by water. They do not reduce the refractoriness of the charge materials and, for this reason, they are used as chemical bonding agents for those mixes that have to be suitable for high temperatures. These are mainly mixes with an alumina content above 60 %. The chemically-bonded bricks give the refractory material a higher resistance to attacking melts, slags and dust. They are, therefore, preferably used for furnaces in the aluminium industry.

One type of bonding is the hydraulic bond. The agents used here are artificially produced mineral powder material, e.g. cement, which consist mainly of compounds CaO with SiO_2 and Al_2O_3 in various proportions. They form a pasty mix which hardens after several hours at room temperature in air and under water thus forming water-containing minerals resulting in a rock-like material which is insoluble in water. When heated up, the greater part of the chemically-bonded water escapes between 500 and 600 °C; the cement loses its bonding properties and the castable its strength if there is no ceramic sintering with the charge material. Aluminous cements are used in temperature ranges above 1,500 °C. The strength of the refractory castables, containing aluminous cement, does not decrease nearly as much in the critical temperature range of 600 to 1,000 °C as in the case of other refractory castables. There is always ceramic sintering above 1,000 °C.

Insulation material

The refractory material coming into contact with the metal or with the hot furnace interior and the combustion products requires good temperature resistance combined with high mechanical strength in certain areas of the furnace. But one other feature is also required and this is good heat insulation behavior. The high density refractory material does not offer the optimal characteristics for heat insulation. This is required, however, to reduce heat loss in the walls and to avoid overheating of the furnace shell with potential danger to the operating personnel. The structure, fulfilling all requirements – resistance to the various forms of attack – as well as good insulation, can be obtained by a multi-layer system comprising a hot face lining and insulation refractory material that is tailor-made for the temperature range as occurring in the temperature gradient from furnace interior to the outside.

For the thermal insulation of furnaces and other combustion equipment, materials with high pore volume are used. Because of the large proportion of pores, these materials contain a substantial quantity of air so that the poor thermal conductivity of air can be exploited for thermal insulation.

There are some basic considerations which apply for the production of heat-insulating materials:

- The lowest limit of thermal conductivity, which can be achieved by air insulation, is determined by the thermal conductivity of the calm air.
- The thermal conductivity of a porous material is lower the finer the pores are in the material. In bricks with large pores, generally permeable to gases, heat exchange by convection already occurs at 500 °C. Above 1,000 °C, heat transfer by radiation must be considered.
- The porosity of a brick alone does not necessarily determine its insulation properties. The structure of the refractory material, whether sintered or fused, is an important factor as well.

Different methods are applied for the production of insulating material. There is first the selection of raw material that could be expanded fireclay, power plant ash or the like. By addition of volatile or combustible materials, such as styropor, coal, pet coke, saw dust, cork or nut shells, the proportion and the size of pores can be determined. Also foaming agents, such as saponin and colophonium soap, are used.

Materials, which are mainly used for thermal insulation, usually have a porosity between 45 and 75 %. According to their application, they can be divided into two groups:

- Heat-insulating bricks for an application (service) temperature of 900 °C, max. 1,100 °C which are mainly based on vermiculite, kieselguhr or the like
- Insulating material with an application (service) temperature above 1,100 °C which are based on fireclay, sillimanite, silica etc.

Because of their low refractoriness, the heat-insulating bricks of the first group are mainly used as rear insulation. In comparison, the refractoriness of the second group is very high. They are generally described as insulating refractories. They can be used in direct contact with hot combustion products, provided that the walls are not subjected to mechanical stress or that there is no slag or dust attack and that the operating temperature does not exceed the respective application limits. Materials, which are resistant to abrasion, have also been developed.

For thermal insulation also products out of ceramic fiber are used. The raw materials, which are pure natural products, such as kaolin or synthetic products of the SiO_2-Al_2O_3-system, are melted and pulverized or sprayed. As long as the Al_2O_3-content is less than 60 % by weight, vitreous fibers are formed with a diameter of only very few µm.

The fibers are used, either with or without a binder, to produce blankets, felt, paper, vacuum-shaped parts, rope and pre-fabricated parts. The fiber materials can be cut to size and they can be attached safely in a simple manner with the aid of pins or anchors of heat-resistant material.

Plate material is produced from vermiculite which is a natural stratified silicate with high water content that swells by a factor 10 to 15 when heated. The plates obtained by pressing and drying are easy to cut. They have good thermal resistance and a heat expansion factor similar to steel which makes them the ideal compound material with steel.

Calcium silicate is a synthetic material made of CaO and SiO_2 that is mixed with filling additives and reinforcing fibers. The mix is pressed or cast and then treated in pressure vessels with hot steam. The chemical reaction results in the formation hydrate of calcium-silicate. The final plates and shapes can be cut and machined easily. Plates out of calcium silicate have good thermal insulation properties and the material is resistant to liquid aluminium. However, it breaks easily; therefore, it is preferably used as rear insulation in refractory linings.

There is always a discussion about advantages and disadvantages of brick linings and monolithic refractory linings and the proper selection of the best suited system.

Refractory mixes solidify and sinter only after they are installed; the temperature gradient in the furnace also causes a gradient in the lining. This fundamental difference between the two products determines the advantages and disadvantages in their application.

A considerable advantage if using refractory mixes is the quick process of installation. However, this is a question of techniques for application and the process of pouring, vibrating and gunning. Quicker heating-up is often possible which means less downtime for the production. This is important when repair or complete re-lining is required. Usually mixes have a better insulating effect than the corresponding bricks and a higher thermal shock resistance because of the sintering gradient but also less densification. With monolithic linings, expansion joints are used for large areas only. This means that there are fewer permeable joints than with brick linings. Monolithic structures are considered to have an improved mechanical resistance to vibration and impact. The anchoring system, together with the property and the temperature gradient in the lining, permits high mechanical compressive stress at high temperatures for short periods.

Complicated shapes can be installed in the furnace, thus avoiding the high cost for producing special shapes. In addition, good compatibility with the shape of the flame is also easier to obtain. Another advantage is seen in the possibility that expansion matches the application. With certain mixes, the expansion at service temperature is compensated by the initial shrinkage, thus providing a practically expansion-free lining. However, considering the varying conditions in our aluminium furnaces, this argument seems to be far away from practical experience.

The disadvantages are obvious. Because essential work procedures, such as installation, drying and curing, cannot be carried out in manufacturer's production facilities, damages may occur due to incorrect methods applied. This applies mainly for the curing procedure. External heating equipment has to be used with connected temperature control. But even using such equipment, it seems to be obvious that an even temperature distribution is very difficult to obtain. Bricks are manufactured according to pre-defined and precisely controlled parameters. Consequently, the critical items of a brick lining are only the joints. These are very small in modern brick linings; they are in the range of 1.5 to 2 mm only. Thus, they are critical but not as much as in a complete lining.

3.5.2.5 Curing

The final bonding of a refractory system in a furnace is always the ceramic bond obtained during sintering at a temperature above 1,000 °C. The other kinds of bonding are required to stabilize the refractory lining during the heating cycle. Once sintering has taken place and the material is stabilized, the refractory lining can be exposed to the varying conditions in the furnace.

It is essential that the installed refractory lining is heated up to the temperature required for initiating sintering. Due to the thickness of the lining, the temperature must be held at the critical temperatures for a period of time to allow a thorough heating of the material. This process is called curing and is required for every type of lining, independent of the type of refractory lining, i. e. monolithic or as brick system. The progress of curing with heating time and holding periods, however, depends very much on the kind of refractory material. Bricks are fired and have reached the sintering stage in controlled conditions inside the firing furnace. Thus, only the mortar has to be cured to establish the bond between the individual bricks. This is different with a monolithic system. Firing has to take place in situ, e. g. in the furnace. This requires very special heating equipment.

Due to the nature of the different material, the heating curve differs between a brick lining and a monolithic lining. Careful temperature control is required through all stages of the curing. To take the temperatures, usually Ni-Cr-Ni thermocouples, as also applied for furnace control, are used. They are easy to handle and give accurate readings.

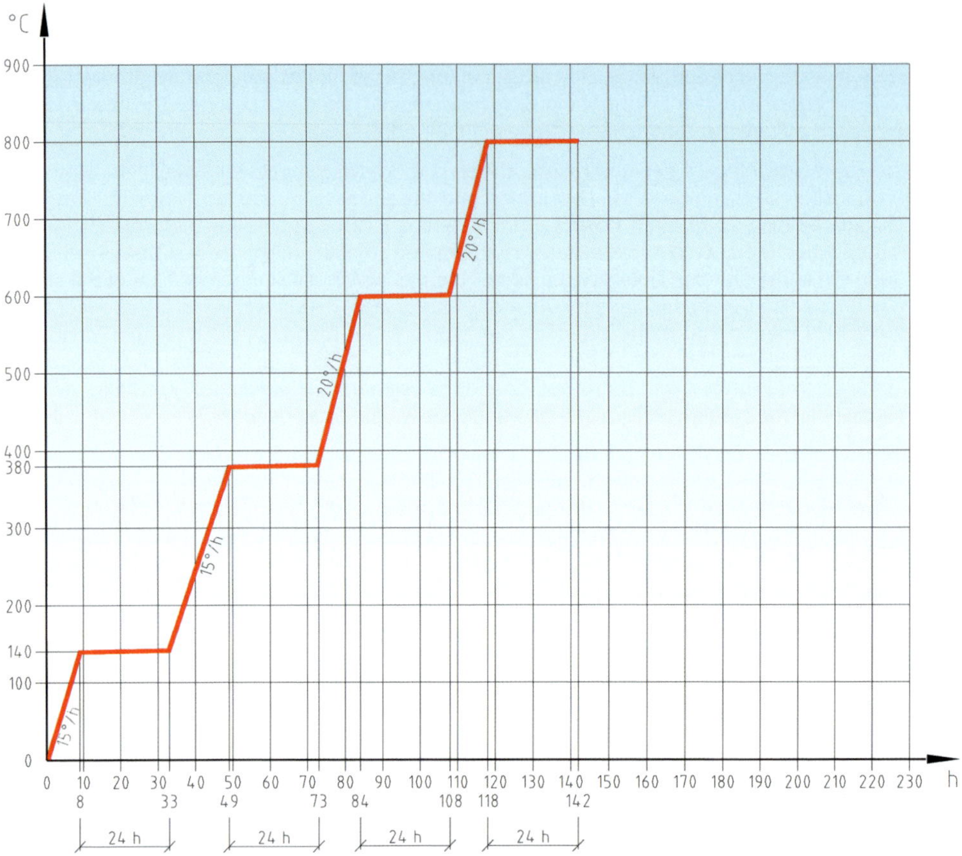

Fig. 3.125: Curing curve for a rotary drum furnace lined with bricks

After installation, the procedure for curing starts after a bonding period of 24 hours. Usually there is a holding period at 150 to 200 °C for approximately 6 h to improve the capillary transport of liquids. At a heating rate of max. 15 °C per hour, the temperature is increased to 400 °C with a holding period of 12 h. The reason is to have slow and controlled dehydration. After the holding period the temperature is increased by 20 °C per hour to 1,100 °C. The furnace is then ready for operation.

Fig. 3.125 gives an example for the heating cycle of a brick lining. This is realized in a rotary drum furnace. The first holding temperature is also approximately at 150 °C. The next stop is at 380 °C which is reached with a temperature increase of 15 °C. The holding period is at 600 °C. Then heating continues to 800 °C. After this time the maximum temperature of 1,000 °C can be reached with a temperature gradient of max. 20 °C per hour.

Generally, the refractory lining comprises layers of different materials. In such a case, for curing the heating curve of the material with the lowest permissible temperature gradient is applied.

Fig. 3.126 shows the heating curve for the refractory lining of a holding furnace using different materials for the hot face lining, including a multi-layer refractory system. Here the different tem-

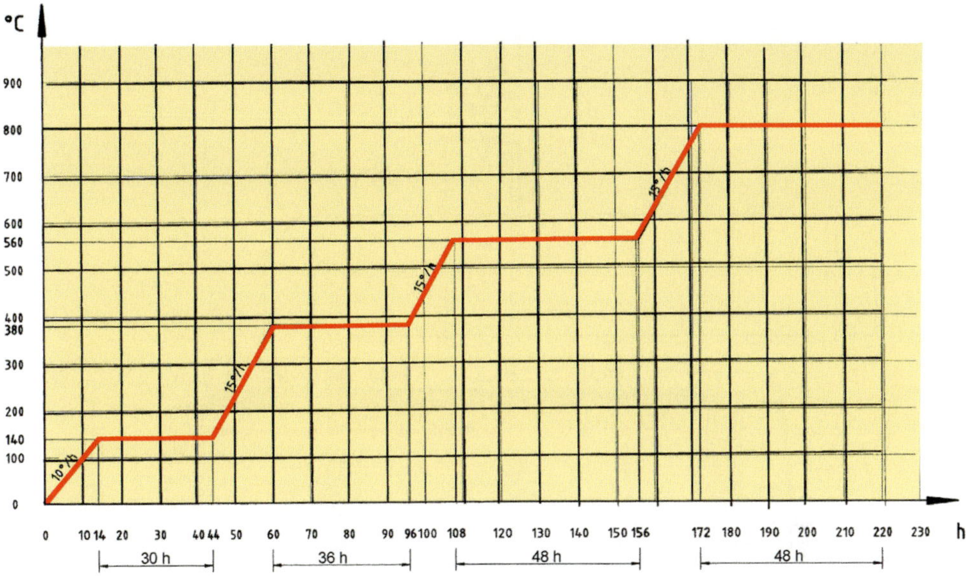

Fig. 3.126: Curing curve for a reverberatory furnace designed as multi-layer system

peratures can only reach temperatures according to the temperature gradient from the furnace interior to the furnace shell. Here again, four holding steps are used for curing.

After completion of the first holding steps all liquids should be removed. If the operators still notice vapor coming out of the refractory lining, the holding time must be extended until no vapor is noticed anymore.

3.5.2.6 Refractory design

In order to avoid the manufacture of special shapes, furnace linings are generally designed as a combination of brick lining and monolithic systems. Apart from this, the one or the other system may offer advantages in certain furnace areas. There is also a difference in refractory design depending on the type of furnace. A rotary drum furnace requires very dense refractory material with good resistance to the treatment salt. Therefore, the content of alumina should be lower than in the case of a reverb furnace. Consequently, the hot face lining of a rotary drum furnace consists of high alumina material with a proportion of Al_2O_3 not exceeding 60 %. Contrary to that, the lining for the bath area of reverberatory furnaces has an alumina content of not less than 80 %.

Let us have a look at different lining concepts of furnaces. The first example refers to a reverberatory furnace (**Fig. 3.127**).

Bath area

The bath area is designed as multi-layer system built of three layers. The hot face comprises a layer of refractory bricks on a bauxite base having an alumina content of 81.5 %. The thickness of this layer is 250 mm. The maximum temperature is 1,500 °C. The brick has a good infiltration resistance, high strength and good resistance to temperature shock. The second layer comprises an insulating refractory castable with special infiltration protection. Its alumina content is 40 %. This type of material is selected to provide a good support for the hot face layer to handle impacts

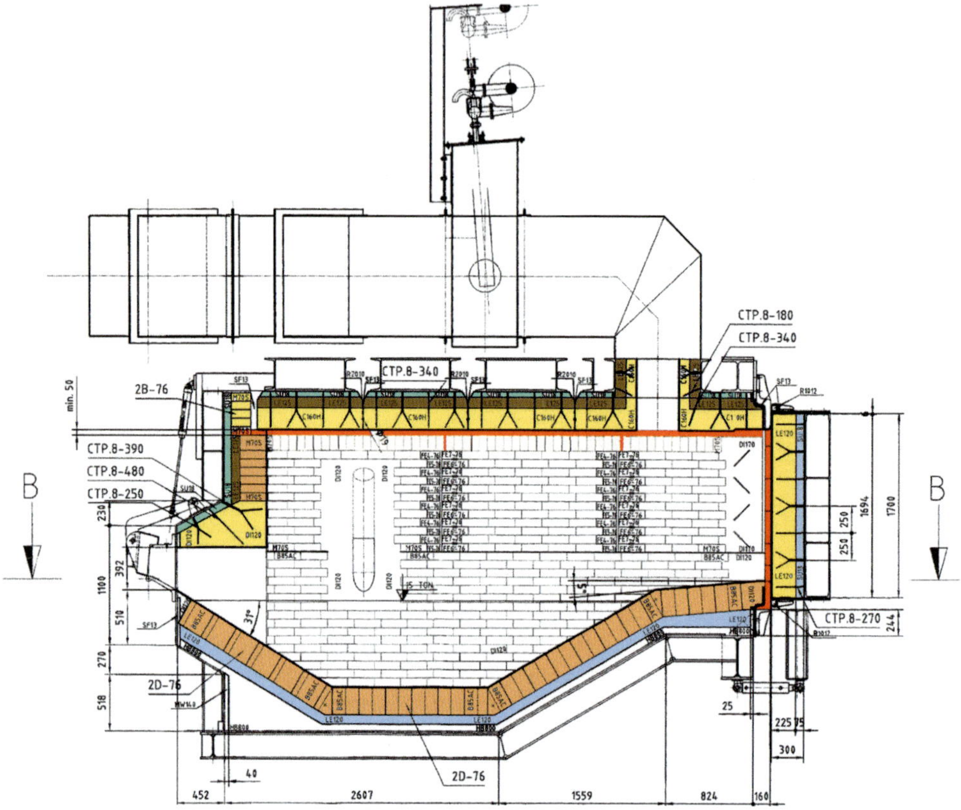

Fig. 3.127: Lining of a reverb furnace

and to provide a blocking system to resist infiltration of aluminium. The thickness is 86 mm. The third layer comprises an insulating layer of calcium silicate. The thickness is 64 mm. For the de-watering of the bath area, ropes of mineral wool are attached to the steel shell prior to the laying of the refractory castable.

Walls

Walls in the lower section of the furnace are identical to the lining of the bath area.

The upper section of the furnace walls is designed as three layer system as well. The hot face consists of high-alumina brick with an alumina content of 73 % which is based on bauxite and mullite. The thickness is 250 mm. The second layer consists of an insulating refractory castable having an alumina content of 26 % and a thickness of 85 mm. The third layer comprises calcium silicate of 40 m thickness. For additional reinforcement 4 anchors per m^2 are installed.

Between front layer and rear lining a thin ceramic fiberboard is positioned in order to prevent molten metal penetration into the rear layers in case the front layer gets some cracks.

Roof

The furnace roof is made of individual slabs which are pre-fabricated on site and, after drying, placed in the furnace.

The hot face layer is made of refractory castable with an alumina content of 91 % and a corundum base. The thickness is 200 mm. The second layer consists of an insulating refractory castable having an alumina content of 26 % and a thickness of 100 mm. The third layer comprises a calcium silicate of 50 m thickness. The pre-fabricated slabs are insulated against each other by ceramic wool expansion joints. The anchors are a combination of steel and ceramic anchors.

Door

The lining of the door is a special challenge. There is where the attack of the full heat of the furnace interior hits. The lining must also be of light weight since the door needs to be lifted for opening. On top of that, there should be no distortion to maintain the required sealing of the door. Therefore, the lining comprises different sections to allow a certain displacement of the individual slabs under the influence of the temperature of the furnace interior. They are separated by expansion joints filled with refractory wool. The hot face comprises an insulating refractory castable having an alumina content of 26 % and a thickness of 225 mm. The final layer comprises a calcium silicate of 75 mm thickness.

Burner brick

The burner brick is made of a high temperature resistant refractory castable on corundum base. In this area there may be impingement of a very hot flame. Thus, very heat-resistant material must be used. This, in general, is corundum-based refractory castable.

There are two more very critical areas in the furnace which are the wall opposite the charging door and the pre-heating ramp. Since the lining comprises dense bricks, no special measures are required for the back wall. Normally, the pre-heating ramp is made of dense bricks; no special measures are required except that the bricks are laid with particular care. Since the area is exposed to mechanical impacts during charging and de-drossing, more intense wear than in other areas must be expected requiring frequent repair of refractory materials. To ease this repair work, the ramp could be constructed as monolithic system, whereby the front layer is armored with steel needles to reduce the wear. Monolithic linings are easier to maintain than brick linings.

The belly band, that is the area between liquid metal surface and the upper furnace, is critical in regard to corundum scaling. However, special measures are required only if metal is held for an extensive period of time at that level. In such cases bricks with a carbon content are installed. In any case, operators should be instructed to keep the sidewalls clean. This is comparatively easy as long as the furnace is hot and operated at short intervals, i. e. every time the furnace is empty the layer of corundum is thin. It becomes more difficult if only cleaned at large intervals with a substantial thickness of the scaling.

Other furnace areas are not that critical. Having made all minor repairs during the frequent annual maintenance periods, a refractory lining, which is carefully designed using proper materials, skilfully installed and cured, will have a service life of five years at a minimum; some of them even operate for 20 years.

It is very important that sufficient expansion joints are provided for in various sections of the refractory lining. The final expansion joint must be located at the door frame in order to avoid expansion forces in this area leading to possible deformation; the door cannot close completely and furnace pressure cannot be maintained. No matter whether the forces are elastic and the door frame comes back to its original position after the furnace cools down, this distortion must be avoided.

The furnace lining, as described above, is designed for a holding furnace. Thus, no special measures are applied for ramp and belly belt. But it is a combination of brick lining and monolithic refractory lining. Instead of the bricks, a monolithic lining could be selected. In such a case, refractory gunning mixes are used instead of the bricks. The mixes have the same material composition as the bricks. In the bath area, high-alumina material with 80 to 90 % Al_2O_3 is applied.

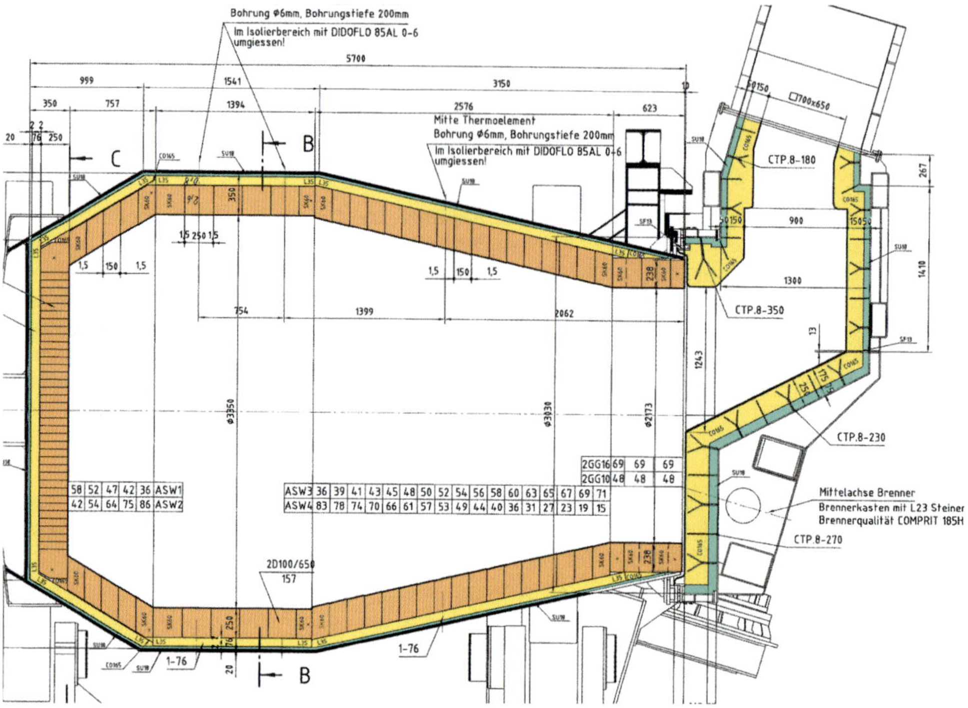

Fig. 3.128: Refractory lining of a rotary drum furnace with bricks

In the upper area the alumina content is less. For holding the lining, so-called "Christmas-tree" type anchors, made of stainless steel, are used which are welded to the steel shell. Additionally, ceramic anchors are used.

Ceiling, door and burner bricks are identical to the design as described for the brick lining.

Another example is the lining of a rotary drum furnace (**Fig. 3.128**).

The lining of a rotary drum furnace is exposed to many heavy attacks. There is first the charge of different materials and, last but not least, liquid metal. The charge also contains a high percentage of salt with an addition of aluminium fluoride and cryolithe which is used to digest bauxite in the Bayer process for producing aluminium oxide. Since refractory material also consists of a high percentage of aluminium oxide, the impact of the aluminium fluorides leads to an extended chemical attack. Therefore, when selecting the refractory material, the Al_2O_3 content should be as low as permissible in regard to heat resistance and also resistance to the other components of the batch. The mechanical forces attack the lining also very severely. Besides the grinding effect of the rotating batch, large metal blocks sometimes bump against the walls of the drum. The burner flame inevitably hits the furnace wall, too. In case the furnace is equipped with oxy-fuel burners, the flame is very hot and its temperature may even exceed the permissible peak value of the refractory material for a very short instance. All these factors reduce the service life of the refractory lining. As a consequence, a complete relining is usually due every two years. Depending on the type of scrap to be processed and with a normal gas/air burner system, it may be sufficient to replace the refractory material in certain areas only but even then it may be that only one complete relining is skipped.

Due to the rotation of the drum, only little rear lining can be installed. Thus, the heat loss through the drum shell is higher than in the case of a reverb furnace. To support the cascading of material in a rotary drum furnace, sometimes the hot face layer is shaped polygonal. This certainly helps as long as the refractory material is new. But due to the excessive wear, the inside will be round again. Therefore, it is more efficient to adjust the rotating speed of the drum to the requirements.

Considering these facts, the refractory lining is designed for the example as follows:

The hot face is made of refractory bricks on andalusite and corundum base having an alumina content of 58 %. The thickness of this layer is 250 mm. The intermediate layer consists of insulating fireclay and has a thickness of 78 mm. The third layer, installed at the drum shell, comprises calcium silicate with a thickness of 27 mm.

The rear wall of the furnace is built identically. The charging side could also be installed as monolithic lining, reinforced with steel needles. It has been proven, however, that the lined charging section with bricks is more durable.

3.5.2.7 Design and installation

There is a difference in design using bricks and unshaped material. Brick sizes, which should be considered for the design, are standardized in ISO 5019. The standard differentiates between

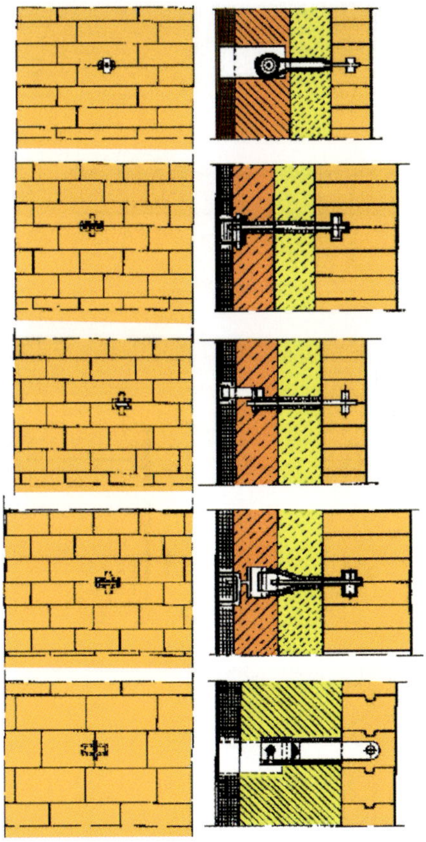

Fig. 3.129: Refractory anchors for a furnace wall

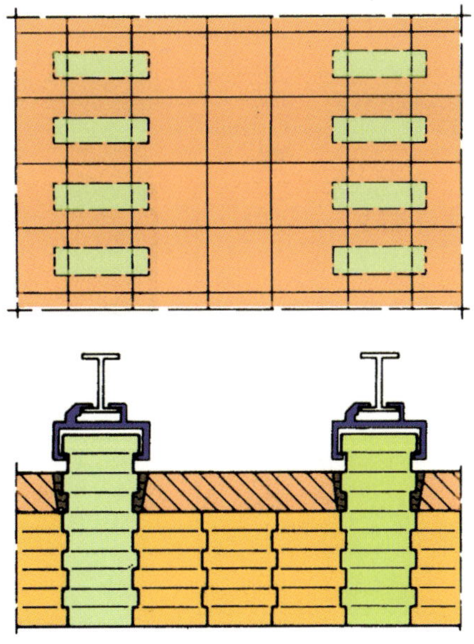

Fig. 3.130: Refractory anchors for a ceiling

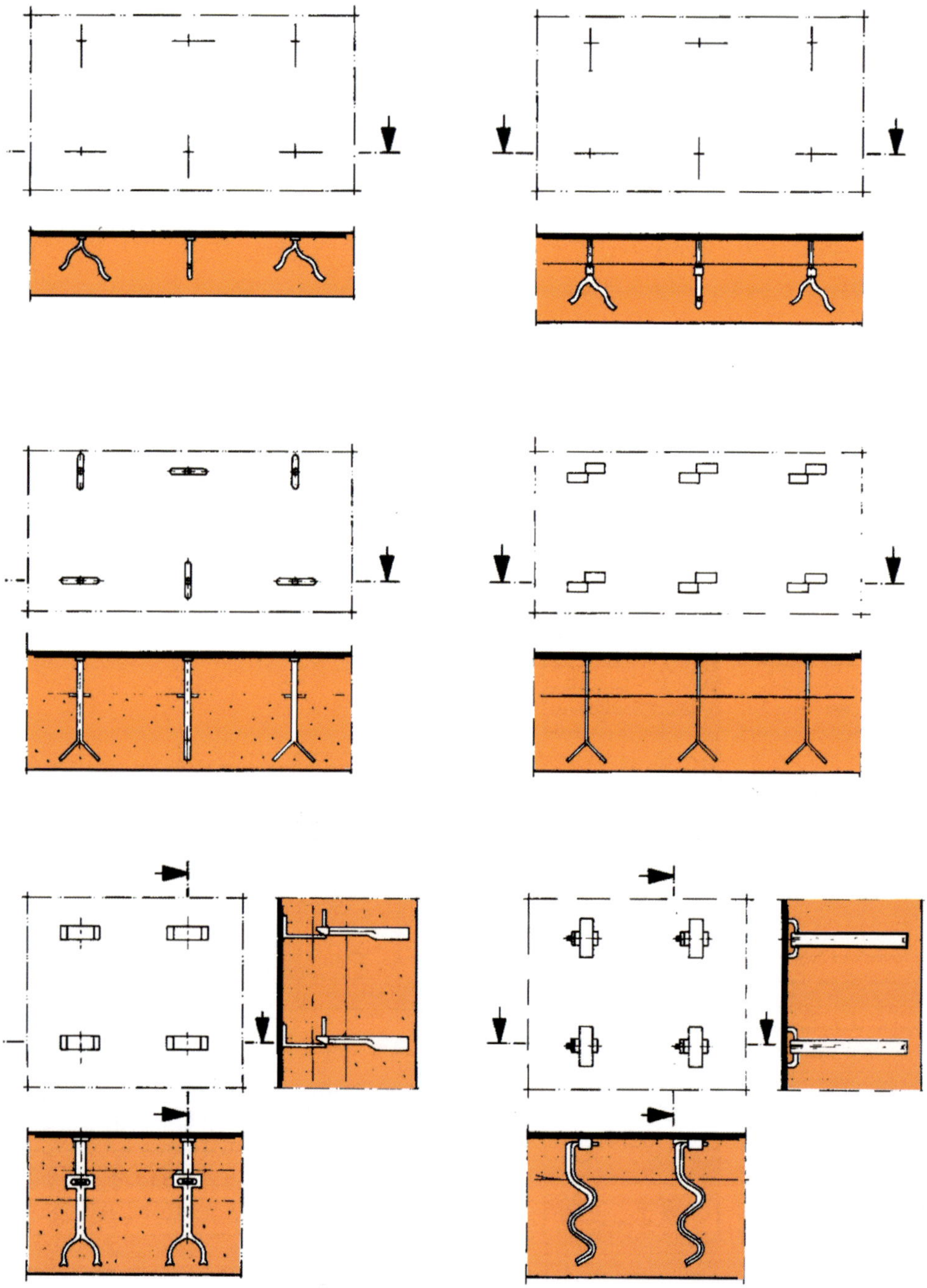

Fig. 3.131: Steel anchors

normal sizes and arch bricks used for the construction of arched furnace roofs. There are also standards for special bricks for suspended roofs and for expansion joints. All other bricks will have to be specially made for the purpose. They are referred to as special shapes. Pre-fabricated parts are also regarded as special shapes. Designers working for the aluminium industry try to avoid special shapes because they are expensive when purchasing and to kept in stock. If a structure in the furnace can not be constructed by using standard shapes, monolithic sections are installed as described in the previous chapter.

Regardless if a brick lining or monolithic lining is installed, anchors are required to stabilize the refractory lining. For bricks this refers to walls only; the bottom does not require anchors in case of the brick lining. Installed refractory anchors must have the same characteristics as the brick material. They are connected to the steel structure by means of steel or cast iron anchors (**Fig. 3.129** and **3.130**).

Due to the nature of the material, more and different anchors are required if installed in monolithic material; they connect the lining with the steel shell. Depending on the design and temperature requirements, steel anchors or refractory (ceramic) anchors are applied. Numerous types of anchors – steel or refractory – are available. **Fig. 3.131** shows some examples of steel anchors. They must be covered with a minimum layer of refractory material to avoid contact with the material in the furnace and no heat bridges to the furnace shell are created. The minimal thickness of this cover is 20 mm. As a general rule, ceramic anchors are used if the application temperature exceeds 1,200 °C. Our aluminium furnaces are just at that limit. Thus, in cases where the anchor may get into contact with metal, refractory anchors are applied. As mentioned before, it must be made sure that the anchor material has the same characteristics as the material selected for the lining. **Fig 3.132** shows an example of the installation of refractory anchors in a furnace wall. **Fig. 3.133** shows the steel structure for roof elements with holders for the anchors welded to it. The refractory anchors for the furnace door are shown in **Fig. 3.134**. It also shows the ceramic fiber rope for the door sealing with their cast iron holders.

Joints are critical components. They form the links between the individual bricks and, therefore, they are filled with mortar or mastics of adequate specification. The thickness of the joints, particularly in the liquid metal area, must not exceed 1.5 or 2 mm. It requires skill and experience combined with care to lay bricks for a good refractory lining (**Fig. 3.135**). Different types of joints are applied in furnaces. **Fig. 3.136** shows some typical joints. Expansion joints are very important.

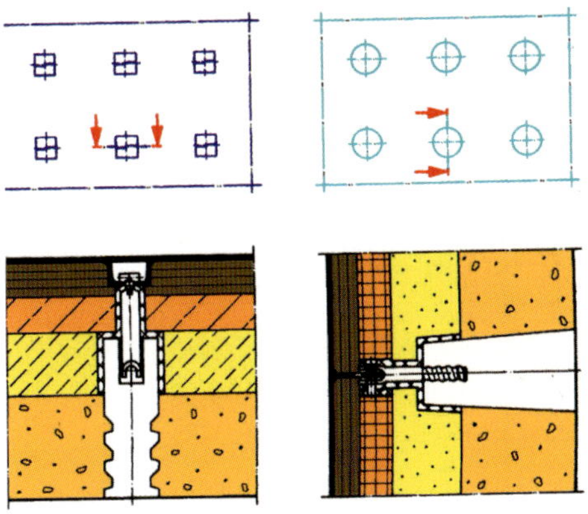

Fig. 3.132: Refractory anchors

Fig. 3.133: Roof element with holders for refractory anchors

Fig. 3.134: Door structure with steel anchors welded to it

Fig. 3.135: Installation of refractory bricks in a reverberatory furnace

They are required to compensate the expansion of the refractory material. If the refractory material expands too much, even a sturdy furnace shell may not be able to bear the related forces resulting in deformation of the furnace or breakage of the refractory material. It is not possible to compensate the expansion by one large expansion joint in a central area of the furnace. It must be expected that the different sections will crack due to the tension forces. Therefore, the entire lining must be divided by expansion joints in various sections. It is very important to have an

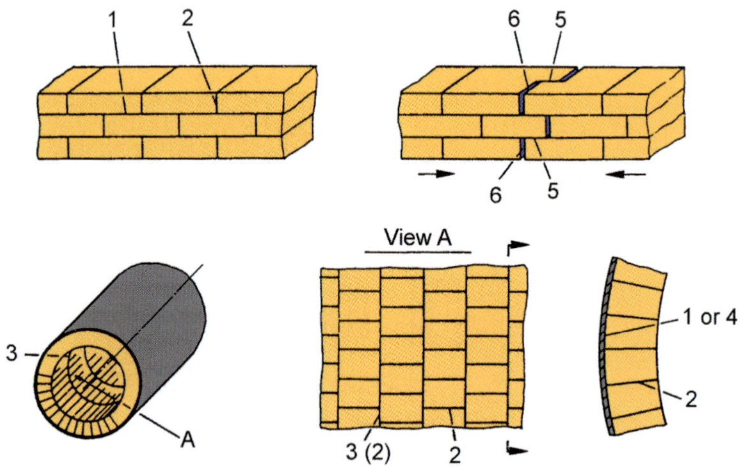

1 - bed joint, 2 - vertical joint, 3 - ring joint, 4 - levelling joint, 5 - sliding joint, 6 - expansion joint

Fig. 3.136: Typical joints of refractory brickwork

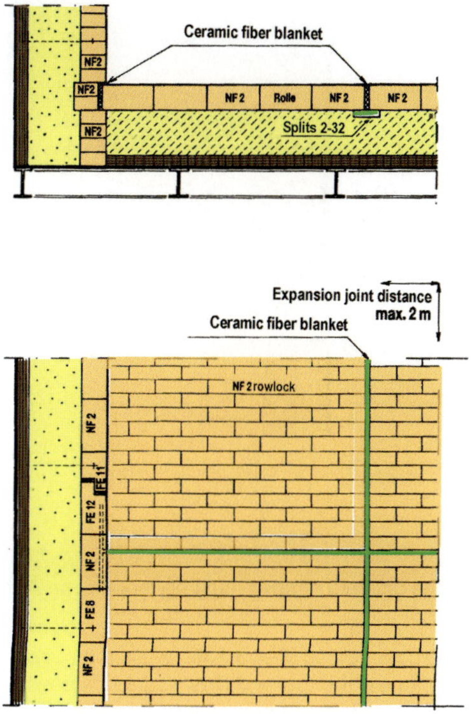

Fig. 3.137: Expansion joints, furnace bottom

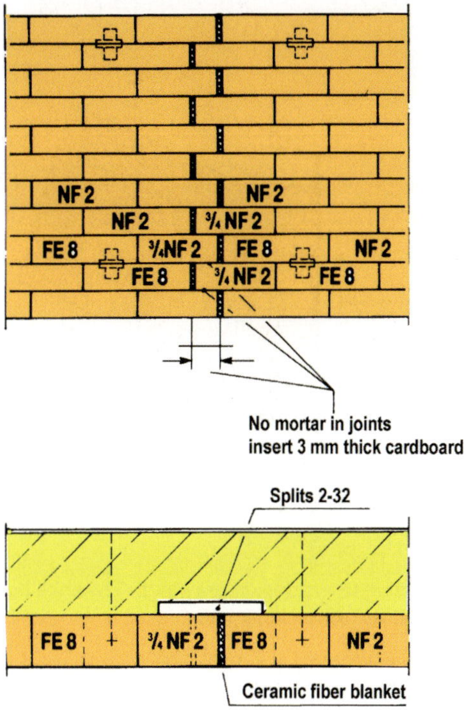

Fig. 3.138: Expansion joints, furnace bottom

expansion joint arranged around the door frame of a furnace and at the outer frame of the furnace door. This is to avoid pressure on the relevant steel structures which may lead to distortion of these important furnace components. Maintaining a positive pressure in the furnace is then very difficult to achieve. The joints are usually sealed by means of refractory fiber, leaving some safety margin for the compression at operating temperature. The arrangement of expansion joints needs much attention on the part of the designer and personnel on site. **Fig. 3.137** and **Fig. 3.138** show the arrangement of expansion joints for the furnace bottom and for a vertical wall for a lining as described in the previous chapter. The arrangement of joints differs between brick lining and monolithic lining. Linings built of mixes require more anchors but less expansion joints. When designing such a lining, the irreversible expansion must be considered. During first heating of certain monolithic linings the expansion is higher than during repeated heating. Bricks have been fired before and do not show this effect.

Shaped materials are installed similar to the experience obtained over ages when building houses, bridges and other structures. However, different materials are used for the refractory lining and the brickwork is to be laid quite accurately. Thus, the persons installing the refractories are skilled specialists although their basic hand tools do not differ much from those used in normal masonry work.

Straight walls and furnace bottoms are constructed in sections following the specification of the design engineer. These sections are generally selected from expansion joint to expansion joint. No hollow spaces are permitted when installing the lining. All joints, particularly expansion joints, must be free of foreign particles. When placing pre-fabricated slabs a correct placement is necessary

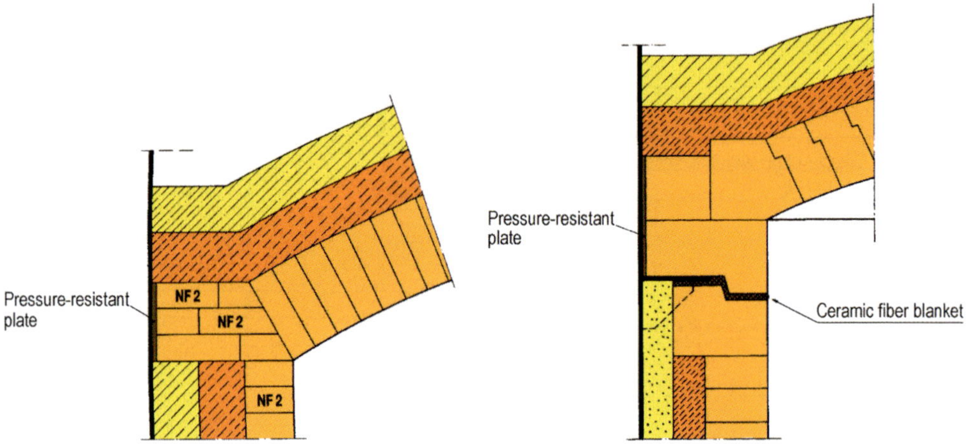

Fig. 3.139: Arched roof construction

according to the design drawings. The links between the different shapes must be properly sealed or built as expansion joints.

In some furnaces, i. e. oval furnaces, arches have to be constructed (**Fig. 3.139**). While the bottom can be installed similarly to the construction of flat bottoms, the roof requires boards. The work then continues section by section. Care has to be taken that the arch is securely supported by the lower structure.

The lining of a rotary drum furnace is a special case. For the construction, a system comprising a wooden beam and a spindle will be used. After placing this tool (**Fig. 3.140**), the work starts by laying the lower part of the drum. Once this part is completed, the furnace is turned and the work continues (**Fig. 3.141**). Additional jacks have to be brought in as the work commences. Depending on the furnace size, two or more jacks are used. After the section ring of the lining is complete, the work can continue by installing the next ring. Although the lining is self-supporting, special care has to be taken to prevent accidents by collapsing refractory lining in the drum. Additional supporting beams may be placed in completed sections. This is very important when removing the old lining during repair of refractory material or re-lining work.

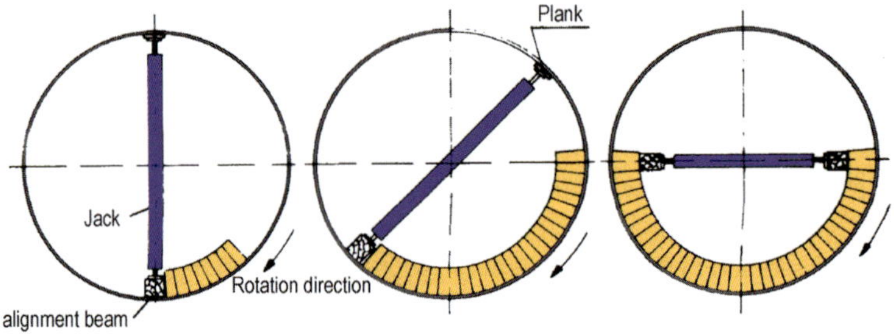

Fig. 3.140: Lining procedure for rotary drum furnace

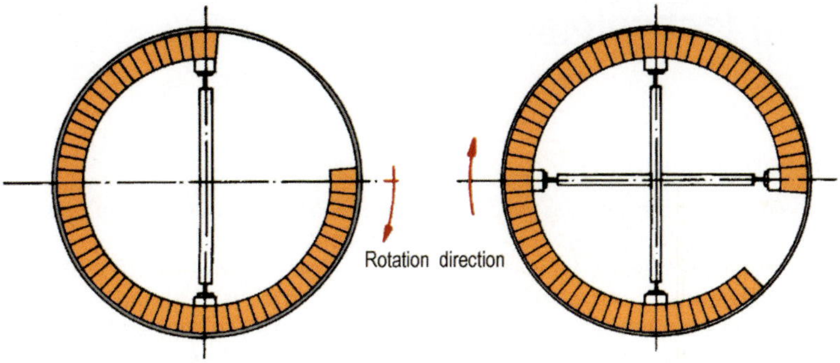

Fig. 3.141: Completion of lining in a rotary drum furnace

Monolithic materials will also be installed in sections. Borders between such sections must be placed between rows of anchors. Refractory castables require molds. After welding of the anchors the material must be brought in swiftly and without interruption of work for the section. The fresh castable must be compacted with vibrators. Installation of ramming mixes also requires boards or molds. The material is compacted by means of ramming tools. In order to open a passage for the water evaporated during curing, small holes (3-4 mm dia.) have to be drilled in the fresh lining at a distance of approximately 150 mm. Some refractories are installed by gunniting. For this work a mobile gunning machine is used. Water is mixed with the dry mix in such a machine and sprayed with the aid of compressed air to the area to be lined. Compacting during this work is mainly due to the impact of the coarse fraction of the mix. A certain quantity of the material rebounds and must be considered as lost.

In order to get at a good refractory lining, work sequence at the site has to be well-organized and all necessary tools and equipment have to be available. Since refractory mixes have to be prepared on site, good mixing equipment is needed. Water should be checked to ensure that the quality is according to the requirements for the mix.

The environmental conditions at the construction site must be considered when installing refractory material. The ideal temperature is in the range of 15-25 °C. At temperatures of more than 30 °C or tropical conditions, faster drying of the material must be expected. In some cases it may be required to work during the cooler night hours and to cover the freshly installed refractory material with wet fabric. Working at temperatures below 15 °C is critical, too. Refractory castables have to be brought to 15 °C and other materials to a minimum temperature of 5 °C before mixing. The work environment should have a minimum of 15 °C. At cold temperatures the continuous progress of curing must be ensured as well in order to ensure proper functioning and to achieve the expected service life of the refractory lining.

4. Casting technologies

Christoph Schmitz

4.1 General considerations

As soon as the alloy is ready and the temperature has reached the desired temperature, the liquid metal must be cast into the shapes as required by the customers. There is a principal difference between refiners and remelters. Refiners produce alloys from scrap with a wide range of alloy composition. Most of them carry a high percent of silicon. It is state-of-the-art that these variations are used for producing casting alloys which will have to be melted again ahead of the production of different types of castings. Consequently, the secondary aluminium recycling industry supplies remelt ingots. Some are used in the steel industry for killing steel. The de-ox alloys are manufactured as cubes or pellets. Refiners process in-house scrap, primary metal or clean wrought scrap. The product is used for manufacturing sheet extrusion, shapes, foils and the like. Basis for this production are billets (extrusion ingots), slabs (rolling ingots) or strip. Although both furnaces used – refiners and remelters – are identical with some exceptions, the casting equipment to obtain the specific product differs substantially.

4.1.1 Product mix

Primary metal is provided by the smelters in the format of 20-25 kg ingots, stacked to bundles with a weight of 1 or 2 tons. The individual ingots are shaped to permit stacking in stable layers.

Secondary ingots are produced as ingots having a weight of 7-12 kg. The individual ingots are also shaped to obtain stable layers within the stacks. These are placed on pallets which are preferably cast as foot pallet of the same alloy and from the same batch this avoids pallet handling since the foot pallets can be melted together with the ingots.

Large ingots of 500-1,000 kg weight, so-called “sows”, are also quite common. They are cast if the type of scrap processed or the plant facilities do not permit production of alloys within narrow tolerances of the analysis. It will also happen that a no-go batch is produced. Instead of going through the entire casting procedure, the complete batch is cast as sows which are then gradually used for the normal production.

De-ox material is manufactured as cubes of different weight or as pellets of different size distribution. Their normal size is 6-8 mm or < 6 mm. Pellets are provided in big bags while cubes are delivered in container boxes.

Ingots for extrusion and for rolling are entirely different products. They go directly to the downstream processing of extrusion shapes on heavy presses or rolling in heavy cold or warm rolling mills. Both are produced on discontinuously operating casting machines which allow a standard casting length of 7.5 m. Extrusions are of round shape and have a diameter that ranges from 120 to 600 mm diameter. The average that is suitable for most extrusion presses is in the range of 200-260 mm diameter. For shipping they are cut to the required lengths of usually 600 mm in the casting plant and bundled to stacks. Crop ends are cut off and recycled.

Rolling ingots are of rectangular shape of different sizes depending on the customer’s rolling facilities. They might have a size of 600 x 2,400 mm. Crop ends are also cut off and recycled. The large slabs are divided into lengths as required by the customers or shipped at full length.

For certain productions aluminium is provided as strip. This is to reduce forming on rolling mills, particularly as starting material for tubes and certain average containers. The material is provided as thin stripes or as strip material up to a thickness of 25 mm and a width of 2.4 m.

4.1.2 Quality requirements

Primary ingots are provided as pure metal as defined for the production of a primary aluminium smelter. In former days, ingots produced by secondary aluminium smelters were regarded as material of minor quality. The common philosophy was that this metal has to be remelted anyhow and the quality is established during the final production. This has changed dramatically in recent years. There may still be recycling plants supplying this material of lower quality but the majority of the aluminium recycling plants produce high quality alloys with narrow tolerances for analysis, hydrogen content and inclusions. The stringent limitations are defined by the automotive industry as part of their quality assurance program which is passed on to the aluminium melting plant as requirement of the Conditions of Purchase for aluminium supply.

Sow ingots are not that critical. Their minor quality is accepted but the producers may have to live with lower prices for their aluminium. Quality requirements for de-ox alloys are not so critical as long as they do not contain elements which may spoil the steel. The production of billets and slabs is very demanding with regard to quality. They are used directly in downstream production and minor deficiencies in the material may result in remarkable defects in the final product. This will have an influence on the production i. e the type of casting equipment. Due to the casting process of the direct chill casting machine, segregation of alloy components is inevitable resulting in a different crystal structure from inside to outside. Casting has to be followed by heat treatment. In case of the extrusion ingots, this requires a separate heat treatment, i. e a homogenizing plant. This is not required for rolling ingots. They are heated to a comparatively high temperature prior to rolling. Surface is also important. Casting processes have been developed to obtain smooth surface on extrusion ingots. Efforts to arrive at the same surface quality for rolling ingots have not been successful so far. Therefore, the products require machining on heavy equipment. Segregation is also a problem at the horizontal position for casting. For primary metal this is less critical. Thus, it is possible to cast anodic bus bars at remarkable size for a primary electrolysis plant. Small extrusion ingots can also be produced on horizontal casting machines as well as short ingots of primary metal.

Strips, as the product of strip casting equipment, can be produced with insufficient quality. The alloys are apparently not very sensitive with regard to segregation during the casting process but casting of hard alloys with high magnesium content is possible with very specific casting machines only.

4.1.3 Casting machines

Before we look at the casting equipment in more detail it may be helpful to have some overview about the technologies involved.

It is very simple to cast sow ingots. Liquid metal is poured into molds of adequate size. In order to avoid splashing of metal during casting, the molds should be free of any moisture. Metal can be poured directly into the mold lined up in front of a crucible. This requires some handling for filling and transport of the crucible and perhaps removal of filled molds. A better solution is the arrangement of the molds in a casting circle. Metal is fed directly from the furnace to a swivel launder at the center of the circle. Filled molds are removed and replaced by empty molds while casting commences. If a complete furnace batch has to be cast, the number of molds arranged in the circle may not be sufficient. The aluminum may still be liquid at the surface. Thus, the molds have to be handled with care! More molds can be arranged with a casting train. As soon as one mold is filled, the train cycles to the next mold without interrupting the casting. For a large furnace the casting train can be designed as double strand unit. The aluminum remains in the molds until it is solidified and can be handled. This may take as long as 24 h. Before casting, special steel anchors are placed in the molds which provide fastening hooks for removing the sows from the mold.

The simplest equipment for casting the standard ingots is the casting circle as well. Usually a number of ingot molds are combined as large casting plates. A number of these plates are arranged in a circle. Metal is fed from the swivel launder as well. This system is suitable for a very small production rate. For higher capacities the circle can be mechanized. The molds circle with the aid of a mechanically driven circle structure. At one point of the circle, the filled and cold molds are lifted and turned to remove the ingots. The empty mold plate is returned to the circle. This system, too, is limited in capacity and requires substantial maintenance. State-of-the-art is the ingot casting machine designed as horizontal conveyor. The molds are arranged on a chain conveyor. They are filled at one end of the conveyor individually and the ingot falls out of the mold at the turning point of the conveyor chain and the molds turn upside down. At that point the ingots can be collected in containers or passed into an automatic stacking device. When collected in a container, the ingots are stored for some time and after that they will be stacked manually on their way from the casting point to the ingot discharge. The aluminium is cooled directly by air, by water spray or indirectly by water-cooling. Air-cooling is based entirely on radiation from the aluminium surface. This takes time and requires a long casting machine. Water-spraying is the most effective means of cooling. It utilizes not only the full temperature difference of water temperature between supply and metal but also the heat of evaporation of water. Cooling water is released as vapor after contact with the ingot. This is to be replaced by fresh water which may be a problem in countries where water is very scarce. But also in industrialized countries, the plant has to pay a fee for the water provided by public sources. This problem is overcome by having the ingot molds pass through a water basin where water is supplied in a closed loop. The water is cooled by means of cooling towers or air / water heat-exchangers. Requirements for fresh water are small and limited to the evaporated loss in the cooling tower. The temperature of the ingots leaving the casting chain is in the range of 200-250 °C. Having that temperature, they are sufficiently solid for handling. If the automatic stacking device is equipped with an automatic strapping device, this temperature is far too high. After cooling to ambient temperature, the straps will be loose. Consequently, the automatic stacker needs to be equipped with a cooling conveyor reducing the ingot temperature to below 80 °C.

The metal feed to the casting machine must be fast enough to fill the mold during its short time passing the pouring point. The standard pouring device comprises a tundish with pouring lips. As soon as the mold is in position, the tundish tilts and metal flows through the pouring lips into the mold. After passing the position, the tundish tilts back. Due to the cascading of metal during pouring, oxides will settle at the pouring lips which then pass into the metal; the quality of the ingots is not too good. The efforts to clean the metal ahead of the casting operation are in vain. The casting wheel is the better solution to avoid these problems. Metal is fed to a casting wheel that is equipped with nozzles at the circumference having the same pitch as the casting molds. These nozzles dive into the molds thus releasing the metal close to the bottom of the mold. This avoids cascading and surface turbulence and results in improved ingot quality. The tundish is still required when casting de-ox cubes on the chain-type casting machine. The casting machine equipped with accurate metal flow control, casting wheel, cooling system and stacking machine is a very reliable high performance unit which permits continuous casting for 24 h/day. It can be designed for any production rate between 3 and 25 tons per hour.

The vertical direct chill (VDC) casting machine is used to produce extrusion and rolling ingots. Metal is poured from the top into molds having shape and dimension of the ingot to be cast. The bottom of the mold is closed by a foot piece (bottom black) mounted on the casting table that moves downwards as soon as the mold is filled with metal. The velocity of the table is just sufficient to allow the formation of a stable skin of metal at the surface of the billet. A number of ingots can be cast simultaneously on one machine. Metal is fed to the different molds by means of a distributing launder system. If the mold is tilted by ninety degrees, the casting machine works horizontally. The newly cast billet is moved by a roller system. The solid metal shipment has sufficient strength to support the liquid metal and cooling needs to be fast enough to avoid melting of that skin. Therefore, the size of the ingot is limited, particularly for critical alloys.

4.2 Heat balance

4.2.1 Casting temperature

Different alloys have different temperatures for melting and, consequently, for solidification. In general, the melting point of an alloy is lower than for pure metal. This is shown in the phase diagram which we know from section 3. The casting temperature should be selected to be not too high above the solidification temperature but must consider the particular conditions of the casting machine. In the DC casting machine, fast cooling is essential so that not too much heat must be removed. The ingot casting machine is not critical in regard to the length of a solidification zone. Furthermore, since the material is melted again in the following production, there is also no problem in regard to segregation. Thus, slow cooling conditions can be established to obtain a good ingot without shrinkage cavities. Upon setting of the metal in the casting furnace, the heat loss during metal transfer and during metal treatment must be considered. The conditions are different from plant to plant and depend on the plant layout and the arrangement of the equipment. It requires skill and experience of the operators to define the correct temperature of the metal and to balance the required parameters to get to the optimal result.

4.2.2 Cooling conditions

The heat to be removed is of the same quantity as the heat required for melting. This heat will be removed by air or water in the casting equipment. Additional heat is required in furnaces to compensate for losses. Also in casting machines the heat cannot be used at an efficiency of 100 %. More cooling medium has to be provided than required if looking simply at the heat balance. Looking at the total energy balance of the plant, the energy regarded as useful during liquid metal processing is removed by the cooling medium and will end up finally in the atmosphere with no chance of recovery. Actually, the same energy is even required to remove the "useful" energy from the metal. This will go to blowers and pump. But unlike for furnaces, efficiency is not important. The comparatively small quantity of energy required, providing more cooling water or conditional air, does not require consideration.

As outlined above, the heat removal from the metal follows the laws of heat radiation, heat convection and conduction for transporting energy form inside of the metal to the outside during the stages of cooling – solidification – final cooling. As during melting, the heat conduction will affect the metal. During the production of re-melt ingots, solidification takes place in the mold. Heat is removed from the surface and must flow from inside to outside allowing sufficient time for conduction, i. e removing energy from the surface in the rate of the heat transfer from inside to outside. Apart from the extensive time required for casting to be provided this way, there are other effects. The alloy is composed of different elements and those with lower melting point will solidify first on the outside. This creates different crystal structures from inside to outside. **Fig. 4.1** shows the crystal structure of a shape. Also enrichment of different alloying components is seen in the cross-section of the ingot. During cooling and solidification, the aluminium shrinks since metal with lower temperature has higher density; it fills less room. To compensate for this, liquid aluminium from the inside is still sucked to the outer area and this effect causes the shrinkage cavities which can be seen on the top side of the ingot. Generally this is not a serious problem

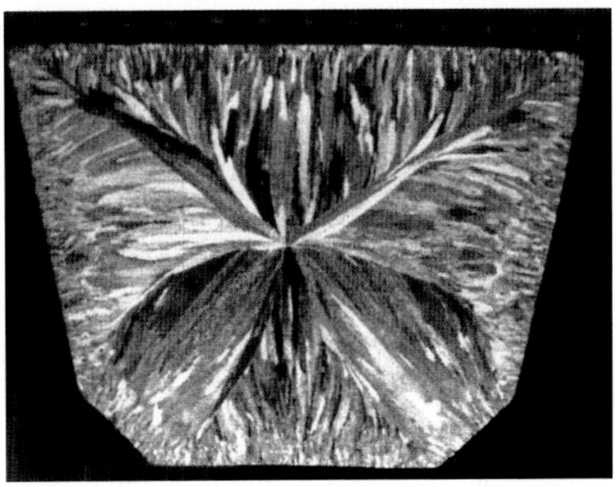

Fig. 4.1: Grain structure caused by segregation during solidification of an aluminium bar

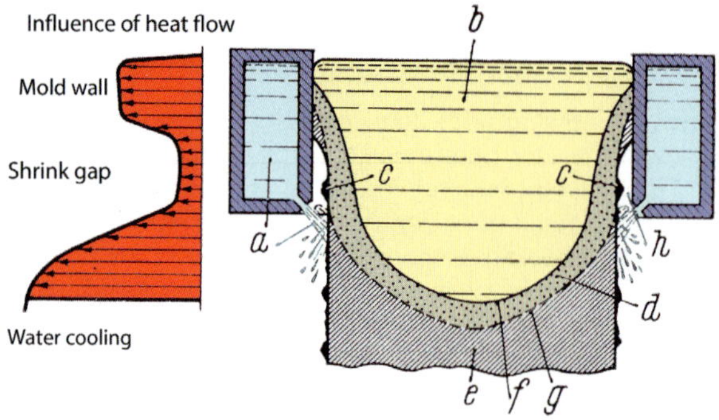

Fig. 4.2: Cooling principle in a vertical direct chill casting machine (Source: VDC)
a - water cooling, b- liquid metal,
c - newly occurring metal bleeding,
d - plastic grain texture,
e - solidified billet, f- liquidus area,
g - solidus area

or at least only an optical one for remelt ingots. The situation is very different for wrought alloys produced on DC-casting machines or strip casters. Liquid metal coming in contact with the cool mold solidifies immediately and forms a skin of solid metal. But then the metal shrinks and loses contact to the mold. Liquid material from inside penetrates through the solid metal by melting it partly, thus causing bleeding. Then at the exit of the mold water-spraying becomes effective and efficient cooling commences (**Fig. 4.2**). The trick during DC-casting is to reduce the contraction of the partly solidified material to its maximum possible extent. This is achieved by the air veil mold system **(Fig. 4.3)**. Aluminum flows from the header into the mold. Contact between mold and metal is prevented by introducing air at the top of the mold. It is important that air pressure is balanced with the static pressure of the aluminum. An air veil is created between mold and metal which avoids harsh cooling and the resulting shrinkage lubrication oil is injected. Apparently the expression lubrication oil is wrong. Since there is no contact of areas, which may cause friction, the oil may assist to obtain a stable veil that softens the cooling. After leaving the mold the water spray is applied to create a water veil and this ensures rapid cooling as in the conventional system. The air veil casting mold was developed by various companies but has been introduced by Wagstaff to the general market.

Although operated successfully for billets, a similar system is not yet available for slabs. Here the floater system, which distributes the metal in the mold and controls the metal flow by a nozzle with plug, is still in use. It is common to all DC-casting systems that a liquid metal swamp appears inside the cast ingot. This swamp solidifies as the ingots commence downward motion. In order to avoid liquation, the swamp should be kept at a low level. Unlike the solidification in the remelt

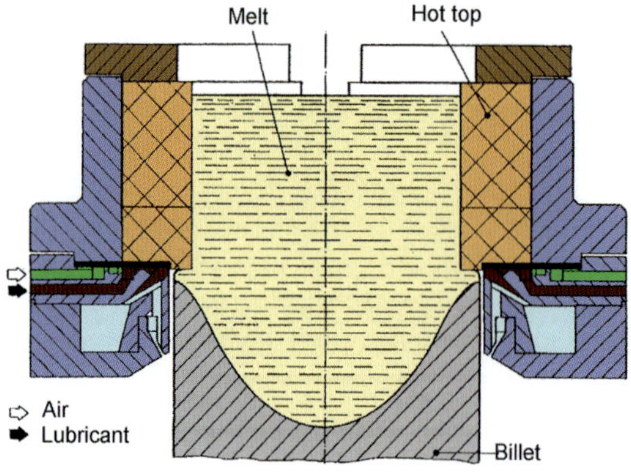

Fig. 4.3: Air slip mold for billet casting

ingot, the liquation takes place at the grain boundary and can be eliminated by homogenizing billets and slabs after casting.

In a horizontal casting machine the conditions are similar. However, no air slip system has been developed so far. The reason might be that the application of that system is supported by the static pressure head of the liquid metal. Thus, the system may be very efficient for primary metal but fails with casting alloys, for instance.

Strip-casting machines are characterized by very rapid cooling (**Fig. 4.4**). The liquid metal swamp is extremely small which causes an intensive grain boundary liquation. This is of advantage for primary metal. For wrought alloys heat treatment of the coils is required. The liquation is reduced if the strip is cast between water-cooled steel belts (Hazelett) or a water-cooled chain comprising individual slabs (Alusuisse/Lauener).

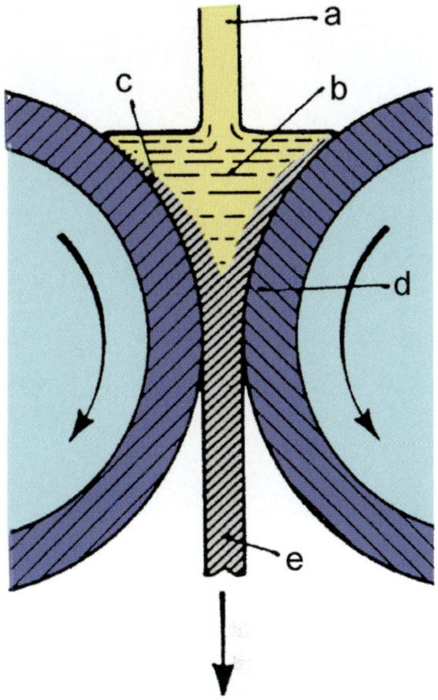

a - liquid metal, b - liquid metal swamp, c - skin of solidified metal, d - rolling of solid metal

Fig. 4.4: Cooling principle for a strip casting machine

4.3 Design of casting equipment

4.3.1 Casting circles

In former days the capacities of aluminium recycling plants were low and the quality requirements were far below the standards required today. For the production of ingots it was quite common to use casting circles and they are still used for small quantities of some special products such as de-ox alloys.

The casting circle comprises a steel structure arranged in a circle (**Fig. 4.5** and **Fig. 4.6**). Aluminium is fed through a swivelling launder to which material is fed at the swivel center. This is supported by a supporting structure equipped with a bearing to allow easy movement. The launder can be tilted or lifted to be able to pass from one pouring position to the next one without spilling of

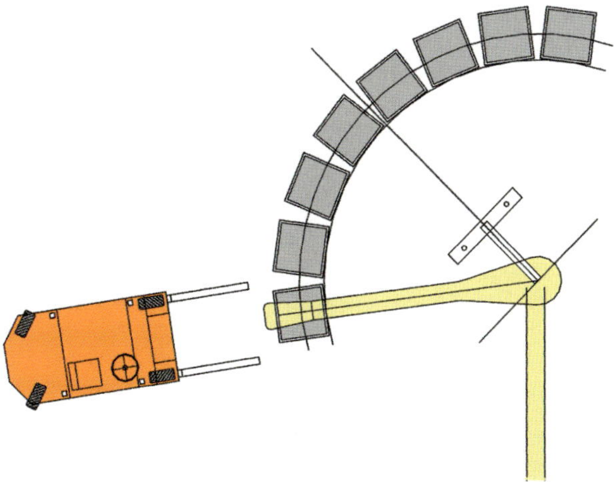

Fig. 4.5: Casting circle for sows (500 or 1,000 kg)

Fig. 4.6: Sow molds in a mold casting circle

metal. Each pouring position is equipped with a pattern of special cast iron for casting a number of ingots. After filling one pattern with metal, the operator moves to the next one by swivelling the launder to the next position. The plates either remain in their position to cool or they are removed after the ingots are solid by means of a jib crane. This is certainly required for producing at least some reasonable quantity. The mold plates rest on a supporting structure which can be equipped individually with a pneumatic lifting system to ease removal of the patterns.

Metal is provided directly from a furnace or by a crucible arranged on a tilting device. The casting circle can be designed as casting machine. The supporting structures of the patterns are arranged on a large rotating table or on a rotating train with chain and sprockets. After pouring the metal in a casting position, the system cycles to the next position. After a number of positions, the ingots are solid and can be removed. To do so the molds are lifted from the carriage by a grab and moved to a discharge position where the pattern is turned and the ingots drop into a container. The empty pattern is then returned to the circular conveyor. The mechanized casting circle can be equipped with a water-cooling system.

This type of equipment is not used anymore for aluminium and has been replaced by the chain-type casting machine.

4.3.2 Sow casting system

Large ingots having a weight of 500 or even 1000 kg, commonly called sows, are cast in heavy cast iron molds. These can be arranged on the operating floor in rows. Metal is poured from a crucible which is transported and tilted by an overhead crane.

But for larger production the sows are cast in a casting circle (Fig. 4.6). The unit is similar to the casting circle described above for casting small ingots. Of course the steel structure is heavier to be able to take the weight of the ingot and instead of patterns the sow molds are placed in position. These molds are shaped with a design that permits transportation by forklift truck. The inside is also slanted to ease removal of the cast aluminium. Additionally, heavy steel clamps are placed on the molds which are used to hook the mold to lifting equipment by means of a steel rope. This is very helpful to get the sow out of the mold. The design of the clamps is very simple. They must be shaped to hold the weight of the ingot and permit easy removal after they are not in use anymore. If not shaped correctly they may jam in the aluminium because of the shrinkage of the sow.

If the capacity of the casting circle is insufficient, the casting mold can be placed on a casting train. The supporting structures are placed on roller carriers moving on rails. They are pulled by means of a chain drive.

Metal is poured at a fixed pouring station from the furnace. As soon as one mold is filled, the whole train cycles to transport the next empty mold to the pouring position. Since the number of molds is not limited, as in the case of the casting circle, the casting train is able to accommodate a larger number of molds. However, there is also a limitation since it becomes more and more difficult to move the whole train with an increasing number of sows. If required, a parallel train could be arranged in addition.

In a well-designed plant the sow casting equipment is arranged in such way that sows can be cast from the melting furnace as well as from the casting furnace. This permits production of sows from no-go batches as well as from scrap with more or less undefined alloy composition. These sows could be sold to a customer at lower price or charged to the normal production at small rates.

One important procedure must always be done prior to casting any mold. All moisture collected in the molds has the potential danger of causing an explosion, particularly if the inside of the mold is coated. Therefore, it is indispensable that every mold is thoroughly dried before it comes in contact with liquid metal.

4.3.3 Ingot casting machine

The standard ingot casting machine (**Fig. 4.7** and **Fig. 4.8**) is designed as a chain conveyor, comprising a transport chain with casting molds, pouring station, cooling system, ingot ejection device, chain heating system and coating device.

The basic casting machine, made of a sectional steel construction, comprises the front end belt deflection and tensioning station, the horizontal cooling section, the upward gooseneck, enabling the connection of a stacking equipment, and the upper deflection and drive station. The deflection stations incorporate chain wheels for driving and deflecting the transport chains.

Two transport chains (collar roller type) are arranged at the right and left side of the strip. The chains run on suitable roller guides in the casting belt structure. The casting molds, made of special cast iron, are bolted to these chains. They are designed to obtain the required shape of the ingots. These are slanted from bottom to top for easy removal from the molds and to be able to have an interlocked stack for transport. They may also have some special shapes to improve the interlocking of the stack.

The molds are arranged so that each mold corresponds to one chain roller spacing, respectively. The chain pitch depends on the mold size used (heavier casting weight = increased chain pitch).

The drive unit consists of a plug-in gear unit with flange-mounted control mechanism which is plugged on the drive shaft of the upper deflection unit.

The torque support is provided to support the drive as required. The speed control of the casting belt is accomplished by means of a remote control unit from the casting operator's desk. The casting belt rate in pigs per hour is shown on the scale of an indicator.

The casting belt tension is adjusted by means of two adjustable disk spring assemblies provided at the lower deflection unit or tensioning the complete deflection axle together with the chain wheels.

Metal is supplied to the casting machine by a feed launder system which is equipped with a level control for liquid metal. The metal level can be adjusted manually or by the automatic level control. The level is maintained by controlling the tilting of the casting furnace depending on the metal requirement of the casting machine. The velocity of the casting chain determines the throughput

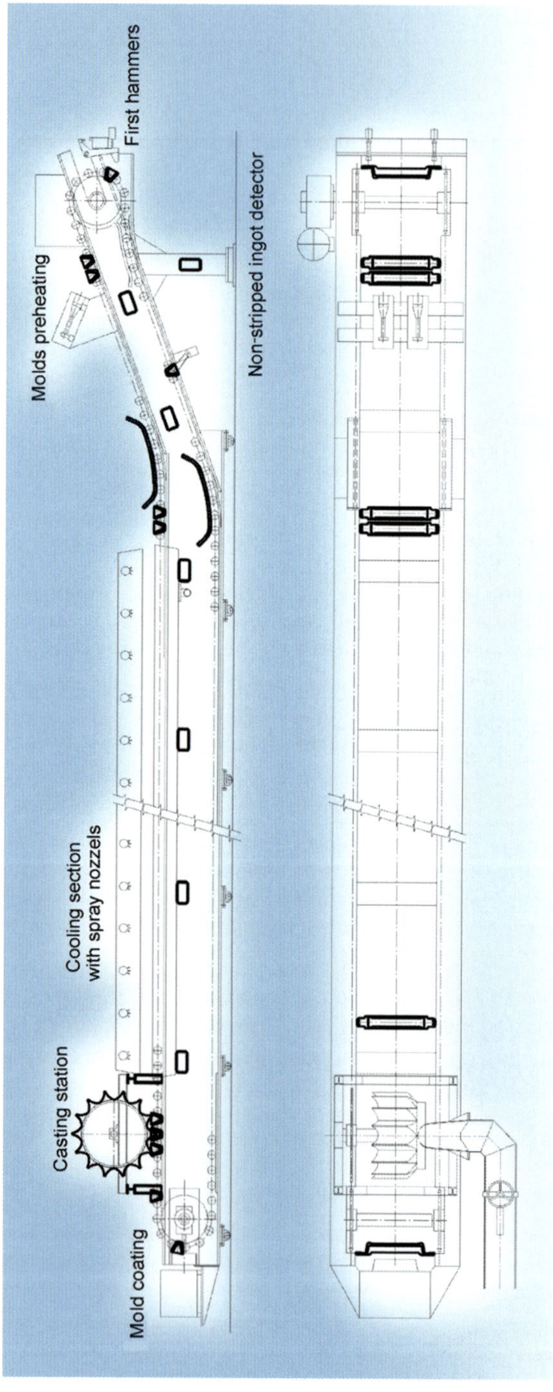

Fig. 4.7: Casting chain for re-melt ingots with casting wheel (Source: Ingotech)

Fig. 4.8: Arrangement of ingot casting machine (Source: Ingotech)

rate. The metal level in the feed launder is maintained at a certain value which means the furnace is tilted to provide more metal at a higher casting rate or less metal if the casting rate is lower. As additional flow control for the pouring system, a regulating gate is installed in the launder just ahead of the casting wheel. This arrangement is very helpful during start of casting. This gate can also be used to open the link between the casting furnace and the casting machine. The metal flow is then controlled automatically by a level sensor in the trough behind the gate while the tilting of the casting furnace is separate also by a level sensor ahead of a gate. This method is usually selected if a casting machine with independent electrical control is installed in an existing plant. As additional aid, a gate to the feed launder system allows drainage of the launder at the end of a cast. Metal in the launder system and the metal can be poured into a sow mold to empty the

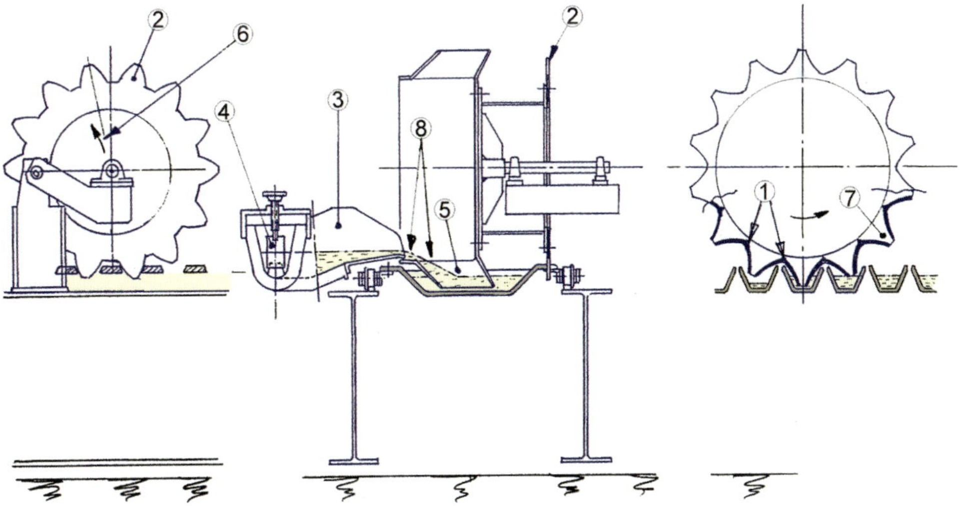

1 - metal guides, 2 - drive sprocket, 3 - feed trough, 4 - stopper, 5 metal level in casting wheel, 6 - lifting device, 7 – casting nozzles, 8 – metal flow

Fig. 4.9: Casting wheel system (Source: Brochot)

system and to avoid that metal solidifies in it. It is always a hard job for the operators to remove solid metal from the launders and the adjacent equipment. This is in addition to the possible damage to the refractory lining. The transfer of metal to the casting wheel must be configured in such a way to reduce the metal cascading to the lowest possible level. All efforts to remove oxides ahead of the casting machine are spoiled if metal gushes from an open launder into the casting wheel. A large-sized nozzle, which supplies the metal beneath the surface of the metal heel maintained in the casting wheel, reduces the intake of oxides. Another solution is depicted in Fig. 4.5. Independent of the fluctuation of the metal level, due to the position of the pouring nozzles at the circumference of the casting wheel, a level in the launder is always maintained to permit metal flow beneath a protective oxide skin.

The casting wheel is designed to equally distribute the liquid aluminium to each individual mold.

Outlet nozzles are arranged at the circumference of the casting wheel. These may be an integrated part of the casting wheel as slot-like shape (**Fig. 4.9** and **Fig. 4.10**) or comprise individual nozzle inserts or nozzles welded to the outside of the casting wheel. The pitch of these nozzles corresponds to the distance of the individual mold on the casting chain. Liquid metal is fed into the interior of the casting wheel at controlled flow rate. A rim at the outer circumference stops backflow of the metal. The other side of the casting wheel is closed. The bearing shaft of the casting wheel is attached to this back wall. The nozzles not only distribute the metal evenly in the individual mold but also prevent turbulence during casting. Since the nozzle dives into the mold to a point close to the mold bottom, most of the metal is poured underneath the surface. Therefore, the amount of oxide inclusion is greatly reduced; this results in a better top surface of the cast ingot as well.

The casting wheel is supported by a free-standing base frame with pivot assembly. Via a hydraulic or pneumatic cylinder, the casting wheel can be lowered into casting position or raised into maintenance position. The supporting structure can be arranged at either side of the casting chain in order to facilitate the plant layout requirements. A sprocket, having the same pitch as the chain, is flanged to the casting wheel shaft thus assuring identical speed for casting wheel and molds. For the production of de-ox cubes a different system is required. This was used for pouring of normal ingots as well before the casting wheel was introduced.

A casting car with the casting pot is arranged at the pouring position of the casting belt. A mechanical linkage system synchronizes the tilting movement with the chain speed. In this way the change of the casting belt speed does not require the readjustment of the casting pot tilting speed. At the end of the casting process, the casting pot returns to its normal position waiting for

Fig. 4.10: Casting wheel (Source: Ingotech)

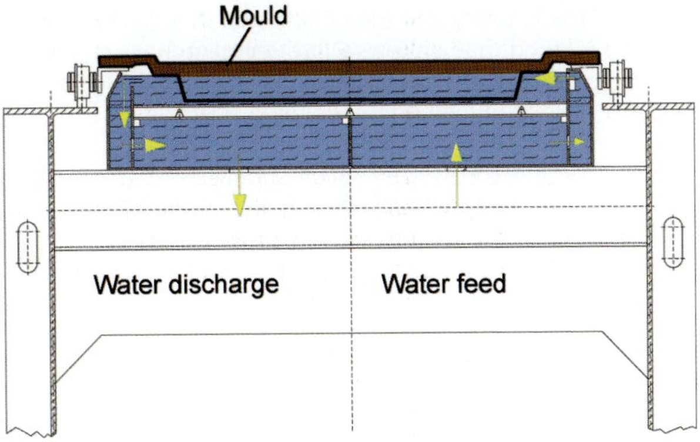

Fig. 4.11: Water basin of the ingot casting machine

the next following catch. For casting de-ox cubes, the casting molds are separated into several sections. The casting pot is equipped with a number of pouring lips to fill each individual cube.

After casting, the filled molds continue their travel in horizontal position until the metal has solidified in the molds so that it can be removed as ingots from the molds.

Bottom and sides of the filled molds are water-cooled by a basin system (**Fig. 4.11**) which is divided lengthwise into a supply section and a discharge section by a separating wall. The molds travel through a water trough. Water flows from the supply section through small slots to the water trough rectangular to the travelling direction of the chain, leaving the trough through small slots to flow into the discharge section of the basin underneath. From here it is discharged via pipes to the water return pipes. Water quantity is adjusted to match the production rate of the casting machine. A different means of ingot cooling could be accomplished by a spray-cooling system. This comprises a nozzle system which provides for water spraying on the hot ingots. A hood is arranged above the entire cooling area to collect the water vapor. A basin underneath the carrying row of the conveyor collects the excess water not evaporated during the spray-cooling. The collected water is directed to a water drain. Each of the individual rows of nozzles is equipped with a control valve to adjust the water flow according to the temperature of the ingots.

The length of the casting belt and its travelling speed are matched to provide the required cooling efficiency. The cooling water generally flows counter-currently to the casting chain. In order to improve the cooling efficiency, cold water is fed at various points along the casting machine while the hot water is discharged. As mentioned before, the temperature of the discharged ingots at the end of the casting machine is usually approximately 200 °C. This temperature is a function of casting chain length, ingot throughput, ingot weight and cooling media. If mechanical processing is desired, a separate cooling device is indispensable to arrive at a temperature below 80 °C that is necessary for strapping.

At the outlet end of the casting belt the ingots are normally released from their mold by gravity and due to their shrinkage by cooling. The ingots are successfully released from the molds only if the mold is of exact casting quality and if the ingot molds are properly faced. In order to ensure that in no case a pig is retained in the mold, an ingot knocking device is provided with a pneumatic hammer to release any ingot from its mold. This device is controlled from the belt deflection unit by means of a cam wheel and an initiator.

The ingots are collected in a container and can be removed from the casting machine for cooling, followed by manual stacking.

The casting belt is lubricated by an automatic impulse lubrication system which is controlled by the chain. By this lubricating system the chain is lubricated from the outside by adding lubricant between track rollers and the chain links. In order to ensure a proper functioning of the chain, care has to be taken to always use the lubricant in amounts as small as possible and to always keep the chain tracks perfectly dry.

Due to the fact that the casting molds must never be cold and moist, they have to be pre-heated and dried after prolonged shutdown periods. For this purpose, the casting belt comprises a heating device arranged below the casting belt which heats the returning molds to improve the heating efficiency. The heating system consists of an in-line burner system of modular construction to allow any combination to suit each and any heating requirement. An injector burner is provided which by itself takes in the combustion air without the necessity of a fan. A gas/air mixer is attached to the burner. The required gas pressure has to be set by the user by increasing or decreasing the pressure.

A mold spray device is used for coating the inside of the molds to ease removal of the ingots at the end of the chain and to obtain a smooth surface. The unit is arranged close to the pouring station and comprises a small tank equipped with a pump and nozzle arranged close to the molds.

A stamping device arranged close to the end of the conveyor is used to mark every individual ingot. The actual stamp is equipped with an interchangeable set of steel letters. Thus, marking can comprise typical production data such as batch number, production date and shift number for later reference in case of a complaint by a customer.

4.3.4 Ingot stacker

Instead of being collected in containers these ingots can be stacked automatically. The ingots removed from the casting machine are placed on a roller conveyor by a pneumatically or hydraulically operated discharge device. Since they are still very hot they first pass into a cooling section. This comprises a closed housing equipped with water spray nozzles. The water quantity is large enough to suppress excessive evaporation. Water is collected in a basin underneath the conveyor and returned to the cooling water circuit. After leaving the cooling sections, the ingots are passed over to another roller conveyor to change their orientation to the transport direction. When discharged from the casting machine, the ingots travel with the small side in transport direction. Now they are oriented with the long side in transport direction. The ingots are slanted from bottom to top; every second ingot is therefore turned upside down to obtain a stable layer. At the next station the ingots are tightly pushed together to assemble a layer. As soon as the layer is complete, a robot equipped with a layer grab takes the layer to a stacking conveyor (**Fig. 4.12**) and places

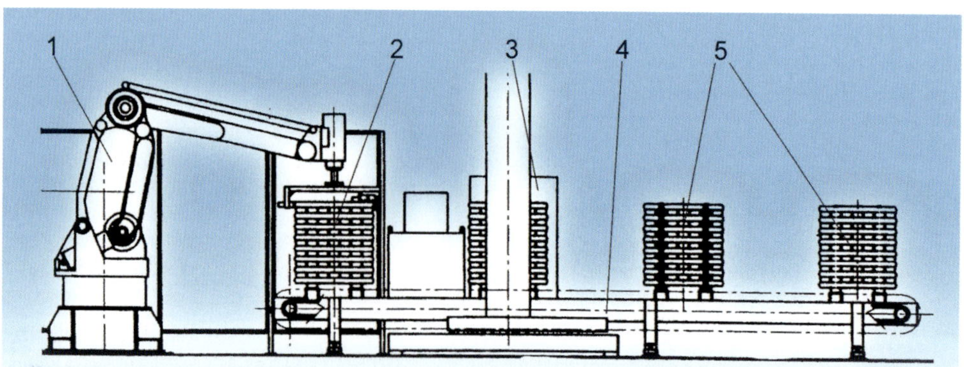

1 - stacking robot, 2 - completed stack, 3 - strapping position, 4 - storage conveyor, 5 - stacks ready to be removed

Fig. 4.12: Stacking unit of ingot 25t/h casting facility (Source: Brochot)

Fig. 4.13: Completed stacks assembled on the storage conveyor

it on the layer already transferred before the layer is turned by ninety degrees. The first layer can be placed on a pallet, either cast from the same alloy or on a so-called Euro pallet. Both can be fed by hand or automatically. The stacking conveyor is usually designed as walking beam. After completion of one stack, the walking beam transports it one station further, either to a strapping position or directly to a storage section (**Fig. 4.13**) From here the complete stacks are removed by forklift trucks.

The stacking unit is a highly mechanized system with substantial capacity. For a casting machine, producing for instance 8 kg ingots at a casting rate of 10 t/h, 1,250 ingots have to be handled at the stacker, i. e. one ingot every 2.88 seconds! If the casting rate is much higher a dual system operating with two robots is required. This may also be necessary if the stacking unit is fed by two casting machines operating in parallel mode.

In addition to the strapping station, that may be operated manually or automatically, a marking device that spays a number code to the stack may also be installed. The ink is usually based on a solution of the pigment in alcohol. This must be considered when supplying such equipment to Islamic countries.

4.3.5 Vertical direct chill casting machine

Ingots for extrusion and for rolling are produced on the vertical direct chill casting machine (VDC) as shown in **Fig. 4.14**. Direct chill means that the aluminium is cooled and solidified by direct contact with water. This is introduced by establishing a water veil after the material passes the mold. In a very short period, the shape of the ingot is established by indirect contact with water in the mold (Fig. 4.2).

The casting machine comprises the mold frame with molds and cooling water supply, the hydraulically-operated casting table with guides, either externally or arranged in the cylinder, the heavy hydraulic cylinder and the water system with pumps and piping.

The molds are sized according to the required size and shape of the ingot. They are designed as housing with internal water cooling and the spray system. Up to more than 60 individual molds for casting extrusion ingots are arranged on the mold frame. The number of molds depends on the size of the ingot. Rolling ingots generally have a larger size which limits the number of molds on the casting table. To supply the large rolling mills, the ingot size may be as large as 600 x 2,400 mm. A specific launder system provides metal to the molds.

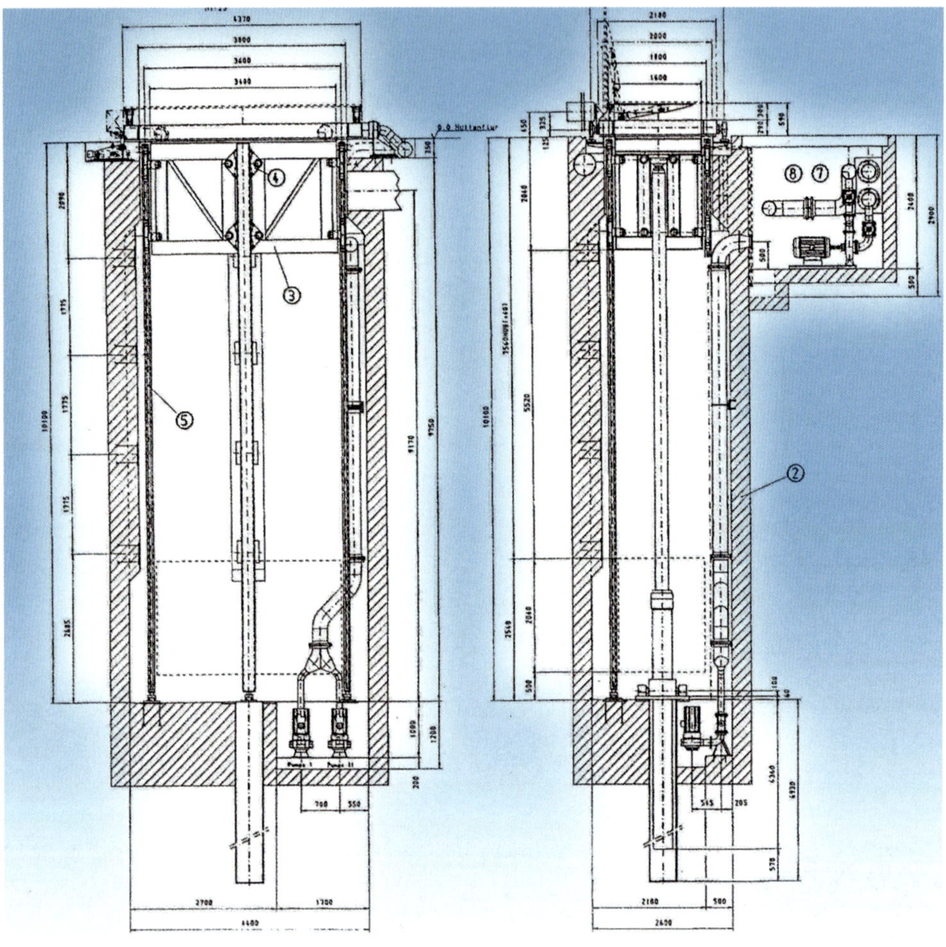

1 - casting furnace, 2 - melting furnace, 3 - grain fining wire feeder, 4 - casting launder, 5 - in-line degassing box, 6 - casting machine

Fig. 4.14: Vertical direct chill casting machine (Source: VDC)

The design of the molds is the key technology for DC casting. For billets very sophisticated mold systems have been developed. Nowadays, the air veil mold system (Fig. 4.3) as described above is used. In former days metal was provided from the distribution launder through individual nozzles to the mold. The quantity of the metal was controlled by a floater made of light refractory material which was equipped with a small cone towards the nozzle opening. If the metal level in the mold was too high, the cone was lifted by the floater into the nozzle to adjust the metal quantity to change the cross-section of the nozzle opening. The floater system requires permanent attention of the operators during the entire cast operation. It is still used for casting rolling ingots. In this case the floater is of box-type design made of steel plate. For casting extrusion ingots the hot top system is very common today. With this method metal is provided from the launder directly to the molds at the same level for the complete mold system arranged on the casting table. The launders are connected to the molds via a refractory ring.

By maintaining the level in the metal feed system, i. e. the distribution launders, the casting process can be automated. At the end of the cast operation, the mold frame must give way to be able to remove the ingots. Therefore, the complete casting frame can be lifted by crane, tilted by a hydraulic cylinder or moved sideways if designed as mold car. This unit is supported on rollers and travels on rails installed in the operating floor. Movement can either be manually or with the aid of a motor drive. The design of the mold car permits the automatic exchange of mold frames for the different sizes of molds. Since the arrangement of the molds depends on the size of the ingots to be produced, it would be difficult to start a re-arrangement on the casting frame for every size. Therefore, different casting frames are already prepared and held in stock. For easy handling the mold frame storage and the mold change procedure can be automated (**Fig. 4.15**). In this case the mold car with the attached mold frame travels to the storage rack where the mold frame is removed and stored in the assigned positioned on the rack. A new mold frame is reclaimed and placed onto the casting car. The mold frame is attached to a supporting structure or the mold car fixed in the casting position.

The casting table carries the bottom blocks of the mold system (**Fig. 4.16**) to receive the multicast molds or the large sizes for rolling ingots. The table must have an adequate size. Standard dimensions are in the range of 2 x 3.5 m. Larger sizes are also being used.

At the beginning of the cast operation the individual molds are closed by a bottom block matching the accurate mold opening. As the ingot is shaped, the casting table moves downwards within the casting pit thus carrying the ingots growing in length. The casting table is a heavy structure made of structural steel members. The load on the table may reach 50 tons with an average of 25 tons depending on the capacity of the casting furnace. The casting velocity is in the range of 5 to 8 cm per minute. Considering a standard casting length of 7.5 m, the casting operation will require 1.5 to 2 hours. During this time the casting table must travel very smooth at precise velocity. Any stick-slip in the motion of the hydraulic cylinder will cause a defect on the ingot surface.

For a good ingot surface the guiding of the casting table must be very accurate. There are two principle arrangements. One is to have a roller system moving on four guide rails attached to each

Fig. 4.15: A mold frame being reclaimed from the storage for complete mold frames

Fig. 4.16: Bottom blocks of VDC casting machine

Fig. 4.17: Removal of billets from the casting pit

Fig. 4.18: Billets lifted on casting table

concrete wall of the casting pit. This requires very accurate alignment of the rails which is not easy to achieve. Checking and readjustment is required in frequent intervals and is usually done during the annual maintenance period of the plant. Another solution is a guided rod of the hydraulic cylinder. This system has proven meanwhile to be very reliable and accurate. In this case it is of advantage to have the molds arranged symmetrically to have balanced load conditions. This is advisable to reduce wear on the rod guides.

The VDC casting machine requires a very deep pit which is sometimes difficult to construct. To obtain a casting length, the depth of the pit must be in the range of 10 m since space must be available for the table and support of the hydraulic cylinder. Additionally, space must be provided for the cylinder. This is realized by having a large steel tube buried in the ground below the bottom of the casting pit. The pit is a concrete structure. It accommodates the foot valve for the water pumps required to remove the water fed for cooling the aluminium. These are arranged in a separate pit close to the casting machine with the possibility of easy access. There is always a stand-by pump installed to be able to continue casting if the water pump or pumps fail. The casting machine can be operated as dry pit, which means with a low water level, or as wet pit which means with a high water level. This is to the liking of the operators which consider different cooling conditions for certain alloys.

The cooling water discharged from the casting machine is fed into the cooling water circuit.

After the cast operation the ingots have to be removed from the pit. To do so, the casting mold system is set aside or tilted and the casting table is lifted into the upper position (**Fig. 4.17** and **Fig. 4.18**). The ingots can now be removed by crane equipped with the aid of rope slings or special grips in the case of the rolling ingots.

4.3.6 Horizontal direct chill casting machine

The handicap of the VDC casting machine is the limited casting length which requires stopping of casting after every 1.5 or 2 hours. It would be convenient to have a continuously operating casting machine similar to those used in the steel or copper industry. Unfortunately, aluminium does not have a plastic phase before it solidifies completely. Thus, the cast ingot cannot bend. Therefore, the horizontal DC casting machine was developed (**Fig. 4.19**).

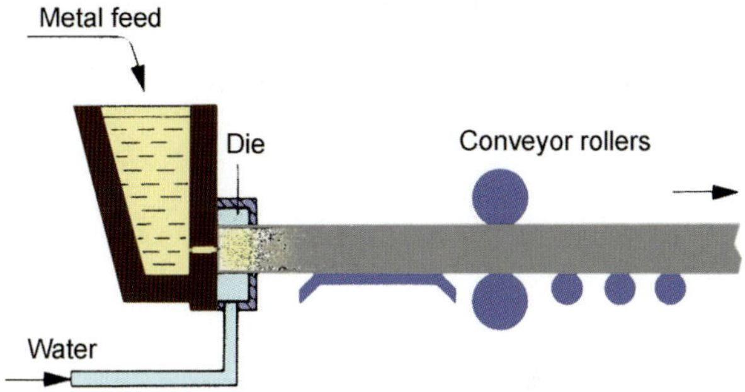

Fig. 4.19: Horizontal casting machine

The molds are turned by 90 degrees into horizontal orientation. Several molds can be arranged in a line. Metal is fed to the individual mold by a large size spout matching the size of the mold which is of similar design as the mold of the vertical machine. However, air slip systems are not possible for the horizontal process. The cast rod is moved by means of a suitable conveyor as they grow in length. For normal application a flying saw is arranged to cut the required length of the ingots. The beauty of the system is to allow continuous billet production with integrated continuous homogenizing, installed in-line. Unfortunately, the system is limited to billet diameters which usually not exceed 150 mm and alloys. However, if long slabs are required, as in the case of the long bus bars for the electrolysis of a primary smelter, the horizontal casting machine is ideal. In some cases remelt ingots of primary metal are produced in horizontal casting machines. One advantage is that different shapes of remelt ingots can be produced by just changing the mold. But, if comparing chain-type machines of state-of-the-art design, there seems to be no apparent reason to have such a production facility.

The exit end of the casting machine is constructed as collecting section from where the cast product can be removed by forklift truck.

4.3.7 Strip casting machine

For the production of strip directly as cast product, strip casting machines are installed. They can produce an aluminium belt of 6 to 30 mm thickness and up to 2,000 mm width. The basic concept of such a casting machine is the roll caster system Hunter or Alusuisse (**Fig. 4.20**).

Liquid aluminium is poured via a slot nozzle between the water-cooled rotating rollers where it immediately solidifies. The distance between the rollers determines the thickness of the strip. The rollers apply a force to the strip thus already causing some rolling and subsequent forward movement. The casting station is designed similarly to a rolling stand of a hot mill. The rollers are supported by heavy antifriction bearings. Their distance can be adjusted with spindles to the gauge to be produced. The speed of the drive can be adjusted infinitely and comprises a drive motor and gear box with coupling. Behind the rollers the strip is taken over by a roller unit which can be adjusted in height by a hydraulically operated swivel arm. This allows adjusting the discharge to the casting conditions during the starting phase. The strip is carried further by a roller conveyor with pinch rolls, a flying shear and scalping station for the edges. Finally, a coiler takes

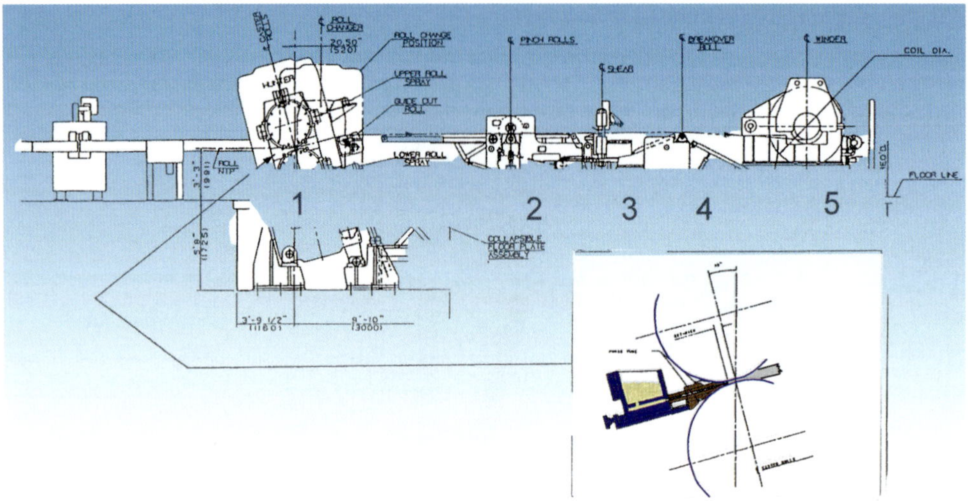

1 - caster, 2 - pinch rolls, 3 - shear, 4 - breakover rolls, 5 - coiler

Fig. 4.20: Roll caster (Source: Alusuisse)

over producing coils of the strip of practical size. The coiler tension provides the pinching directly up to the nip of the casting machine. As soon as a coil is complete, the strip is cut by using the flying shear. The pinch roller will provide the necessary tension to the strip during this phase. After strapping, the complete coil is removed by overhead crane or forklift without interrupting the casting process and transported to the storage area. The maximum magnesium content in the alloy is 2.5 %, production rate is 1.5 to 2 tons per hour and the strip dimensions are 6-10 mm thickness with a strip width of max. 2,000 mm.

It is also possible to have the strip fed directly into a small hot mill to produce sheet material for different applications or to feed the strip to a station for producing round punchings for the manufacture of tubes.

The rapid cooling causes segregation at the grain boundaries. Therefore, the strip material must be homogenized after casting. The range of alloys, as well as the size of the strip, is limited. Larger sizes can be produced on a different type of strip caster. Liquid aluminium is poured into the gap between two circulating belts. The solidification phase is extended and boundary segregation is reduced to some extent. The casting station comprises the belt system with drive unit and belt roller and deflection unit for the upper and the lower belt. It is very important that the belt tension is assured to avoid differences in thickness of the cast strip. Therefore, the belt unit comprises a belt tensioning system with adjustable supporting rollers for both belts. The belt is water-cooled by spray nozzles and a water film is maintained at the belt. The casting speed is adjusted by changing the circulating speed of the casting belt. The critical item in this technology is the steel belt which has to be straight and distortion-free during the entire casting process. It requires many efforts to arrive at a reliable procedure for the manufacture of the belt which has to be changed frequently.

The block caster developed as caster II by Swiss Aluminium and Lauener (**Fig. 4.21**) avoids this problem. It uses individual chilling blocks which are closely fit together. These blocks form a caterpillar-like double chain system. Two sets of blocks, rotating in opposite directions, form a casting cavity into which the liquid metal is fed through a nozzle system. The liquid metal, upon contact with the chilling blocks, is cooled and solidified. The strip of metal travels during the cast-

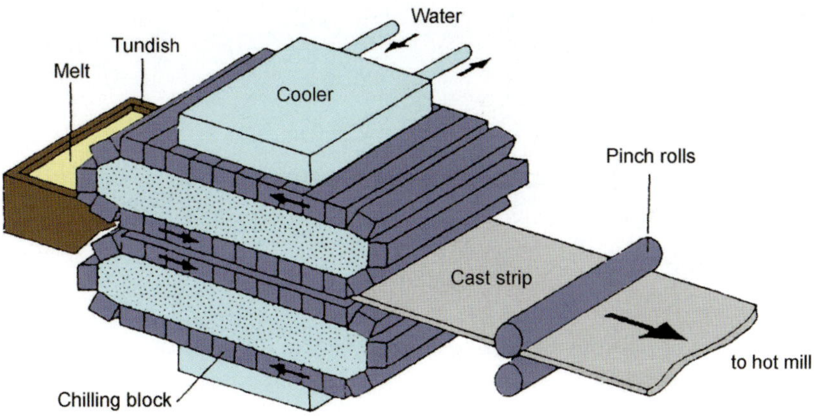

Fig. 4.21: Strip caster, model Alusuisse caster II

ing period along with the chilling blocks until the strip is sufficiently cooled and the blocks lift off to start their return path. The heat stored in the blocks is removed by an external cooling device and the blocks enter the casting cavity once again at the desired temperature.

The block caster is a development of a high capacity strip casting machine which is not limited to the casting of pure aluminium and soft alloys but can also successfully cast hard alloys including alloys for body stock and lid. The thickness of the strip of 20 mm is reduced to a coilable thickness of 2 to 4 mm through a hot tandem rolling mill arranged in-line with the caster (**Fig. 4.22**).

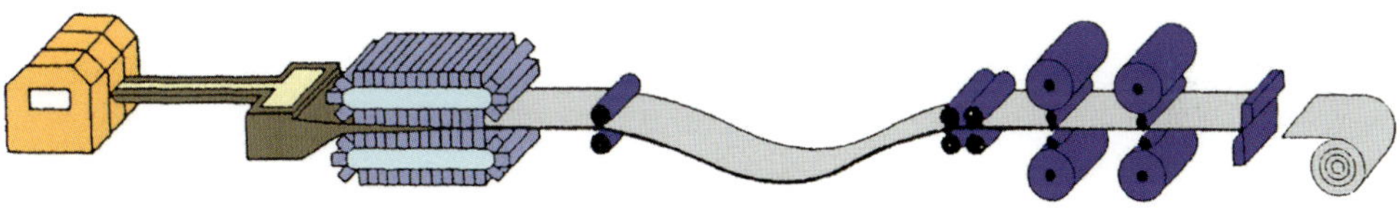

Fig. 4.22: Tandem arrangement of strip caster with rolling mill

4.3.8 Rod casting machine

The rod casting technology is known throughout the world as Continuous Properzi process. The casting machine is equipped with a casting wheel with a diameter of 1.4 to 3.2 mm (**Fig. 4.23**) according to the required production rate. The casting wheel carries a copper rim which has a cavity shaped for the product required. The cavity is closed by a steel belt (**Fig. 4.24**). The belt circles as the casting wheel rotates and can be tensioned by means of a deflection wheel. Material can be provided to the cavity horizontally or vertically by a special nozzle. After cooling and solidification, the metal leaves as continuous bar and is directed to a bar straightener through an extractor and pinch rolls. After straightening, the bar passes a rotating shear which may be used to cut the rod into uniform lengths which are supplied to an automatic stacking device. The advantage of this type of ingot production is the very clean surface, free of surface oxides and shrinkage cavities.

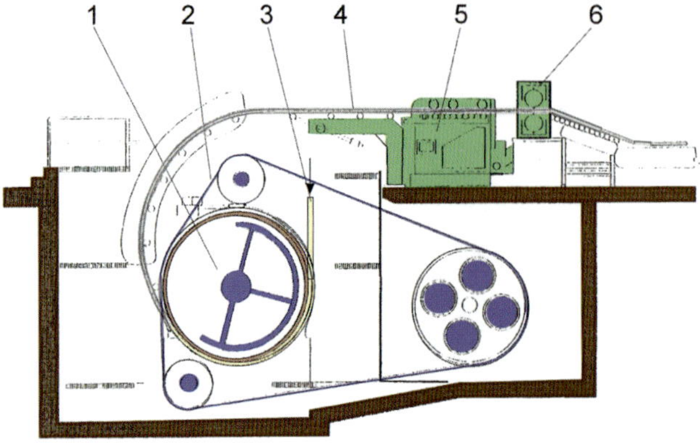

1- casting wheel, 2-steel belt, 3-pouring nozzle, 4-cast rod, 5-support rollers, 6-pinch rollers

Fig. 4.23: Properzi rod caster

The traditional Properzi system is used to produce wire in different sizes for cable production. Instead of being cut the rod continues to a small rolling mill (**Fig. 4.25**) with individual rolling stands reducing the cast rod to round wire of 7.2 or 9.5 mm diameter. To produce the 9.5 mm wire, the last two rolling stands are removed. A coiler plant, designed as twin coiler or as double basket coiler, is installed in-line of the rolling mill. A flying shear cuts the wire after completion of one coil and the system switches automatically to the second coiler. The completed coil can be strapped and taken to the storage area. If a rolling mill unit is installed, the casting wheel is usually arranged for vertical casting. The rolling mill can be used for producing de-ox pellets. This only requires that the flying shear is designed to cut very small peaces of rod. For higher capacity a double shear system may be required. Considering the comparatively high cost for casting machine and rolling mill, the production of de-ox pellets on this machine appears not to be economically justified. Considering the high production rate, compared to other pellet production equipment, could change the situation.

The plant is small even for high capacity production. If the rolling mill is installed additional space must be provided for the emulsion system.

Fig. 4.24: Properzi copper casting wheel with steel belt

Fig. 4.25: Properzi rod casting plant with rolling mill and twin coilers. The required emulsion and lubrication system is located in the space underneath the operating floor

The capacity of the rod caster ranges from 3 to 20 tons per hour. The alloys produced are mainly wrought alloys although alloys having higher silicon content have been produced on a Properzi machine.

4.3.9 Pelletizing table

De-ox pellets are usually produced on pelletizing tables (**Fig. 4.26**). These comprise a large water-cooled rotating table. The surface of that table is machined. Metal drops from a special tundish via a row of nozzles on the table where it cools and solidifies. The pellets travel on the rotating table to the discharge position and will be removed.

An inclined belt conveyor receives the discharged pellets and feeds them to a sorting drum to classify the product according to size. The final product will comprise different sizes that are filled into big bags for delivery to the customers. The rotating table is supported by a large bearing and

Fig. 4.26: Pelletizing table for de-ox pellets

a sturdy supporting structure. The drive consists of a frequency-controlled motor and a gear box. The cooling water is sprayed onto the surface of the table from underneath so that there is no contact of aluminium and water. It is then collected in a basin and returned to the closed water circuit by gravity.

4.4 Water treatment

In order to reduce the water consumption, a closed circuit cooling water system is usually installed. It comprises the water basin, the pumps and filters, the water cooler, the feed piping system and the water return system. Water discharged from the equipment by gravity is collected in an auxiliary basin or – as in the case of the VDC casting machine – in the casting pit. It is returned to the hot well of the main water basin by pumps with sufficient flow rate. In case the full capacity of the cooling equipment is not required, cold excess water flows into the hot well from the cold well. From here the water is pumped to the cooling equipment and, after being cooled, reaches the cold well. A pumping system now supplies the cooling water to the casting machines. A small water treatment section of the basin allows water treatment to suppress formation of weed or adjust the degree of hardness. Possible dirt is collected from time to time as mud and supplied to the public sewage system.

The water basin can be constructed out of concrete or steel. The return water basin is divided into three sections. The water first reaches the return chamber. One large section is used as settling and mixing tank and one as pump intake. A small treatment chamber is attached to the mixing tank. Fresh water is fed through the treatment chamber. The overflow of the cold well is connected to the return chamber of the mixing section.

The pumps are equipped with intake filters. There should always be a stand-by pump which can take over as soon as another pump fails. For keeping pumps in functional condition, all pumps, including stand-by units, should be operated in regular sequences. The individual pumps can be separated from the other pumps by hand or motor-operated valves.

Different systems can be used for water-cooling. The standard heat exchanger with air-cooling requires large energy-intensive blowers but otherwise can operate in a satisfactory manner. Problems in the water/air heat exchanger may arise in a hot climate where the temperature of the outside air that is required for cooling the water is close to the return water temperature i. e ambient air temperature < 30 °C, required cooling water temperature < 30 °C). Simpler solutions are provided by cooling towers which are able to operate in almost any climate, provided the unit takes care of these conditions. They operate on the basis of evaporation. Thus, some cooling water will be lost in the unit and has to be replaced by fresh water.

5. Metal treatment

Christoph Schmitz

5.1 De-gassing systems

With the ever-increasing demand for defect-free aluminium castings, particularly by the automotive industry, optimum methods for aluminium melt treatment are becoming a necessity. Removal of inclusions by flotation is one of the most commonly used treatment methods in the aluminium foundry. In order to optimize this process, it is very important to understand the basic mechanism underlying it. The quality of cast aluminium and product largely depends on treatment of the molten metal prior to casting. A variety of methods has been developed and is being used to treat the molten metal. Gas bubble flotation, one of the effective ways of treatment, has been used in the casting industry for a long time. In this process either the reactive gas or an inert gas or a combination of both is purged into liquid metal. The most commonly used reactive gas is chlorine and the most commonly used inert gases are argon and hydrogen. While the gas in the form of bubbles rises to the surface it comes into contact with the inclusions and dissolved hydrogen that may be suspended in the melt (**Fig. 5.1**) and carries them to the top slag where they are skimmed with the dross. The inclusion removal efficiency largely depends on the contact time between the bubbles and the metal. It also depends on the ratio between the interfacial gas metal surface and the volume of gas injected into the melt. Therefore, the gas bubbles must be finely distributed within the melt to provide the necessary surface to offer a large reactive area (**Fig. 5.2**).

Secondly, the residual time for the contact between gas and metal is achieved by a long distance for the gas bubbles to travel. During their movement to the surface the pressure in the liquid metal will decrease. This means the gas bubbles will expand and could finally agglomerate. This effect is not wanted. Thus, there is a limitation to gas flow quantity and also to the height of a treatment system. To overcome this problem the gas treatment system can be separated in various stages.

There are various technologies and methods developed to fulfil the required conditions. The simplest system is to use chemicals in the form of powder or pellets introduced into a liquid bath in the furnace combined with severe stirring, either manually or by adequate equipment. This method is not very environment-friendly since the excess gases will leave the furnace through the door and represent a hazard to the operators and to the environment in the vicinity of the furnace. Treatment in the furnace can be improved by introducing gassing lances from top or from the side into

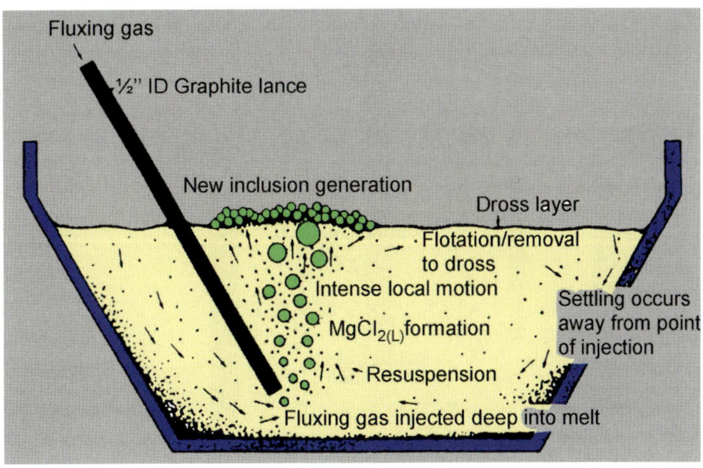

Fig. 5.1: Principle of gas fluxing

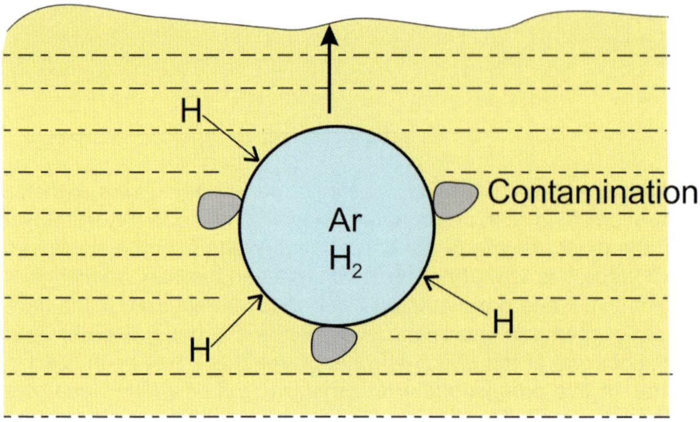

Fig. 5.2: Mechanism of removal of hydrogen and mechanical inclusions

the metal bath. Treatment gas, such as chlorine or argon or nitrogen, is introduced and creates bubbles in the gas which then act as flotation agent to the liquid metal. These activities take time and immediately require subsequent further processing such as casting.

Inline treatment systems are more effective when the relevant unit is located ahead of the casting machine. It usually comprises a degassing unit and a mechanical filtration system. There are different methods for introducing the fluxing gas into the inline filtration unit. They all have the target to obtain a very good distribution of the gas bubbles. One method comprises the introduction of the gas from the bottom via porous bricks or special carbon blocks with a number of fine holes drilled into them and connected to a gas control system. Metal is forced to flow from the top through an opening at a separating wall which is arranged on the bottom of a filter box thus passing a curtain of very finely distributed gas bubbles. This static system is very reliable and only a little maintenance is required. However, it is not possible to obtain a very fine dispersion of the gas in the melt, particularly if the metal height in the unit is selected too large. A better distribution is obtained by rotating systems. (**Fig. 5.3**) These comprise an impeller through which the purging gas is introduced into the melt. The impeller causes the metal to move and at the same time introduces the gas while the larger gas bubbles are broken by the mechanical effect of the impeller.

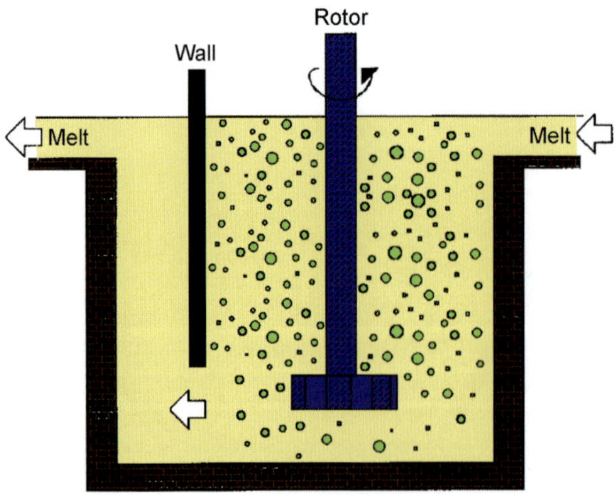

Fig. 5.3: Principle of an inline degassing unit

There are various systems available on the market and it can be said that almost every aluminium company has developed their own technology. The target is to have an even better distribution of the gas bubbles. The actual advantages of one system compared to another are, however, difficult to determine sometimes.

A system for introducing gas into a furnace is obtained by feeding gas through the impeller of a liquid metal pump or to the system of the electromagnetic pump. However, in these cases the reaction time is extremely short and gas bubbles will disappear as soon as the larger bath area is reached. Thus, the efficiency of such systems is somewhat limited.

The in-line degassing is usually followed by a mechanical filtration. This can comprise porous filter inserts of different mesh-size or porosity or can also comprise the gas injection system from the bottom. This will support the flotation effect and result in good pre-removal of inclusions already. The furnace can also be equipped with porous bricks at the bottom through which gas is injected. This will allow having the furnace completely closed and the excess treatment gases can leave through the normal flue gas removal system. If aggressive gases are used in the furnace it has to be ensured that no equipment, such as recuperators or regenerators, are effected.

5.2 Some basic considerations

There are non-metallic contaminations in the aluminum comprising:

- Gaseous inclusion, mainly hydrogen.
- Solid reaction products of gases such as oxygen, nitrogen, hydrogen with solid or liquid metal. This will form oxides, nitrates, and hydrates.
- Inclusions of not desired trace elements, such as carbon, aluminium-carbide, calcium-carbide, sulfur and phosphor.
- There might also be inclusions from sodium which is present during the melting of the scrap metal. This is actually a metal contamination but because of its influence on the non-metallic contaminations it mostly has a negative effect.
- These contaminations have a negative influence on the quality of the metal. The included gases create blisters during the heat treatment and also cause the material to generate fissures if hot-rolling. The solid inclusions appear also as blisters in the material and seed for the generation of bubbles.

The gases, which are dissolved or trapped after solidification, comprise mainly H_2. Traces of CO_2, CO and CH_4 can be neglected.

Main source for the hydrogen in aluminum is the humidity having contact with the metal. There are different possibilities:

- humid atmosphere (high air humidity)
- humid or corroded scrap charge (the corrosion product of the aluminium is a water-containing oxide)
- humidity from the furnace lining (this in particular after the installation of a new refractory lining)
- cleaning flux with high moisture content, humid furnace tools
- water from the combustion products

Upon contact of water vapor with the liquid metal, besides aluminium oxides, atomic hydrogen will be formed.

$$2Al + 3H_2O_{steam} \rightarrow Al_2O_3 + 6H \tag{5.1}$$

Products of reaction are absorbed or dissolved as contaminations in the liquid metal bath. The solubility of the hydrogen in the aluminium depends on the partial pressure of water vapor which is in contact with the aluminium, the metal temperature and the metal composition. The dependency on the pressure is expressed by Sievert's law.

$$H = k \cdot \sqrt{\rho_{H_2O}} \tag{5.2}$$

This means that the hydrogen quantity H is proportional to the partial pressure of the water vapor acting on the metal bath; k is a temperature-dependent factor.

The actual condition, as noticed in practice deviates from the theoretical values. The erection with water vapor may result in higher hydrogen partial pressures. In spite of that, the tendency remains that the hydrogen absorption in aluminum is largely dependent on the temperature. Thus, for instance, newly tapped primary metal from the electrolysis with a temperature of 850 °C to 900 °C could contain more than 1.00 cm^3 of H_2 / 100 gr. Al. At the solidus point, the hydrogen solubility is reduced in a sharp step by an exponential factor of 10.

Temperature and pressure relation for the water solubility can be defined:

$$\log S = 0.5 \cdot \log \rho - \frac{A}{T} + B \tag{5.3}$$

With S = dissolved quantity H_2 in cm^3 / 100 g Al, p = pressure, T = temperature in K. A and B are material factors.

Alloying components do have a different influence on the solubility of hydrogen. Si and Cu reduce the solubility while Mg results in an increase. This may be explained by the fact that Mg has a higher solubility for hydrogen than aluminium. Na acts only indirectly on the solubility. Due to its rapid oxidation at the metal surface, it supports a less dense oxide skin which has less resistance to the action of water vapor on the metal below. Besides the acceleration of the hydrogen absorption, the hydrogen solubility of Na in aluminium increases the absorption of hydrogen drastically.

The content of hydrogen in the aluminium melt ranges, depending on the melt treatment, between 0.8 cm^3 and 0.5, mostly however in the range 0.12 to 0.25 cm^3 H_2 / 100 g Al. During solidification there are different possible mechanisms for the presence of hydrogen. These depend on solidification velocity and have, subsequently, different effects on the final product.

An affinity of aluminium and its alloys to oxygen results in a very tight oxide skin on the surface. The oxide quantity increases with temperature, as well as with increasing content of water vapor in the atmosphere. During melting of scrap with high specific surface area and due to the movement of the melt (particularly during melt treatment such as alloying, stirring and skimming), the oxides can get into the melt and finally be found in the casting. As mentioned in one of the previous chapters, aluminium oxide appears in the aluminium in two modifications which are α Al_2O_3 and γ Al_2O_3. The γ modification can be transferred by local superheating into the very stable a modification (corundum). This transformation is accelerated in the presence of magnesium. Since magnesium has a very high exothermic energy, which will be freed during oxidation of the magnesium, this modification can be influenced by other foreign oxides. Aluminium oxide can be found in a very fine but also in a very coarse dispersed form in the aluminium. The very fine modification is particularly critical since it cannot be removed by the cleaning processes available today.

To obtain the requested quality, which means a very low content of inclusions and hydrogen, there are primary methods and also secondary methods.

The primary methods try to protect the metal from the contaminations of the furnace atmosphere and to keep water vapor in the furnace atmosphere as low as possible. Secondary methods are to remove the dissolved hydrogen und to clean the metal from solid inclusions. Primary methods comprise the application of protective fluxes which comprise in principle chlorides of potassium and calcium, as well as chlorides of potassium and sodium. It is also important to maintain a low metal temperature and to avoid superheating beyond the level required for casting.

There are different methods for the secondary treatment. These comprise holding the melt for a certain period of time, treatment of the melt fluxing gases, such as Cl_2 and argon, or mixture of these gases, gas removing salts, in-line treatments and filtration. Treatment by holding at constant melt temperature and constant atmospheric conditions can have an influence only on the mechanical inclusions by having them settled and collected at the bottom and walls of the furnace.

Due to tendency of aluminium to absorb hydrogen, the time between treatment and casting should be as short as possible. Treatment in the furnace requires comparatively low investment but may not be as efficient like a well-selected in-line treatment system, unless metal such as sodium or magnesium has to be removed. It also requires time that may be lost for the production. The equipment for in-line treatment, on the other hand, must be sized to achieve the required contact time between gas and aluminium. It is typical for both systems that the contamination cannot be reduced to zero. Only a reduction of the components is technically possible. Thus, always a primary treatment in the recycling plant should be considered and to reduce the casting time to the lowest possible extent.

The gas quantity required is based on experience with the different treatment systems. As an indication, the gas consumption ranges from 0.1 to 0.8 m^3n per ton of aluminium. It is standard practise to obtain a hydrogen content as low as 0.1 cm^3 / 100 g Al.

5.3 Inline degassing units

There are different technical solutions available to fulfil the requirements as stated above. As examples we will use some typical designs.

As representative of the static systems we will look at the DUFI system (**Fig. 5.4**) which was developed by Alusuisse in the early seventies. Treatment gas is injected by means of porous bricks from the bottom of the unit. The chamber is divided into two sections: one for introducing the metal and

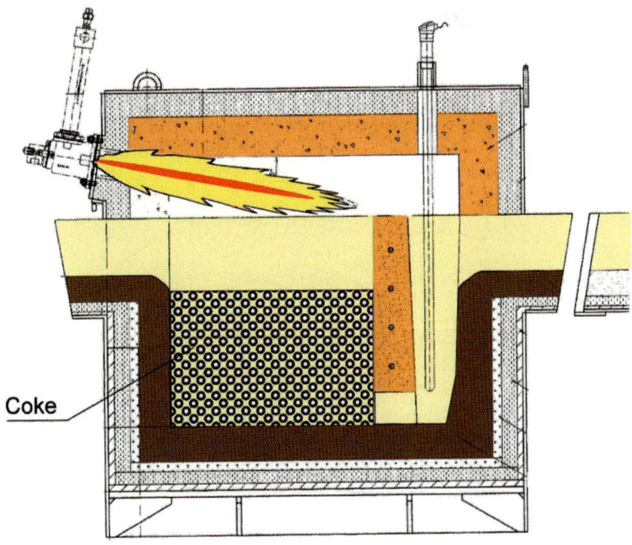

Fig. 5.4: DUFI-Filler (Source: Alusuisse)

then, after passing openings, on the bottom of a separating wall for charging metal. The section where metal is introduced is filled with coke. Aluminium passing this coke bed is cleaned by the deep bed filtration effect and also by flotation. The flotation process is characterized by components in the melt having different border tensions to the gas bubbles. By means of this physical effect, hydrogen and also some inclusions are carried to the surface of the melt and from there they can be skimmed off. The cleaned metal rises through the discharge section and continues to the down-line equipment such as additional filtration and casting equipment. After every cast – no matter how long it takes – the unit has to be emptied and the coke filling must be deposited. If the alloy is not changed, the content of the unit can be kept hot by means of a heating lid and is available for the next cast. The DUFI is equipped with a heating lid that is equipped with a burner or electrical heating elements. Charging side as well as discharge port are designed as normal launder ports and can be connected to the casting launder system. There are different means of getting the unit emptied. Hooks and swivel point can be attached to the filter vessel which allows tilting the vessel by means of an overhead or a jib crane. The unit can also be placed onto a tilting device which is arranged rectangular to the flow direction of the metal. First the liquid metal is discharged via a spout into a sow mold. The sow mold is then replaced by a normal container which receives the coke filling. After the coke is replaced by fresh coke, the unit can be heated again and is then available for the next charge. The purge gas to the porous bricks at the bottom of the unit is provided via a control system that enables the operators to set the gas flow or a mixture of gases comprising argon or nitrogen or, if required, even chlorine to the melt. Gas connections are made by flexible hose pipes via quick-changing couplings attached to a common manifold.

Units equipped with an impeller operate with a different principle but the target of distributing small gas bubbles within the melt is the same. Liquid metal is supplied to a closed box and passes through openings at the bottom of a separation wall to the exit. The treatment chamber accommodates one of more impellers. The fluxing gas reaches these impellers through the bath. Additionally to the fine dispersion of gas bubbles, the liquid aluminium is moved by the rotating action of those impellers. This ensures a very thorough contact of metal with the gas bubbles. The metal now flows at the bottom of the unit to the exit chamber of the filter system. As in the case of the DUFI, the removal of inclusions and mainly hydrogen is due to flotation and lowering the partial pressure of the hydrogen in the melt causing the hydrogen to penetrate into the gas bubbles. The unit comprises a refractory-lined box which is closed by a lid that carries the drive units for the

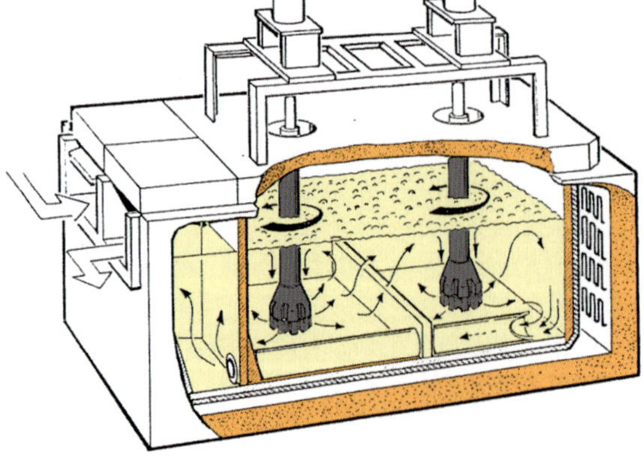

Fig. 5.5: In-line degassing unit (SNIF)

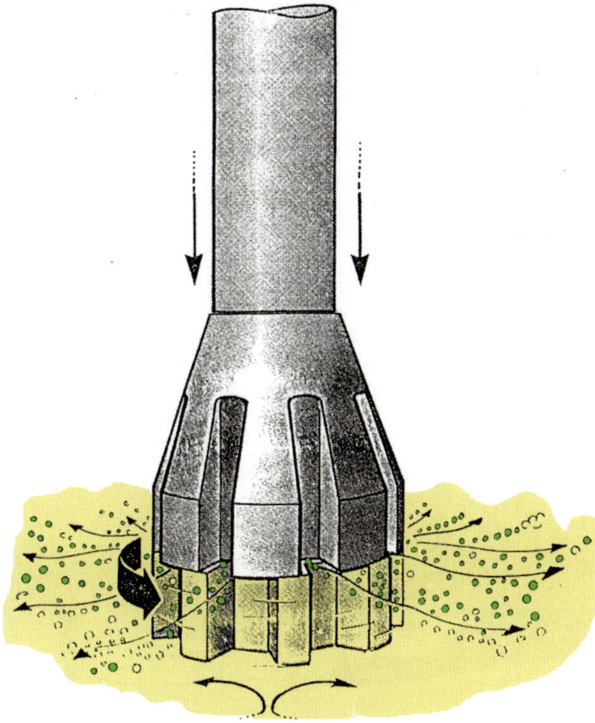

Fig. 5.6: Impeller (SNIF)

rotors. The lid is also equipped with burners to avoid heat losses. Gas generated by these burners leaves via a flue gas exit accompanied by the hydrogen and argon discharged from the liquid melt.

The unit may continue operation until casting is stopped. If alloy change is required, the treatment box has to be emptied by means of a tilting device. Before discharging remaining metal the heating lid has to be lifted. The lid of the treatment furnace may be lifted by means of an installation on the treatment furnace or by the overhead crane or jib crane. After pre-heating, the inline degassing unit is ready for operation again. The complete unit contains a substantial amount of metal. Thus, it is advisable to have a certain quantity of metal passing through the unit before alloy change is intended.

Inline degassing systems, as provided as SNIF unit or as ALPUR (**Fig. 5.5**) system, are very efficient and provide a very good separation of hydrogen. They may be redesigned for application in a refining operation. This will be different in a re-melting installation where a DC-casting unit is installed to produce ingots for immediate use in extrusion or rolling plants. If the metal is quite clean, a small box having just one rotating impeller may be sufficient for treating the metal inline.

The complete in-line treatment unit is quite complex considering the essential elements such as heating lid, impeller (**Fig. 5.6**) with drive, treatment box, tilt system and electrical equipment.

As mentioned before, one of the problems is to generate very fine bubbles evenly distributed in the melt. The bubbles generated by porous bricks or also by a rotating system may agglomerate, particularly, if the metal level in the treatment unit is too high. A system developed for the copper industry may overcome this problem. The treatment gas is introduced via jet nozzles on the bottom of the treatment box. A row of the nozzles is arranged in the bottom of a comparatively small

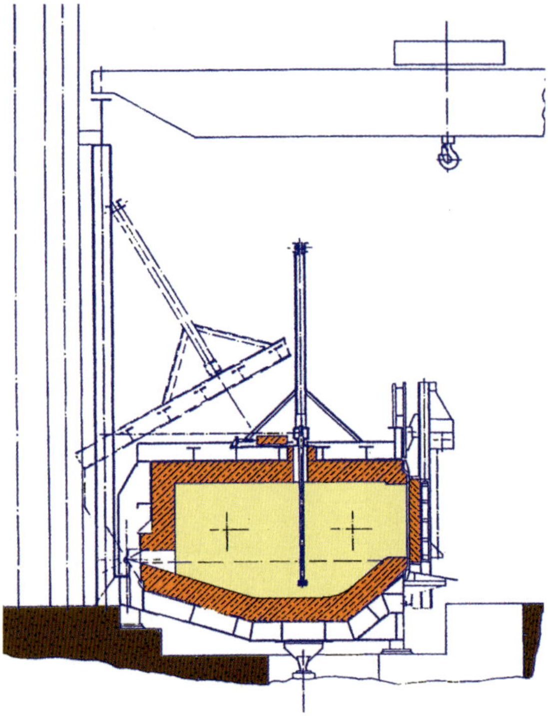

Fig. 5.7: Furnace-mounted gas fluxing system

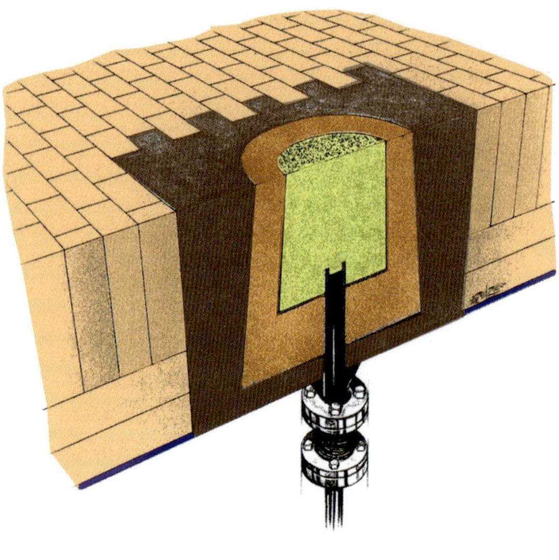

Fig. 5.8: Gas injection via porous bricks

launder. While passing through this launder the metal comes into contact with the bubbles generated by these nozzles. This appears to be a very efficient system. A problem might be the comparatively high pressure required for the gas jet.

5.4 Furnace-installed systems

There are different methods which can be implied in the casting furnace after the melt is prepared and adjusted for casting. We have already mentioned the treatment with chemicals, which actually is not representing the state-of-the-art treatment with fluxing gases, since this can be performed out in the closed furnace. Lances for introducing gas can be lowered into the metal bath from the furnace roof (**Fig. 5.7**). To do so, openings have to be provided in the furnace roof through which the lances can enter. A lifting system is installed above the furnace. This is connected to the gas supply system. After treatment the lances are lifted out of the furnace and the openings are closed. This system requires that access is provided on the top of the furnace. The system injecting the gas could be equipped with rotors for good distribution of the gas in the melt. However, the better the system is to treat the melt, the more complicated it is to design, to operate and to maintain.

A very efficient and simple system is to equip the bottom of the furnace with a number of porous bricks (**Fig. 5.8**). The gas flow from these bricks not only acts to remove hydrogen and inclusions but also to move the bath in the furnace for good heat exchange.

For certain alloys it is convenient to use the installed degassing system. It allows the removal of excess magnesium or sodium. The treatment can also be extended over a lengthy period of time which may be required if the comparatively short treatment in the in-line degassing does not suffice. Fluxing in the furnace is supported by holding the metal for some time.

5.5 Metal filtration

During degassing mechanical inclusions may be generated in the melt or distributed in the liquid metal. These inclusions can be removed by a metal filtration system. In the simplest case it can comprise only a narrow meshed system of heat-resistant porous brick. But normally the metal passes through a filtration unit. This comprises a box where the metal is introduced and forced to pass through special porous filtration inserts made of refractory material. The pores of this insert may be larger than the grain size of the inclusions. The filtration effect is due to adhesive forces in the porous material (**Fig. 5.9**). The unit comprises a refractory-lined box with a suitable opening to receive the porous refractory filtration insert. As long as casting commences these plates need not to be replaced. But after every cast, metal will solidify within the refractory filtration which has, consequently, to be replaced as soon as a new casting commences. But since the insert is comparatively small and inexpensive, this is acceptable. The box is equipped with a lid with integrated burner to maintain a certain temperature level. Every time casting stops the lid has to be removed.

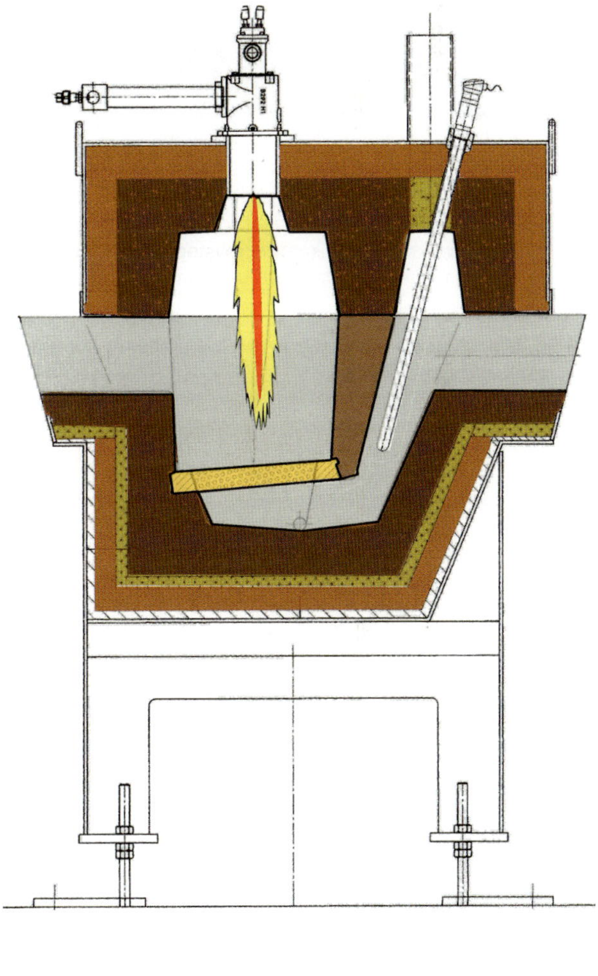

Fig. 5.9: In-line filtration

Depth filtration

- Flow
- Porosity
- Particle diameter vs. pore diameter

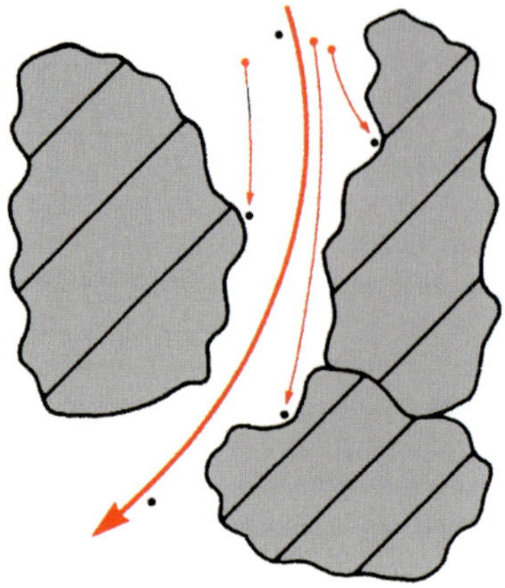

Fig. 5.10: Deep bed filtration

The unit is emptied from liquid metal and the filtration insert must be replaced. For this reason the filtration box must be in the operating range of a crane or at least of a jib crane.

For high performance filtration a deep bed filter should be applied (**Fig. 5.10**). This provides a longer passageway for the metal with a resulting higher filtration efficiency (**Fig. 5.11**). The filtration effect is also based on adhesive forces at the filtration medium, i. e the filling of tabular alumina or coke. This type of filtration is applied for producing high-quality extrusion or rolling ingots. In the secondary smelters, which mainly produce re-melt ingots, the normal filtration box, on the basis of a porous filtration inside, is effective and fully sufficient. The filter medium comprises a crushed tabular alumina or coke, similar as used in the DUFI filtration system as described above.

Metal filtration is always arranged downstream of the degassing unit since its purpose is to remove mechanical inclusions trapped in the metal during the melting process but also during degassing.

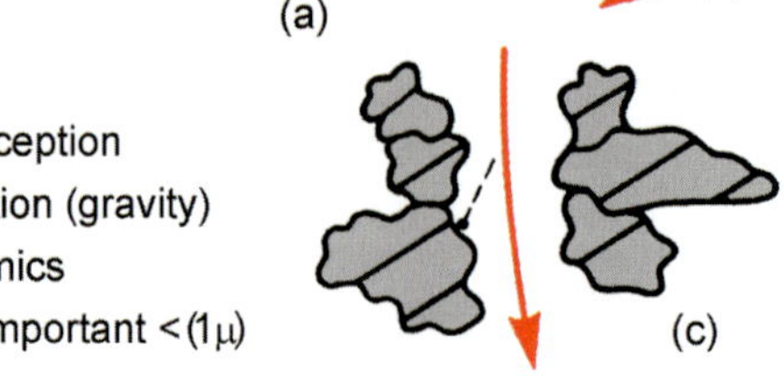

- Inertia
- Direct interception
- Sedimentation (gravity)
- Fluid dynamics
- Diffusion (important <(1μ)

a) particles settle by gravity, b) fluid dynamics, c) adhesive forces

Fig. 5.11: Filtration mechanism

6. Homogenizing

Christoph Schmitz

As detailed in the previous section, recycling plants are operated as refining plants, which produce casting alloys, and remelting plants processing in-house scrap and primary aluminium. Billets and slabs are used directly in the own rolling or extrusion plant. The billets require "homogenizing". This heat treatment at comparatively high temperatures is required to reduce internal stress and to obtain the grain structure for the extrusion. This typically is a batch type unit or – if production rate and product mix permit – a continuous process. The batch type process provides more flexibility for serving individual requirements of different customers. Therefore, most of the plants are batch type plants.

6.1 Metallurgical considerations

Before DC ingots are fabricated into semi-finished forms, it is necessary to homogenize at a high temperature to reduce segregation and to remove non-equilibrium, low melting point eutectics that may cause cracking during subsequent work. These effects have their source in the comparatively rapid cooling during DC casting. The undisturbed growth of crystals starts around elementary cells (seed grains) within the melt. **Fig. 6.1** shows the different stages until the final grain is complete. Starting from the elementary cell, more and more other cells attach to this nucleus until

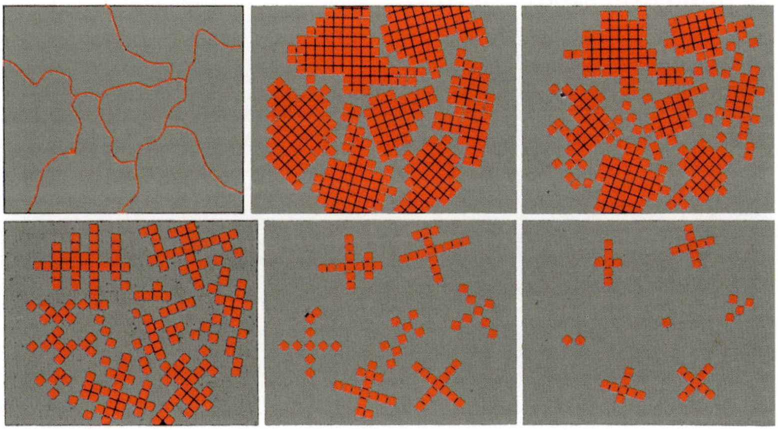

Fig. 6.1: Development of a grain structure starting from a nucleus grain

they finally form a grain. This effect takes place at various points within the melt until all melt is consumed. The different grains can increase in size until they reach the borders of neighboring grains thus forming the crystal structure. The growth does not always result in little ball-like or cylindrical grains. The heat transfer at the edges of the newly formed grains causes a more rapid growth velocity. This results in the formation of so-called "denditric" crystals (**Fig. 6.2**). During DC casting the cooling takes place at a fairly rapid velocity which decreases towards the center of the billet (see also section 3.4.1). The related temperatures are shown in **Fig. 6.3**. Consequently, crystals are small towards the mold, followed by a zone of "columnar grains" and finally small irregular grain in the center (**Fig. 6.4**).

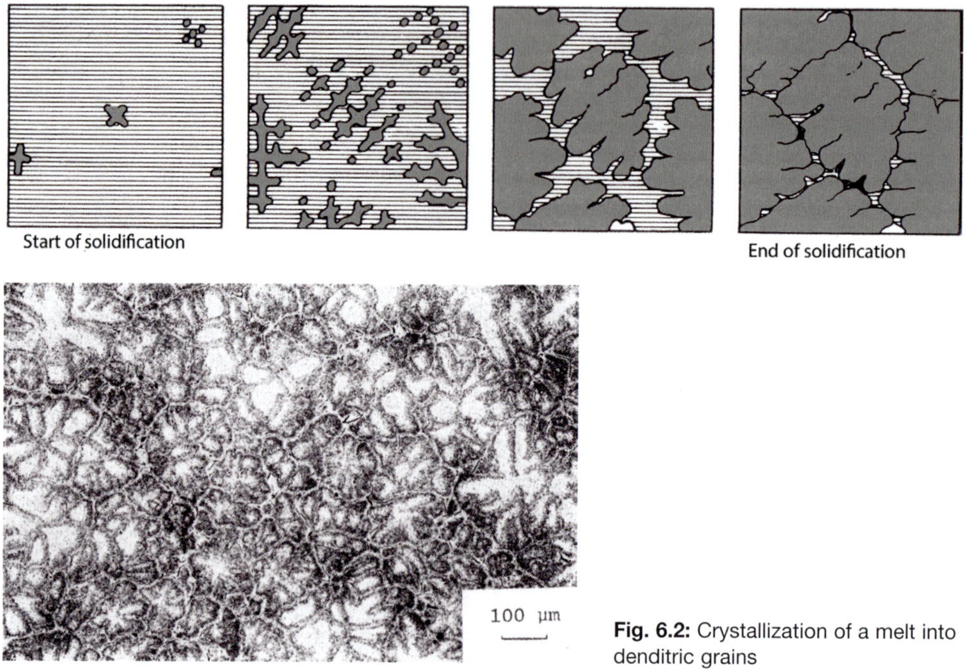

Fig. 6.2: Crystallization of a melt into denditric grains

But other processes take place during solidification of a melt since alloying metals are involved. Looking at an alloy that may, for example, consist of aluminum and magnesium it can be determined from the phase diagram at the solidus temperature that 17.4 % of the magnesium can be dissolved in the aluminum. The resulting crystal melts already at 449 °C. Above this temperature a

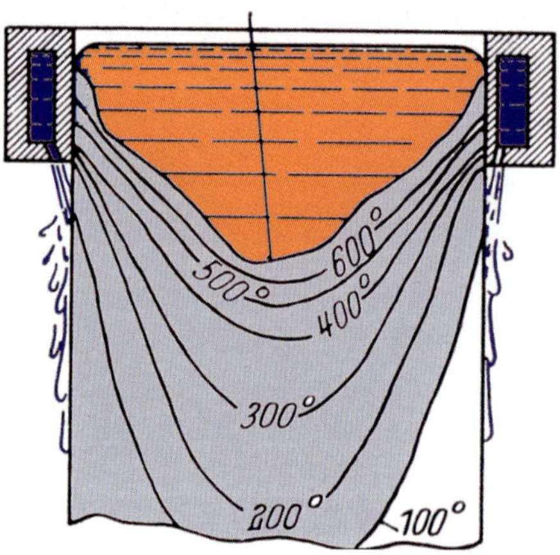

Fig. 6.3: Temperature in a billet during DC casting

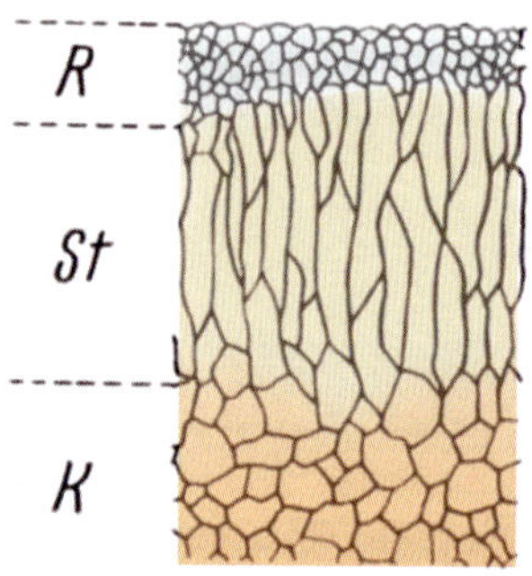

Fig. 6.4: Growth of grains from outside to center. The photo rolling was obtained by etching the surface of a cast rolling ingot

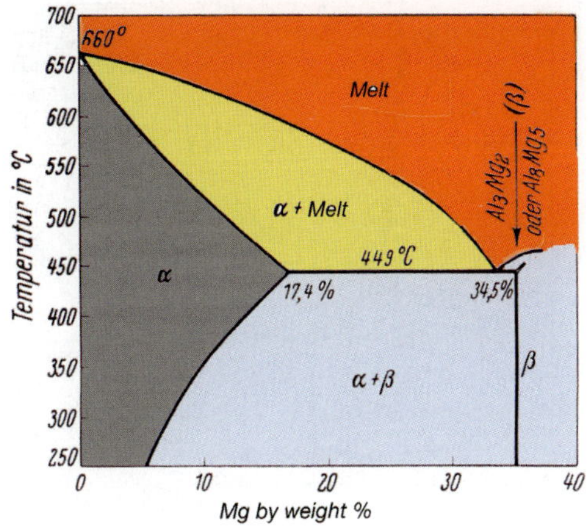

Fig. 6.5: Phase diagram of the system AlMg showing the maximum solution of Mg in aluminum of 17.4 % Mg at a temperature of 448 °C

mixture of melt and AlMg crystals are present (**Fig. 6.5**). Below this point at the solidus temperature aluminum is solid. The alloy is now considered as being "homogeneous." If we look at a melt which may have – as an example – a magnesium content of 5 %, some interesting processes occur during solidification. First, grains will have a Mg content of 1 % (**Fig. 6.6**) at time D. At time E the Mg content is 3 %, at time D 10 % and finally reaches 30 %. The result is that crystals with the high Mg content settle at the grain borders (**Fig. 6.7**). The process is called grain segregation or heterogeneous segregation and their extent depends very much on the cooling time of the melt. With the DC casting process the cooling time is different over the cross-section of the billet depending on the temperature of the melt. The result is that we find not only different grains but also a variation of mixed crystals and solid solutions. Since the atoms are still very movable the segregation can be corrected by heating the alloy to a comparatively high temperature. The structure is homogenized.

The homogenization is essentially a thermal treatment to remove the undesirable features of the cast structure. The variables are the heating and cooling rates and hold time and temperatures. In this regard, the homogenization time at a particular temperature has been found to be inversely related to the square of the dendrite arm spacing in the ingot. Homogenization is particularly important for the higher strength alloys as it also serves to precipitate and redistribute the submicron intermetallic compounds of the transition metals such as $MnAl_6$, $Al_{12}Mg_2Cr$ and $ZrAl_3$. These

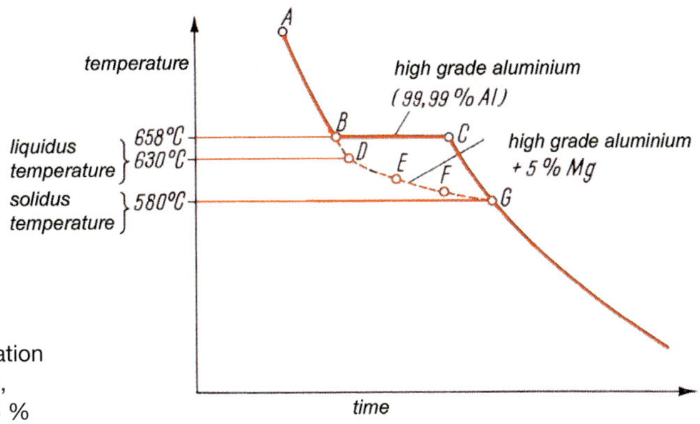

Fig. 6.6: Time-dependent solidification of a melt. B – C primary aluminum, B – G alloy with a Mg content of 5 %

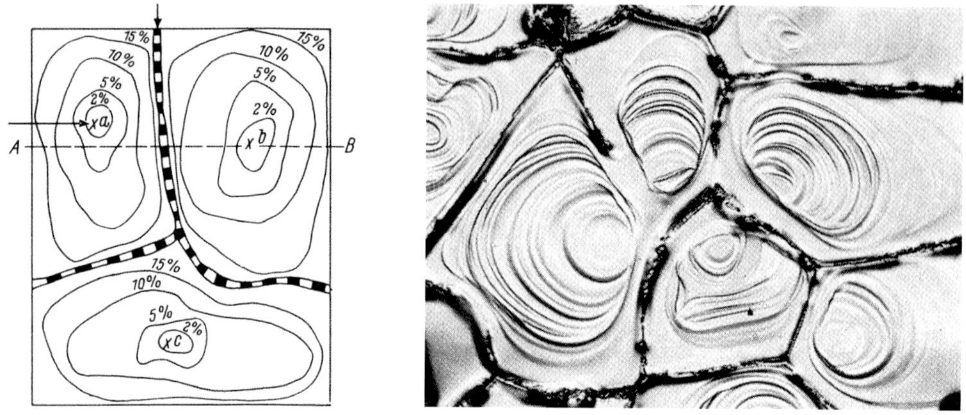

Fig. 6.7: Grain segregation of an alloy with 5 % Mg. Grain borders with segregation of Al-containing grains. These border crystals can be seen in the magnification photo

transition metals may supersaturate in the aluminum during the relatively rapid cooling of DC ingots and it is then necessary to promote precipitation of uniform dispersions of the compounds in order to control grain structure. In addition, it is now realized that they may have a marked influence on various mechanical properties through their effects on both the subsequent response to ageing treatments and on the dislocation substructures formed by deformation.

Regulation of these various functions requires a careful choice of conditions for homogenizing ingots of different alloys. If precipitation of these compounds is involved, both time and temperature are significant and the rate of heating to the homogenization temperature is of crucial importance. Relatively slow rates, e.g. 75 °C/h, are necessary to promote nucleation and growth of a fine and uniform dispersion of the compounds. It has been found that the compounds are actually nucleated heterogeneously at the surfaces of precipitated particles which form and grow to relatively large sizes during the slow heating cycle. Once formed, the submicron compounds remain stable at the homogenization temperature, whereas the precipitates are re-dissolved. This effect is illustrated for an Al-Zn-Mg alloy in **Fig. 6.8**. In this case the precipitate η-($MgZn_2$) has provided an interface for the subsequent heterogeneous nucleation of MnAl6 in the ingot that was

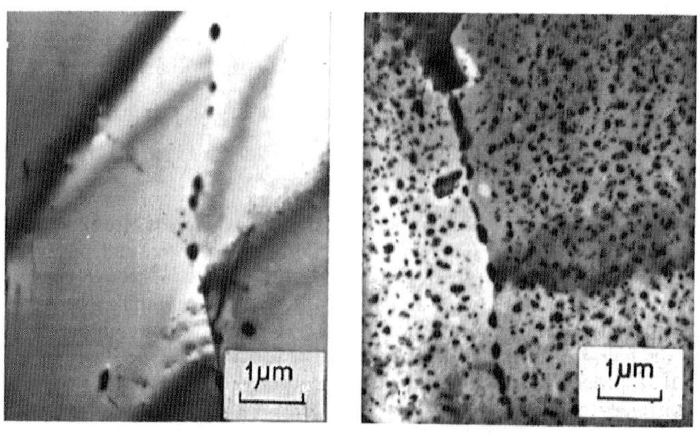

(a) (b)

Fig. 6.8: Transmission electron micrographs showing formation of submicron particles of MgAl6 in an Al-Zn-Mg alloy containing 0.3 % Mn. The ingots were homogenized at 500 °C for 24 h following: (a) a fast heating rate (500 °C/h); (b) a slow heating rate (50 °C/h) [6.1]

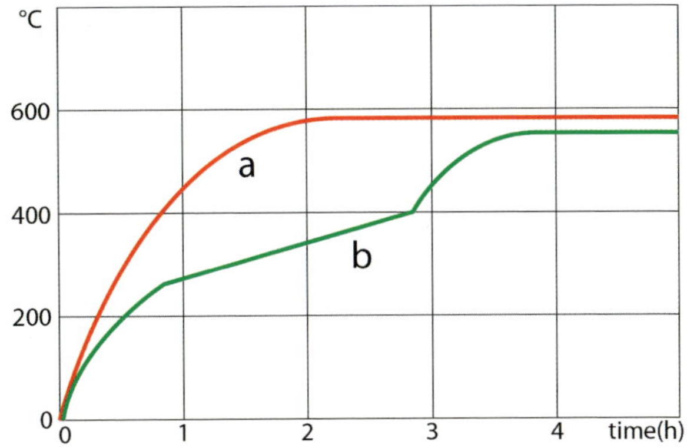

Fig. 6.9: Heating regime of a homogenizing process (Source: Hertwich)

heated at the slower rate. Further modification to the size of these submicron particles is possible by control of the rates of cooling of ingots after homogenization is completed [6.2].

Since during homogenization several processes are ongoing, specific temperatures during treatment are needed for a specific chemical composition of the material. Mainly the temperatures are near to the solidus temperature. The homogenization takes several hours, normally between ten and twelve hours (**Fig. 6.9**) because the diffusion process is running not very fast. For specific alloys a treatment at lower temperatures at the beginning is necessary to avoid melting of ternary eutectics. For several alloys the cooling rate after homogenization is important. These alloys are, for example, some of the 6,000 series (billet alloys) to maintain an advantageous microstructure for the extrusion process. The equipment design must allow a setting the process parameters as required [6.3].

6.2 Batch-type homogenizing plant

The cast billets are sent to a heat treatment plant (**Fig. 6.10**) which operates automatically. This includes billet handling like stacking and de-stacking and stack transfers, heat treatment and cooling. Heat treatment of billets is done in a furnace with air circulation. This requires a specific

Fig. 6.10: Batch homogenizing plant (Source: Hertwich)

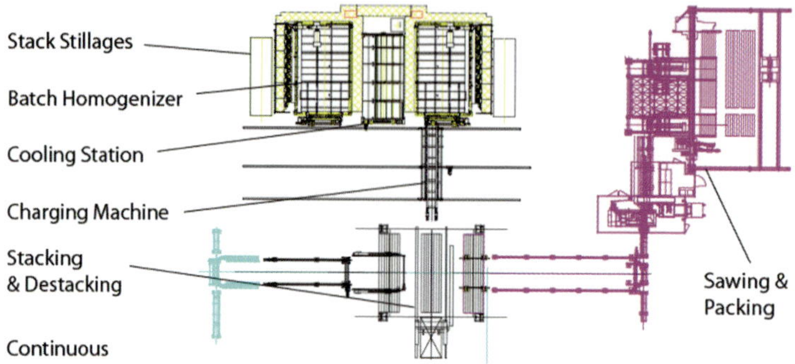

Fig. 6.11: Layout of a batch homogenizing plant (Source: Hertwich)

stacking of the billets for improved heating. The billets are taken out of the casting pit by overhead crane and are placed on a storage table. From here a certain amount of billets, the number, which represents one layer in the furnace, is separated onto the combiner – a table arrangement which provides space for one layer of billets. By separating them from the billets on the storage table, the specified number of billets is combined for one layer.

Next the billets will be taken by an automatic stacker and put onto the stacking stand (**Fig. 6.11** and **6.12**). Between two layers of billets, spacers are placed which provide for enough space where the air can circulate around the billets for optimum heating of the billets. When a stack is stacked up, it is taken by a charging machine and placed in a furnace. The charging machine runs on rails along the furnace line and the preparation stands. The charging machine is placed in front of the loading position, where a billet stack is waiting. While the stack is on the charging

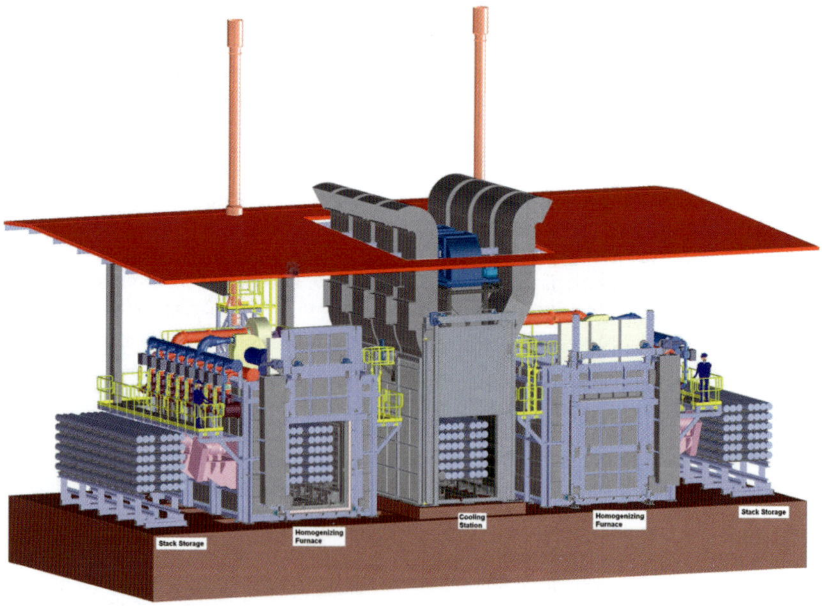

Fig. 6.12: Homogenizing plant (Source: Hertwich)

machine, the machine moves along the furnace line to an empty furnace. The furnace door opens and the stack is moved into the furnace and placed inside. After the charging machine has left the furnace, the door closes and the heating process starts. Now the charging machine is free to take a heat-treated batch from another furnace and move it into the cooling chamber or take a batch from the cooling chamber to the unloading station.

Once the stack is placed on the unloading station, the stack will be disassembled automatically. The billets are taken from the stack and put onto the preparation stand for the ultrasonic testing and for cutting while the spacers are placed in a storage area or directly onto the new stack of billets.

Ultrasonic testing is done to ensure the quality requirements.

After the test shows the accepted result, the billet might be cut to pieces of customer-pecific length. After that the billets are packed and prepared for transportation.

6.3 Homogenizing furnace

After being placed in the homogenizing furnace the billets are heated by convection. For this the furnace is equipped with an air circulation system. The air is propelled by fans and guided by channels and distributed for covering the whole side of the stack of billets homogeneously.

Heating by convection requires a homogeneous flow of the air through the layers of billets. Therefore, the inner casing of the furnace is configured with channels on the sidewall which distribute the air to the channels created by the spacers in between the layers of billets.

6.3.1 Furnace structure and concept

The homogenizing furnace consists of a steel structure, which contains the insulation, air guiding channels, burners for heating and the fans for air recirculation. The furnace inside is equipped with the supports where the batch is placed on and the rails for the charging machine. The furnace front contains the charging door which is installed as a lifting door. The door seals tight against the furnace front door frame and is guided in vertical rails.

Door design is very important. Because of the very even temperature distribution no tramp air from outside can be tolerated. A sealing consisting of a ceramic rope installed in the door blade is pressed by pneumatic clamps against the door frame. If the door travels the blade is lifted in order to avoid damage of the seal.

The air is propelled by fans which are installed in the roof area. From the roof area the air flows into the sidewall channel. This channel contains a flow distribution system which directs the air onto the stack sidewall for the even distribution of the airflow to the billet layers. On the opposite side, the air is taken in by the same distribution system which also ensures the even collection of the airflow without deviating the flow to one direction. In this channel on the furnace sidewall the air flows back to the roof and is sucked in by the recirculation fan. On the way to the fan the flue gas from the burners is mixed into the airflow which effects heating of the air.

Heating of the furnace is done by gas burners or electrical heating elements which are installed in the furnace roof. Those burners are recuperative tube burners for reducing the energy consumption of the heating process. The furnace shell comprises a steel plate structure adequately reinforced by steel members. A lightweight refractory lining minimizes heat losses.

The furnace has an inner casing which serves for guiding and distributing the air to one sidewall and collecting the air on the opposite sidewall. Furthermore, the inner casing supports even mixing of hot flue gas into the circulating airflow. Efficient heating and temperature distribution of the billets are achieved by the reverse flow which is installed in the furnace (**Fig. 6.13**). This means that the flow is first running through the batch from left to right and after a certain temperature differ-

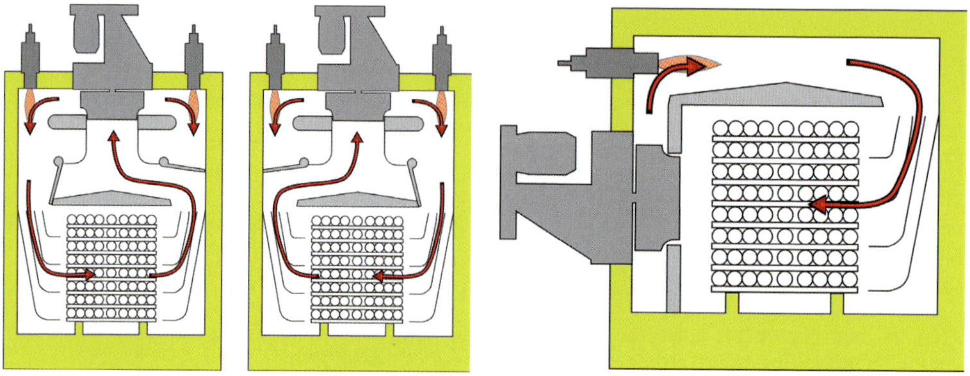

Fig.6.13: Air distribution in a batch homogenizing plant (Source: Hertwich)

ence between inlet and outlet has been achieved, the flow turns and passes the batch from right to left. By this the heating is improved, the homogeneity of the heating is more even and the heating time is reduced. Control of the reversing cycles is done by the control system and the temperature probes in the airflow inlet and outlet as well as by the material temperature testing devices.

6.3.2 Control system

The control system of the furnace adjusts the burner power according to the needs of the specific process step. The inlet temperature to the furnace heat treatment area is controlled by thermocouples. Likewise the outlet temperature is measured and taken into consideration for controlling the power of the heating system. For exact control of the billet heating, the material temperature of the billets will be measured by special testing devices which are installed on the sidewall of the furnace. These devices consist of a thermocouple which is pressed against the billet for measuring the billet temperature. In order to avoid any influence from the circulation air on that measurement, the device is equipped with a special arrangement for insulating the testing area around the thermocouple.

6.4 Cooling chamber

The cooling chamber is installed in line with the furnaces and the preparation stations. The charging machine serves it in the same manner as the furnaces and the preparation stations. Cooling also must follow a certain program which may differ as desired for different alloys (**Fig. 6.14**). The cooling chamber consists of a steel casing for guiding the cooling air through the billet layers. Mainly the cooling chamber is of the same design as the furnace, but the casing is not equipped with insulation. The cooling is effected by fresh air, which is sucked into the cooling chamber by the air fans and released to the ambient surroundings after passing the stack of billets. The cooling chamber is equipped with a jalousie instead of a door plate, because there is no insulation needed on the closure of the cooling chamber. Charging of the cooling chamber by the charging machine is done in the same manner as for the furnaces. The lower area for the cooling chamber, containing batch supports and rails for the charging machine, is of the same design as the lower furnace area.

The cooling chamber as well can be equipped with a reverse flow system for improved cooling of the billet stack.

After the cooling reaches a sufficient temperature, the batch can be taken out of the chamber by the charging machine and sent to the unloading station.

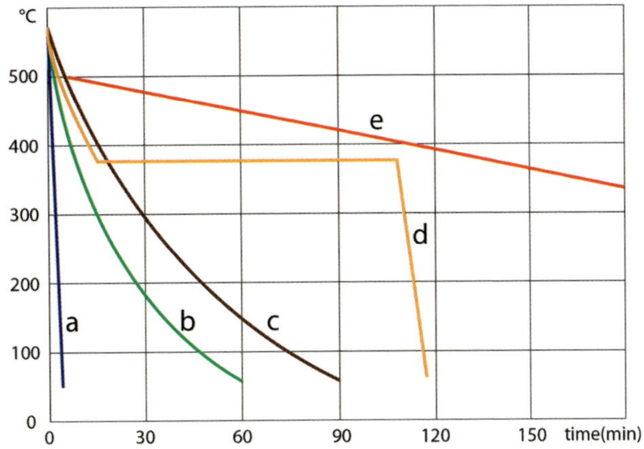

Fig. 6.14: Cooling curves for homogenized billets. The different cooling graphs depend on the alloy to be processed (Source: Hertwich)

For the fresh air supply to the cooling chamber a channel, which extends to outside of the building and allows sucking in of cold air, is installed on top of the chamber. The air outlet is installed parallel to the inlet and releases the warm air to the ambient surroundings. Because there is not any contamination produced or connected to the cooling chamber and its batch, the air can be released without further treatment.

One cooling chamber can serve several furnaces (Fig. 6.12).

6.5 Charging machine

The automatic charging machine serves all the stations and chambers of the plant. It takes the stacks from the loading station into the furnace, from there to the cooling chamber and afterwards to the unloading station.

The charging machine consists of two frames, the lower and the upper frame. The lower frame carries the whole machine and contains the wheel sets, which are running on the rails. The upper frame moves on the lower frame and can be lifted for taking the stack and carries the billet stack. For taking the billet stack from the loading position and for placing it in the furnace, the cooling chamber and on the unloading position the upper frame is lifted or lowered respectively. The lifting is effected by a lever system and a hydraulic drive. Moving the charging machine is effected by electric drives on the wheel sets. The upper frame moves under the stack, and upon reaching the end position it is lifted, which also lifts the billet stack. Next the upper frame, including billets, moves back onto the charging machine.

For letting the upper frame move into the stations and chambers these are equipped with rails on the bottom. These rails support the upper frame when leaving the lower frame of the charging machine. The upper frame runs into the furnace and, once reaching the end position, the frame and the stack are lowered which places the stack on the supports in the furnace. The supports are fixed installations on the furnace bottom.

All movements of the charging machine are controlled by the transport control system. Suitable sensors monitor the positions of the machine.

6.6 Stacking and de-stacking station

An automatic stacker does the stacking and de-stacking. The stacker station consists of the billet layer combiner, the stand where the stack is located on and the automatic stacker handling.

The stacker station serves for stacking and de-stacking of the batches. The stands are used for stacking and de-stacking.

The billets, coming from the casting machine, are fed to the layer combiner. This unit serves for combining a sufficient number of billets for one layer and offers them to the stacker. The first layer on the stacker stand is the layer of the basic spacers. On these spacers the whole stack will be located. These spacers are left in place after de-stacking.

The stacker takes the first layer of billets from the combiner and puts them onto the basic spacers. Next it takes a layer of spacers and puts them onto the layer of billets. These spacers can be taken from the stack that is de-stacked at the same time. Afterwards the next layer of billets is taken from the combiner and placed onto the stack. This is repeated until the stack is complete. The number of spacers placed in one layer is chosen according to the weight of the billets placed on it in total. Parallel to the stacker stand, a second stand serves for de-stacking. On the de-stacker stand the heat-treated and cooled stack is placed for de-stacking. The layers of billets are sent to the preparation stand for ultrasonic testing and the layers of spacers are placed on the new stack on the stacker stand.

The stacker-handling device consists of a crane unit, which can move above the de-stacking station. The stacker stand can take, lift and lower the billets and the spacers. A grip unit, which contains the clamping devices for spacers and billets, is attached to the cables of the crane unit. **Fig. 6.15** shows a typical layout of a batch homogenizing plant.

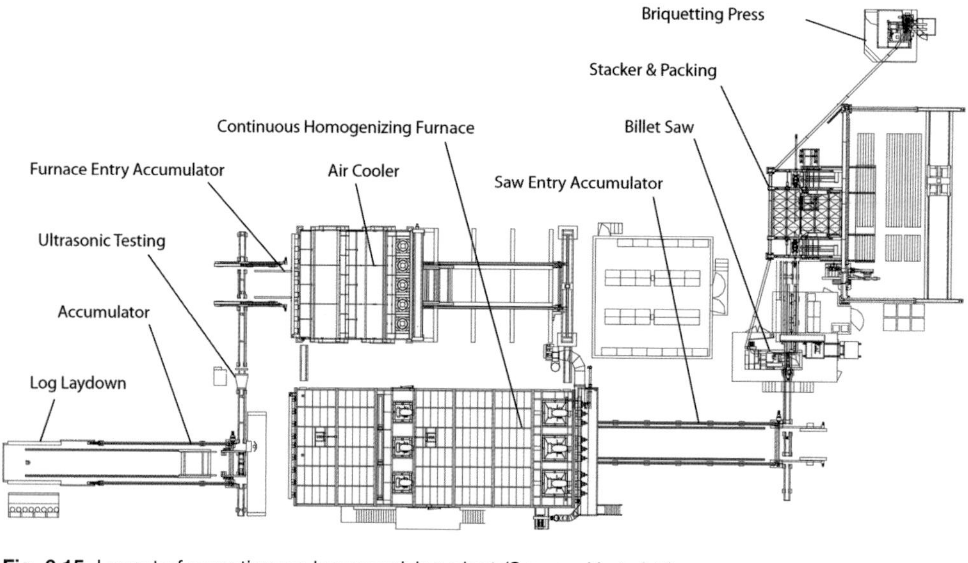

Fig. 6.15: Layout of a continuous homogenizing plant (Source: Hertwich)

6.7 Continuous homogenizing plant

The continuous homogenizing plant is generally designed for a continuous operation with large capacities and limited variations in billet sizes (**Fig. 6.16**). The equipment is usually operating 24 h per day. This description is considered to be information in order to complete an overview about homogenizing technologies. According to Hertwich, the continuous homogenizing plant offers some significant advantages compared to the batch homogenizing plant:

Fig. 6.16: Continuous homogenizing plant (Source: Hertwich)

- Perfectly uniform metallurgical properties of the billet due to precise and reproducible temperature regime during heating, holding and cooling,
- Lowest labor costs because the largely automatic operation requires only few operation personnel,
- Perfect straightness of the billets,
- Sophisticated control software,
- Low energy consumption.

6.7.1 Operating description

The continuous homogenizing plant serves as a heater with adjoining continuous homogenizing for round extrusion ingots. Heating – holding – cooling follow at every step automatically (**Fig. 6.17**). For homogenizing these are fed by crane into the entry magazine. They are then con-

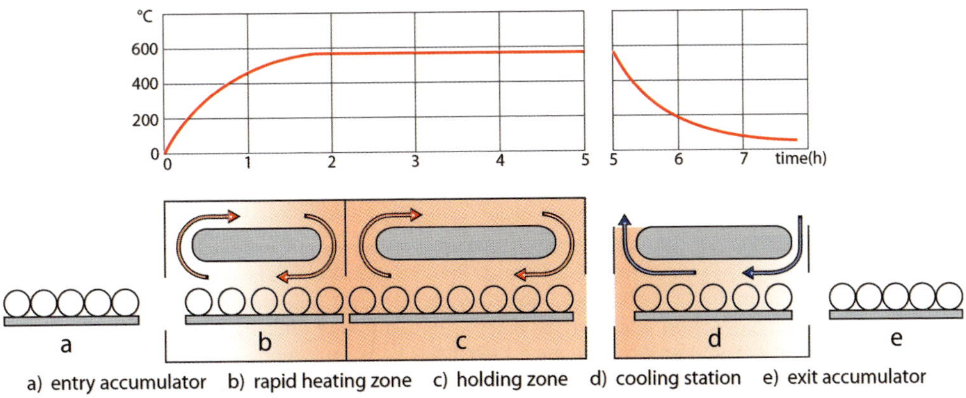

Fig. 6.17: Heat flow in a continuous homogenizing plant (Source: Hertwich)

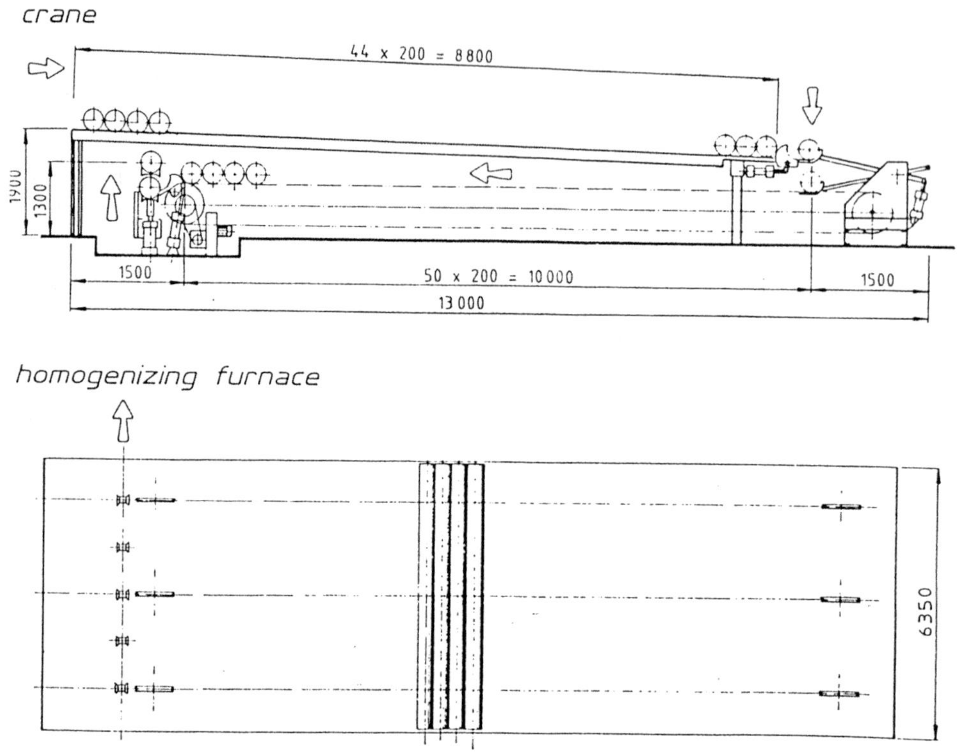

Fig. 6.18: Feed magazine and billet transfer (Source: LOI)

veyed through the entry magazine and charged into the rapid heater followed by entry into the holding zone (**Fig. 6.18**). Rapid heater and holding furnace are designed as one unit (**Fig. 6.19**).

A preheating furnace is comprised of three sections and the holding zone of two sections. The input magazine is arranged just in front of the furnace section.

After the signal "furnace position 1 free", the next billet is fed through the hydraulically operated charging door at one side of the furnace and taken over by the furnace roller conveyor (Fig. 6.19). A measuring device picks up the billet diameter in its position. As the billets are moving through the furnace the entrance becomes empty and the next billet can enter. The correct position of the billets is controlled by limit switches.

The billet is now conveyed through the furnace step by step by a specifically designed walking beam conveyor (Fig. 6.19). At every step through the furnace the billet is turned partially. As the billets pass through the furnace they are heated by air flow from top and bottom. The air nozzles are arranged to prevent deformation of the billets. By using high velocity flow rates and equally distributed air nozzles, efficient and uniform heat transfer is ensured. The heating furnace may receive as much as 90 billets. The quantity of billets determines the heating time and thus the furnace capacity. Temperatures are controlled in the individual heating zones within a temperature difference not exceeding 5 °C.

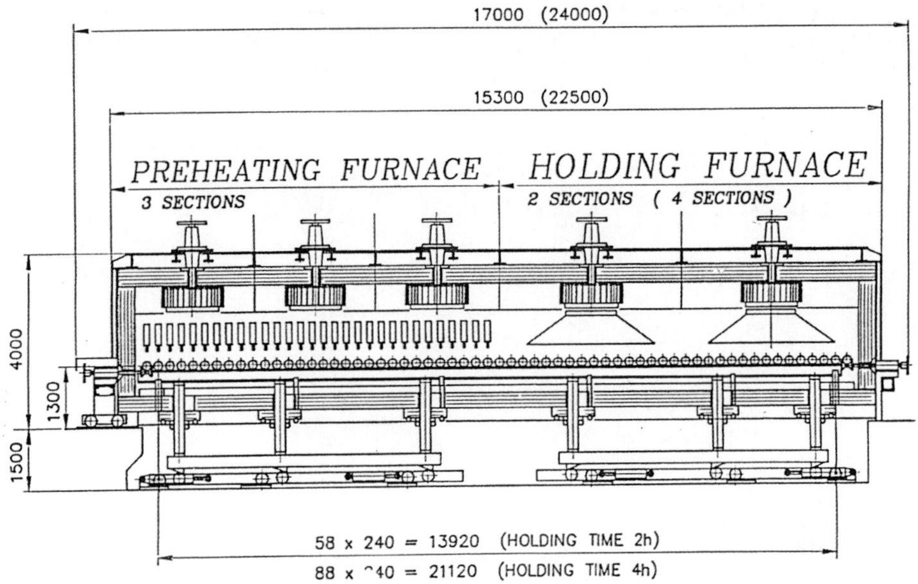

Fig. 6.19: Preheating and holding furnace (Source: LOI)

The holding furnace section may hold up to 60 billets in order to ensure the required holding time. As the billets are passing through the holding zone they are heated from above by hot air flowing down through the billets. The air flow is directed in such way that it is not disturbed by the walking beam conveyor or the furnace base construction. It is essential to maintain a temperature difference of not more than ± 2.5 °C.

The homogenized billets exit through a hydraulically operated side door to the air cooling section followed by the water cooling section (**Fig. 6.20**). Here the billets rotate and are exposed to a water spray until they reach ambient temperature. A cutting saw can be installed after the homogenizing plant (**Fig. 6.21**).

The complete cycle including transport, preheating, holding and cooling as well as the

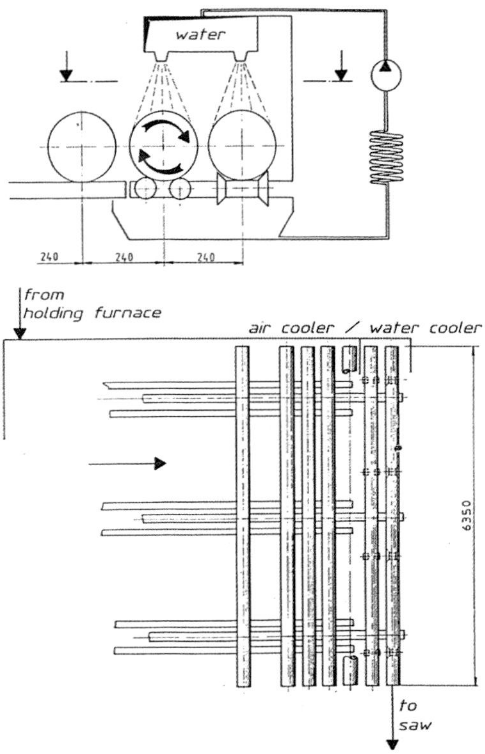

Fig. 6.20: Cooling section (Source: LOI)

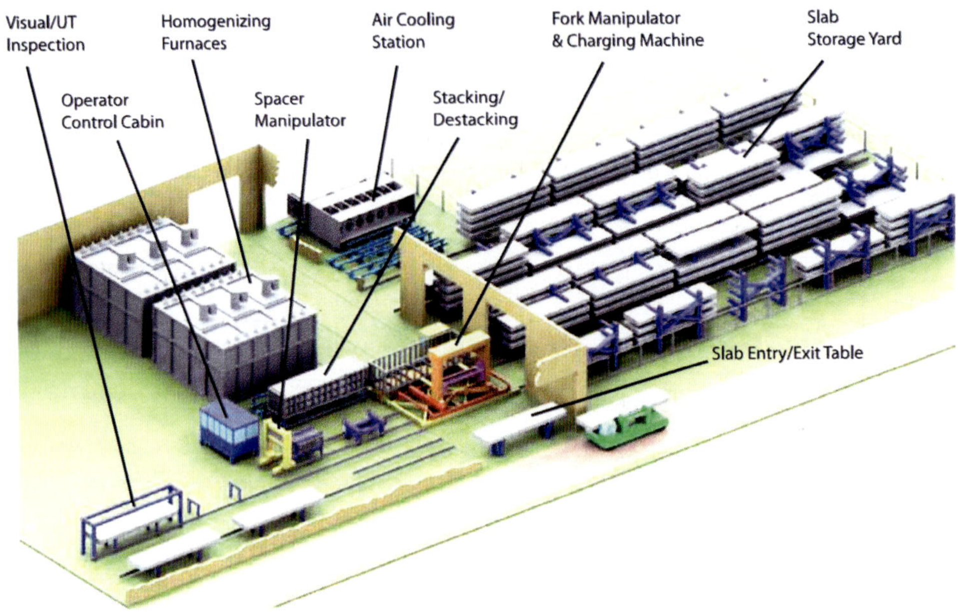

Fig. 6.21: Layout of a continuous homogenizing plant (Source: Hertwich)

optional cutting unit are controlled by control loops and inter-lockings. The entire process is directed by an overall process management system.

7. Environmental control

Christoph Schmitz

Efforts to reduce the impact of the "carbon footprint" have resulted in design improvements in equipment design and operation practises. Focus is also directed to the so-called "diffuse emissions" of a melting plant which means that the fumes occurring at all stages of the production process must be collected or at least to their absolute minimum.

A production plant and also the aluminium recycling plant cannot be regarded as stand-alone system. It is interacting with the surroundings and its effluents do have distinct impact on the environment. It is impossible for any production facility to operate without such interaction but influence must be kept as low as possible. Toxic or aggressive effluents must be neutralized within the plant. Where this is not possible, safe collection and transport to specialized treatment facilities or safe deposit must be assured. Most critical effluents of an aluminium recycling plant are

- production waste such as slag or mechanical contaminations of the scrap which have been separated during the production process,
- gaseous emissions,
- filter dust,
- contaminated water,
- used oil from vehicles and equipment,
- noise.

The plant management will be responsible for taking care of the emissions and to minimize the quantity. The performance within the limits set by the environmental regulations are monitored and supervised by the authorities. The public is very conscious of what type and which quantity of harmful effluents are emitted by a plant. The generation of such effluents should be reduced to the absolute minimum in the processes applied. The quantity finally released from the processes must be treated and neutralized to the greatest possible extent, although this is costly. More and more stringent regulations will be issued as time goes by. For the construction of a new plant it should be a rule that the current regulations should not only be fulfilled but the target for processes and waste treatment should be to arrive at even better performance than required. This should be a rule also in countries which do not – yet – have issued stringent regulations for the protection of the environment. We all should have in mind that it is the duty of any industry today to conserve the environment for the benefit of future generations.

7.1 Design criteria

The most critical effluents are the gaseous emissions which require thorough treatment and handling within the plant. Some of the equipment and processes are customized for the secondary aluminium plant. They result not only in the release of clean gases to the ambient air but also some filter dust requiring safe deposit or treatment in special plants will be the result of these efforts. We will look into the technologies in more detail. Production slag also occurs in large quantities. Its generation must be reduced as far as possible but special treatment is required to recycle salt and aluminium and arrive at a sellable but at least harmless residues. Treatment of this effluent has to be done in specialized plants.

Water is used for casting machines. The treatment of the contaminated water is no challenge and standard methods are applied. This also refers to the small quantity of used oil from vehicles and equipment. It can be burned in the plant.

7.1.1 Gaseous emissions

During the operation of aluminium recycling furnaces gaseous emissions will occur which do have different compositions of contents depending on the type of furnace. In state-of-the-art plants these waste gases are cleaned, i. e the harmful chemical components and dust are removed from the waste gases based on the standards required by the regulations for environmental protection. Furthermore, the generated combustion products, such as CO_2 or NO_x, cannot be removed but their volume has to be reduced to the lowest possible level by optimal combustion conditions and good energy efficiency during their generation.

Depending on the type and application of the furnace, waste gases of different composition will occur. Some problems in operating a production plant for aluminium recycling could arise from diffuse emissions. The waste gases of the actual process equipment are collected more or less efficiently. But there are areas which generate emissions to the operating area. One major source is dust escaping from furnaces and handling equipment. This can be noted by a dust layer in the plant area which may cause hazards. Also smoke and fumes escaping due to imperfect sealing or collection contribute also to these emissions. A very large area is the treatment of the melt with chemicals. This is done while the furnace door is open and the dust collection hood may not be able to catch all the fumes flowing out of the furnace. External heating systems for launders and casting equipment add their share to the diffuse emissions as well. Dust will escape when filling the charging equipment and during charging of the furnaces.

The simplest answer to the problem of the diffuse emissions is to provide a casing for all critical areas in order to be able to catch all products escaping into the operation building otherwise. But is this really a solution? All external casings create obstructions within the production facilities. The better solution appears to be to have a proper design to catch all emissions at the very point of their generation. This again is a challenge to the designers to care more about the environmental control of their equipment.

7.1.1.1 Combustion products

Combustion products, consisting of carbon oxides, water vapor and oxides of sulfur and nitrogen, are generated during any combustion. Water vapor is not harmful to the environment. Carbon in the form of CO_2 is not toxic but its impact on the atmosphere by increasing the green house effect is common knowledge. Its quantity can only be reduced by energy savings. Sulfuric oxides depend on the type of fuel used for the firing of the furnaces. It is mainly generated by using oil as fuel, particularly heavy fuel oil. Part of the SO_x generated is absorbed and neutralized by the additives used in the waste gas scrubbing system and will finally go to the filter deposit. The neutralization of NO_x is a sophisticated process which is not applied in the aluminium recycling industry. But the limits set by the environmental regulations are low and it must be expected that these values will be lowered furthermore in the future. The generation of NO_x must be controlled during the combustion process. Formation of NO_x requires a high energy of activation and occurs mainly at high combustion temperatures. These occur mainly by excessive pre-heating of the combustion air but also when using oxy-fuel. In this case actually no generation of NO_x should happen since the combustion takes place in the absence of nitrogen. In fact, at operating plants the expected decrease of NO_x generation could not be noted when operating oxy-fuel burners. The reason may the unavoidable ingress of air into the furnace. Here the furnace designers are challenged to improve the sealing of the furnace and the adjacent systems even more to avoid contact of combustion products with ambient air. Limitations are set, however, when a substantial amount of air is introduced during charging and with the material charged. NO_x generation can also be reduced by having a reasonable pre-heating of the combustion air in recuperators and certainly also in regenerators.

Another source of combustion products is the pre-heating of launders and casting equipment. The gases are not collected to be treated in the waste gas system and contribute to the diffuse emissions.

7.1.1.2 Products of evaporation

When contaminated scrap is charged to the furnace the temperature is usually quite low and the organic components smoulder. The products removed by this process step contain toxic components and, therefore, it is essential that all smoke generated during this pyrolytic process is captured and passed to the waste gas scrubbing system. The captured products are treated in the scrubbing system. After charging, the treatment is much more efficient either by the internal post-combustion in the furnace or by an incinerator system.

7.1.1.3 Carbonization products

The target of any thermprocessing is to arrive at complete combustion. This cannot always be realized. Under normal conditions the combustion system operates efficiently and with some excess air. Thus, complete combustion is assured. But particularly when charging the furnace with scrap of not so well-known content of organics or in case of maintenance problems or lack of data or sufficient knowledge of the operating personnel, combustion may be incomplete. This will result in the formation of unburned components such as CO or free carbon. To reduce these problems the combustion conditions must be monitored carefully by better training of personnel and more sophisticated process control. In state-of-the-art plants this is assured already. Still there might be some uncertainties which are compensated for by post-combustion in the furnace or by the process equipment. Final contaminations are neutralized by the additives of the scrubbing system.

7.1.1.4 PCDD/F

Polychlorinated dibenzodioxins and polychlorinated dibenzofurans, in short PCDD/PCDF are present everywhere in the surroundings. The chemical structure is depicted in **Fig. 7.1**. In processes and products these chemicals appear as unwanted by-products. These are processes for the production of paint, herbicides, in the paper industry and the recycling of used oil. They are also generated during the incineration of household waste and in the metal recycling industry during combustion of organic substances in the presence of chlorine. There are 75 kinds of PCCD and 135 types of PCDF. They appear as colorless needle shape material and can be attached to the dust or turn up in the flue gas as particles. Studies have shown that only very small quantities are present as vapor or gas.

PCDD and PCDF always appear as mixture. The individual modifications are of different toxicity. Amongst them is the very toxic 2,3,7,8-tetrachloro dibenzodioxin, commonly known as the Soweso poison which is a very hazardous substance and may cause different health problems such

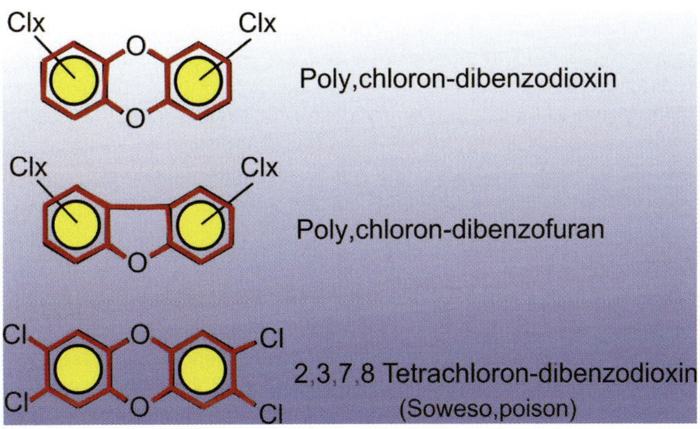

Fig. 7.1: Structure of PCDD and PCDF

as irritation to skin and eyes, headache, weakness and vomiting, damage to liver and pancreas as well as to the nerve system. It is also very carcinogenic.

In the aluminium recycling processes dioxins and furans are mainly generated in the rotary drum furnace process where plastics and other organics, including fuel, are burned with salt. Unless there is salt treatment, only very little PCDD and PCDF need to be expected in holding/casting furnaces or in melting furnaces for clean scrap.

Dioxins and furans are destroyed at temperatures exceeding 900 °C. This will always be in incinerators and mostly in the flue gas of the furnaces. But they will appear again when the waste gas cools slowly and thus we will find them finally in the clean gas leaving the stack. The re-formation of PCDD/F, called de-novo synthesis, can be suppressed if the waste gas is cooled very rapidly. This can be achieved in the quenching chamber or by timely and fast addition of cold air. Rapid cooling occurs also in the heat exchanger bed of a regenerator. However, results published in this connection appear somewhat misleading. This type of equipment can be used only in a fairly dust-free operation. This means for melting of more or less clean material there may be contamination with oil. Therefore, the PCDD/F generation is also quite low. The particles attached to the little dust quantities are held back in the regenerator bed and are, consequently, not measured. For cleaning of the regenerators all dust is removed either by washing with water that goes to the public waste disposal or dust removal that goes somewhere as well.

7.1.2 Clean gas requirements

The environmental regulations, such as the German TA-Luft, define the permissible emissions for the different components of clean gas. These are measured generally as daily averages. The technologies applied for process and waste gas cleaning are in general below these values.

The relevant values are listed in **Table 7.1**.

Table 7.1: Waste gas emission achieved and required by the environmental standards

Emission	Values	
	TA-Luft [mg/m³n]	Obtained [mg/m³n]
dust	20	5 - 20
fluoride as HF	5	< 1 - 5
chloride as HCl	30	< 5 - 30
tot. C	50	< 10 - 50
CO		< 100 - 50
PCDD/PCDF	0.1 ng/m³n	0.1 ng/m³n

7.1.3 Water

Water is used as cooling medium for casting machines. This is done as spray-cooling or in a closed water circuit. For some waste gas cleaning systems, such as the quenching chamber, water is also injected into the waste gas. The quantity of water used for spray-cooling of metal is evaporated to the atmosphere and is to be regarded as lost. The same applies for the water for waste gas injection. A closed water circuit returns the used water to the process. Only the losses unavoidable, due to normal evaporation or in the cooling tower or in the equipment, must be replaced. The cooling water basin of a closed loop system will collect some contaminations due to abrasion in the equipment and as precipitation products originating from water-conditioning. These contaminations collect in the basin and have to be discarded form time to time. Since they are not toxic or otherwise harmful to the environment they are usually passed on to the public wastewater treatment plants.

7.1.4 Noise

Different noise sources are noticed in an aluminium recycling plant. The noise generated by the operation of equipment such as machinery, pumps and other components is very well within the required limits. Blowers are equipped with silencers to keep their noise level low. Burners also generate some low frequency noise but since they are operated in the closed furnace the level is low, even with units having a capacity. High momentum burners develop a higher noise level but they, too, operate in a closed furnace. In total, these noise sources are mostly found within a closed building and generally no special noise abatement measures are necessary, even at nighttime. A very high noise level occurs during charging of the furnaces. Metal is dropped into buckets and charging machines with the noise level exceeding the permissible level of 50 dB (A) at night and 60 db (A) at daytime. These emissions occur over a short time and may be accepted within the plant battery limits. At nighttime some special measures may be required such as special wall panels or enclosures. Some noise emission can be reduced by good organization of the work in the plant area. The loading and handling activities in the open plant area can, for instance, be restricted to daytime.

7.1.5 Production waste

Major production waste is slag generated in the rotary drum furnace. This material is collected and supplied to special treatment plants. If such plants are not available, the slag must go to an intermediate storage area and from there to landfill as agreed with the public authorities. In the long run, slag treatment plants should be installed to recover the salt as well as the aluminium still trapped in the slag. The final waste is usually neutralized during the recovery process and could be used for road construction or the refractory industry.

Filter dust is also collected and must be supplied to specific deposits as defined by the public authorities. Efficient treatment processes have not yet reached the production scale status but may be available in the near future.

Auxiliary material, such as packing material and floor dust, sweepings and other refuse, may be supplied to public deposits.

7.2 Fume collection and health precautions

The operating personnel is exposed to some extent to gases, heat and dust emitted by the equipment, particularly by furnaces during metal treatment. The state-of-the-art design must ensure that such emissions are reduced to the absolute minimum. Where necessary, adequate enclosures must be provided or safety clothing must be worn during specific operations. The plant management is obligated to maintain clean and dust-free surroundings by good housekeeping. All diffuse emissions, that cannot be collected or treated by the existing equipment, must be kept to their absolute minimum. With increasing efficiency of the waste gas treatment systems, the share of these diffuse emissions in the total emissions has increased and, therefore, more attention must be directed to this factor.

7.3 Waste gas treatment

The waste gas system in a recycling plant comprises the gas collection system and the waste gas treatment. Injection of additives to provide absorbents for the neutralization of acidic components as well as trace contents of the waste gas is a very important stage in the system. The final stage of the waste gas treatment is the filtration plant. The different technologies applied by industry are shown in **Fig. 7.2**. Only some of them are used in aluminium recycling plants. These are marked in the graphic. The following will refer only to these technologies.

7.3.1 Casting furnaces

Metal without contaminations attached will be charged and processed by these furnaces. The resulting waste gases comprise combustion products of the fuel used, i. e. CO_2, H_2O, N_2, NO_x and, from case to case, SO_2. During not optimal combustion conditions, incompletely burned products, such as CO, C and other forms of carbon hydrates, may be present. The content of SO_x is low since even when firing light fuel oil the sulfur content will be < 1 %.

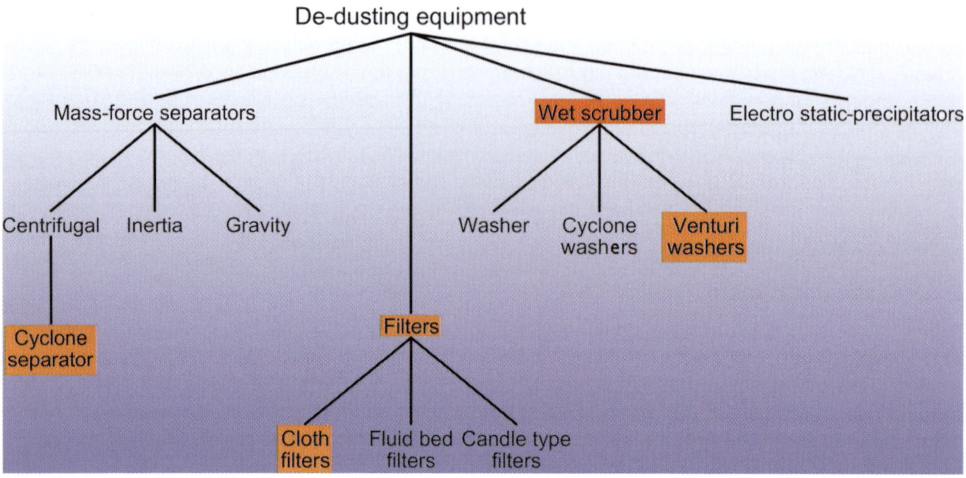

Fig. 7.2: Technologies for waste gas cleaning

Waste gases exiting these furnaces do not require cleaning since their content of harmful components is far below the values permitted by the environmental regulations, i. e. the German TA-Luft. Additionally, the dust-collecting hoods arranged above the furnace door reduce the percentage of the harmful gases. It is even recommended to release these waste gases directly to the ambient air instead of mixing them with waste gases from also present melting furnaces. This would thin out the waste gases coming from these furnaces, thus reducing the efficiency of the waste gas scrubbing system and increase its size.

In many cases, however, liquid metal treatment with chlorine-containing gas mixtures or salts will be done in these furnaces. The waste gas will then contain small quantities of hydro-chlorine or reactive free chlorine and also some quantities of HF. In spite of the small quantities, waste gas treatment is recommended. It is pointed out, however, that waste gas treatment in state-of-the-art furnaces is done by using argon or an in-line filtration ahead of the casting equipment with external equipment. Here again no waste gas treatment is necessary.

As consequence, it can be stated that furnaces processing clean metal only do not require waste gas treatment. This is generally the case in the downstream industry and in primary aluminium smelters.

7.3.2 Melting furnaces

In these plants aluminium is melted by using different kinds of scrap. Due to the contaminations and compound materials, mostly plastics, paint and oil, the waste gases carry, besides the combustion products, fumes of PCDE/PCDD (furans and dioxins). Additionally, the waste gas contains, due to the application of fluxes, HF in small quantities and HCl as well as picked up salts, such as NaCl, KCl and CaF_2. There is a certain amount of so-called "spray aluminium" which partly oxi-

Table 7.2: Contents of contaminations of raw gas and clean gas after passing the scrubbing plant

	Raw gas	Clean gas
Dust	50000	5 mg/m^{3}n
HCl	300	30 mg/m^{3}n
Cl_2	10	3 mg/m^{3}n
HF	60	5 mg/m^{3}n
NO_X	500	100 mg/m^{3}n
SO_2	250	200 mg/m^{3}n
PCDD/F	11 to 40	1 pg 1-TE/m^3

dizes to Al_2O_3. When processing dross the waste gas carries a large quantity of alumina as dust. **Table 7.2** summarizes the contents of the flue gas and its quantity as specific value per m^3 of combustion products.

7.3.3 Waste gas system

All gases and dust-loaded air are directed to the waste gas scrubbing system (**Fig. 7.3**). This comprises a dust filter, preferably a cloth filter. Wet filters are mainly used in shredder plants because the air exhausted from this equipment contains water and sometimes a large amount of oil. Cloth filters would clog immediately.

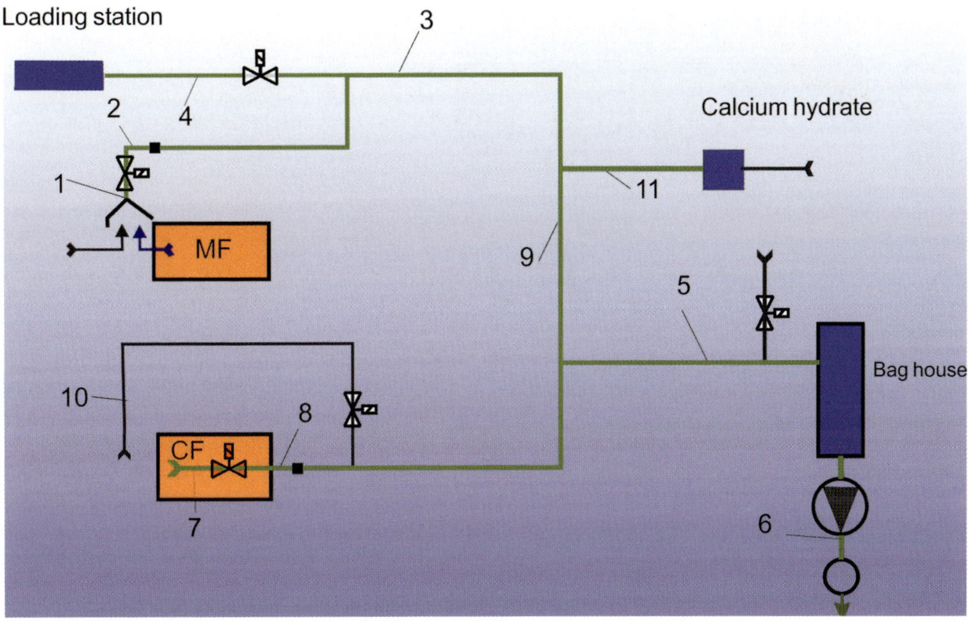

Fig. 7.3: Principle arrangement of a waste gas cleaning system

Dry filters (**Fig. 7.4**) are the most common technology in state-of-the-art aluminium processing plants. They are designed as tube or pocket filter. For neutralizing the acidic components, calcium hydrate is added to the flow of waste gases by means of a special reactor. Together with the dust, the reacted lime hydrate is passing the filter to be cleaned. Filter dust stripped from the cloth filter collects at the bottom of the filter housing and is then discharged by means of a screw conveyor and finally filled into big bags.

The conditions for a filtration system for a single furnace are easy to define. Only different stages of the furnace cycle need to be studied. If more furnaces are connected to the scrubbing system, many variations of waste gas quantities, temperatures and gas condition require attention. For design and operation of the plant, extreme conditions can be only evaluated. Once covering these conditions all other stages can be handled. The concept shall be explained by an example

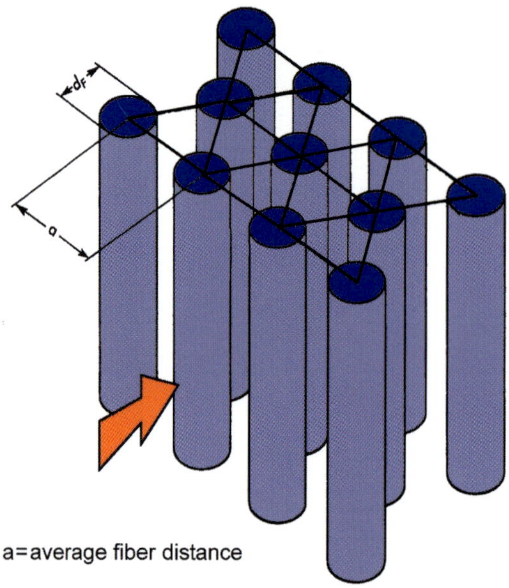

Fig. 7.4: Gas flow through the fibers of filter cloth

describing a plant consisting of a tiltable rotary drum furnace and two reverberatory furnaces used as casting furnaces. The extreme values possible during operation are listed in **Table 7.3**. The conditions define extreme values. All furnaces may be in operation at full power of all burners with an exhaust temperature of 1,173 K. A condition, where two furnaces are in operation at low fire and a combustion temperature of 973 K is also possible. This would mean that one of the furnaces is being loaded or tended while the other furnace is in holding position. The highest dust load must be expected when a furnace is at full power, even at a lower flue gas temperature. This will be as soon as a furnace is loaded with fresh material and starts the heating and melting sequence. Particularly in the melting furnace this will happen at the beginning of the furnace cycle when the batch is still cold with the resulting good heat transfer from the combustion products to walls and furnace batch. The filter designer must consider these extremes when sizing the unit. Further

Table 7.3: Operation condition of a complete plant considering the sequences of different furnaces

	Intake [%]						
Melting furnace 1)	100	100	100	0	20	20	0
Loading 2)	0	0	100	0	0	0	0
Charging 2)	0	0	0	100	0	0	100
Casting furnace 1 1)	100	0	100	100	20	20	20
Charging/tending 2)	0	100	0	0	0	0	0
Casting furnace 2 1)	100	100	100	100	20	0	20
Charging/tending 2)	0	0	0	0	0	100	0

Notes: 1) refers to the full burner capacity, i. e. full waste gas quantity
2) refers to the de-dusting air requirements

Table 7.4: Fabrics and their main characteristics used for cloth filters

Band name	Polyacrylnitril Dolanit	Polyester Trevir, Diolen	Polyamid Nomex	Polyphenylen Ryon
acid resistance	good	moderate	moderate	good
caustic resistance	moderate	moderate	moderate	good
temperature resistance [°C]	120	150	220	200
thickness [mm]	1.8 - 2.8	1.1 - 6.0	1.5 - 3.1	1.1 - 2.1
weight per m^2 [g]	350 - 650	250 - 960	300 - 650	380 - 650

aspects requiring attention are the dew point of the flue gases under these extreme conditions and the service temperature of the filter cloth which is dependent on the type of material selected.

The temperature resistance of the different filter material is limited (**Table 7.4**). A heat exchanger is required if the maximum temperature level of the raw gas is higher than the temperature resistance of the filter material. Various heat exchanger designs are available. Due to the sticky nature of the dust from aluminium melting furnaces an automatic cleaning device for the cooling surfaces is mandatory. The cooling air temperature at the heat exchanger inlet shall reach a level of about 40 °C all year. If shock cooling is required, as for instance in case of suppressing re-formation of dioxins, a quenching chamber (**Fig. 7.5**) may be the best solution. In such a unit the flue gases are passing through a chamber where water is injected by means of spray nozzles. The heat of evaporation of the water is used to cool the gas stream. Since the temperature remains safely above 100 °C water will be present as steam.

The standard cooling system comprises a heat exchanger system whereby ambient air passes through a series of tube bundles or rectangular channels while the hot gases flow along the outside (**Fig. 7.6**). For better efficiency, particularly if a larger temperature difference is required, the cooling air is recycled in different stages through the system (**Fig. 7.7**).

Heat exchangers ahead of the filter plant are generally costly inh regard to investment and operation. It appears to be more convenient to use the de-dusting air from the dust collection hoods (Fig. 7.3), although the throughput of the bag house increases, requiring larger filter area. Intakes of de-dusting and smoke collection are located above every furnace door. These air intakes are contaminated to a certain extent and must pass through the filter anyway. The hoods are usually

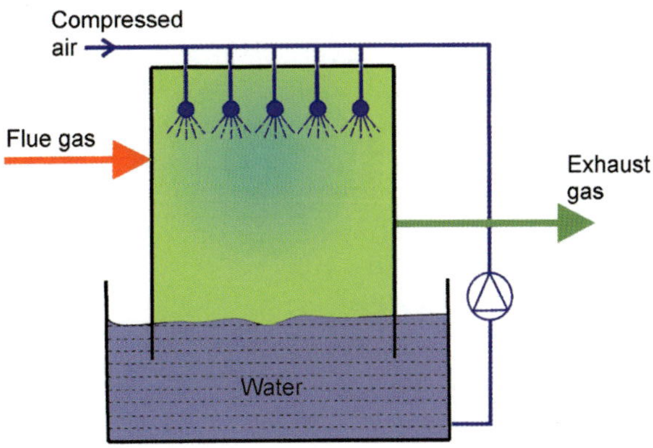

Fig. 7.5: Quenching chamber

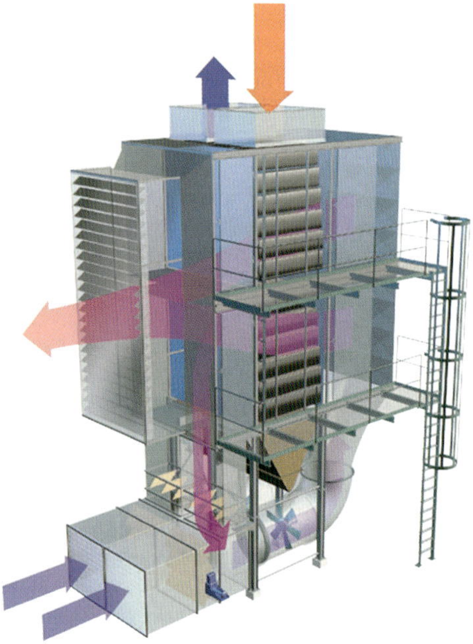

Fig. 7.6: Waste gas cooler with air circulation (Source: Nedermann)

in operation once required, i. e every time the doors are opened for charging or furnace tending. They are also in operation to remove diffuse emissions from the plant building. Such a situation may not match entirely the requirements for cooling. It is possible, however, to use one or more of the de-dusting intakes for adding cooling air to the system although they are not required for their intended purpose. Using the de-dusting air intakes for a lengthy period of time will remove air from the production building. This may have a very nice effect in summertime or in a hot climate. During wintertime or in a cold climate it will chill the working surroundings. An air intake from outside of the building would be much more convenient.

Every intake, be it gas or air, must be equipped with a motor-operated valve for adjusting pressure and flow rate. This is of particular interest for the de-dusting points. The valves are controlled

Fig. 7.7: Single flow waste gas cooler (Source: Nedermann)

by the overall pressure conditions of the system. For effective dust pick-up the air must reach a velocity that is higher than the down float velocity of the individual particles or the upward draft when carried by the hot gases. This is necessary to reduce settling of dust in the dedusting pipes. Taking this into account, the air velocity should be in the range of 20 to 25 m/s at the intake duct connected to the collecting hood. There are two conditions to be fulfilled at that point. Firstly, the pressure difference between the outside and the duct must be sufficiently high to accelerate the air to its velocity and, secondly, the required air volume must be assured. In systems having multiple entry points, most of the air will be sucked in at the point of the lowest pressure drop on the way to the bag house. This requires that the different ratios have to be adjusted with the aid of valves. However, the conditions change permanently. Thus, manual setting to one specific set-point is insufficient. The system requires a kind of "slave control" to follow the fast changes. Usually, the valves at the point in operation open fully and the other valves at intake points close one after the other – or simultaneously – with specific settings as required to obtain the necessary pressure at the operating point.

The pressure required to obtain the necessary pressure to accelerate the air to the velocity required can be calculated with the Bernoulli equation

$$\Delta p = \zeta \cdot \frac{1}{2} w^2 \cdot \rho \qquad (7.1)$$

Δpp is the pressure difference required in kg/m², w the gas velocity in m/s, ρis the density in kg/m³ and ζ the pressure loss factor. This factor depends on the size ratio of the outside duct. If the air intake is from the open surroundings, ζ will be 0.43.

Example:

The suction air volume at a fume-collecting hood above the door of a casting furnace shall be 20,000 m_n^3/h and the temperature 100 °C equal to 373 K. The density r for the temperature of 100 °C is 1.3 · 273/373 = 0.97 kg/m³. The air velocity shall be 25 m/s. Which pressure is required in the duct and what diameter is required? (see also below)

The actual air volume is

$$V_a = 20000 \cdot \frac{373}{273} 27326\ m_o^3/h = 7.59\ m_o^3/s$$

This results in a duct diameter of

$$d = \sqrt{\frac{4 \cdot 27326}{\pi \cdot 25}} = 0.622\ m$$

The required pressure difference is

$$\Delta p = 0.43 \cdot \frac{1}{2} 25^2 \cdot 0.97 = 130\ kg/m^2 \text{ eqal to } 1300\ Pa$$

The calculated pressure differences must be summarized for all intake points to arrive at the total pressure relevant across the system and must include all line losses, too.

For mixing the hot gases with the cold air, the duct section is designed as mixing chamber with tangentially entering of one of the two components. Thorough mixing, to obtain a temperature which is preferably below 200 °C of all gas components, avoids the de-novo-formation of dioxin.

To have this effect, mixing should take place close to the furnace with very little heat loss to have a thermo-shock of rapid cooling of the gases.

The hot gas section of the piping must be lined with refractories. This is not necessary for the transport of gas mixed with air since its temperature is sufficiently low.

During charging, dust will escape if material like dross is handled mechanically. For other material the generation of dust is not critical. But due to the natural draft of the air passing through the furnace, due to the temperature difference between furnace and surroundings, some dust from the furnace, present due to the oxidation, will escape as well. This is also the case during furnace tending which means stirring and skimming; dust and gases released will leave the furnace. Treatment with chemicals requires open doors as well. This will result in a large volume of gaseous emissions. Finally, some of the combustibles charged into the furnace will fume-off already and leave the furnace together with dusty emission released from paint. Part of these emissions leave the furnace through the door and will be collected by a hood arranged above. Contrary to the hot waste gases, the temperature of the fumes collected by the hood is comparatively low and will not exceed 200 °C as long as no combustibles start burning. This is actually desired, however, to reduce the content of unreacted organics. For a short period of time, the gases collected by the hood carry all contents of the flue gases from the operating furnace.

To calculate the cooling air quantity, the equations developed in section 3.1 are helpful. Temperature and volume of the flue gas change permanently on their way to the filter. Only the weight of the different gas fractions remains unchanged.

$$G = V_0 \cdot \rho_0 = V_a \cdot \rho_a \tag{7.2}$$

According to the first law of thermodynamics, also the energy of the system remains. Thus, we are able to define

$$H_{AG} = H_A + H_G \tag{7.3}$$

with H_A enthalpy of the air introduced and H_G the enthalpy of the waste gas and H_{AG} the enthalpy of the gas/air mix. The enthalpy of the components is

$$H_A = c_{pA} \cdot G_A \cdot T_A \tag{7.4}$$

for the air and

$$H_G = c_{pG} \cdot G_G \cdot T_G \tag{7.5}$$

for the waste gas. For easier reference we replace the weight by the volume at mean condition

$$G = V_o \cdot \rho_0 \tag{7.6}$$

and with this we can express equation (7.3) as

$$c_{pAG} \cdot V_{0AG} \cdot \rho_{0AG} \cdot T_{AG} = c_{pA} \cdot V_{0A} \cdot \rho_{0A} \cdot T_A + c_{pG} \cdot V_{0G} \cdot \rho_{0G} \cdot T_G \tag{7.7}$$

The index 0 refers to the mean conditions (T = 273 K, P = 1,022 hPa). If we now look at the values for ρ_0 of gas and air as well as at the value of c_p for both, we notice that they are very close

ρ_{0A} = 1.30 kg/m^3_n

ρ_{0G} = 1.34 kg/m^3_n

c_{pA} = 1.32 kJ/m^3 K

c_{pG} = 1.38 kJ/m^3 K

(Please note that we have used the common expression m_n^3 for the volume at mean conditions). Since the air – gas mixture will also have the values close to them for air and gas, we can simplify the equation to

$$(V_{0A} + V_{0G}) \cdot T_{GA} = V_{0A} \cdot T_A + V_{0G} \cdot T_G \tag{6.8}$$

With this equation we are in position now to calculate air quantities to be added and the temperature of the air/gas mix.

Example:

A furnace has burners with a total rating of 3,000 kW. The volume of natural gas required will be 3,000/10 = 300 m^3/h. With an excess air ratio of 5 % this results in a volume of the combustion products of 300 · 12 · 1.05 = 5,780 m_n^3/h. The temperature of the flue gas will be 900 °C equal to 1,178 K.

Now, what is the temperature of the air/gas mix if 2,000 m_n^3/h of air is added at a temperature of 20 °C ‰ 293 K?

With equation (7.3) we obtain

$$T_{AG} = \frac{3780 \cdot 1173 + 2000 \cdot 293}{3780 + 2000} = 869\,\text{K or } 596\,°\text{C}$$

Most of the filter cloth material can be used up to 200 °C. Thus, the gas temperature is far too high. Consequently, we must increase the air volume to arrive at an acceptable temperature. The new value shall be 15,000 m_n^3. To be realistic, the volume of the 15,000 m_n^3 for dedusting of a furnace is too low anyhow; we would need at least 20,000 m_n^3. But for our example we calculate with the lower value again:

$$T_{AG} = \frac{3780 \cdot 1173 + 15000 \cdot 293}{3780 + 15000} = 470\,\text{K or } 197\,°\text{C}$$

We notice that the total volume going to the filter is much higher than the volume of the combustion products. Cooling of flue gases by means of a separate heat exchanger would definitely result in a higher cooling air requirement since the efficiency of the heat exchanger requires consideration. As mentioned above, cooling by means of a quenching chamber is also possible but needs a different approach due to the cooling water requirement. On the other hand, a smaller gas quantity would have a higher concentration of dust and other contaminations per m^3 of gas. The permissible filter load (we will come back to this later on) must then be lower for proper functioning of the cleaning system which in turn increases its size. Thus, we may not fully benefit from the lower flow rate.

For the sizing of the components of the treatment system we need the actual flow rates of the gas.

From section 3.1 we know the general gas equation

$$\frac{p_0 \cdot V_0}{T_0} = \frac{p_a \cdot V_a}{T_a} \tag{7.9}$$

Although temperature and volume will change in our waste gas system, that means actual values and not values related to mean conditions for calculations, the system pressure does not change so that we can assume isobaric conditions. Therefore, we have

$$V_a = V_0 \frac{T_a}{T_0} \tag{6.10}$$

And thus we obtain the actual waste gas volume for the example to be

$$V_{aG} = 3780 \frac{1173}{273} = 16242 \text{ m}^3/\text{h}$$

The actual air volume is slightly lower compared to that of standard conditions:

$$V_{aA} = 15000 \cdot \frac{293}{273} = 16100 \text{ m}^3/\text{h}$$

The actual gas volume arriving at the filter plant will have a temperature of 470 K

$$V_{aAG} = (15000 + 3780) \cdot \frac{470}{273} = 32382 \text{ m}^3/\text{h}$$

This is now the design capacity of the filter.

Most plants have a minimum of two furnaces. One is a melting furnace and the other one a casting furnace. We have to incorporate this setup into our calculation. Coming back to our example, we define that the second furnace willl be a high performance melting furnace with a burner rating of 6,000 kW. The relating waste gas quantity will be twice as much as for the other furnace amounting to 3,780 · 2 = 7,560 m_n^3/h. All temperatures will be the same. We definitely require much more cooling air now and define this quantity to be 50,000 m_n^3/h. Now we calculate again:

$$T_{AG} = \frac{3780 \cdot 1173 + 7560 \cdot 1173 + 50000 \cdot 293}{3780 + 7560 + 50000} = 456 \text{ K equal to } 183 \text{ °C.}$$

Our estimate was good and we arrive at a filter size of

$$V_{aAG} = (3780 + 7560 + 50000) \cdot \frac{456}{273} = 103458 \text{ m}^3/\text{h}$$

If the plant is equipped with more furnaces, the equation becomes complex and the variations of the parameters to be evaluated according to Table 7.3. will likely become a time-consuming task. In order to consider all possibilities of the plant operation, please feel free to expand the table. For evaluating all or at least the extreme possibilities, an excel spreadsheet is very helpful. **Table 7.5** shows an example. Burner input and excess air are variables which can be adjusted as the present case requires, also the input of cold air. The effect on the temperature at the filter intake is then shown immediately. Under the provision that the burner rates are set at this instance, the air quantity can be put in as required. The numbers in column "location" refer to the schematic Fig. 7.3.

The collected waste and dedusting gas is now sent to the filtration plant. The ducts can be designed as steel tubes installed in the steel structure of the building. Experience teaches that

Table 7.5: Gas flows due to burner input, cooling air requirements and total air flow. This spreadsheet is very helpful for the analysis of different operation conditions

Location	Temperature (°C)	Gas quantity [m^3/h_{actual}]	[m^3/h_n]	Air quantity [m^3/h_{actual}]	[m^3/h_n]
1	900	28964	6741		
4	30			44396	40000
3	155	73360	46741		
11	30			1110	1000
9	153	74470	47741		
5	187	125631	74482		
7	900	28964	6741		
10	30			22198	20000
8	249	51162	26741		

a flow rate in the ducts should not be below 12 to 15 m/s in order to avoid settling of dust. To allow heat expansion and to compensate for inaccuracies in construction, expansion joints are integrated in the ducting and the duct supports are designed to permit movement due to heat expansion. The preferred cross-section is a round shape but rectangular ducts are also quite common. A steady flow of gases should be ensured over the entire length of the ducting. At junction points the cross-section must be widened to accommodate the additional gas or air volume. The changeover to the new diameter is designed as tapered connection tube.

When flaps or gates are integrated, easy ladders and platforms must permit access to these elements for maintenance. At regular distances doors for frequent dust removal are necessary. This is very important on elbows since dust collects very rapidly at points with heavy turbulences combined with dead corners.

Sometimes the ducting to the filter plant is designed as underground smoke flue constructed out of refractory material. The advantage is that no upward ducts obstruct the plant area for cranes or other equipment. But investment cost is higher than for steel ducting. Large access tunnels on both sides of the flue lined with refractories permit easy access. The tunnels are aerated to remove the heat escaping from the refractory walls of the duct. To enable regular cleaning, doors permit access to the flue. Due to the construction requirements, the flue duct will have a substantial size. Thus, the construction of underground ducting is generally limited to plants with large quantities of waste gas.

7.3.4 Additives

Dust is removed from the waste gases by the filter cloth which is not able to remove gaseous components. Therefore, absorbents are injected in order to neutralize the acidic components by chemical reaction and absorption. Calcium hydrate (Ca $(OH)_2$) is the most common additive but also calcium carbonate ($CaCO_3$) is used in some cases. The reactions for the different acidic components are

$$2HCl + Ca(OH)_2 \rightarrow CaCl_2 + 2H_2O$$

$$2HF + Ca(OH)_2 \rightarrow CaF_2 + 2H_2O$$

$$SO_2 + Ca(OH)_2 \rightarrow CaSO_3 + H_2O$$

Free chlorine reacts via the intermediate stage of formation of hydro chlorine. Calcium hydrate $Ca(OH)_2$ is a proven absorbent for the acidic components of the flue gas. Carbon carbonate $CaCO_3$ is also used but its main application is in wet filtration systems where sometimes the even more reactive NaOH is used.

Compared to wet scrubbers, dry scrubbing systems have the advantage of better absorption of components in the waste gas which are present as traces only but consist mainly of very poisonous chemical substance such as PCDD/PCDF (dioxin) and in some cases even heavy metals such as Hg. Removal of PCDD is very important since traces of this extremely poisonous and carcinogenic substance are harmful to human beings. Calcium hydrate and calcium carbonate are not sufficient to the remove the toxic organic and inorganic contents of the waste gas. From this follows that one more sorbent is required

Brown coal (lignite) dust and active carbon have shown very good properties for this application. They can be added separately or mixed with the other sorbents. Carbon additives have proven well in many scrubbing plants for garbage incineration and also in aluminium recycling plants. There are two potential risks when using carbon dust. There is first the danger of glowing combustion and, secondly, the potential of shock combustion. The danger of glowing combustion cannot be avoided completely if the carbon content of a mix exceeds 3 %. But by careful handling, this hazard can be very much reduced. The potential of shock explosion can not entirely be excluded if the carbon content is over 35 % and oxygen for reactions as well as a source of ignition is present. This danger is very much reduced with the carbon content in the mix not exceeding 20 %. Even in an additive mix, glow burning could occur due to the formation of local clusters of carbon. To suppress this effect, filter systems are equipped with a flushing system using nitrogen. As soon as the temperature at critical points within the filter increases, the filter housing is automatically flushed and any further reaction with oxygen is suppressed immediately.

The danger of burning can be avoided by using an inorganic absorbent, as for instance developed by Rheinkalk, Wülfrath, called Minsorb. This is a material that has a fine powdery characteristic. It comprises a mixture of calcium hydroxide, calcium di-hydroxide and magnesium silicate hydrate. Due to a special production process it has a high specific surface area (BET) of 183 m^2/g. This is, however, much less than that of carbon powder. Therefore, it requires two to three times the quantity of Minsorb than that of carbon. The material is also very aggressive and harmful to health. Thus, careful handling is required. For this reason ready mixture with the required calcium hydrate should be purchased from the supplier.

As mentioned above, there are two ways to apply the sorbents; they can be introduced separately or as mix with calcium hydrate. The separate injection permits exact adjustment to the actual condition of the waste gas. The question is, however, how can we determine this condition? In-line measurement of acidic components can already be a challenge and the continuous measurement of trace elements is virtually not possible. In a plant with a more steady state condition, such as a garbage incineration facility, separate addition of sorbents may be a possibility. In case of aluminium recycling plants, the mix seems to be the more workable solution.

Additives are charged in over-stoichiometric quantities to compensate for incomplete mixing of waste gas absorbents, as well as for uncertainties and fluctuations of the waste gas condition. A good mix generally comprises 500 – 1,000 mg /m_n^3 of calcium hydrate and an addition of 50 – 1,000.2 mg/mm_n^3 of carbon powder or 150 – 200 mg of Minsorb. Application and quantity of carbon depend on the scrap basis of the plant. The additional installation may sometimes be a high additional cost item.

The quality of the neutralization by absorption and chemical reaction of the additive depends on the thorough mixing with the waste gas. During the injection, the waste gas must enter into an intensive turbulence but not into a circular flow to avoid a cyclone effect, separating the additive particles from the gas again. In a wet scrubber the additives are dissolved in water and the suspension is then injected by spray nozzle.

The reactor is arranged in some distance ahead of the filter feed point but after the cooling stage to allow sufficient time for the reaction. It comprises a reactor vessel and the additive feed and injection system. This comprises one or more feed nozzles arranged at the circumference of the reactor vessel with interior vanes to create turbulence. The absorbent is provided by a screw conveyor to a small surge bin and charged at controlled feed rate via a speed-controlled rotary lock. A pneumatic feed system is also used instead of a screw conveyor. The additive is stored in a large bin providing sufficient capacity over a lengthy period of production time (**Fig. 7.8**). The storage bin is loaded by means of a bucket elevator or a pneumatic system that can discharge directly from the special transport truck. At the bottom of the silo the additive is reclaimed via a rotary lock to the pneumatic conveyor feeding the surge bin of the reactor.

From the reactor, the flow of waste gas arrives at the filter. The typical cloth filter comprises a series of tubes or pockets made of a special fabric. Dust particles, even smaller than the distance of the individual fibers, are held back by the cloth. The dusts, together with the reacted additive, do have a positive effect on the separation efficiency of the filter. It creates a dense phase bed with very small passageways for solid particles. The gas also comes in intensive contact with the additive again for a final absorption and reaction.

The filter dust, comprising the dust removed from the waste gas and the additive fed to the system, is collected at the lowest point of the filter housing and filled into big bags for transport to a central deposit. This dust contains all toxic chemicals and other components removed from the waste gas. Central treatment or at least proper deposit is essential.

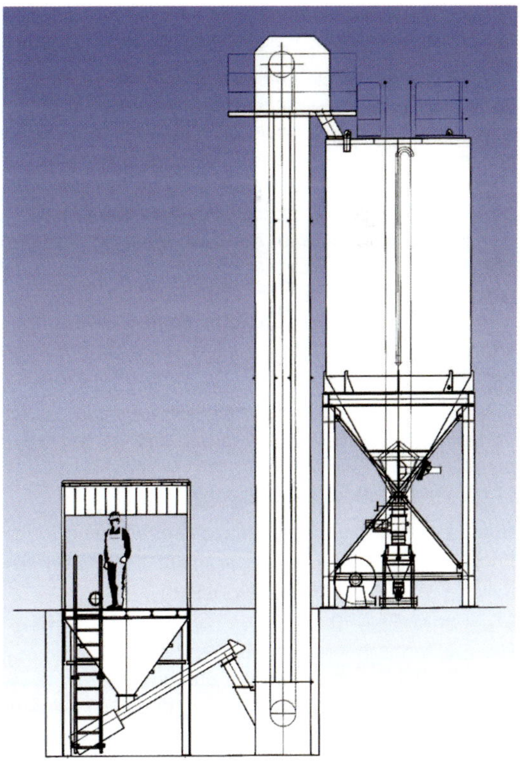

Fig. 7.8: Feed bin and additive storage

Regarding ore, to improve the separation efficiency filter dust will be recycled. A typical circulation load is 500 %. A quantity equal to the fees rate of sorbent is taken out of the circuit and loaded to big bags. Recycling of filter dust requires skilful control of the material flow in the plant as well as good control of the furnace operation. In the beginning of any melting cycle, the dust content in the waste gas is high. It decreases after a short period of time. The same is true for other contaminations which mainly have the source in the organic and non-metallic contaminations of the scrap. After the initial period of the furnace cycle, the calcium hydrate remains fairly clean and the addition of fresh material can be reduced or the filter dust is good enough for recycling and provides sufficient media reaction and absorption. For these procedures interlocking with the production plant is necessary.

The combustion products CO_2 and H_2O, as well as the nitrogen, cannot be removed and are released to the surroundings. Other combustion products, such as SO_x and NO_x, will be reduced with limited efficiency. It is must be a target for all process development steps to reduce the generation of waste gas in the equipment to the utmost possible extent by reducing energy consumption, temperature control, treatment technologies and mechanical processing prior to feeding the scrap to the furnaces.

7.3.5 Cloth filter

Cloth filters are the most effective dust-separating method. Separating efficiencies of up to 99.9 % can be achieved which means that 99.9 % of the dust reaching the filter is separated. Particles down to a size of 5 mm are held back, although the pore width is much larger. The theoretical basics of the separation in a cloth filter are very complex since they depend on many influencing variables. The theory of filtration is still incomplete. Furthermore, the correlation between theoretical calculation and measured values show substantial differences. One of the most important factors for this difference is the formation of dust layers on the filter cloth which varies with time.

For the theoretical evaluation a simplified model is used by assuming that a filter cloth is designed for a collection of single fibers of identical diameter arranged vertically to the direction of the gas flow (Fig. 7.4). It must also be assumed that the particles are evenly distributed in the gas and interaction between different particles is excluded and that these particles adhere to the fiber as soon as they come in contact. This all permits to define a mathematical model for gas flow and particle movement and, consequently, the separation mechanism can be developed. In order to have an overall model for the filter, the model must be applied for the entire filter cloth. This is not possible. The reason is that the filter material is not exactly definable. Fiber diameter and orientation with regard to the flow direction, fiber cross-section and porosity can hardly be determined. A model for the separation of a cloth filter is principally possible only when based on empirical factors or statistical probabilities. Furthermore, such model is valid only for filter cloth not yet coated by the dust.

The most important factors playing a role in the separation are based on inertia effects, on the influence of Brownian movement and the effect of additional electrical charges. Besides this, mechanisms, responsible for the movement of the particles toward the fiber and adhesive forces at the fibers, are to be considered.

Such a mathematical model is of limited value for the practical dimensioning of a cloth filter since it does neither permit to define the pressure drop for a certain cloth material nor does it permit the calculation of the separation efficiency.

The permissible filter load is one of the governing design criteria. It defines the volume of gas passing through 1 m^2 of the filter cloth and defines the size of the filter area for a given flow rate. The gas volume is the actual gas volume at the gas temperature passing through the filter. For secondary aluminium processing, this value is in the range of 50 – 100 $m^3/h \cdot m^2$. The lower figure refers to a plant processing mainly high dust-generating material such as dross while the higher figure refers to processing almost clean metal.

Example:

How much filter area is required for the gas volume of the plant comprising one melting furnace and two casting furnaces? The maximum filter temperature is to be 183 °C or 456 K.

We have calculated in the example above for the total gas flow a volume of 103,458 m^3/h. The melting furnace is to be a rotary drum furnace which can also process dross, thus generating a fairly high dust load. On the other hand, the casting furnaces process clean material only. Therefore, we select, as permissible filter load, a value of 60 $m^3/h \cdot m^2$.

This results in a filter area of

$$A_{filt} = \frac{103458}{75} = 1725\ m^2.$$

The pressure drop across is a factor for sizing of the blower, together with the line pressure loss and the pressure required for the various intake points. This defines the total pressure difference required for the blower. In standard cloth filters, the pressure loss across the filter is in the range of 20 hPa. The other factor is the gas volume to determine the power requirement. This makes the blower with 150 – 250 kW the largest single consumer of electrical energy in an aluminium recycling plant.

The standard values for the permissible filter load are basically valid only for the clean filter. After the first contact of the filter cloth with dust, the pressure drop cannot be maintained since even after intensive dust blow-off from the filter medium not all of the dust particles are removed. Experience shows that the pressure drop increases gradually with every cleaning and will finally settle at a constant value (**Fig. 7.9**).

Filter materials are selected according to their application. One criterion is the temperature resistance. In the secondary aluminium industry two major groups are used as filter cloth: material up to a temperature of 120 °C and material up to a temperature of 200 °C. Cloth for a higher temperature is more expensive. However, the flue gas is cooled by adding cold air to the waste gas. If a higher temperature is permitted, the total quantity is less for the same permissible filter load. The higher temperature, on the other hand, increases the gas volume by a factor 200 + 273 / (120 + 273) = 1.2 which has to be considered when comparing cost for filter cloth. All filter cloth used in the aluminium recycling industry is made of synthetic fiber. Characteristics of the most common materials are listed in Table 7.4.

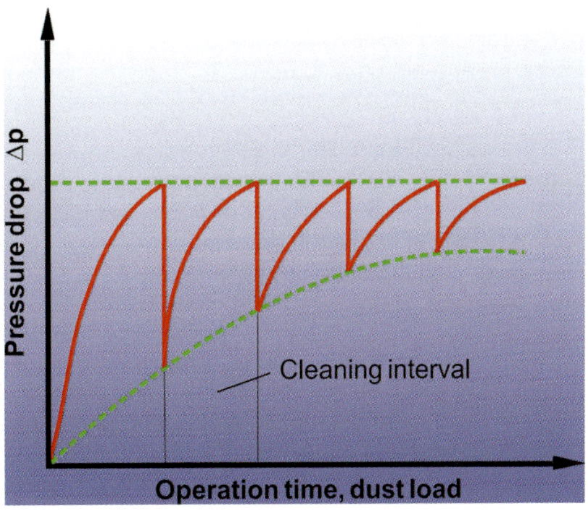

Fig. 7.9: Cleaning intervals

Fig. 7.10: Cloth filter unit (dust inlet on top right, clean air outlet at the bottom)
(Source: Nedermann)

Another criterion is the resistance to undershooting the dew point. This would result in water collection at the filter cloth and, consequently, lead to clogging. The dew point is calculated on the basis of the water content in the flue gas which comprises the water introduced with the combustion air and the water generated as combustion product. Additional water is added as soon as a quenching chamber is installed. For details of the calculation please look at the vapor pressure tables and the Mollier diagram. Usually there will be no problem involved if the temperature of the waste gas is above 100 °C. If sulfur-bearing fuel is burnt, the dew point of SO_x needs to be considered. In such a case the temperature of the gas after the filter should be above 160 °C in order to avoid dew point corrosion in the steel components.

Cloth filters are designed as bag filter or pocket filters. Bag filters are equipped with hoses made of the selected cloth. This is the oldest type of the cloth filters. They have a very good collecting efficiency and are easy to maintain. Pocket filters have a smaller size and can be integrated into a plant layout without much space requirements.

Filter housing (**Fig. 7.10**) is designed as a sturdy welded construction with dirty and clean air compartments, accessible trough large-sized housing doors and fitted with filter hoses or filter bags or filter hoses. The bags or hoses are supported by steel wire cages (**Fig. 7.11** and **7.12**). After entering the filter, the gas velocity is reduced and the dust is collected on the surface of the filter bags creating a so-called “filter cake” consisting of dust particles The raw gas passes into the hoses from the outside and clean gases leave the filter through the bottom of the clean gas chamber. The filter cloth has to be cleaned at certain intervals to remove the filter cake collected at the surface of the filter bags (**Fig. 7.13**). To do so, cleaning air or compressed air is provided either by reversed air fan or by compressed air nozzles. The reverse air filters require less energy in case of large filter plants (> approximately 200 m^2 filtration surface). To improve the removal of the filter cake, either certain rows of filter bags or complete filter chambers are taken out of operation which is called “off line” (**Fig. 7.14** and **7.15**). In case of reverse air filters this can be achieved by means

Fig. 7.11: Supporting cages for filter hoses

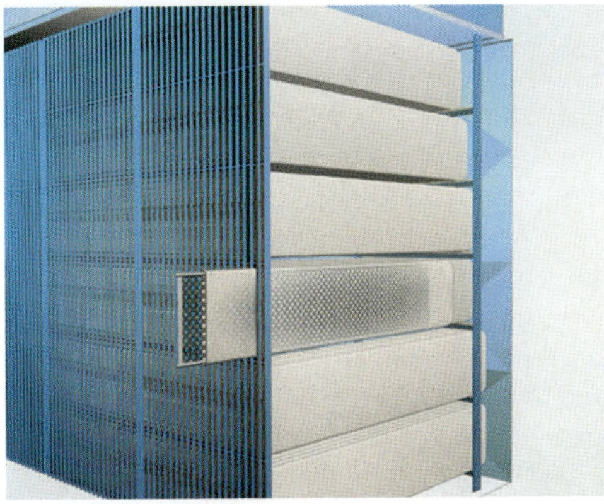

Fig. 7.12: Supporting cages for filter bags (Source: Turbofilter)

of a carriage (**Fig. 7.16**) and in case of pulse jet filters by means of valves in front and after each individual filter chamber (**Fig. 7.17**). The dust removed from the filter bags is collected at the bottom of the filter housing (**Fig. 7.18**) which is shaped as a sloped bin. A screw conveyor removes the collected dust and feeds it via a rotary lock into big bags. To prevent coarse, glowing particles from entering the filter, a pre-separator is required because these

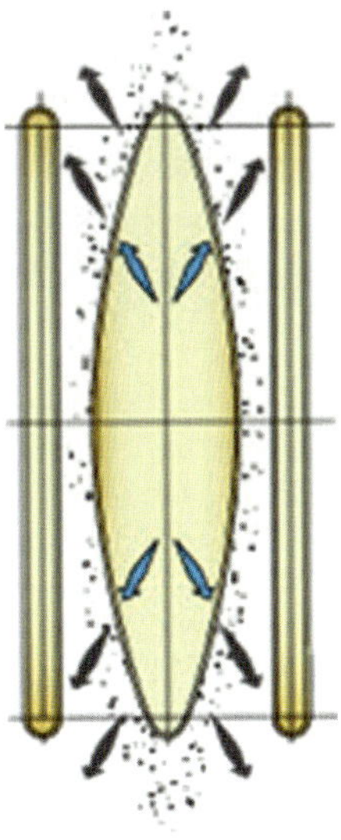

Fig. 7.13: Removal of filter cake (Source: Nedermann)

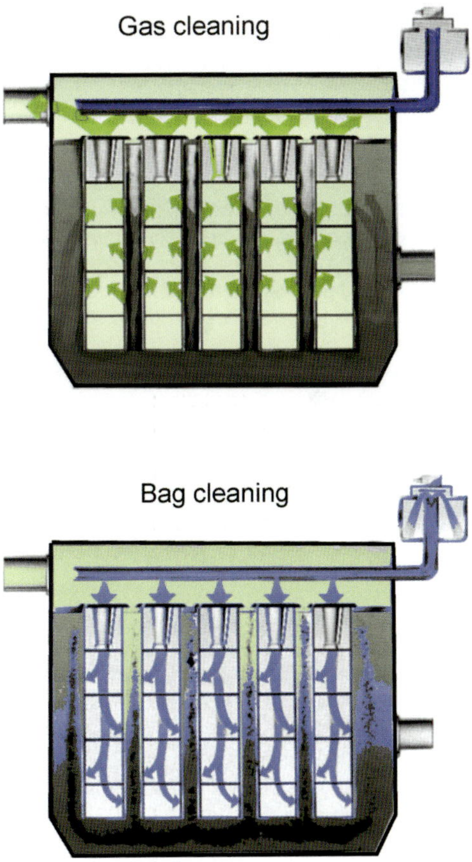

Fig. 7.14: Stages of “off line” operation

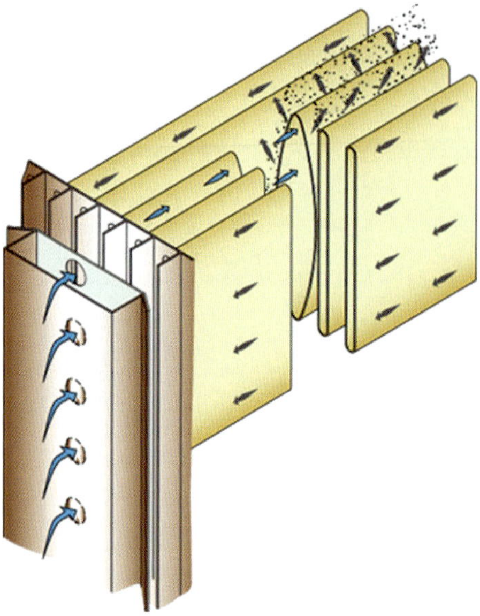

Fig. 7.15: "Off line" filter operation (Source: Nedermann)

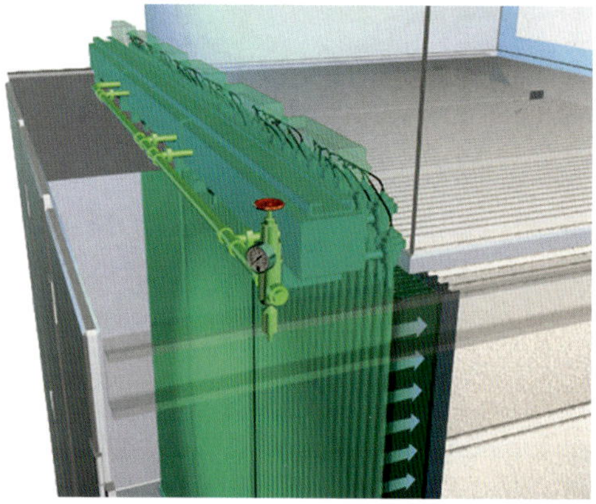

Fig. 7.16: Compressed air nozzles for bag cleaning (Source: Nedermann)

particles would damage the filter material and could even cause a fire. Simple chambers, baffle plates or chain curtains are usually not sufficient. Cyclones or similar units are much more reliable (see chapter 7.3.7). This is achieved by a temperature-controlled cooling air recirculation (Fig. 7.12). The radial blower, as in the case of the pocket filter, by a special blower arranged at a trolley comprising drive and deviation station with timing generator and limit switches. Different sections are cleaned while the others still operate normally (Fig. 7.12). The progress of cleaning is automatically controlled by the cleaning air vent that is activated by magnetic valves (Fig. 7.13).

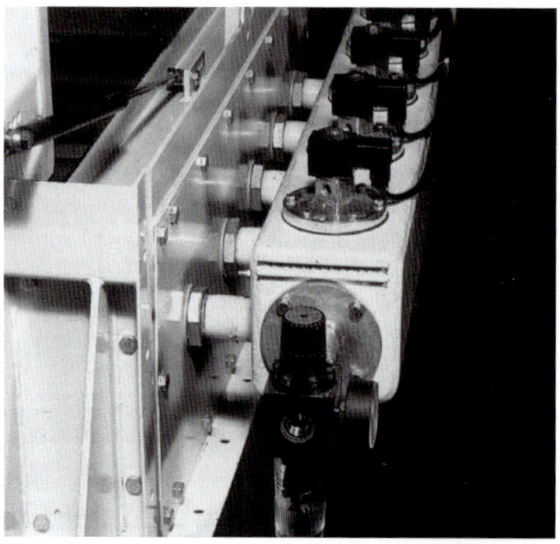

Fig. 7.17: Valve arrangement for compressed air cleaning of air nozzles (Source: Turbofilter)

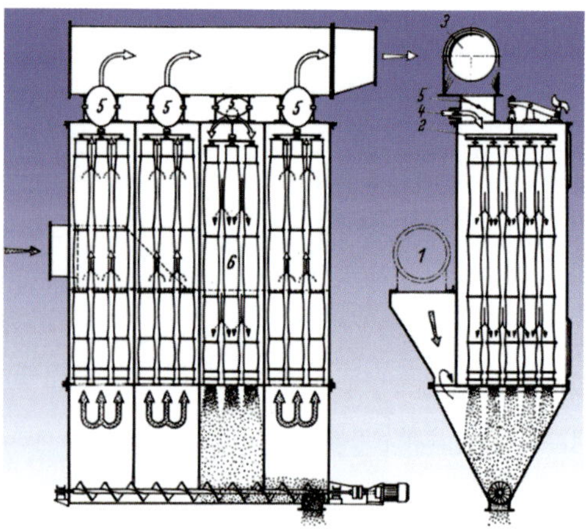

Fig. 7.18: Sequence for bag cleaning

From time to time the filter cloth needs replacement. At the hose-type filter the complete casing is lifted requiring sufficient space above the filter housing. This may be a problem. The bag-type filter system requires less height since the bag units are taken out to the side of the filter.

The dust is collected at the bottom of the filter housing which is shaped as sloped bin. A screw conveyor removes the collected dust and feeds it via a rotary lock into big bags.

If the filter is arranged close to the furnaces, a baffle plate is arranged at the filter entry which works as gravity separator. A chain type baffle system or a pre-separating cyclone can also be arranged ahead of the hydrate reactor. These measures are required to separate coarse particles arriving from the furnace still glowing.

A service platform along the cleaned air area provides access to the doors at the filter housing for maintenance. The radial blower, for providing the necessary draft in the entire waste treatment system, is arranged close to the filter housing. In order to protect the fan against overload or over-current when it is started, the fan runs when handling pressure with closed throttle valve.

Noise abatement must be designed according to the local standards and regulations. In Germany, this is DIN 45 635, requiring for an extraction fan the sound pressure level to be 85 dB(A) at a distance of 1 m. Furthermore, the noise at the stack will be reduced to not exceeding 85 dB(A) at distance of 1 m by installing an absorber with baffles between fan and stack.

Thermal insulation is required for the filter housing to maintain a temperature above the dew point. This will be including the insulation of the exterior surfaces by means of insulating mats and a casing made of galvanized not lacquered metal sheets.

The under-construction consists of a stable steel structure, braced torsion-free and equipped with base plates, siderail for cover rim, ladder with back protection, loosening device by means of compressed air, stop valve.

7.3.6 Wet scrubber

Wet scrubbers are not commonly used in aluminium recycling plants. Occasionally they are installed in auxiliary plants such as decoating or chip-drying plants. The reason is that their separation behavior for trace components, such as heavy metals or, and this is very important, for PCDD, is comparatively poor. In industrialized countries wastewater has to go to public treatment plants and the collection and treatment systems are very much developed and are able to clean the effluent water. But even there the public plants may not be fully equipped to handle toxic waste as generated in a scrubbing plant for the aluminium recycling industry. In plants located in a cold climate, the low temperatures also create problems in regard to the handling of water; while in a hot climate, water may be very expensive.

But still, there may be applications where wet scrubbers might be the best choice. This could be in shredder plants where sometimes wet material is to be processed which may cause clogging of dry filters.

The basic principle of the wet scrubber is to bring the solid particles in contact with water, attach them to the droplets of the liquid and separate them from the gas. The process within a wet scrubber can be classified into four steps:

- Add water to the mixture of dust and gas,
- bring dust in contact with the liquid,
- attach dust particles to the water droplets,
- separate the dust/water mix from the gas.

Studies have shown that the wetting ability of the dust particles is not very important since the dust particles will penetrate into the liquid where they settle. In any case, the adhesive forces between liquid and dust are sufficiently high to attach the solid particles to the liquid, even in case the solid particles are not wettable. Generally speaking, there is always a bonding between the two phases.

The principle criterion for the efficiency of a wet scrubbing system is apparently that all dust particles come in contact with water. This can be achieved by various measures.

- Establishing large surfaces of the liquid by creating thin film layers, i. e. on surfaces of solid bodies, creation of veils of liquid or establishing numerous fine droplets by injection nozzles or other distribution elements. These methods also provide for a large surface between liquid and gas.
- Rapid replacement of reacted liquid by fresh liquid, obtained by flow conditions for liquid and gas. This can be realized by sudden change of direction by baffles, change of velocity by changing the cross-section of the flow. In short: creation of intensive turbulences.
- Change of temperature to initiate condensation.
- The simplest method appears to be to have the gas bubbling trough the water. Experience shows, however, that the separation efficiency of such a system is comparatively poor. It is more efficient to mix the water into the gas flow to be cleaned.

To obtain this mixing effect many methods have been developed.

The simplest type of scrubbing system is the washer (**Fig. 7.19**). Gas and water are mixed by direct injection of water by means of spray nozzles. The gas enters the unit directly or tangentially to obtain a good turbulence with the resulting intensive mixing. Baffles or other obstructions support the mixing effect. Depending on the gas temperature, some of the water is absorbed in the gas. The contaminated water settles at the bottom of the washer where the contamination is removed, i. e. the water is purified.

In the cyclone washer the gas flow hits the water with the effect of a pre-separation of the dust particles. In the rotating water flow, particles are in

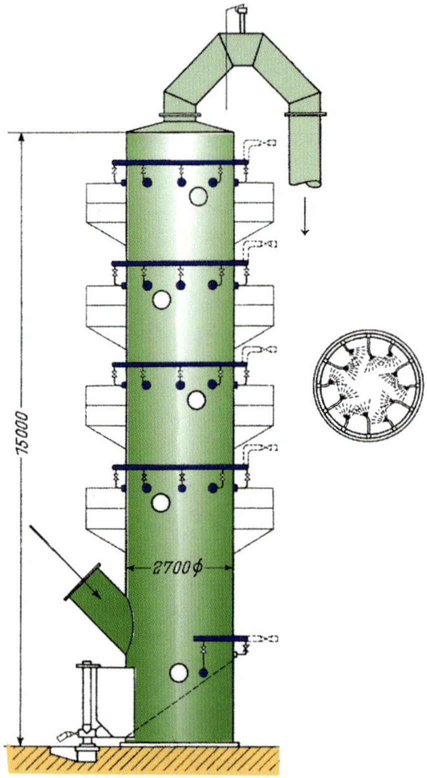

Fig. 7.19: Gas washer

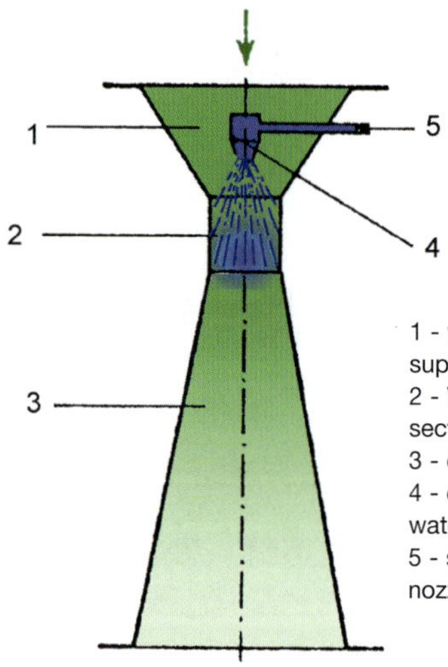

1 - waste gas supply duct,
2 - Venturi section,
3 - diffuser,
4 - cooling water feed,
5 - spray nozzle

Fig. 7.20: Venturi washer

intensive contact with the water and the final separation takes place in a cyclone - like inertia separator. This type of washer requires no water circulation system but its capacity is very small.

The Venturi washer (**Fig. 7.20**) is a high performance system with cleaning efficiency that is comparable to cloth filters or electro-static precipitators. The gas is accelerated in the Venturi section and mixes intensively with the water injected just ahead of this section. Due to the interaction of gas and water at high velocity, a fog-like liquid veil with high specific surface area is established which causes the solid particle to settle. In the diffuser the velocity decelerates and larger dust-loaded water droplets are created. These will be separated from the gas flow in inertia separators. The Venturi washer is able to process large volumes of waste gas. It is of uncomplicated design but is characterized by a high pressure drop.

The rotation washer (**Fig. 7.21**) is characterized by a mechanical rotor through which the cooling water enters the gas flow. The rotor creates a veil of water through which the gas must flow passing the dust particles to the liquid droplets. The washer may comprise different stages. It is characterized by a very good cleaning and separation behavior. The advantage of the technology is that the energy for creating the droplets is not provided by the gas flow. Therefore, the separation efficiency is independent of the gas flow and the scrubber is very suitable for alternating gas volumes.

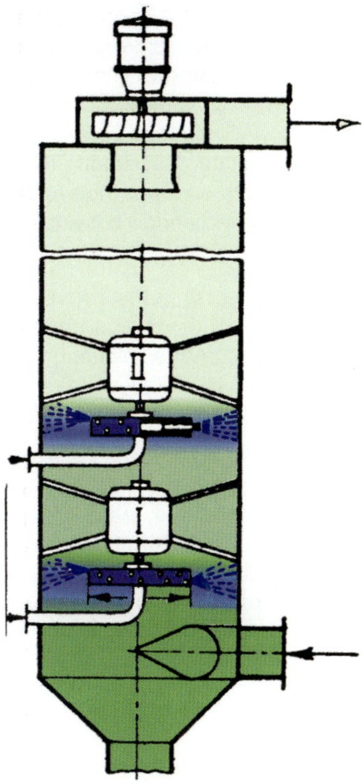

Fig. 7.21: Rotation washer

Fig. 7.22: Wet scrubber unit

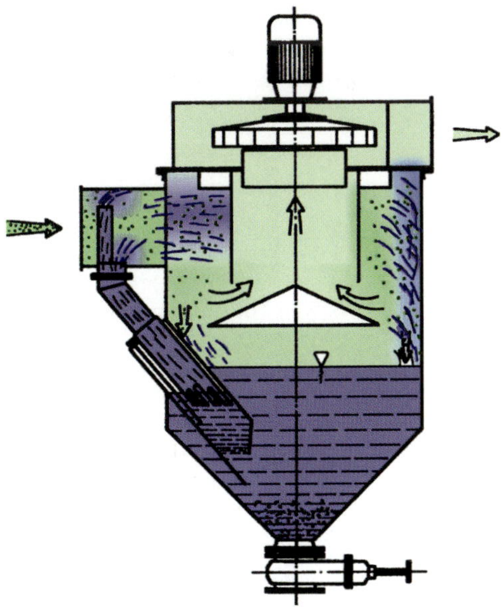

Fig. 7.23: Wet scrubber principle

Fig. 7.22 and **7.23** show a combination of Venturi washer and inertia separator. The gas enters the separator vessel tangential after it has passed the Venturi section. Due to the high velocity difference between gas and water, the dust particles are absorbed by the water droplets. The water injection works without pump only due to the pressure drop in the Venturi section. The separation of water and gas takes place in the cyclone section in the center of the unit. Separated and contaminated water flows along the sidewalls down into the collecting basin. Part of the water passes the gas entry section, thus forming an additional water veil. The cleaned gas leaves the unit via the central immersion tube. Some water is lost due to evaporation. Therefore, the water level in the collecting tank is measured. Water losses are automatically replaced.

Wet scrubbers are very convenient to use if the waste gases contain liquids such as water, emulsion or oil and the temperature is too low for having these contaminations burned. This is mainly the case in shredder plants.

7.3.7 Centrifugal separator (cyclone)

The aluminium recycling industry uses cyclones for the separation of comparatively coarse particles. They are usually applied ahead of the cloth filter to improve the separation efficiency of that filter. If the filtration plant is close to the furnace, still glowing particles may reach the filtration plant. This would be a dangerous situation, particularly if carbon powder is used as additive in the filter. Cyclone separators (**Fig. 7.24**) remove these particles efficiently before the waste gas reaches the additive reactor and the filter.

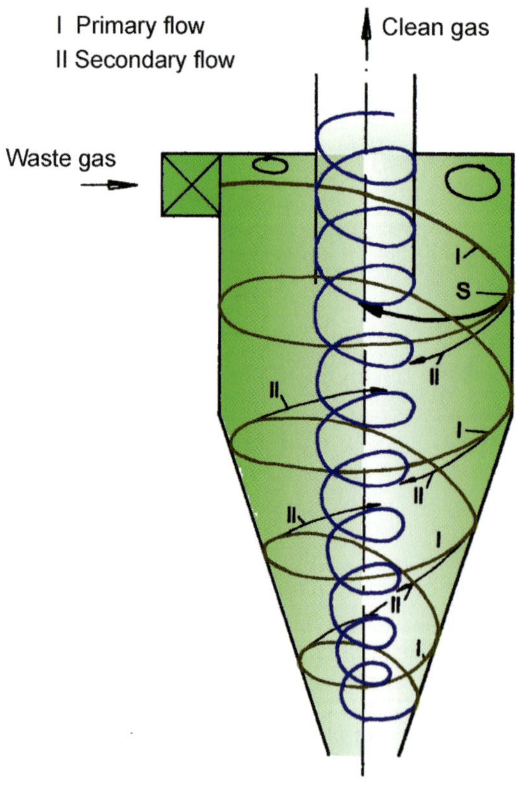

Fig. 7.24: Cyclone separator

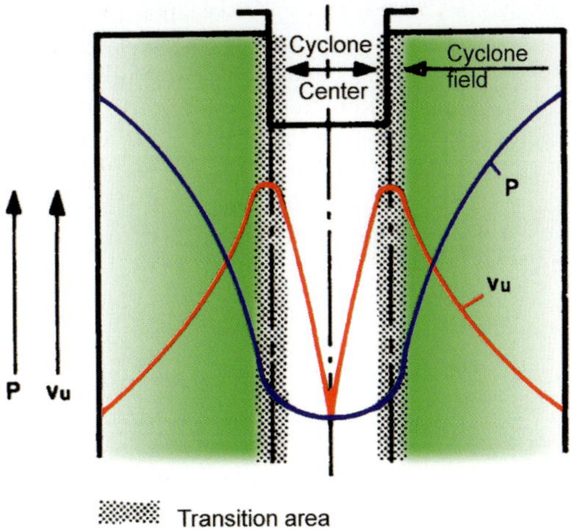

Fig. 7.25: Pressure and velocity distribution across a cyclone separator

The waste gas passed to the cyclone, shaped as vertical cylinder with conical bottom, flows tangentially which forces the gas/solid mix to enter into a radial flow. The dust-loaded waste gas is forced to travel down to the bottom in a spiralling movement until it is stopped by the solid bottom and travels now upwards in a second cyclone to the exit (**Fig. 7.25**). Dust travels due to the centrifugal force and is thus separated from the gas in combination with the gravity. From the bottom the collected dust is discharged via an airlock required to avoid intake of dead air.

The physical conditions of a cyclone are similar to that of the circular flow, following the equation

$$v \cdot r = \text{constant}$$

where v is the gas velocity and r the radius of the cyclone. But the conditions within a cyclone are very complex. Thus, the equation should read

$$v \cdot r^m.$$

The correction factor m is in the range of 0.5 and 0.7. The velocity increases according to this equation hyperbolically from outside to the inside and, consequently, according to the Bernoulli equation the pressure decreases. Pressure P can be determined according to Euler's law

$$\frac{dP}{dr} = \frac{\rho_G}{2} \cdot v^2 \tag{7.11}$$

with ρ_G being the density of the gas. The distribution of pressure and velocity is shown in Fig. 7.25.

The principles of cyclone separation appear to be very simple. But again, the conditions within a cyclone are much more complex. The different flows interact with each other. Thus, only some simplified calculations are possible which are based on similarity considerations. These take into account the drawdown velocity and grain size distribution of the dust particles, the velocity and viscosity of the gas, friction coefficient, and the desired separation efficiency. There are also some empirical factors involved as well as the type of cyclone. This is then related to typical dimensions of the selected type of cyclone. The separation efficiency of a cyclone ahead of the filter plant of aluminium recycling production is not very critical since the cyclone is more like a safety unit.

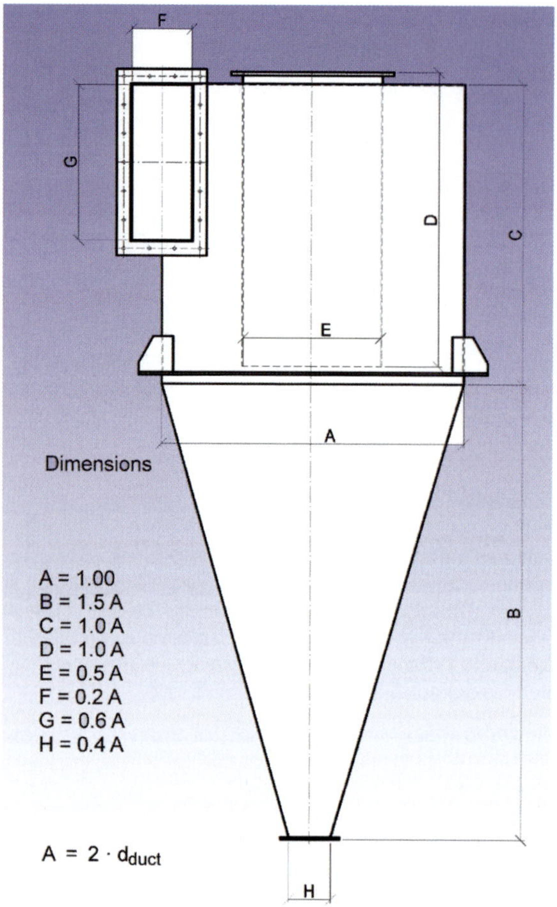

Fig. 7.26: Typical cyclone dimensions

But it can help to reduce the dust load on the filter during more dust-generating phases of the furnace cycle.

As good approximation for the dimensioning for a cyclone, the data in **Fig. 7.26** can be used. These are based on good experience in operating plants. All dimensions refer to the diameter of the connecting pipe which is determined by the value of the gas velocity which should be in the range of 20 to 25 m/s, calculated with actual waste gas volumes.

7.3.8 Quenching chamber and incinerator

Quenching chamber and incinerator are not scrubbing systems since no separation of dust and gases takes place in this type of equipment. But there is of course some treatment since the flue gases are conditioned for final treatment.

The incinerator is used to destroy still reactive components leaving the furnace process. Carbons, as well as organic elements, are converted into CO_2 and water. But also trace components and particularly the very poisonous dioxins are destroyed at high temperature before they leave the furnace area. Design details of the incinerator are described in the previous sections.

The hot waste gases leaving the incinerator and the furnaces have to be cooled before they enter the filter plant. There are different reasons to do this. Firstly, the cloth filter cannot process waste gases having a temperature exceeding 200 °C. In this case the filter cloth would be destroyed. The other reason is that dioxin will be formed in the so-called "de-novo synthesis" again, if the waste gas cools too slowly.

The quenching chamber (Fig. 7.5) permits cooling of the waste gas very rapidly. It works on the principle of a water injection into the waste gas.

It comprises a housing equipped with spray nozzles. With the aid of compressed air the water is atomized to very fine droplets to cause rapid heat exchange. The water evaporates immediately and the temperature settles at a lower level due to the heating up of the water. Some dust may also be collected at water droplets first. However, these will evaporate. The dust cannot be held in the water particles. To compensate for the efficiency of the system, more water is fed as can be utilized. Excess water is collected in the basin at the bottom of the housing to which fresh water is added to compensate for the quantity evaporated and leaving with the waste gas. A feed pump supplies the spray nozzles with water.

The waste gas is introduced at the lower part of the quenching chamber and flows countercurrently to the waste gas exhaust at the upper section. The quenching unit can be designed as rectangular box or as round vessel. The velocity in the unit shall be very slow to provide sufficient time for evaporation. A good value is a velocity of 5 m/s or less.

The required water quantity is calculated on the basis of the required temperature difference and the energy removed by the water.

Example:

A quenching chamber is to be operated with water having a temperature of 20 °C. A volume of 15,000 m_n^3/h of waste gas shall be cooled from 1,200 °C to 250 °C. How much water is required for quenching? The average specific heat cpG for the temperature range of the flue gas is 1.4 kJ/m^3.

Fig. 7.27: Installation of a filtration plant in a secondary aluminium smelter (Source: Nedermann)

The quantity of heat to be removed is

$$Q_G = 15000 \cdot 1.4 \cdot (1200 - 250) = 19.95 \cdot 10^6 \text{ kJ/h}$$

$$q_W = 4.19 \cdot (100 - 20) + 2235 + 2.1 \cdot (250 - 100) = 2.87 \cdot 10^3 \text{ kJ/kg h}$$

We can obtain the same result if we would use the steam tables.

With this value we calculate the required water quantity as

$$G_W = \frac{19.95 \cdot 10^6}{2.87 \cdot 10^3} 6950 \text{ kg/h or } 6.95 \text{ m}^3\text{/h}$$

This is quite a quantity which we would have to add to the flue gas every hour. With the aid of the steam tables, we need to calculate the dew point temperature considering the water volume generated during combustion and the water injected as humidity with the combustion air. The temperature of the flue gas in the filter must always be above the dew point temperature. In operations without quenching chamber a temperature of above 100 °C does not present any problem.

Fig. 7.27 shows filter equipment in an aluminium recycling plant. It comprises a bag house with additive silo. As usual, the unit is located close to the production building.

8. Process control

Christoph Schmitz

Looking at the requirements of pollution control and at the so-called "carbon footprint", in particular process control has become a key factor in the operation of a modern aluminium recycling plant. Besides the efficient design of components of the process equipment, the operation must be reliable and safe but must also provide safety for the personnel and all this with the optimum efficiency.

The many different functions in an aluminium recycling plant require permanent settings and controls. This refers not only to the individual equipment, such as furnaces or casting machines, but also to the interlocked operation of the entire plant. Temperatures have to be monitored and recorded. Water flow, gas qualities and electrical energy require recording and supervision. Proper operation of the individual components and alarms, in case of malfunctions, is required to alert the operation personnel to take the necessary action. Interlockings are defined to eliminate failures in operation or to initiate automatic operation sequences. To optimize plant operation, the best solution is to automate the entire production process. This, however, is only possible for certain activities. In many other cases interference and action of the operators are required. For instance: Loading of the charging machine requires the action of the forklift driver to pick up the batches that have been prepared manually as well. As soon as the container of the charging machine is filled with the correct batch, the forklift driver initiates a signal and the interlocked sequence can commence automatically. The burner is switched off, the furnace door opens, the charging machine moves to the furnace, the rotation speed is reduced and the charging starts. Now the operator's input is required again. Since different materials require different charging times, the operator gives a clearing signal as soon as the batch is in the furnace. Now the automatic sequence starts over again. The charging machine retreats, the door closes and the burner starts firing.

Similar interfaces between operators and plant equipment exist in other plant and production areas. For these activities the operators must possess information about the status of the equipment and when their input is required. Best overview is provided by visualization on the display of a computer screen. Although there is an interlocking of all plant operations, it will certainly happen from time to time that there is some trouble with the equipment. In such a case all other equipment must be able to function without trouble. The requirement, as outlined, will lead to a specific control architecture.

The process control system can also communicate with the commercial plant management system and with other sources such as the scrapyard via radio – and the quantometer for metal analysis. Definition of control components (control architecture) for the process control is set up for various levels (**Fig. 8.1**).

The first level comprises the control of the individual equipment. All controls, interlocking and safety equipment as well as the motor control are designed as stand-alone unit. Furthermore, the power feed is designed to allow isolation of the individual unit without interrupting the power supply to the others. The unit is able to operate individually and is controlled by the operator's panel at the switchboard. All control loops and Interlockings, based on a programmable logic control (PLC) system, are accommodated in the cabinet. A good example of such PLC system is "Simatic" offered by the Siemens company. Starting with the Simatic S5, a milestone in the history of digital control, the modern Simatic S7 (**Fig. 8.2**) is not limited to digital signals but is able to process continuous signals required for automatic control loops with feedback.

The individual controls communicate via a bus bar system with each other and with the PC-based process control system. Data from the individual equipment controls are collected and recorded. The status is visualized on the computer monitor. From the process control units the data for the

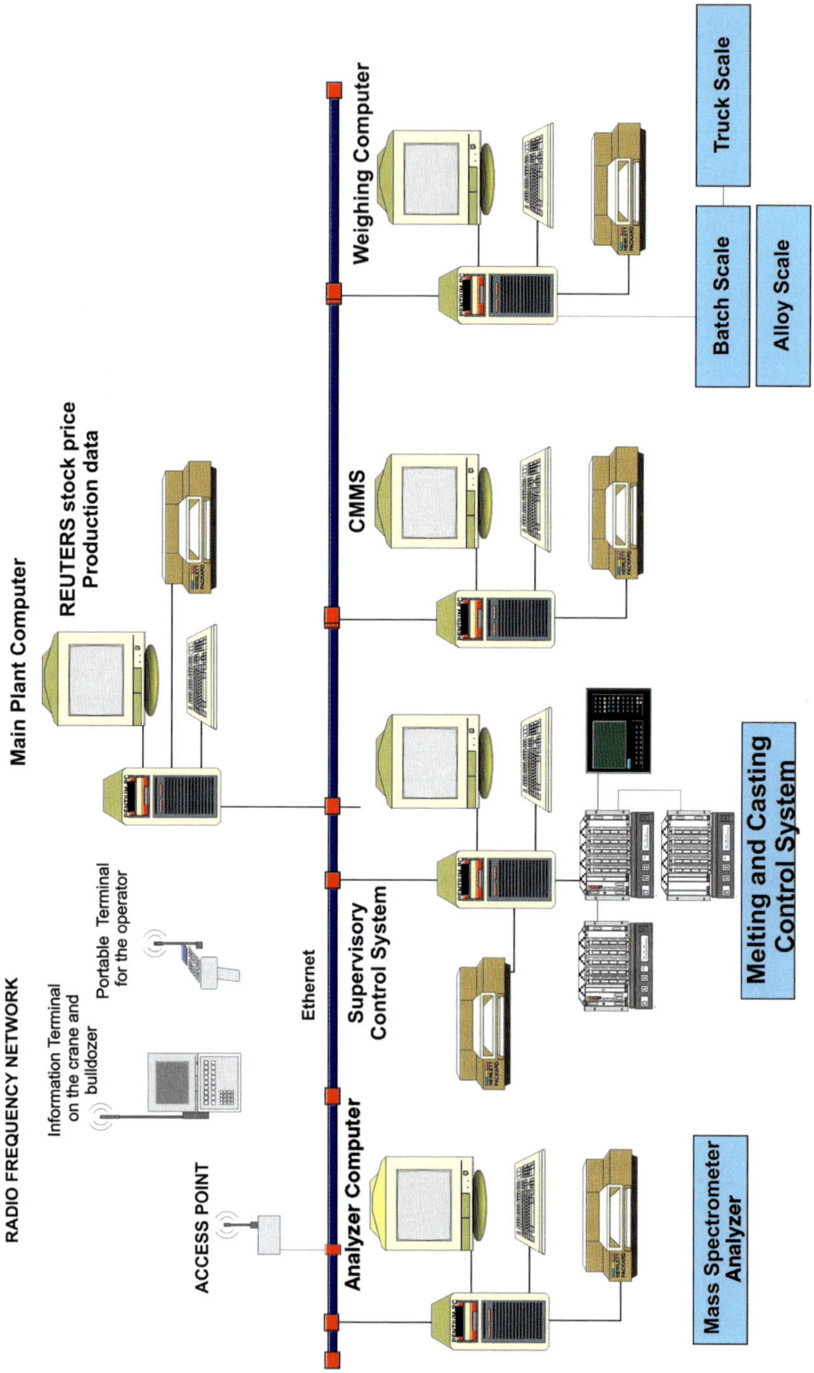

Fig. 8.1: Control architecture (Source: Novatherm)

Fig. 8.2: Simatic S 7 PLC control system

process can be set as far as they are required to run the process. No input can be made, however, to change safety settings, i. e temperature limits or interlockings.

Production data, such as alloy shift number, quantity, material charged, final analysis, energy consumption and the like, are stored in the computer system. Alarms are printed out and the time of action is recorded as well. To simplify production procedures and recipies, they are defined and can be recalled from the PC. Some of the equipment settings will be made automatically as soon as the procedure is selected and the required action by the operating system is indicated. The process control system can also communicate with the commercial plant management system and with other production areas such as the scrapyard (via radio) and the metal analysis computer (quantometer).

There is quite a quantity of data that can be collected and made available. Here a selection and restricted access is required for meaningful utilization of the data.

8.1 Sensors and actuators

8.1.1 Sensors

Basic elements of any control are sensors giving signals for position, temperatures, flow rates, pressure and status of the various equipment. This information is then processed within the control loops or in data processing.

Position is defined by limit switches. Various kinds of activation elements are available for the different applications. They are activated mainly via a roller system. Very commonly used is the proximity switch. There is no direct contact required. The signal is initiated by a steel piece that is attached to the moving part of the equipment. Also limit switches with direct contact to the equipment element are used.

Another type of touchless switches is the photo-electric switch. It comprises a light source and a photo cell receiving the light. If the path of the light is interrupted, a signal is generated. Photo-electric systems are used for counting purposes, i. e. counting of ingots at the feed to the ingot stacker (**Fig. 8.3**) or to identify hindrance in the path of moving equipment such as a charging machine, for instance. They are also used where the installation of contactors is not possible or subject to damage as it happens in connection with limit switch actuators in concrete floors.

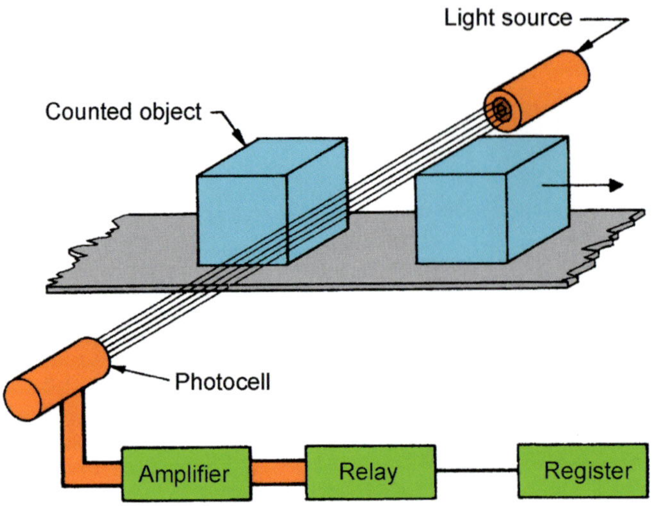

Fig. 8.3: Counting system with a photocell

The limit switches used in the equipment in an aluminium recycling plant must be accommodated in a sturdy housing that resists severe conditions such as dust, chemical attack, mechanical impact, temperature and the like. The designer must take care to place them in a protected location to keep away at least some of the impacts from the surroundings.

Special types of limit switches for pressure and level are used for instance in hydraulic circuits. They are of particular design to meet the requirements.

Flow meters and flow indicators are integrated in the relevant circuits. Flow measurement is required for different applications within the plant. The air and gas quantities in burner circuits are required to determine the air / gas ratio. Cooling water quantity of a casting machine is important to know as well as the volume of fluxing gas during metal treatment. The flow rate can be obtained by a diaphragm combined with pressure measurement. Basis of this method is the principle of conserving mechanical energy through conversion of fluid velocity to pressure. If the fluid changes its velocity, the pressure will also change. The relevant pressure difference can be measured and converted to signal the flow rate. The principle is also used for the Pitot tube which is used to measure the flow velocity of the gas manually. For precise measurement of the flow, the turbine type flow meter is very suitable. Depending on the flow rate, the turbine will have a different speed that can be converted into an electrical signal.

For any pick-up of flow it is important that the measuring unit is placed in a turbulence-free area where pressure drop, due the flow obstruction, can not influence the flow conditions in the pipe. As rule it can be defined that any flow meter, no matter if installed permanently or temporarily, should have a distance to the sensor of 5x pipe diameter as straight section. Small gas flows are measured by a rota meter where the position of a metering float in a conical tube gives an indication of the flow rate.

Pressure can be obtained by measuring the displacement of a spring or specially shaped tube or a bellow. The size of the displacement gives the indication of the pressure.

Furnace pressure control requires the measurement of a very low pressure difference between the hot furnace and the surroundings. For taking the interior pressure a small tube is inserted trough the refractory lining so that the pressure value can be measured at the outside of the furnace (**Fig. 8.4**).

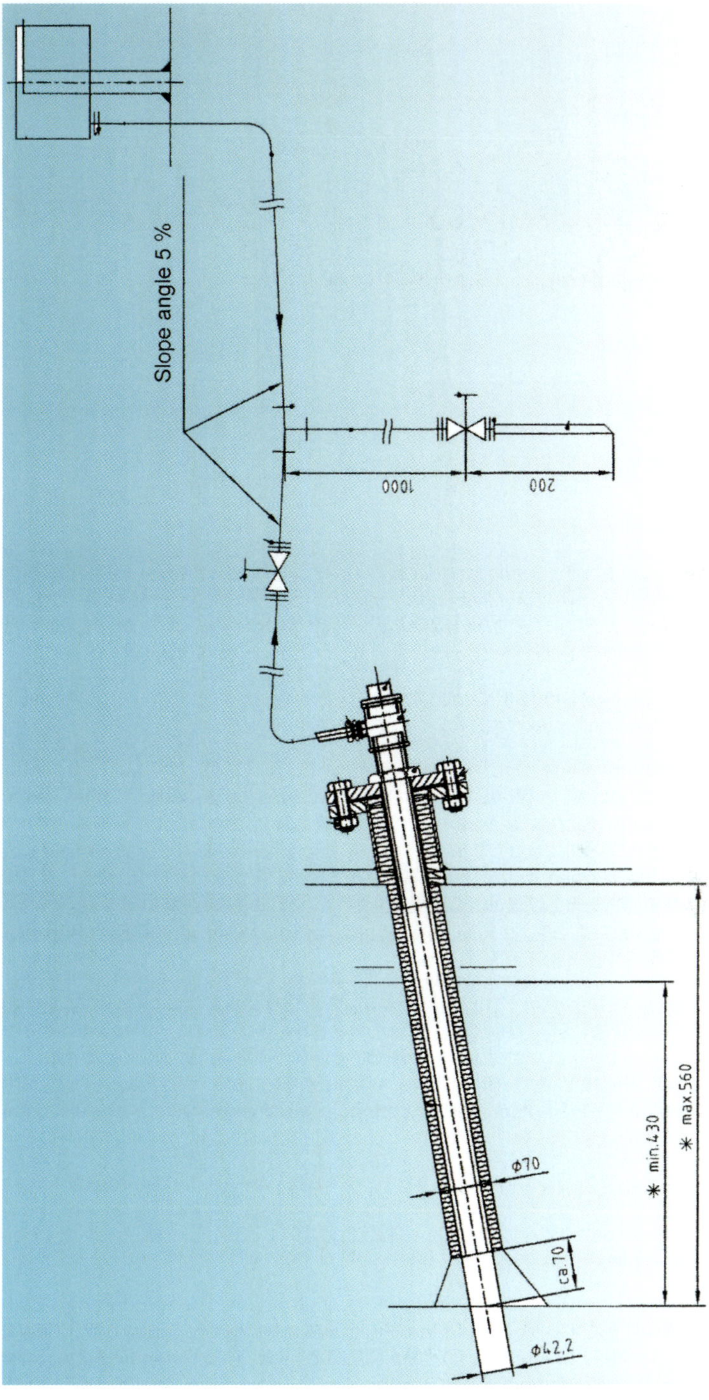

Fig. 8.4: Furnace pressure system

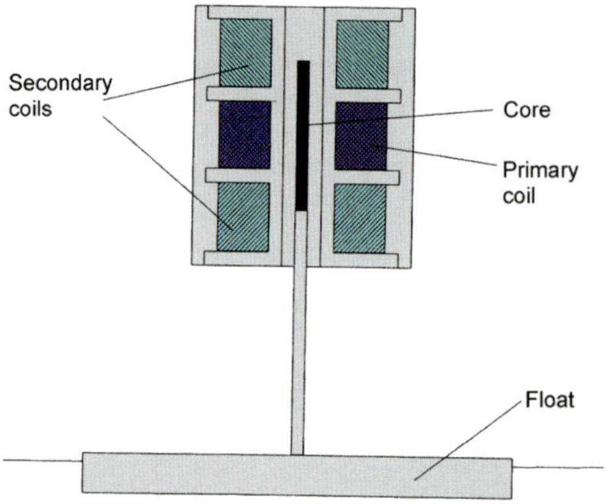

Fig. 8.5: Inductive level sensor probe for liquid metal

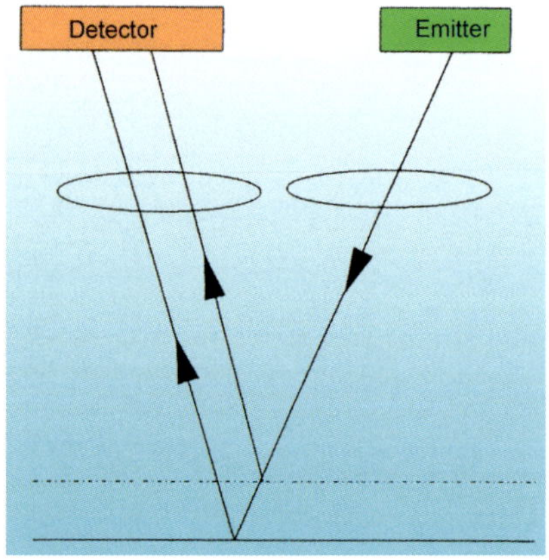

Fig. 8.6: Laser level control principle for liquid metal

For the determination of the level of a liquid in a tank or container, a system with floater is quite common. The displacement of the floater is measured via a chain or a rod. The buoyant force of a floater can also be used in a pressure system. The change of pressure proportional to the volume of a pneumatic displacer can be measured by comparing the balanced air pressure in a bellow that results in a displacement of the bellow.

Measuring the liquid metal level in a launder is essential for a precise flow control. Temperature and the harsh surroundings require very specific methods. The unit should also interfere with the operation as little as possible. With the conventional method a ceramic float rests in the metal and is mechanically connected to a transducer generating a signal due to the change in inductivity or capacity (**Fig. 8.5**). Ultrasonic systems have also been tested but the results have not been satisfactory. Trials have been made using Gamma rays. A source is located at one side of the launder and a scintillation tube at the other. Rising metal level reduces the rays picked up by the tube. However, the use of even a low level source of radiation brings up a number of practical, procedural and emotional headaches.

The laser method is meanwhile very common. Laser devices transmit a light beam at an angle to the metal surface (**Fig. 8.6**). The position of the reflected beam varies with the distance from the device and the receiver is capable of detecting where the reflected beam hits it. This measuring principle is not affected by temperature. Another principle using laser rays is based on the measurement of time for the reflection that changes with the distance between metal surface and measuring probe.

Temperature is measured with the aid of thermocouples. Two different metals are joined and will develop a voltage difference between the two at a value depending on the temperature. The voltage difference at the cold junction is measured. The hot junction is at the point where the temperature will be measured (**Fig. 8.7**)

The circuit is completed at the cold junction in the instrument. The temperature-dependent value is in the range of mV. The thermocouple does not give an absolute reading but only the difference in the temperature of the hot junction and the cold junction. Thus, there may be different readings depending on the ambient temperature. For this reason compensation is applied (**Fig. 8.8**). Once adjusted, the temperature compensation is effected automatically. The connection of the thermo-

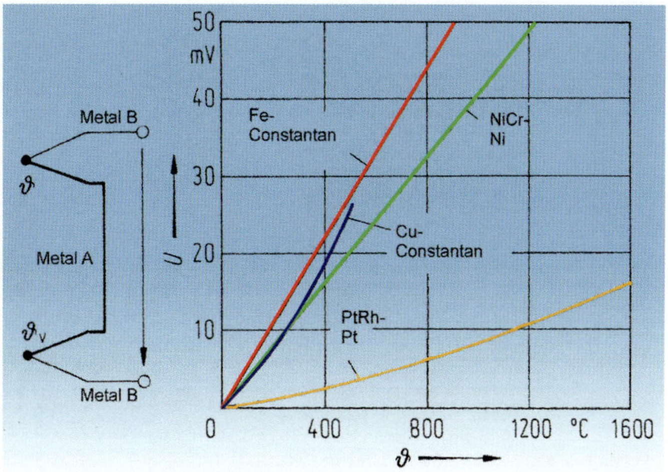

Fig. 8.7: Thermocouples

couple wire through different junctions will also create new thermocouple junctions. Therefore, the wiring between thermocouple hot junction and cold junction introduces additional junctions which may cause faulty temperature readings. It is common practise to connect the hot junction thermocouple with the instrument or the mV/ mA converter with special lead wire which may be of the same material as the thermocouple itself. This ensures that the cold junction will be inside the instrument case. The combination of the bi-metal depends on the application temperatures. The most common thermocouple in the aluminium industry is the Ni – Cr – Ni combination. These thermocouples are placed directly at the location where a temperature reading is required, i. e in the refractory material of the furnace roof. For the installation it must be assured that the reading gives the correct count at the point of placement. For instance, if the thermocouple is placed in the refractory material of the roof it must be placed close to the furnace interior. If located in some distance within the refractory material, it may give the temperature inside the refractory lining but not the roof temperature of the furnace. Where the thermocouple is in contact with metal it must be protected by a tube of cast iron or special refractory material. If a thermocouple fails, the control

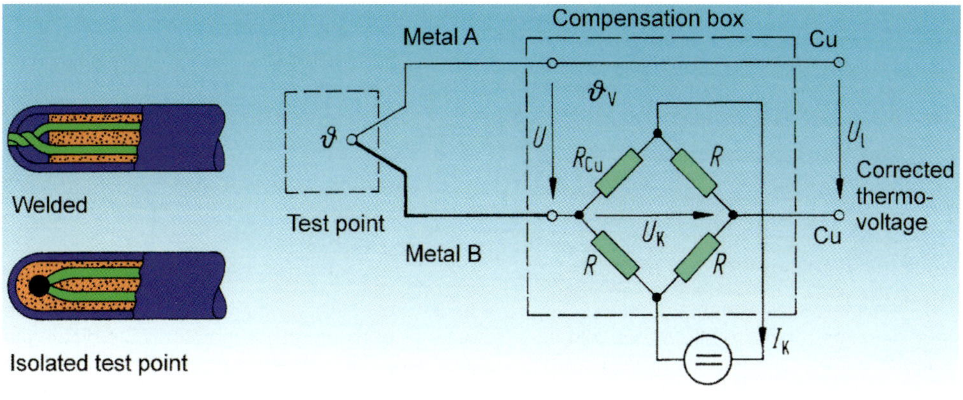

Fig. 8.8: Thermocouple and compensation

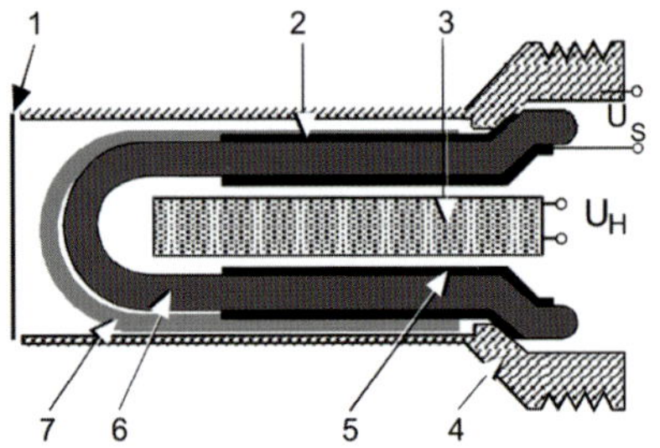

1 Gas intakte, 2 Working electrode,
3 Probe heating, U_S Probe voltage,
U_H Heating voltage, 4 Housing,
5 Reference electrode (Pt),
6 Zirconium oxide ceramic,
7 Protective layer

Fig. 8.9: Oxygen probe with zirconium oxide (Source: Lamtec)

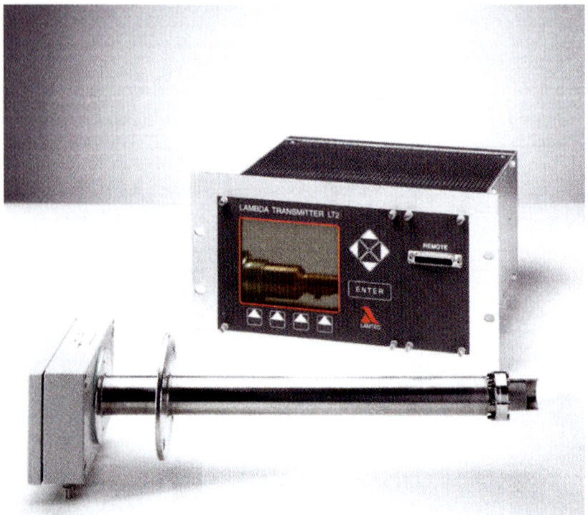

Fig. 8.10: Probe extension for protection against high temperature (Source: Lamtec)

parameters must go automatically to a safe setting. For instance, a breakage of a thermocouple in the circuit of the burner control would then give a signal to increase power input. This could damage the equipment. Thus, in this case the burner must be at low fire.

For some melting processes, in particular if controlling of the combustion is required, it will be necessary to measure the content of oxygen and unburned combustion products in the flue gas. The standard λ- probe uses the behavior of zirconium oxide to change the resistance for electric current depending on the presence of oxygen (**Fig. 8.9**). The probe can be operated by measuring the electric current or the voltage difference. This signal is processed to obtain an indication of the oxygen content of the flue gas. The probe is arranged in the flue gas duct. Since working temperature of the measuring cell is low, the standard probe permits a maximum gas temperature of only 300 °C. For higher temperatures – as is the case in our aluminium melting furnaces – part of the flue gas will be taken out of the duct by an extended probe tube and directed to the measuring cell (**Fig. 8.10**). Modern probes are designed to measure the content of uncompleted combustion usually expressed as CO_e (**Fig. 8.11**). The advantage of such a system is shorter response time

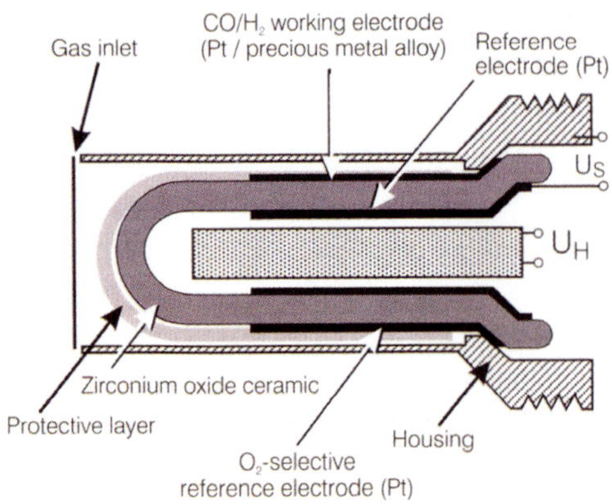

Fig. 8.11: Combination probe for oxygen and CO (Source: Lamtec)

and better controllability of the firing system. Other measuring principles are based on optical systems ("Fire Eye"). These systems are not very common in the aluminium industry.

The rotating speed of the drives must be controlled in some applications. This could be the r.p.m. of the drum of rotary drum furnace or the casting speed of the ingot casting machine. To simplify things, the actual speed is not measured but the set point of the speed control is used as indicator. In most cases this practise is acceptable but we must be aware of the fact that we do not get the real value and slips between control and actual motion are not taken into consideration. If more precise reading is required or the speed is to be used as control parameter speed, indicating units, such as tacho-generators, must be used.

8.1.2 Actuators

It is very convenient to have the status of the plant and a reading of the important data. To have them processed the equipment has to be actuated. In most cases this will be realized by electric motors. The rotating speed of an electric AC motor depends on the frequency of the power supplied to it. By controlling the frequency, the speed of the motor can be varied. The frequency can be adjusted by a thyristor. All the complicated systems of infinitely adjustable gear boxes or electric frequency converters are replaced by the thyristor control. Only in very specific cases the controllable hydraulic drive may offer advantages.

Typical applications for frequency controlled motors are rotary drum furnaces or casting machines or the large electric motor of the waste gas filter.

Linear movements are either actuated by hydraulic cylinders or by pneumatic cylinders. Both types are operated from a control valve rack or a hydraulic kit. Hydraulic operation permits, in connection with the relevant electric control, very sensitive movement of equipment components.

8.2 Control architecture level 1

The basic control systems comprise the interlockings and sequence control as well as the automatic process control.

8.2.1 Interlockings

The sequence controls with the relevant interlockings are based on a certain status and position input which is provided by field instruments and limit switches. Manual input may also be a criterion. All sequences and interlocking follow logical statements such as AND, IF, NOT, OR and the like: For example, the charging machine can move forward IF the door is in open position AND the burner is switched off AND material is in the charging machine. The latter was an input of the operator using a push-button. Other options are linked to an OR option. The furnace can be filled from the operator's panel OR a push-button station near the launder. But in case no operator is available or the furnace door is closed, operation NEITHER from the push-button station NOR from the operator's panel is possible. The interlockings generally involve safety settings which cannot be changed by the operating personnel as it may be possible for operation data and settings.

8.2.2 Automatic control loops

While these functions are good for interlockings and sequences they are not suited for automatic control loops although they may be integrated. The major control loop in an aluminium recycling plant is the temperature-dependent firing of the burners. We could simply control this by an on / off method but we all know the result. The burner will fire with full power. As soon as the set-point is reached, the burner shuts off. Due to the inertia at the furnace – the refractories are very hot – temperature still raises above the set-point and having reached a peak it comes down to

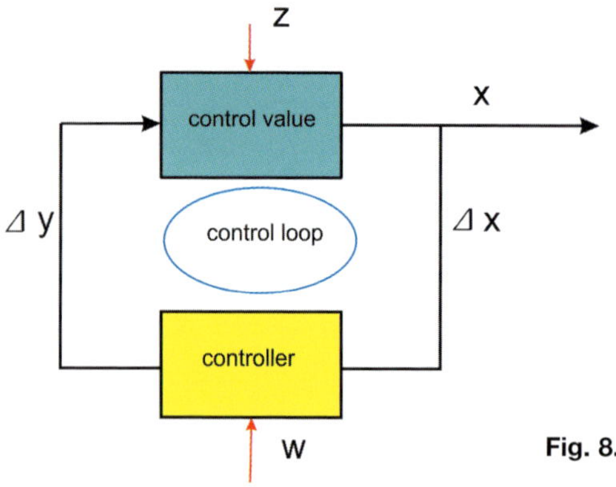

Fig. 8.12: Automatic control loop with feedback

the lowest permissible level. The burner starts firing again and the whole story is repeated. What we need is a control system with feedback where the deviation of the set-point has influence on the setting of a continuous controller. **Fig. 8.12** shows the principle of such a closed control loop.

For the definition of the control loop the following designations for the parameters are in use (**Table 8.1**).

Table 8.1: Designations for parameters

	Reference	Example
w	Set point, command variable	Required temperature
x	Actual value, control value	Actual temperature
y	Control output, correcting variable	Opening of gas valve
z	Disturbance variable	Temperature fluctuation
Δx	Deviation from set point	

The function of the control loop comprises different elements (**Fig. 8.13**). The operator defines a set point w which is processed by a controller or the software of the PLC. This generates a signal y1 which is submitted to the actuator, i. e a control valve. The actuator now submits the correct-

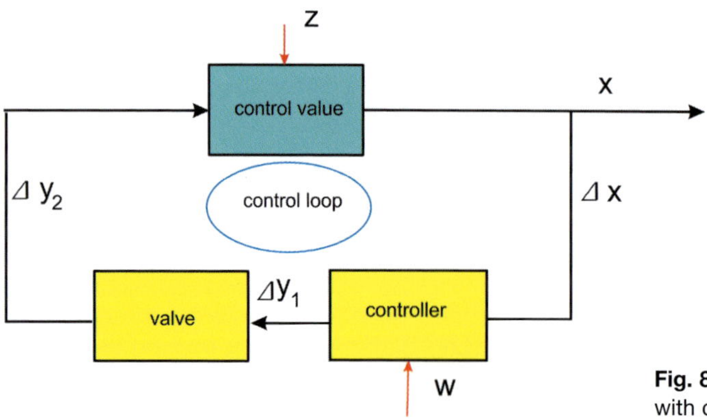

Fig. 8.13: Closed circuit control loop with controller and control valve

ing signal y_2 to actual process unit. The resulting x value is from a sensor and returned to the controller for processing. The deviation from the set point of the process is called Δx or e. Thus, the loop acts in the sequence

> Δx ⇒ w ⇒ x, in words: deviation from set point ⇒ set point or command variable ⇒ actual value.

This shows that the change in valve position depends on

- the deviation from set point: Δy~ (w-x) ⇒ Δy~Δx
- the product of deviation and time: Δy~ (w-x) · t ⇒ Δy~(Δx) · t
- the change of deviation per unit of time Δt: Δy~ (w – x)/Δt ⇒Δy ~ Δx/Δt

The control of the temperatures in the furnaces is done by the most important control circuits. Other loops refer, for instance, to hydraulic oil temperature or the tilting of a furnace.

The deviation from the set point will give a signal to the controller which will try to compensate for this deviation. In the case of our furnace, the controller will give a signal which may react with a time delay to the gas valve to open or to close. There still might be an over-shooting of the temperature but, due to the closed loop, the oscillation will flatten and come to an end very soon (**Fig. 8.14**).

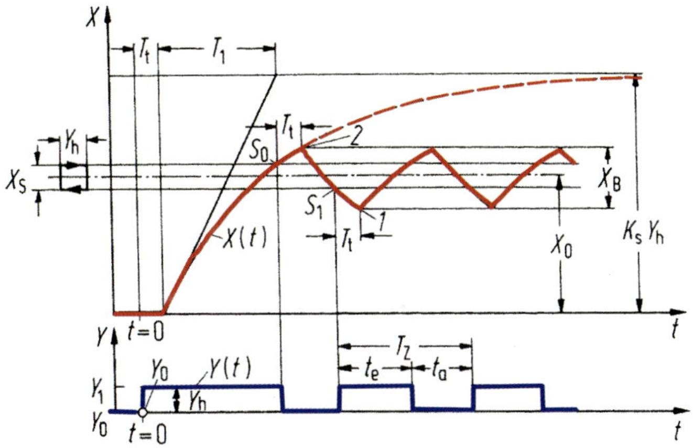

Fig. 8.14: Typical on-off control

Different controller systems are applied in industrial processes.

The two-point or on-off controller is the simplest system for controlling temperature. Since setting must be between a maximum and a minimum temperature and this causes a permanent oscillation (**Fig. 8.15**). It will also take some time for the temperature to rise, expressed by the residual time TR which is defined as the time for the controlled variable to reach a specified value. TR will, for instance, be much longer for controlling the roof temperature than for the flue gas temperature.

Better control can be achieved by the so-called “three point step controller”. While the two point controller knows only two positions such as, for instance, “heating” and “off”, the three point controller knows a third position, for instance “cooling”. With this characteristic a quasi continuous control can be achieved that may be sufficient for simple control loops.

The time-dependent behavior of the controller and the resulting response of the control loop depend on the specific characteristics of the controller. For defining the controller characteristic a possible deviation is defined by a sudden change of the controlled parameter, i. e the x-value.

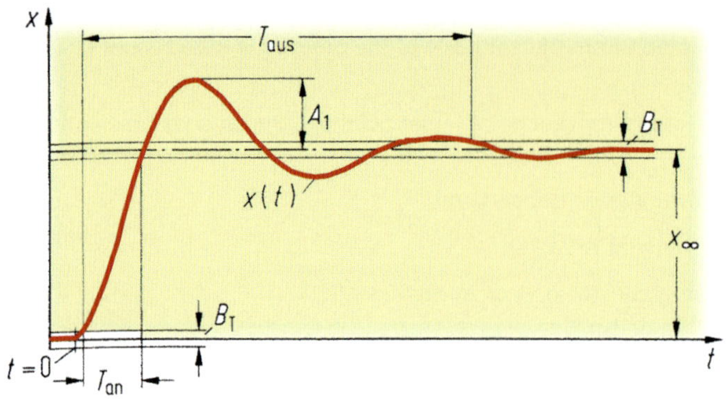

Fig. 8.15: Control behavior

This cannot be obtained in real life but it can be simulated in the software.

The P-controller will respond with a signal x that is proportional to the deviation y. Thus, the basic function is

$$y = K_p \cdot \Delta x \tag{8.1}$$

The factor Kp is called the amplification or proportionality coefficient which is responsible for the reaction of the value of the deviation of set point and actual value. If the value of K_p is too large, the control loop tends to generate oscillations. The advantage of the P – controller is the fast reaction upon occurrence of a deviation. If K_p is selected carefully, the controller provides a very stable loop. The disadvantage is an unavoidable permanent deviation since the controller can only react upon the presence of a change of deviation.

Another characteristic is the floating controller, called I-controller. It follows the equation:

$$y = \frac{1}{T_N} (\Delta x) \cdot t \qquad \text{or } Ki = 1/T_N \tag{8.2}$$

$\frac{1}{T_N}$ · is called the response time and K_i is the integration coefficient. Both define the velocity of the corrective action. Thus, the I-controller is characterized by the velocity of the corrective action x. In other words, the velocity of the corrective action x is proportional to the deviation y. Since y will change during its adjustment, the controller will sum up (integrate) the response value until the set point is reached. Therefore, the controller is called integral Controller I. The advantage of the I controller is that there is no remaining deviation during the steady state condition. However, with a large T_N the loop is very slow while a small T_N may cause the control loop to oscillate and even may result in an instable situation.

The D-controller is different again. The corrective action is proportional to the velocity of the change of the deviation following the equation

$$y = T_v \cdot \Delta x / \Delta t \tag{8.3}$$

T_v is called the idle time. This is the time passed before the controller starts acting.

The D-controller (derivative controller) cannot be realized technically. It is only possible in combination with the P-controller. It will reduce the overshooting but still a permanent deviation, however much smaller than with the P-controller, will remain.

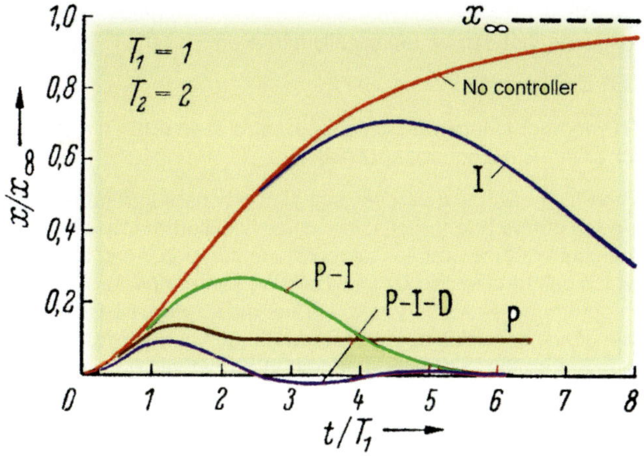

Fig. 8.16: Control characteristics of different controllers

The different characteristics bear a distinct influence on the quality of the control loops. There will always be a first overshooting of the set point and the setting will enter into a damped oscillation. The amplitude and settling time depend on the parameters of the control loop, specifically on the time factors (**Fig. 8.16**).

For technically meaningful control, a combination of the different controllers is combined. The P-I controller combines the fast response of the P-controller with the accuracy of the I-controller. The PID-controller will slow down the controller settings to obtain a steady state of the control loop. **Fig. 8.17** shows some typical responses of the controllers involving a sudden controller input

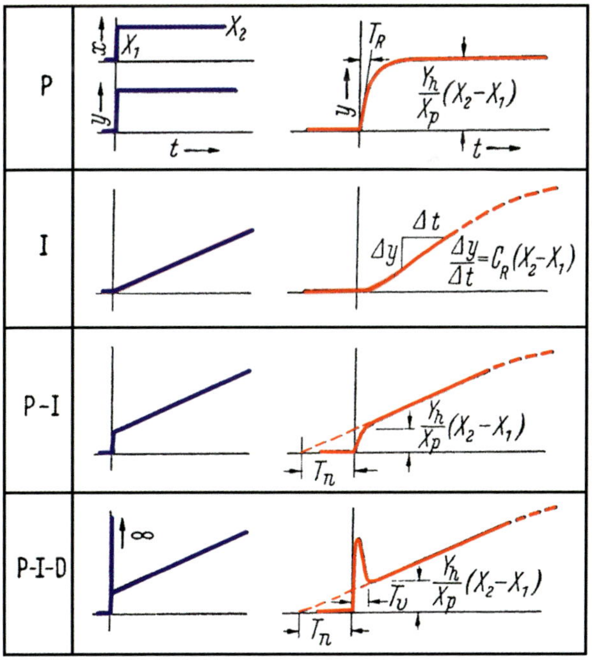

Fig. 8.17: Controller response

(deviation). The left column indicates the theoretical response while the right column shows the response that can be technically achieved.

Let us summarize:

A Proportional-Integral-Derivative control loop (PID) is a standard feedback loop component used for all industrial control applications.

The PID measures a process signal of a sensor and controls an "input" with the goal of maintaining the output at a target value which is called the "set point". An example of a PID application is the control of the process temperature although it can be used to control any measurable variable which can be affected by manipulating some other process variable. For example, it can be used to control pressure, flow rate, air/fuel ratio, force, speed or a number of other variables. Automobile cruise control is an example of an application area outside the process industries. The basic idea is that the controller reads an input signal of a sensor (i. e furnace roof temperature). Then the controller subtracts the measurement from the desired "set point" to determine an "error".

The error is then treated in three different ways simultaneously:

Proportional – to handle the present. The error is multiplied by a negative proportional constant P and sent to the output. P represents the band where the controller's output is proportional to the error of the system. For example for a furnace, a controller with a proportional constant PR of 100 and a set point of 700 °C would have an output of 100 % at 70 °C, 50 % at 350 °C and 10 % at 630 °C. Adding the change to the output makes the output self-adjusting. For example, if the burner were to get dirty, decreasing the heater's efficiency, the base line output would just drift upwards a bit and then re-stabilize.

Integral – to handle the past. The error is integrated (or averaged or summed) over a period of time, then multiplied by a constant IR, and added to the proportional output. It represents the steady state error of the system.

Using the proportional term alone, it is not possible to reach a steady set point temperature. Real world processes are not perfect and are subjected to a number of environmental variables. As these variables are often constant, they can be measured and compensated for. Using the proportional example above at 19.9 °C, the controller output is 1 % and at this temperature environmental losses by way of heat transmission are 3 %. In this scenario the system controller will never be able to reach set point. However, it can be corrected by introducing an integral term which will attempt to remove errors that last for some time. In practice, the integral term of a controller only considers a relatively short history of the controller.

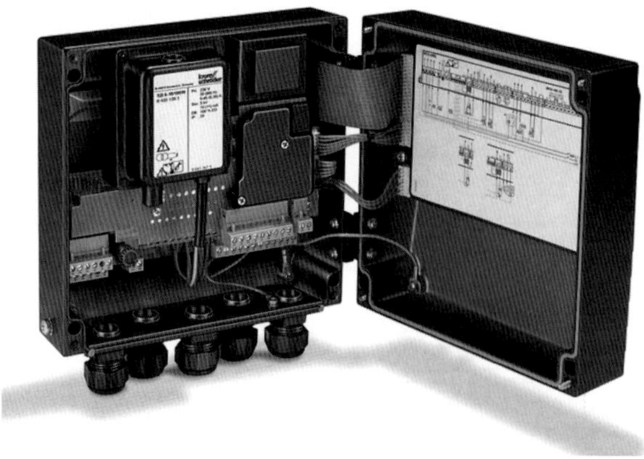

Fig. 8.18: Independent temperature control and safety device for burner control

Derivative – to handle the future. The first derivative of the error (its rate of change) is calculated in respect to time, multiplied by another constant D and summed with the proportional and integral terms. The derivative term is used to govern a controller's response to a change in the system. The larger the derivative term, the more rapidly the controller will respond to changes in the process value. It is a good thing to reduce when trying to dampen a controller's response to short term changes.

For setting the control parameters it is helpful to use the procedure defined by Ziegler/Nichols. It starts with P-control which only means D-control and I-control are switched off. Now the controller is set manually until the control loop enters a harmonic oscillation. Period time T_{crit} and the manually set proportionality factor K_{crit} are noted. By using these values the parameters K_P, T_N and T_V are calculated by using the experience-based equations.

P-control	$K_P \approx 0.5 \cdot K_{Pcrit}$
PI-control	$K_P \approx 0{,}45 \cdot K_{Pcrit}$
	$T_N \approx 0.85 \cdot T_{crit}$
PID-control	$K_P \approx 0{,}45 \cdot K_{pcrit}$
	$T_N \approx 0.85 \cdot T_{crit}$
	$T_V \approx 0.12 \cdot T_{crit}$

Using these values and a bit of experience of the process engineer, it is certainly possible to arrive at a stable and fast reacting control loop. The parameter can be set at the controller device (**Fig. 8.18**) by turning the dedicated knobs or in more recent plants by using the Simatic software.

8.2.2.1 Controls for combustion system capacity

Not all of the control loops can be fully realized over the entire range of an operation on the principles of automatic control. One typical example is the temperature control of furnaces which we will discuss in more detail. In metallurgical processing of aluminium, heat is one of the most important factors. Heat control – that means temperature control – is decisive for metal processing and energy efficiency. This explains the focus on temperature control during the entire process control.

In a rotary drum furnace the only temperature parameter available is the flue gas temperature. The wall temperature is normally approximately 100 °C lower: The flue gas temperature now controls the burner input.

The combustion system capacity is controlled by means of a temperature controller which uses as input signal the temperature of the furnace roof, the waste gas temperature or the temperature of the melted metal. The temperature controller acts on the gas or air regulating valves and together with the gas/air ratio controller adjusts the capacity of the combustion system to the value required for furnace operation. With electronic controls, if the system capacity changes, the air flow signal is predominately increasing capacity and the gas flow signal is predominate if the system capacity is decreasing. In that way it is ensured that the system during the capacity change stays in the air excess range. When using a pressure type gas/air ratio control, the temperature controller acts on the air control valve. The gas is then controlled by the zero pressure or difference pressure regulator according to the air pressure signal. With preheated combustion air, the air amount measurement is corrected accordingly to the actual air temperature. With hot air systems and a pressure-type control system, the measuring orifice for generating a signal to the gas regulator is placed in the cold air pipe.

In aluminium melting furnaces, the roof temperature is used as the input signal in the first part of the melting cycle when solid metal is in the furnace. After the charge is melted down, the temperature control switches to the bath thermocouple and the metal temperature is used as the input signal.

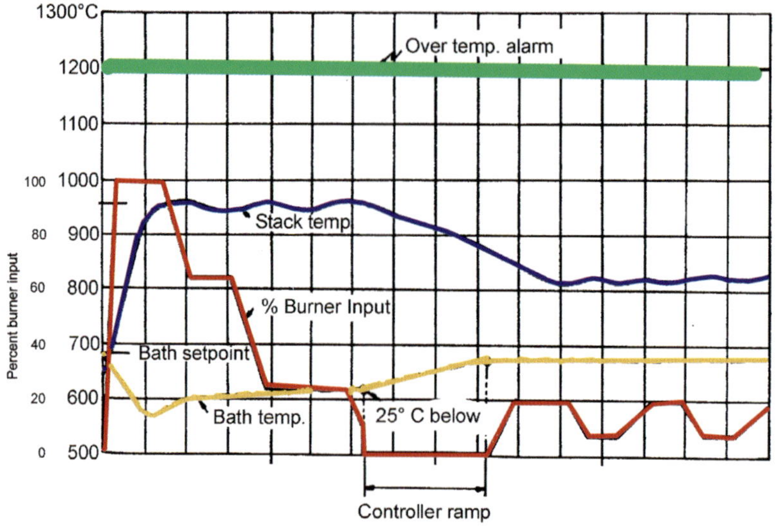

Fig. 8.19: Temperatures in an aluminium melting furnace

There are generally the following methods of changing the burner system capacity:

- modulating control (gas and air flows are controlled continuously)
- ON/OFF (or ON/LOW/OFF) control (the burners are working at full capacity; the capacity change is achieved by the changing of firing or breaking time intervals
- cascade control which is a combination of modulating and ON/OFF control (the burners are controlled down to a certain capacity and then the control occurs in the ON/OFF mode).

Referring to the furnaces used in the aluminium recycling industry we need a closer look at the controls.

Fig. 8.19 shows the furnace's typical bath and wall temperature and percent burner input profiles for a complete melt cycle using a single temperature controller controlling with the flue gas temperature. At some time into the cycle, the bath temperature begins to rise. At this point the flue gas temperature approaches 900 °C and continues to rise until the bath thermocouple senses a bath that nears final set point. Furthermore, at this time the solid metal is going through a phase change and will not readily accept more heat input. Therefore, from a certain period on, practically all heat input is exhausted out the flue, thus wasting fuel. Depending on the amount of heat input required, the burners are operated between 0-100% of their firing rate. Stack gas temperature is limited to 900 °C.

The burner control sequence is as follows:

1. The doors are opened and the furnace is charged, at this time the burner is on low fire.
2. The doors are closed; the operator enters the program number for the particular alloy. This also determines the final set point for the bath temperature.
3. The controller, acting on flue gas temperature, takes control of the burners and controls the flue gas temperature to 900 °C. When this set point is reached the one burner is reduced to low input.

Once bath temperature is assumed to have the correct temperature, the flue gas controller cascades the burners to hold the bath temperature.

All burners require close attention to maintain the proper fuel/air ratio. The burners in melting furnaces are to be checked and adjusted. A step-by-step manual is used to aid the operation.

A stack analysis is frequently done periodically and an effort is made to keep the burners between 8 and 12 % excess air.

In general, the major sources of heat transferred in a flat bath aluminium furnace come from radiation. The sources are re-radiation from the refractory material, from the gas blanket (POC) and from the flame envelope. The magnitude of the heat transferred depends on the temperature difference between the surface areas of the source and receiver. In an aluminium holding furnace, the area of the receiver is limited to square footage of bath surface.

As the bath temperatures increase, the amount of heat transferred to the metal begins to decrease. To maintain heat transfer levels, higher and higher differential temperatures are required between the radiating sources and the bath surface. The major contributor to heat transferred to liquid metal is the re-radiation from the refractory material to the bath surface.

The goal is to bring the refractory material to operating temperature as uniformly as possible. This will ensure maximizing the heat transfer to the entire bath surface available without local hot spots.

Radiation from the flamed envelope to the bath is appreciated. But when the heat release produces areas of intense heat transfer, localized overheating occurs. The surface under the flame is raised to a temperature well above the average surface temperature. In addition, it is the flame envelope that contains free oxygen. Therefore, the areas most subject to oxidizing have the greatest source of available oxygen and potentially H_2 pickup.

The goal of the combustion system is to raise the temperature of the refractory material rapidly and uniformly to provide maximum and uniform heat transfer to the bath.

If the bath can accept the heat transfer uniformly, the bath will be uniform. In actual operation this is difficult or impossible to achieve. Conductance of heat from one area to another is necessary to achieve uniformity. However, if the combustion system can provide uniform refractory material temperature without areas of significant higher temperature on the metal surface, it has produced an environment subject to uniform heat transfer.

As discussed before, it is often necessary to have a combustion system that, when needed, raises the bath temperature rapidly to casting or tapping temperatures. This is an added burden to the system design. It requires burners with greater input than necessary for holding but with the turndown to provide efficient holding, minimizing oxidation (on ratio) and maintaining uniformity. The best option to raise the metal temperature rapidly is high velocity burners.

The problem here is now to maintain the benefits of high velocity burners while providing adequate on ratio turndown and maintain furnace temperature. With high velocity burner turndown, the outlet velocities decrease and the overall effectiveness decreases. Using the high velocity burners, in conjunction with a modified pulse fire controls, eliminate this problem.

Pulse fire control is a system that fires the burners at high fire on a time basis that depends on the difference between the set point and the process variable (PV). If the furnace temperature is considerably lower than the set point, the on time of the burners is reduced for the stop time period. As an example, if the stop time period is one minute then the control system will address the burner on time every minute. If the set point and process variable are far apart, then the control system tells the burners to stay on the full 60 seconds or one minute. If in the next minute the set point and P.V. come into the control range, the controller will decrease the burner on time for the next minute. Let us say the control stays 90 % at on time and then the burner will fire only 54 seconds during the next minute. At 80 % only 48 seconds and so on to the minimum on time of 5-7 seconds. Below the minimum, the burner stays in the off condition (air and gas are off).

The effectiveness of the burner to circulate gases and maintain uniformity requires on time without overheating. The input requirements, holding versus heating, are dramatically different requiring two input rates. By reducing the input rate as the set point is achieved, increases on time without overheating or temperature peaks. High velocity burners lose some effectiveness between a 100 % design input and 50 % input. Modifying the pulse fire system to reduce input, as the set point is changed from 100 % to 50 %, increases on time and reduces overshoot in temperature. Uniformity is maintained and the heat is still released in the furnace rather than the outlet of the tile.

Once the system is in holding mode, it is important to avoid overheating at the tile outlet. Releasing heat here is not effective in heating the furnace and this causes overheating of the metal in contact with the refractory material. This can cause increased build-up of dross in that area and, depending on the alloy, premature refractory material failure.

When controlling a furnace in holding condition a major problem is control of the heat-up rate without overheating if the furnace is boosted in its temperature when the metal temperature is raised quickly. What happens to the heat built up in the refractory material once the metal reaches temperature? Two scenarios are possible with standard types of control system:

- Metal temperature control – The burner system responds only to metal temperature. The metal comes to temperature faster but there is much heat built up in the refractory material that even with the burners shut off the "fly wheel" effect occurs. Heat from the refractory material transfers to the metal and the metal surface exceeds temperature subsequently increasing dross and metal loss.
- Roof temperature control – The burner system responds only to the roof thermocouple. The roof set point is depressed to avoid overheating of the metal. The time to heat the metal to temperature increases.

Cascade control is a simple system using metal temperature and roof temperature to determine the burner firing rate. **Fig. 8.20** shows a 2-loop cascade control system.

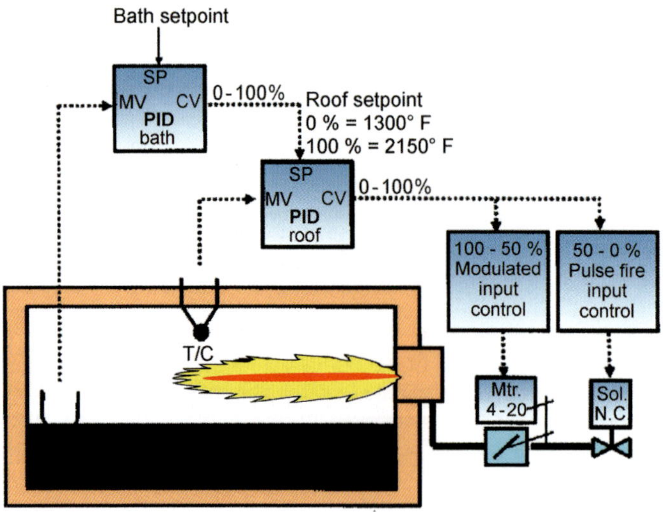

Fig. 8.20: Cascade control of an aluminium melting furnace
(Source: Bloom Engineering)

Control loop 1 – monitors the metal bath temperature and varies the 4-20 mA output based on the difference between the bath temperature and the set point. If the metal is at or above set point, the output is 4 mA. If the bath temperature decreases, the output increase to a maximum of 20 mA is necessary.

The 4-20 mA output from loop 1 goes to loop 2 as the set point for loop 2. The 4-10 mA signal is scaled in the loop 2 controller for a set point range generally around 704 °C to 1,176 °C. This means that if the input is at 4 mA, the roof set point is 704 °C. At 20 mA, the set point is 1,176 °C. The input varies depending on the bath temperature.

Loop 2 monitors the roof temperature and adjusts the output 4-20 mA to the firing rate controller based on the deviation of the roof temperature from the set point.

In a pulse fire system, a 20 mA signal would hold the burners at 100 % capacity for the entire phase time period (burners are on constantly). At 4 mA the burners would be at 50 % capacity and the minimum on time (5-7 sec. of every minute).

With bath to roof cascade control, a rapid increase in temperature can be achieved while minimizing the bath surface via temperature condition.

If it is necessary to increase the metal temperature, the set point on control loop 1 is raised and the output drives up from the holding condition to 20 mA. This in turn drives the roof set point on loop 2 to 1,200 °C. The burners are driven to 100 % of capacity and 100 % on time. The burners then modulate to hold the 1,200 °C temperature.

As the metal temperature increases, the control output decreases and, subsequently, decreases the roof set point temperature of loop 2. The output of loop 2 will begin to decrease the firing rate of the burners to maintain the set point. Once the burners reach 50 % input, the burners stop modulating and begin stepping. Any further decrease in the firing rate will start depressing the on time of the burner within the pulse cycle.

As the metal temperature increases, the roof temperature decreases and with proper adjustment of the PID parameters the bath temperature can be raised efficiently and in a timely manner without suffering overheating from "fly wheel".

Controls for regenerative burner system

The regenerative burner systems have, on the gas and combustion air side, similar controls like the conventional burner systems – see also Fig. 3.34 – (the air measuring is on the cold side of the regenerator). Additional waste gas flow control is needed which is responsible for suction of a proper waste gas amount through the regenerator ("pull back": approx. 90 % of the actual amount of the generated waste gas). Also a system for the switching of regenerators has to be installed. The switching valves (one for the air, one for the waste gas for each regenerator) are operated in reverse mode: each 40 – 90 seconds once the switching occurs.

In case that the waste gas temperature behind the regenerator is higher than 200 °C, the switching occurs earlier. In the short time, when valves are switching (1-2 seconds), the controls (gas, air and waste gas flow control, furnace pressure control) are "frozen".

This means that the controllers do not "realize" the switching process and after this short period the controllers are working further continuously.

The application of recent high-performance PLC systems allows the design of very efficient controls for regenerative burners.

The requirements in regard to reducing the carbon footprint control of combustibles in the flue gas become more and more a required task

Scrap processed in aluminium melting furnaces is very often contaminated by organic waste products. Their complete combustion must be assured in order to fulfil environmental requirements. These contaminations are occurring in variable quantities. It is, therefore, essential to control and monitor the combustion process. Besides the fuel efficiency of the furnace, it is important to ensure that no unburned components are released (**Fig. 8.21a**) into the environment. These non-burned components are generally CO and H_2. During incomplete combustion carbon monoxide and hydrogen occur together in the flue gas. These combustibles are measured by an oxygen probe. The signal is processed to control the excess air of the burner. As soon as the oxygen content in the flue gases increases, the air quantity is reduced. However, measuring the O_2 content only cannot provide a clear indication of the complete combustion. Information on the proportion of H_2 and CO is also required. Lamtec has developed a concept to take care of this problem. It was necessary to develop a new probe that can detect O_2 and CO simultaneously. The sensor is designed as a zirconium dioxide probe. The system is complete with a sophisticated control system designed as compact unit that can also communicate with the overall process control system of the plant (**Fig. 8.21b**).

8.2.2.2 Fuel ratio control

In order to obtain the control loops, as described above, the energy input to the furnace has to be controlled. This is realized by controlling the flow rate of fuel to the burners according to the temperature in the furnace whereby the relevant controller acts, as described above, in the defined control behavior. Since oxygen is required for combustion, the air (or oxygen) flow must follow

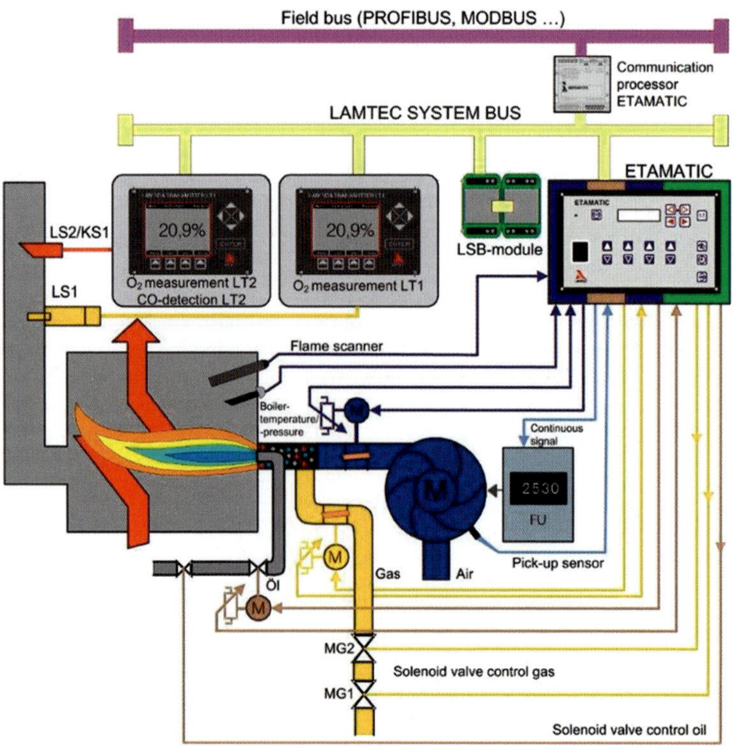

Fig. 8.21a: Lamtec combustion control system for O_2 and CO

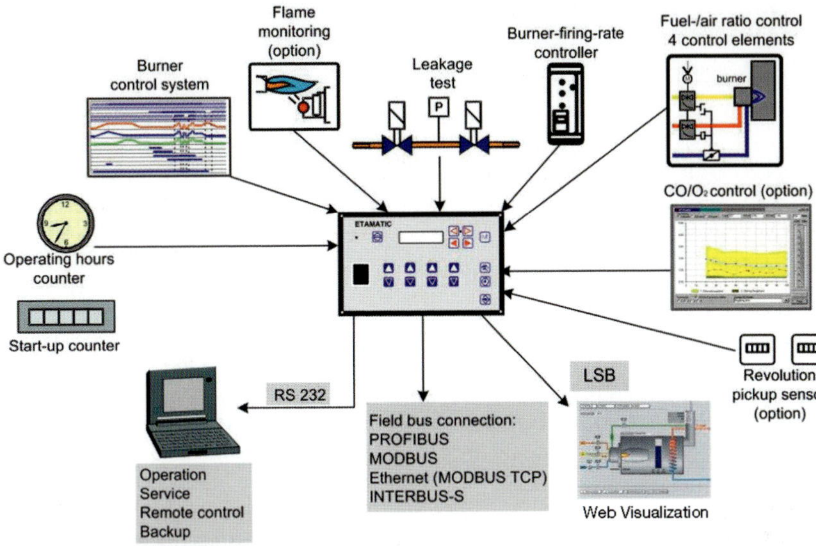

Fig. 8.21b: Lamtec burner system with remote control

the settings of the gas in the required ratio. The burner control comprises, therefore, controllable gas valves as well as controllable air valves. For safety reasons the control must ensure that no excess gas is present in the furnace. When increasing the energy input, the air valve is set first, followed by the gas flow in the required ratio. The process is reversed when the energy input must be reduced. Then the gas flow is reduced followed by the air flow, as required. The defined gas/air ratio depends on the furnace cycle. Consequently, more air volume is required in the de-coating phase of a melting furnace. The ratio, as required, can be set manually as well as by the process control system defining the necessary steps in the furnace sequence. The actual flow rates are measured by flow meters for gas and air in the feed lines to the burner. The efficient gas/air ratio control is an important factor for a cost-efficient and safe operation of combustion systems. The controls have to ensure that the combustion system works with a proper air excess. Too low air amount can be dangerous for the furnace operation. A too high air excess increases the energy consumption (an increase of air excess by one point increases the energy consumption by approx. 2 %) and can increase the metal loss (a too high content of O_2 in the furnace atmosphere causes an increase of a dross generation).

There are two main methods of the gas/air ratio control in combustion systems in aluminium furnaces:

- controls by means of pressure type regulators
- electronic controls by means of a gas and air amount measuring and a controller which acts on the gas and air regulating valves

The gas/air ratio by pressure-type regulators can be realized by means of zero pressure regulators or difference pressure regulators. The principle of such a control system is shown on **Fig. 8.22**.

With such systems it is important to use higher gas/air pressure in a burner (at least 40 mbar is recommended), especially if the capacity of the burners is planned to be controlled in the range 1:5 or higher (the pressure in a burner decreases in square potency of a flow). Usually, the ratio control is realized by setting the gas flow as required. The movement of the gas valves is con-

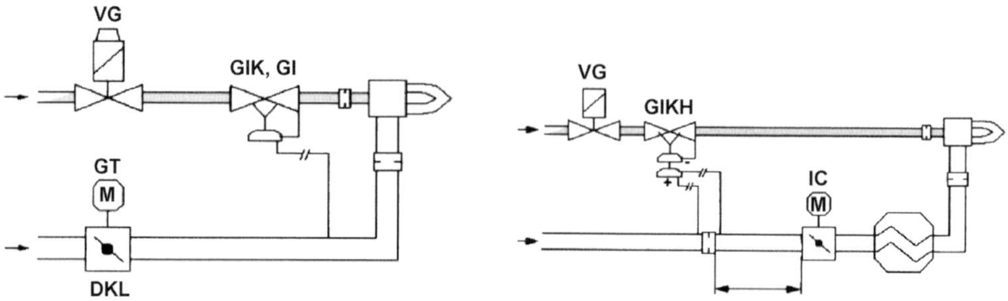

Fig. 8.22: Burner control systems for constant pressure and independent control of gas and air

nected to the air valve by a mechanical system. The λ- value is set permanently and cannot be changed during operation, as may be required. A modern electronic system is much more flexible. Setting of λ is also realized permanently by a curve for the complete firing range. But the control unit allows for different curves for different λ values which can be addressed during operation even automatically, if required.

The other way to control gas/air ratio is an electronic way. The flow of gas and air is measured by means of orifices with difference pressure transmitters. The signals from the transmitters go to the controller which compares the signals and controls the flows by means of gas and air regulating valves so that an adjusted set point for the gas/air ratio is achieved. This method offers much more flexibility for setting air quantities during operation. This is required in most cases considering the different organic components of the scrap to be processed. It also permits adjustment of the air fuel ratio during different phases of the furnace cycle.

8.2.2.3 Burner safety

For the operation of burners very precise regulations have to be applied. This also applies to the burner starting sequence which in some countries must not be realized in the PLC control but requires a separate burner sequence unit. Prior to the start of any burner it must be assured that no combustibles are present in the furnace room which may cause an explosion as soon as the burner is ignited. Thus, for start-up, blower air is supplied for some time to remove any combustibles. This may take as much as close to one minute in some cases. Then the pilot burner is ignited and supervised by a photo or UV cell (**Fig. 8.23**). After a short period of time, the gas

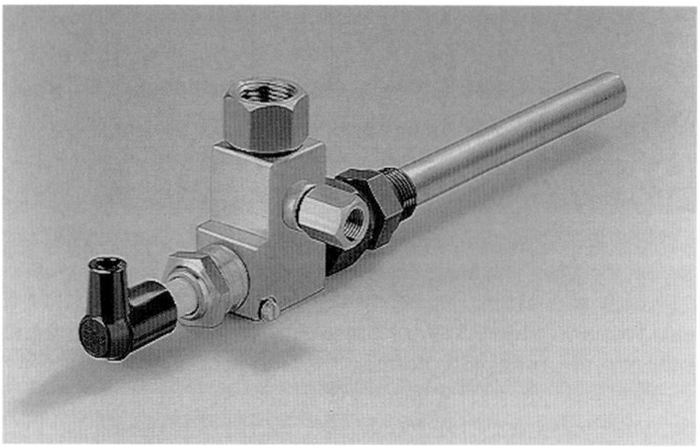

Fig. 8.23: Flame detection unit on ultraviolet basis

valve opens and the burner is ignited by the pilot burner. After a short time, the pilot burner is switched off and a photo cell now monitors if a stable flame is obtained. If not, the gas valve is closed and the whole starting sequence is repeated, including the rinsing with fresh blower air. The start sequence will be executed automatically as soon as the start signal for the burner is initiated. The timing for the different steps of the sequence must be planned upon considering the particular conditions of the installation. For instance, if the time for the operation of the pilot burner is too short, the burner may be shut-off before the gas pipe to the burner is filled with gas and the operating pressure is obtained. Flushing with fresh air is not required if the temperature in the furnace exceeds 900 °C. In some operations flushing could be a problem. If metal is tapped and slag is discharged, the temperature in the drum is usually below the mentioned 900 °C. Flushing would cool the furnace furthermore and the time required to start the burner is lost for production. If the temperature drops below 900 °C during charging, the burner sequence starts again for every charging. Again the result will be loss of energy and operation time. The temperature inside the rotating drum cannot be measured, neither during operation nor at a stopped rotation. This will add one more problem. Here a compromise must be accepted by having the pilot burner in operation even with an open door. The excess gas, which may have reached the furnace in a very unlikely instance, would have been ignited and burned by the combustibles charged with the scrap anyway. Consequently, after the door is closed the burner will be switched to normal operation immediately with increasing energy input and modulating control. There are other safety devices installed in the valve rack. One is the automatic leakage control. This gives an alarm signal if the gas pressure drops during an automatic test procedure. A shut-off valve close to the burner ensures that the gas flow is stopped immediately as soon as the burner receives the stop signal. Thus, no gas remaining in the pipelines can flow into the hot area.

8.2.2.4 Motion control

To be able to operate the equipment the different motions require control. These may be very precise, more general or following the normal principles for drives and actuators.

The rotation speed of electric motors is infinitely controlled by frequency control. The r.p.m. of the drum of a rotary drum furnace requires control in order to adjust the furnace to the specific process steps. This also requires the change of direction of the drum rotation. When changing the speed and the inertia of the drum, a time factor will occur. This is not critical for the process and can be accepted without problems. As soon as the drum stops, a motor brake will hold the position. The speed can be set either by the process control unit or manually from the operator's panel.

All other drives in the rotary drum furnace are operating without speed control but brakes are required to either stop the motion, as it is the case with the unbalanced drive motors of the charging trough, or the motion is stopped to hold the position at the forward drive of the charging machine.

The casting machine requires a frequency-controlled drive. The forward movement of the casting conveyor must be adjusted to the casting velocity during start and stop of casting. Also fine-tuning is required to stabilize the casting speed during operation.

The speed of the large blower of the waste gas filter must be adjusted to adjust the system to the varying air requirements of the plant operation. Therefore, in this case a frequency-controlled motor is required as well.

The control of speed for furnace tilting is very important. As soon as the batch in the furnace is ready it must be poured into the casting furnace or into the casting machine. The different actions require different pouring velocities. To empty the melting furnace it has to be lifted with fast speed into the pouring position and then slow motion is required to adjust the metal flow as required for metal transfer. This requires a proportioning valve in the hydraulic circuit that is by-passed by a fast motion valve. A special electronic card is required to control the proportioning valve. The tilting velocity of the casting furnace requires very precise motion control. After the furnace is lifted into casting position, the proportioning valve is controlled to maintain a constant flow of metal. The control parameter is the degree of filling in the casting launder by a laser probe or other means of level measurement, as described before. The tilting motion during casting is very slow. Precise tilting over the entire casting period is required which may last up to four hours per furnace.

After casting or metal transfer the furnaces tilt back at fast speed into the starting position. The speed for the backward motion is controlled by pre-set hydraulic valves. This motion is also required if a power failure occurs. The furnace must then tilt back in order to arrive at a safe position where no metal overflow can occur. The relevant hydraulic valve must have a spring-operated return position with no electric power required.

An automatic sequence control is initiated for opening furnace doors. In a reverberatory furnace the door is first tilted from the closed position and then lifted by the hydraulic or mechanic system. The entire movement should commence as long as the push-button near the door is pressed. If it is released, the door must stop movement and tilt in closed position and then remain in the lift position. The top position, as well as the conditions in the lowest position, are controlled by means of limit switches. Stop and go motions are softened by special hydraulic valves with delayed shifting action. No electric control is required.

This is different for the door motion of a tiltable rotary drum furnace which is characterized by having a heavy door with high inertia swivelled into open or closed position. To be able to handle the forces caused by the inertia, a proportioning valve controls the flow of hydraulic fluid via a ramp in the electrical control circuit. The ramp is initiated by limit switches in relevant positions.

8.2.3 Control hardware

The controls and control circuits are generally realized today by using free programmable controllers (PLC). The control loops and interlockings are programmed with software and a programming unit, usually by the supplier of the hardware. A wide range of pre-programmed controllers is available. Thus, all varieties of circuits can be realized based on the program defined by the electrical engineer for the specific equipment. The communication between program, controllers, field instruments and power switches is usually based on a 4-20 mA level. 20 mA is the maximum level for all settings while 4 mA is the zero point. This helps to define zero very accurately and also provides a safety value. If, for instance, a thermocouple breaks, the signal will be 0 mA and not 4 mA. This can be used to initiate an alarm. The control loops are more and more based on fuzzy logic systems. This logic does not follow the definition “yes” or “no” but also with expressions “almost correct” and “almost wrong”. Most of the processes work on these principles which are more adapted to real life. The result is a faster response to the deviations regarding the set point.

The PLC system (Fig. 8.2) is accommodated in a control cabinet. Attached is the cabinet for motor and actuator control with the necessary switches and indicators. A small indicating panel permits to read all data and every status of the equipment. Each piece of specific equipment has its own panel. These units are located in one common switch room. This room should be air-conditioned to ensure proper functioning of the electronics. It is common to have cabinets arranged on a raised floor. This makes it easy to interconnect all panels as required. The cables to the equipment leave the control room at floor level and run to various equipment via tubes embedded in the concrete floor or via cable racks. It is common practice to have control cables and power cables fitted separately. Hardware interlockings at the equipment should be avoided as far as possible. However, sometimes they are required due to safety regulations.

The equipment control is designed and installed so that any equipment can be operated independent of any other equipment. This is necessary in case of trouble involving a furnace or a casting machine. In this case the remaining plant must still be able to operate without problems.

8.3 Control architecture level 2

The individual equipment control system communicates via a local area network with the process control system (Fig. 8.1). This comprises a system for cast house control and a production optimization system. The cast house control monitors and directs the operation of the plant. It gives the status of the equipment, the range of temperatures during the last period, operation time for defined process steps such as melting, metal preparation, casting and the like. A summary of gas and electrical energy consumption as well as the names of operators are recorded. A very important feature is the signalling of alarms. This requires by all means immediate action to avoid damage to equipment and production interruption. The alarms are individual on different levels. During night shifts they may be sounded in the gate house and simply indicate the area of trouble. In the production plan a flashing light and an acoustic signal may indicate that there is some trouble. At the level 2 computer, the exact location and the kind of trouble, i. e excess temperature in casting furnace 2, is indicated. The general alarm is maintained until the alarm is acknowledged. The flashing value signal on the computer screen changes to steady red – the normal data may be green. To acknowledge the alarm the operator must enter the level 2 program with his password. Time and access are recorded. Thus, the system gives an exact indication what has happened in the plant, not only by the alarm system but also by the indication of process parameters and their trend over the past hours. The entire process is visualized to be able to read this indication and also to assist the operator. The computer screen will show a graphic display of the process elements with the allocation of equipment, their status and their setting. It may also indicate set points and actual values. The process control unit has a defined access level which can be entered with an assigned password only. The first level is for operators to acknowledge alarms or to confirm actions of the operating personnel, for instance charging is completed. It is also used for initializing actions of the equipment, such as start or switch-off burner. This means the operators are able to set all data for the commencement of operation. The second level can be entered by plant management only. It comprises pre-set production data, such as shift numbers, material to be charged, metal treatment, alloy to be produced etc. The third level is for maintenance purpose only and is used to set certain values for testing or to operate the plant in manual mode.

8.3.1 Process optimization

The standard control system guides the process in accordance with predefined recipe and set points. But it does not adapt the process to any change in the raw material supply, variations in material composition, unexpected situations during charging and melting of the material. Therefore, a process optimization system can be introduced to optimize the following main tasks in the recycling process:

- Optimization of material input, material handling and material flow
- Optimization of the melting process
- Charging and material flow

The following problems are typical for the aluminium recycling process:

- The identification und proper loading of scrap boxes
- The chemical analysis of the scrap which is delivered from the market is not automatically taken for further processing
- Best combination and selection of different scraps und primary material for economic production of the desired casting alloy
- The charge mix for a casting alloy is more or less based on the experience of the cast house operator or supervisor
- Supervision of proper loading of the melting furnace and weight supervision

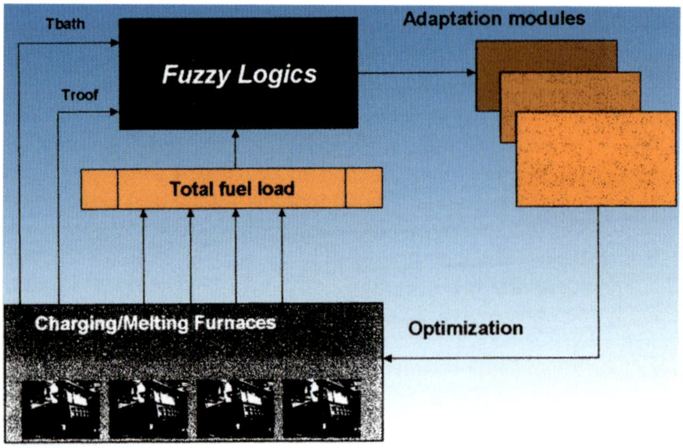

Fig. 8.24: Structure of the furnace optimization system (Source: Inovatherm)

During the loading of the melting furnace it is difficult to supervise which type of scrap has been loaded by the operator. The material identification is done manually and also the sequence of loading is set by the operator. Also the total weight of the metal loaded into the furnace is not known. Additionally, it is difficult to judge which type of scrap has to be loaded in an economic way because the paid value of this scrap is not known. The value fluctuates daily according to the value on the stock exchange.

Thus, the production is more or less based on maximum values to keep the alloys within the tolerances. The introduction of a material charging optimization system is shown in **Fig. 8.24**. The charging optimization system will be connected via an Ethernet LAN (Local Area Network) or by

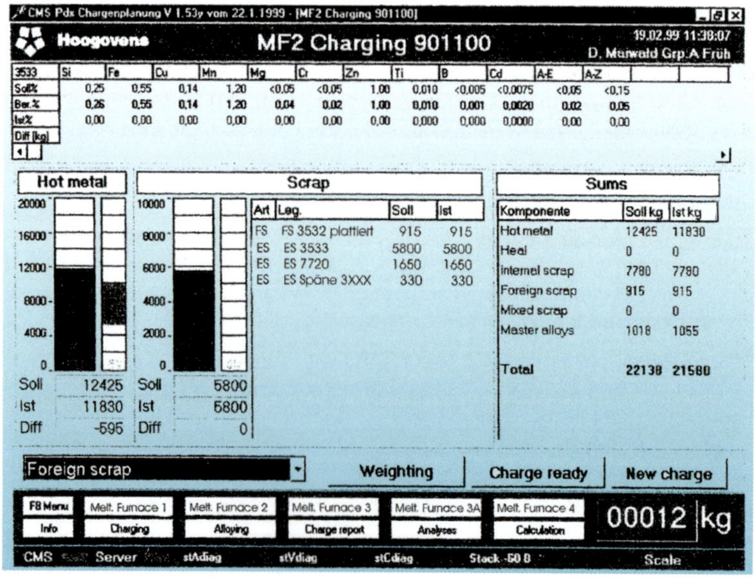

Fig. 8.25: Typical menue for the furnace optimisation system

Ethernet to be the existing plant business system. Weighing scales such as the truck scale, the scrap scale and the master alloy scale and the spectrometer analyzer will also be connected via the existing data network. Further connections to the existing transport vehicles in the scrapyard can be realized by a wireless data LAN to exchange data and transfer commands to operators required for proper scrap handling and furnace loading according to the defined charges.

As an essential task, the charging optimization system executes the planning of charges for each furnace including the charge calculation, the control and assignment of the actual data from the weighing scales and analyzing devices. The planning, preparation and optimization of each charge is initiated automatically by using iterative alloy calculation algorithms and re-assign the calculation each time a new analysis or weight is received.

As a result, the operator is exactly instructed of the weight balance of each furnace and the next action to be taken. A typical menu from this charge optimization system is shown in **Fig. 8.25**. The main features of the charging optimization system can be summarized as follows:

- identification of all types of scrap including analysis
- actual scrap stock database on-line
- charge planning for each furnace
- charge calculation on-line
- supervision of the metal analysis according to quality standards
- weight balancing of each furnace
- operator guidance for furnace charging and post-alloying
- complete production history and documentation according to ISO 9002 standards

8.3.2 Melting optimization

The basic automation and control system guides the process to a dedicated set point. But it does not contain any “intelligence” or expert knowledge to optimize the melting process.

A major problem of the process is, for example, the exact definition of the melting time which is directly related to the energy costs for the melting cycle. Another problem is the formation of dross which must be minimized by cyclic skimming to get the maximum heat into the aluminium bath.

Thus, the main tasks of an optimization of the melting process are the reduction of melting time, dross and energy costs.

For this purpose the basic automation is enhanced by a melting optimization system. This optimization system is connected to the furnace PLC system to get the actual status of the roof temperatures, the bath temperatures, the flue gas temperatures, the actual fuel consumption, O_2 content of the flue gas and the production cycle time.

The melting optimization system handles these data as an input for a fuzzy logic model. From this data the melting optimization system calculates the necessary target set points for the bath and roof temperature and the best production cycle including the remaining melting time by fuzzy control algorithms. Additionally, it generates information for the operator and supervises the action connected to certain operating procedures such as necessary stirring, skimming, sampling or, finally, time of complete melting.

Thus, the operator will be informed just-in-time if a new charging or melting cycle has to be initiated. Furthermore, he will be informed to start and execute necessary stirring and skimming procedures to avoid unnecessary oxidation scenarios.

The supervision results in optimization of the melting time and saving of energy. A higher level of production automation and transparency will be achieved. The optimization of the melting process leads to dynamic set point adaptations of the furnaces which result in lower process temperatures and shorter melting cycles. These facts result in less energy consumption and less dross formation. Experience gained in the past shows that the melting cycle can be shortened by 8 %. The energy savings and shorter melting cycles can reach 10 %. The dross formation can likewise be reduced.

9. Quality assurance

Christoph Schmitz

9.1 Quality management

Quality assurance is essential for an aluminium recycling plant that intends to operate in the metal market. Customers request the product they would like to use in their specific manufacturing line without testing the incoming aluminium in their own facilities and with their own methods and procedures.

Besides these customer requirements, aspects of conformity with the relevant laws, protection of environment and, last but not least, the economy of the recycling plant are important factors for the quality assurance. The relationship with the suppliers of raw materials and other commodities necessary for plant operations is one aspect of quality management.

A plant can only achieve good product quality results if quality management is embedded in plant operations. This applies for the entire organization from management to the operators. All employees must have the motivation to do their best to obtain the high standard concerning the quality of a product. They must be supported by clearly defined procedures for all kinds of work including correction of unavoidable failures and accident prevention. Furthermore, personnel should be encouraged to look critically at everything in the plant and what might be possible to improve. The technology applied for production should be suited best for the range of production and the type of raw material to obtain the product economically at best quality.

The production route from raw material to final product must be supported by intelligent computer software. The result of every production step is tested by suitable methods and monitored up to the final product. The impact of the plant operation on the environment is a part of quality management that should not be underestimated. The generated waste materials, such as dirt and slag as well as the emissions and contaminations of water and air – including noise – must be included in the system. Particularly the generation and deposit of hazardous waste is a factor that may bring about the attention of the public whose acceptance of the operation helps to avoid many problems. Energy is not only a factor having an impact on the efficiency of the operation. With increasing shortage of this resource even the location of the plant may become critical if the production is not able to operate energy-efficient.

To summarize, the quality assurance concept is a combined effort of material testing, technology applied, computer-supported process control, co-operation of management and personnel, customer and public relations as well as supervision of environmental impacts. The requirements for quality management forms, with increasing importance, the basis of supply contracts. Particularly the automotive industry requires a certification of production according to the standards of DIN ISO 9000ff. Apart from the audits of independent institutions, they frequently carry out their own quality audits. The handbook of quality management is a good basis for the instructions to the plant personnel and may also be a good argument during contract negotiations with a potential client. It is a very time-consuming effort to go through a certification and it is also not easy to maintain the status achieved once certified. The final stage of a quality management system is the "Total Quality Management" (TQM), according to DIN EN ISO 8402 which defines the target as follows:

"TQM is a management system for an organization to aim with cooperation of all their members for a long-lasting success of their business operation by having their focus on quality and satisfaction of their customers to the benefit of the members of the organization and the public."

Quality management is certainly required for a production plant. Sometimes these efforts may be exaggerated and the management must be careful that these efforts do not result in an extensive bureaucracy causing cost that is in no relation to the intended effect.

Quality monitoring covers the following:

- Incoming testing of raw materials and additives
- Process monitoring regarding product quality, energy consumption, recovery, consumption of additives, production time, reproducibility of process results
- Testing of outgoing products with regard to product quality, quantity and conformity with sales contract
- Monitoring of environmental impacts such as emissions and outgoing waste, residues from mechanical preparation, slag, filter dust and used refractory material

9.2 Incoming material

9.2.1 Incoming scrap

Incoming material will be tested for various reasons. For further processing it is essential to have a precise knowledge as to what material is supplied and of which elements the aluminium may consist. Furthermore, what type of contaminations and quantity may be contained in the scrap. Furthermore, for a fair pricing of the scrap, the content of aluminium, and the percentage of contaminations should be checked. Therefore, it is essential to have a sample tested representing the entire batch supplied. The persons evaluating the sample must have comprehensive practical experience and must also have the required knowledge about the material to be evaluated. There are defined procedures and regulations. Please refer for more details to Taschenbuch des Metallhandels, Hüthig Verlag, Heidelberg.

The target of sampling is:

- Checking of conformity with the purchasing specification
- Evaluation of composition, shape, recovery and percentage of non-metallic contaminations
- Assignment of the material to a specific storage area.

The most reliable method for sampling of aluminium scrap comprising mix of different alloys, shape and contamination is to melt the entire batch. This is generally not possible. To obtain a reliable picture of the supply a selection of material representing the batch is collected and mixed manually. A second sample is taken in case of doubts. This procedure requires thorough knowledge and sound experience. Out of this mix a sample is melted in a sampling furnace (**Fig. 9.1**) which is generally a small rotary drum furnace with salt and a scrap/salt ratio of 1:1. Clean scrap is easier to evaluate. The proportion of the different alloys can be estimated based on the type of material. The content of contaminations is also estimated. A good method is to

Fig. 9.1: Sample furnace for scrap

take samples frequently during offloading which afterwards are melted in the sampling furnace. All results are recorded and stored in the relevant computerized material management system. It is used to define additions to the scrap in order to obtain the required product as well as the treatment method including the required additives or treatment with gas as well as type of gas and duration of treatment.

9.2.2 Additives

Salt is the most important additive used in a secondary aluminium smelter. Salt is purchased from well-known suppliers on the basis of long term contracts. The quality of the products is not tested since the plant relies on the quality testing of such suppliers. There are some critical contents mixed to the salt which still require frequent tests of the incoming material. Aluminium fluoride is an example. The supplier of salt relies in turn on the company providing aluminium fluoride. Occasional testing of the actual content of aluminium fluoride is advisable since this is a critical item in regard to recovery of metal. The moisture content of the salt may increase during transport and storage of the salt. A frequent test of this parameter should be done.

Other additives, such as master alloys or additives used for metal treatment, are not tested for the quality of the product. This would require sophisticated equipment that is usually not available in a recycling plant.

The incoming material is, therefore, only tested in regard to conformity of quantity and type of material as per purchasing specification.

9.2.3 Products

The metal quality and the analysis of metal are tested during various production steps. For final testing samples of the product are taken on a random basis or on the basis of a defined procedure.

For ingots, a sample will be taken and analyzed at the beginning of the casting operation and at the end – if wanted at the middle of the casting period. The ingots will be cut and analytically tested by an optical emission spectrometer. This includes the visual inspection of the surface. Segregation will be tested by cutting a slice out of the middle of an ingot followed by grinding and etching to examine the grain structure.

For billets and slabs samples can be taken by cutting a sample slice where deemed appropriate. If required by the customer, samples will be obtained by drilling. The metal will then be melted, cast into sample molds and tested like a standard metal sample.

For other products, such as des-ox cubes or granular material, the sampling of the liquid metal in the furnace will be recorded. The same applies for sampling of the product supplied as liquid metal.

Sampling may deviate from the standard procedures following the request of individual customers.

Frequent testing during the various production steps is standard procedure. Basis for all testing is the exact batch calculation which is prepared on the basis of the data recorded for incoming material and the required product. The first sample is then taken after completion of melting. The analysis is basis for correction of the batch calculation and adjustment of the alloy composition in the holding/casting furnace. After melting of the added material, a new sample will be taken. If the result is not as expected, an additional test is required. A final sample will be taken shortly before the start of the casting process or before the metal is filled into transport crucibles. All test results will be recorded for later reference, if required.

There is a test for the chemical analysis of the metal. Another test focuses on the hydrogen content which is required before treatment and after treatment of the metal. Since aluminium tends to pick up hydrogen very easily, frequent hydrogen tests during casting are advisable. Just as for in-line metal treatment there are also continuous testing methods available which provide an excellent monitoring of the production process. This also applies to the content of inclusions. Here in-line testing methods are also available.

All test results, also for intermediate tests, are recorded and become part of the documentation supplied to the customer on delivery of the material.

For internal reference or for documentation requested from different government authorities, other tests are required. These focus on testing of gaseous emissions on the basis of frequent or continuous measurements and the status of wastes, water effluents and slag. Utilities consumption is recorded for seeing the total usage of fuels and electrical energy as well as fresh water provided by an external source. Consumption monitoring is also performed for individual plant equipment such as the specified furnaces or casting equipment to obtain a good monitoring of the efficiency of the equipment and possible changes over the time requiring specific attention.

A very important factor is the recovery of metal obtained during processing of the scrap. If this recovery changes – which is in general a change to the worse – action of the management is required to investigate the reasons. These could be decrease of the quality of raw material thus calling for action by the purchasing department, failures in equipment performance, requiring action by the maintenance department or changes in operation procedures, requiring action of production management.

For the different tests numerous testing equipment and procedures are available. Some of them are typically used in the aluminium industry. Testing methods for gaseous emissions and contamination of water are based on standard principles.

9.3 Testing methods

9.3.1 Metal analysis

In order to determine the chemical analysis a sample of the liquid metal is taken. The temperature of the metal must be sufficiently high in order to ensure that all alloying components are dissolved. The sample is taken by means of a sampling spoon. It must have at least the same temperature as the metal when the sample is taken in order to avoid solidification of metal in the spoon. Liquid metal is then filled into a sampling mold in order to obtain a sample as required for the testing. The surface of the sample must be machined before it is placed on the testing equipment.

The optical emission spectroscopy (**Fig. 9.2**) is the most reliable method and simple for testing the aluminium alloy. It is commonly named quantometer. The system is easy to use, provides results very fast at sufficient accuracy and data can be put directly into the computerized plant management system. The unit works on the following principle: An electric arc is generated between an electrode and the metal surface. A small quantity of metal erodes and atomizes and ionizes. This material emits light at a typical spectrum for each of the elements. The intensity of the spectrum depends on the quantity of the very specific element. As with any spectral analysis, the light emitted by the eroded material is directed to a grid which splits it into the different spectral lines due to the frequency-dependent refraction. Slot channels, arranged in a circle at the specific angle, receive the light of the frequency typical for the elements and the intensity can now be measured, giving a reading for the quantity of that alloying element present in the sample which is directed to slot channels arranged in a circle. The measurement takes place over a defined period of time and the quantity of photons emitted is integrated over this period and fed to the computer. The measuring unit is installed in a closed housing which is designed as vacuum chamber or filled with argon.

The quantometer is easy to handle and does not require much skill of the personnel. It works fast so that results are available to production personnel in a very short time. A further development of

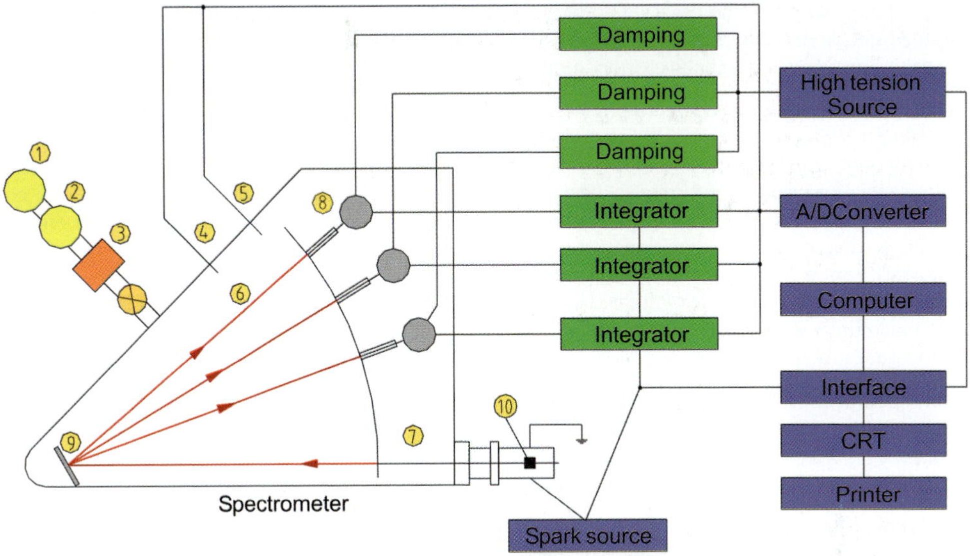

1–3 - vacuum system, 4 - temperature pick-up, 5 - vacuum measuring probe, 6 - optical ray, 7 - primary optical ray, 8 - photo multiplier, 9 - optical grid, 10 - sample

Fig. 9.2: Optical emission spectrometer, principle (Source: Spectro)

the optical emission spectrometer is the system with plasma stimulation. The different elements in the plasma will emit electromagnetic rays of different wavelengths. The liquid sample is atomized and then stimulated in inductive-stimulated argon plasma. Optical system and data processing are similar to the standard emission spectrometer.

The x-ray fluorescence analysis is not so much used in an aluminium recycling plant. It uses the emission of x-rays with typical wavelengths of the individual elements for the analysis. It is also easy to handle but quite expensive.

The atomic absorption spectroscope is not so much of advantage for a melting plant. It operates on the basis that the spectrum of a defined light source shows dark lines, so-called emission lines which are typical for specific elements. The distribution and intensity of the emission lines are a measure for type and quantity of the elements present in the sample.

For the analysis of the samples taken from production, certified samples are required to calibrate the optical emission spectrometer and the unit for x-ray fluorescence analysis. This is not necessary for the atomic absorption spectrometer and the optical emission spectrometer working with plasma. These methods can, therefore, be used for defining the standard samples for the other measuring principles.

9.3.2 Hydrogen content

Hydrogen contained in the aluminium can be tested by different methods.

The simplest procedure is to have 100 g of aluminium poured into a flat mold comprising insulating refractory material or graphite, pre-heated to a temperature of approximately 200 °C. The diameter of the sample will be approximately 60 mm, height 10 to 25 mm. The hydrogen bubbles leaving the metal during solidification create a number of small bumps at the surface. The number of bumps on the surface and the number of bubbles in the cut sample indicate the hydrogen content. This method requires substantial experience of the tester but still only gives a rough indication.

A more reliable method, called melt tester, was developed from this procedure. A probe of liquid aluminium with a weight of 75 g is poured into an electrically-heated steel mold which is placed in a vacuum chamber. At constant metal temperature the pressure inside the vacuum chamber is lowered at a velocity of 8 mbar/s. The surface of the melt is now visually observed through an inspection glass. As soon as the first bubble appears, temperature and pressure are recorded. Based on this data and considering the contents, alloying elements, hydrogen content can be obtained from a nomogram, supplied with the testing unit. The melt tester can also be used to determine the content of non-metallic inclusions in the melt. To do so, the pressure is lowered to the maximum vacuum of 0.5 mbar. The inclusions become visible on the surface of the probe and at the gas bubbles and can be used for an estimate of the quantity of inclusions.

Testing principles on the basis of density measurements are difficult to handle and they are, therefore, not generally used in production plants. Their principle is based on the relationship of the density of a sample solidified under a defined vacuum, which creates a de-gassing effect, compared to a sample that has solidified under atmospheric conditions.

Methods using the principal of the de-gassing with nitrogen determine the partial pressure of hydrogen from the heat conductivity of the hydrogen-bearing nitrogen (Telegas system). A probe is submerged in a liquid metal bath and nitrogen is introduced into the melt. The small bubbles generated in the melt collect hydrogen and will be passed through a ceramic filter once they have surfaced. The cleaned gas is then recycled to the melt until the gas mix reaches equilibrium which is obtained as soon as the partial pressures of hydrogen in the nitrogen is identical to the partial pressure of hydrogen in the melt. The change of heat conductivity of the nitrogen determines its hydrogen content. Taking the metal temperature into considertion, the hydrogen content in the melt can be calculated. This method is expensive since the measuring probe has a very limited life span. It is, therefore, used only in plants producing large metal quantities.

The Alscan method is a further development of the Telegas method. The difference to the Telegas system is the probe which now consists of a porous ceramic plate for collecting the gas leaving the melt. The probe has a much better life span and is, therefore, more useful for use in an aluminium recycling plant.

9.3.3 Non-metallic inclusions

The content of non-metallic inclusions can be determined by metallographic procedures. A defined quantity of metal is poured on a copper plate. The solidified metal is then machined by scalping and the content of inclusion can be determined visually after the sample has been treated by anodic oxidation.

A more precise metallographic testing method uses a ceramic filter of defined number and sizes of pores. A metal quantity of 2 kg will be pressed through the filter. The percentage of contaminations at the surface area of metal and filter cake is measured under a microscope by an optical analyzing system. The system also permits determination of the type of contamination by a cathode ray system. The weight of the metal, which has passed through the filter, can be compared with original metal and thus the exact amount of contaminations can be determined. This PODFA (Porous disc filtration analysis) process is used in plants producing slabs and billets to determine the efficiency of their in-line filtration system.

A very common technique is the LIMCA (liquid metal cleanliness analyzer) method (**Fig. 9.3**). It uses the different electrical conductivity of metal and inclusions. A probe made of steel is submerged into the liquid aluminium. A second probe contained in a tube of quartz glass will also be

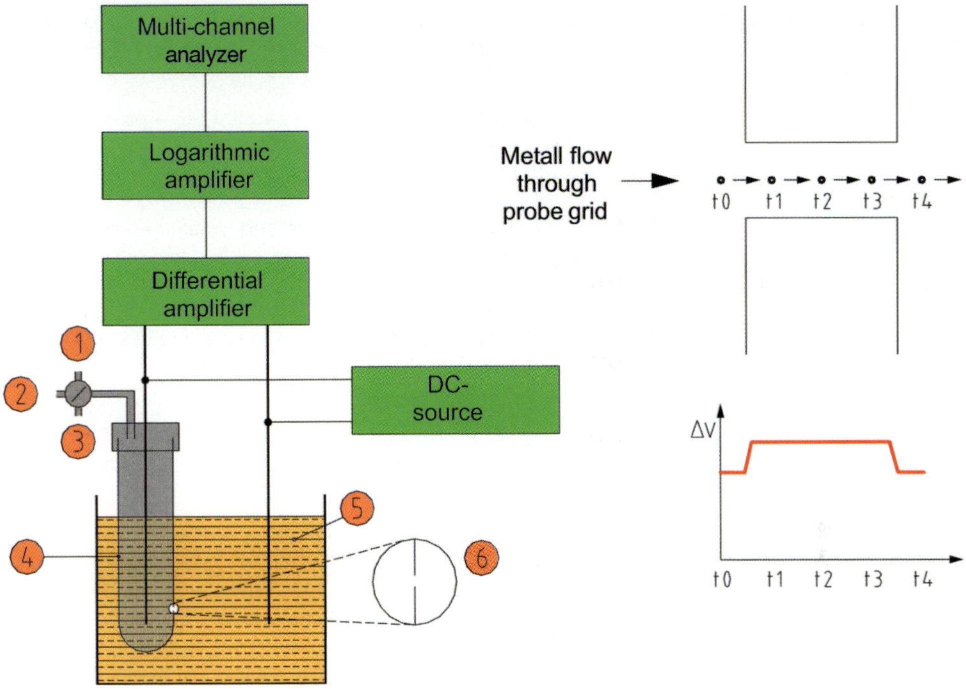

1–3 - pressure and vacuum measurement, 4 - measuring probe, 5 - liquid aluminium, 6 - probe gap 300 mm

Fig. 9.3: LIMCA system

submerged in the metal. The enveloping tube has a defined opening of 300 mm. Both electrodes are electrically connected via this small opening due to the electrically conductive melt at a voltage of 6 volt. Since a low pressure is established in the tube, liquid metal flows through the small opening in the tube. Every non-conductive inclusion reduces the cross-section in the opening with the result of a short increase of resistance between the probes. Number and value of the pulses determine quantity and size of inclusions.

After the tube is filled with metal it is automatically emptied and the test starts again. A complete test lasts approximately 2 min. The method is very reliable but has the disadvantage of high investment cost.

Therefore, the simple to perform Qualiflash method is quite standard. A defined quantity of metal passing through a filter is used to determine the quality of the melt. The testing unit comprises a mold having a bottom made of a replaceable ceramic filter. The metal passing the ceramic filter is collected in a mold comprising 10 sections. Metal is poured at a temperature defined for different alloys into the heated mold. This is continued until no metal passes through the filter. The number of filled sections of the collecting mold is a measure for the contamination of the metal.

9.3.4 Additional testing methods

The testing, as described above, covers most of the requirements in an aluminium recycling plant. Other testing methods are applied to determine physical characteristics and castability or heat treatment behavior. There is a wide range of testing methods for various applications.

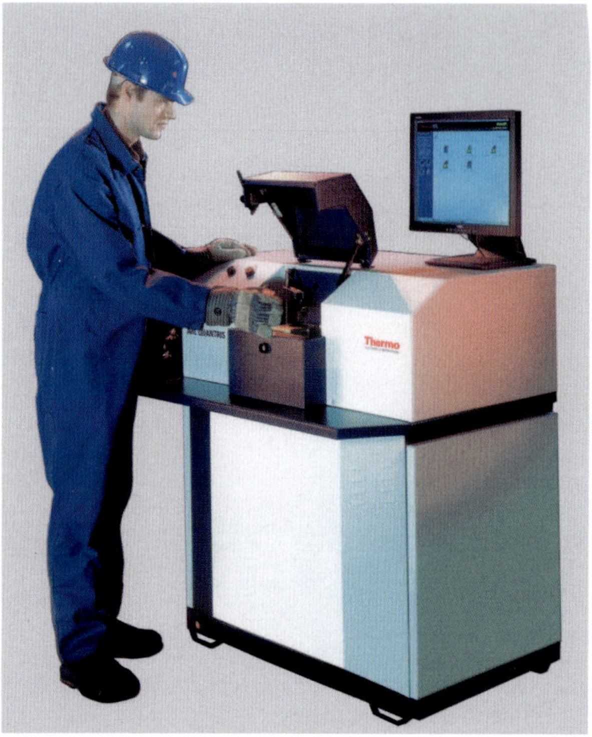

Fig. 9.4: Optical emission spectrometer, production-type unit (Source: Spectro)

However, only a limited number of tests are performed in a secondary aluminium plant. Quality management and material testing is a substantial cost factor. Thus, the test being performed should be limited to the actual requirements, which means to ensure the product quality as requested by the customers, and to investigate reasons and rectification in case of complaint.

It is obvious that all tests and their results are recorded to follow up the entire line of production. This can be done as hand-written reports by using defined forms for reporting and procedures. The better solution is to make use of state-of-the-art computer technology and use a professional process management system (**Fig. 9.4**).

10. Safety

Christoph Schmitz

10.1 General safety aspects

The general rules for accident prevention in an industrial plant also apply for an aluminum recycling plant. Some particulars require very specific measures. Within the plant liquid metal has to be handled at various production steps. This requires care and particular procedures to protect the operating personnel and also the equipment.

Furnaces may also cause some hazards if not operated properly or in case of malfunctions. Some safety measures are integrated in automatic control loops but often in industrial plants operators sometimes try to override or eliminate such safety circuits if production requirements put pressure on them.

Due to the nature of the raw materials to be processed, some hazards may exist. Most critical is the humidity of the scrap. Wet material charged into liquid metal may cause a severe explosion. But also oil or some plastic charged into the furnace can result in stack burning and cause fire attack on the surroundings.

10.2 Personal safety

Apart from the measures, as applied in any industrial plant, special attention is required if handing liquid metal and combustible gases. The general safety codes are valid in the secondary aluminum recycling plant as well and must not be repeated. For details one may refer to the local regulations. Some particular aspects shall be discussed briefly.

Proper clothing is very important. Unlike steel droplets, aluminum does not solidify immediately once hitting the cloth. Due to its low melting point aluminum remains liquid and may pass through the cloth to the skin. Therefore, special fabric must be used for the protective clothing and even underwear.

Safety helmets with eye protection must be used when working in front of the furnace. Operators working in front of the furnace during metal treatment with chemicals must use, additionally, respiratory protectors to counteract the effects of smoke, dust and dangerous gases. Management personnel and visitors must also wear safety helmets and eye protection googles. Safety shoes are worn by all personnel working in production areas. This may sometimes be required for office personnel and visitors, too. Care is also required when handling aggressive chemicals. The personnel must be aware of the nature and dangers of the chemicals. This is in particular valid for chlorine which is sometimes used for metal treatment. Operators must make themselves familiar with the correct first aid treatment in case of poisoning or contact with the material used.

Drivers within the production area must observe the vicinity and people walking around. Correct speed is essential. When operating forklift trucks the driver must ensure that the weight and the load center are within the capacity limits of the truck.

10.3 General plant safety

A fundamental factor of effective accident prevention is the provision of ample ground space at the plant site to reduce crowding of equipment, congestion of plant traffic, unusual fire hazards, unsafe yard conditions, etc.

If a plant is located near a main line of a railroad, consideration for safe access by employees and vehical traffic should be given serious attention. Safe passageways from one building to another are also important. Blind corners, doorways opening to yard railway tracks, etc., should be avoided as much as possible, and, where such conditions must necessarily exist, safety railings, gates, signs, or sound systems should be installed to warn pedestrians of danger.

Because of the inherent hazards of fire and explosion, the storage, handling, and utilization of flammable liquids should be carefully controlled. It is necessary in any event to observe the requirements of any applicable state and local codes.

Buildings, in which dusty operations take place, should be designed to offer a minimum area of projections, ledges, and resting places for dust accumulations.

Probably the most important factor to be considered in connection with floors, stairways, etc., concerns slipperiness. Floors and stairs should be free of projecting nails, bolt heads, etc., as noiseless as possible, wear well, and be strong enough to carry safely any static or moving load. The weight of modern industrial machinery and material handling equipment should be carefully considered if checking floor-load calculations. Floors and stairways should be kept free of unnecessary obstructions over which workers may trip.

Spilling of oil, water, acid, liquid aluminum, should be prevented to eliminate hazards. Splash guards, drip collectors, etc., can be designed in many instances to reduce slippage. Excessive spillage of dusty materials onto the floor is sometimes taken care of by installing floor grates beneath which pits or conveyor systems are located to collect the falling material.

Ample aisle space is very important, especially in foundries where workers carry ladles with molten metal and where there is considerable shop traffic involving power-driven trucks, etc. Aisles should be clearly marked to assist in keeping them clear. One-way traffic is often advantageous.

Stairways should be provided with handrails on both sides. Nonslip treads on stairs are desirable. Stairways should be adequately lighted.

As far as practicable, all doors should open outward or with the natural direction of exit; they must not block passageways from other floors or parts of the building.

For building sections, i. e electrical control room, not less than two possibilities of exit should be provided on every floor, including basements, all buildings or sections; these exits should be separated in such a manner that they are not able to be cut off by a single local fire.

Adequate lighting has a definitive bearing on the prevention of accidents. Workrooms should be well-lighted to reduce eye strain and the possibility of permanent eye impairment and also to remove any danger of employees falling over obstructions or being caught in machinery in darkened areas.

A lack of adequate ventilation in a workroom tends to bring on fatigue and reduces the alertness of workmen thus making them more susceptible to accidents. Where aggressive dusts or noxious vapors are encountered, it is necessary to provide for their removal by the installation of adequate local exhaust systems.

It is desirable that a plan for identifying the contents of various pipelines be adopted so that in case of an emergency it will be possible to determine quickly the service of all pipelines involved.

The logical time to install safety devices is when new machines are being built, while general construction work is being done or when alterations or repairs are being made; results can be accomplished with a minimum of expense and delay at the time plans and specifications are being prepared.

In order to make sure that the question of safety will not be overlooked, it is good to have all plans, specifications, and drawings checked for safety, making special provision for this in each set of specifications and in the remarks on each drawing.

10.3.1 Electrical equipment

In considering electrical equipment for industrial establishments, it should be kept in mind that the danger to human life increases with increase in voltage, transmission wiring and copper parts and the use of high-voltage motors. For small motors, lights, and general service inside industrial plants, installations of 230 or 400 V are recommended.

All switches, fuses terminals, starting rheostats, motors, etc., located within working platforms, should be enclosed or guarded in such a way to prevent accidental contact with live parts, irrespective of voltage. Switches should be arranged so that they can be locked in the open position to guard against a switch being activated accidentally while workers are at work on the lines or equipment that the switch controls. All metallic cases, frames, and supports of such equipment should be permanently grounded; foundation bolts should not be depended upon for this purpose, but substantial ground conductors should be used. It is preferable to have these ground wires accessible for inspection.

The most logical time to consider the safeguarding of a machine is during its design. At this stage, features of safe operation can be incorporated so that there will be a minimum of specific guarding required on the finished machine. The following points are fundamental considerations concerning accident prevention which should be taken into account in machine design.

Care should be taken in arranging clearances of moving parts to avoid shearing or crushing points in which hands or other parts of the operator's body might be caught or injured.

Arrangements should be made so that adjustments, inspections, and manual lubrications can be done safely.

Machines should be designed so that operators are not required to stand in an uncomfortable position, reach over moving parts or exert themselves in awkward positions.

Machines should be designed so that there will be little danger of the operator tripping over parts of the frame or striking against projecting parts during normal operation movements.

Careful attention should be given to the strength of all parts whose failure might result in injury to an operator.

All guards, covers or enclosures should be designed strong enough to prevent the possibility of giving way and permitting an accident in case the operator should fall or be thrown against them.

The point of operation of a machine is taken to be that zone where the work of the machine is actually performed and where the operator, by manipulating the material being processed, is exposed to a hazard by moving parts of the machine. Guards for point-of-operation protection are placed on the machines as additional equipment. The first requirement for a successful guard of this type is that it must be convenient and not interfere with the operator's movement or affect the output of the machine. The following statements describe several basic principles which may be utilized in point-of-operation guarding.

Where possible, the danger point should be completely covered by a barrier or enclosure before the dangerous operation of the machine begins. This may be accomplished, for example, on a treadle-controlled machine by having the treadle operate the guard which, once in proper location, will in turn operate the machine.

Feeding devices may be used so that material to be processed is placed in a feeding mechanism at a place where there is no exposure to moving parts. Special holders or feeding tongs can also be used to better handle work in dangerous positions.

Electronic controls, which operate by the interruption of a light beam or other energy source to protect the danger zone, may be used to start and stop machines.

Electrical interlocks on guarding devices may be utilized in operating circuits so that, unless a guard is in a proper position, the circuit is open and no current will flow until the machine is safely protected by the guard being brought into proper position.

Automation has minimized the hazards associated with the manual handling of stock and has eliminated the need for repetitive exposure at the point of operation, but the urgency of making repairs or adjustments introduces the need for special precautions covering maintenance work such as locking out to power source.

10.3.2 Occupational disease prevention

In any industrial operation, where a toxic material is being processed in such a manner that those persons engaged in or working near the operation are exposed to appreciable quantities of dust, fumes, vapor, or gas, it is important that adequate control measures be adopted. The following statements cover the major considerations involved in the implementation of effective control of industrial occupational disease.

The physical and chemical characteristics of a contaminant should be known. In the case of dusts or fumes, the chemical nature, particle size, solubility, etc., should be determined. For gases or vapors, the composition, vapor pressure, flash point, etc., are important factors. Concerning all atmospheric contamination, the quantity of material in the worker's breathing zone must be known before the degree of hazard can be evaluated. The chemical characteristics are important for the selection of materials to be used in the construction of any control equipment where corrosion, etc. might be factors.

A careful investigation should be done to determine accurately the sources from which the contaminant is being produced or from which it is being dispersed. The most common types of dust-producing operations are crushing, screening, grinding, polishing, etc. Dispersion of dust is encountered in practically all dry-handling operations of fine materials. Vapors and fumes are produced by chemical processes and reactions and are most commonly found in connection with the use of solvents. Removal by means of exhaust hoods, enclosures, etc. located to prevent the escape into occupied areas of any appreciable amount of contaminant at is source is the most effective method of control.

Natural ventilation has a limited application in industrial occupational disease control. It is important with a natural-ventilation system to maintain close supervision of adjustments as required by changes in temperature, wind directions, etc. Regular tests of air content should be made to check on the degree of dilution being obtained. This method of control is not recommended for exposures where severe hazards exist due to extreme toxicity or high concentrations.

Under certain circumstances, the isolation of a hazardous operation, physically, or at point of time, is indicated. For example, a hazardous operation may be done in a separated room in which all contamination can be confined or it may be done outside of regular working hours when no one except the persons engaged in the operation will be present. By isolating an operation, the number of persons exposed to any connected hazard may be reduced to a minimum.

In addition to exposure to toxic gases, vapors, dusts, fumes, etc., there are several physical conditions such as abnormal pressures, temperatures, and humidity as well as radiation (including ultraviolet; infrared; x-ray, a, b, and y rays from radioactive substances; and radiation from handling radioactive isotopes), all of which may, in case of excessive exposure, prove detrimental to the health of exposed persons. Evaluation depends on physical methods of measurement. Protective measures include shielding from radiation and control of exposure time rather than ventilation as in the case of atmospheric contaminations.

10.3.3 Fire protection

The profitable use or availability of facilities is the aim of all business, whether industrial, mercantile, professional, scientific, or educational. Destruction of or damage to the facilities cripples the attainment of the given purpose.

Fire brigades are essential for the development and maintenance of an effective fire protection program at every job site. With a prompt, immediate response and notification of the local fire department every effort should be made to bring the fire quickly under control during the early minutes of an outbreak. The immediate availability of the correct fire protection and extinguishing equipment is essential.

Fire protection engineering involves the application of sound engineering principles to ensure the reduction of loss by fire and related hazards. The Society of Fire Protection Engineers has established a well-defined scope for fire-protection engineering practice, both for building design and for safe operating practices.

Plant construction and the protection of buildings and their contents against fire are frequently governed by local building codes and ordinances and by insurance standards. When new construction or modifications are planned, the property owner should have the advice of a competent fire-protection engineer and should consult local authorities and insurance providers to avoid delay and the possibility of expensive modifications later on.

Fire-resistant types of construction refer to types that resist considerable fire without serious damage such as reinforced concrete or protected steel. Noncombustible refers to any construction that contains no elements of burnable material but which may be structurally damaged by fire such as unprotected metal. Combustible means structures entirely of combustible materials or having combustible elements of such character and distribution that a fire can spread and contribute fuel so that severe damage results.

Fire resistance measures the susceptibility of materials to damage by exposure to fire and is usually measured as the time period of exposure, without significant damage, to a standard fire exposure as specified by a standard fire test. Hydraulic fluids shall be used for all equipment.

Protection of buildings or other structures against fires in nearby vicinity must sometimes be provided. A practical barrier against a conflagration is the presence of a fire-resistant shield along the exposed side. Usually the most severe exposure is localized and protection against it may be provided by a blank brick or concrete wall, by wired-glass metal-frame windows or by open sprinklers alone or in combination.

Special types of protection are adapted to the control of unusual hazards such as flammable liquids. Equipment should be obtained from makers specializing in the form of protection required. However, such use does not eliminate general building protection by automatic sprinklers.

A dense, strong spray of water from suitably designed nozzles is effective for controlling fires in flammable liquids of moderate hazard, for unusually flammable solid materials and for surface fires of ordinary combustible materials. Such a system may be most appropriate for the protection of transformers and other oil-filled electric equipment and for systems handling fuel or lubricating oil under pressure. It must by no means be used when contact with liquid metal is to be expected.

For (1) flammable liquids; (2) electric equipment, such as large enclosed electric generators, and (3) hazards where a space-filling effect by an inert atmosphere is needed or where a no wetting extinguishing agent is desired. Carbon dioxide for fire extinguishing is available in small cylinders for manual use, or in banks of large cylinders, or in refrigerated storage tanks for pipe-extinguishing systems.

Dry chemical extinguishers and extinguishing systems are mainly used for flammable liquids and electrical fires. They are effective also on surface fires in combustible fibers. Multi-purpose dry-chemical extinguishers have special ingredients which make them suitable for fires in ordinary combustibles.

Hand-operated extinguishers with small hose are effective when employees are on hand to fight fires immediately after discovery. They are frequently used to put out the final vestiges of fires brought under control by automatic sprinklers. Common types of hand-extinguishers are the fire pail and water; the hand-pump tank, and extinguishers called soda-acid, non-freeze, foam, carbon dioxide, dry chemical, and vaporizing liquid. Dry compounds, applied by shovel or scooped from bulk containers, are available for fires in combustible metals such as magnesium, aluminum, zirconium, sodium and potassium.

10.3.4 First aid

Trained personnel must be available for immediate action. Personnel must be instructed with regard to the nearest telephone, whereabouts of the first aid equipment including bottles with eye-rinsing liquids and emergency safety showers. If a person operating with chemicals feels unwell he should go out into fresh air and get help.

A first aid station must be located in every plant.

10.3.5 Furnaces

Gases remaining in the furnace are mixed with air once the burners are switched off on purpose or in case of trouble. As soon as combustible gas reacts with air, which may be present due to incomplete combustion in the furnace chamber, there is a risk of explosion. One of the factors having an impact on this risk is the composition of the furnace atmosphere. The percentage share of each constituent of the atmosphere in the furnace chamber does not necessarily correspond to the composition of the gas introduced into the furnace, as the share of combustible components may increase, e.g. by vaporization of or reaction with plastic grease and rolling oil residues on the charge surface.

In an enclosed furnace chamber, ignition of a combustible gas/air mixture will result in an increase in pressure and temperature. Depending on the forces released, the result will be a deflagration and explosion or a detonation. For the sake of simplicity, only the term “explosion” will be used in this description.

No explosive gas/air mixtures can be formed in enclosed furnace chambers, if the temperature of the interior wall is higher than 750 °C throughout because at this temperature level mixing and combustion will take place simultaneously.

This minimum temperature of 750 °C is referred to in the following as safety temperature. Even if this safety temperature of 750 °C is not reached but at a few locations, an explosive protective gas/air mixture may build up.

The gas emerging at specified points must be disposed of safely to prevent the occurrence of hazardous concentrations. This may be achieved by

- drying and removal of combustibles,
- evacuation and
- ventilation in the furnace.

Failure of the source of ignition must trigger an alarm. For removal of the protective gas by evacuation or ventilation, forced-flow equipment must be employed and care must be taken to ensure that this will not cause new hazards anywhere in the plant. The equipment employed must be provided with a failure alarm device.

If there are any materials in the furnace chamber or in the vestibules, which may release combustible components, an explosive gas/air mixture may form if air is enters. To take care of such a case, adequate safety measures must be taken to prevent ignition.

If the furnace chamber is operated at a temperature below 750 °C, the furnace must be supplied with a sufficiently large volume of air to ensure that no air can enter at any point.

Danger can only arise if the supply of gas to the furnace is inadequate and the furnace pressure is lower than the ambient pressure. In such an event, an alarm must be triggered. Hazardous conditions will also occur if the heating system fails and the temperature in the furnace drops thus resulting in a vacuum pressure in the furnace chamber due to the compression of the gas. If this happens, an uncontrollable volume of air will enter which may lead to the formation of an explosive mixture which is sparked off by any of the sources of ignition.

The following can be sources of ignition:

Surface with elevated temperatures such as

- inner walls of the furnace chamber
- electric heating elements
- radiant heating tubes
- charge having a higher temperature than the furnace chamber temperature when brought into the furnace

Flame flash-back occurring, for example, at

- flares
- flame curtains of charging doors

Sparks due to impact or friction caused by

- moving parts or recirculation fans
- contact of projecting parts of the charge with structural parts or the furnace during charging or discharging operations, lifting the muffle or the protective covers

External sources of ignition may be

- naked flames
- protective gas flares
- nearby firing equipment
- not extinguished cigarettes

Infrequent sources of ignition may be

- substances having a pyrophoric or catalytic effect, e.g. soot, sulfur compounds, finely dispersed metals
- electrostatic charging
- triggering or acceleration of the reaction of oxygen with gas which is usually unpredictable
- even at room temperature – entails the risk of local overheating.

The purging of furnace chambers or similar spaces is not a simple process of displacement such as filling a vessel already containing air with water. On the contrary, even with gases of different densities, there will always be a (slow or rapid) mixing process.

Principally, non-combustible gas must be used for purging. If non-combustible purge gas (air) is used for the displacement of air from the furnace, the purging cycle must be continued until the gas concentration has been reduced to less than 1 % by volume. At this point, the furnace may safely be filled with protective gas or air. The purging cycle must be continued until the concentration of combustible constituents has been reduced to less than 5 % by volume. Then the furnace may be safely ventilated.

If, in exceptional cases, – for operational or process reasons – combustible gas is used for purging, adequate measures must be taken to ensure that any source of ignition is safely and reliably excluded.

A complete change of atmosphere from air to combustible gas or from combustible gas to air requires several times the volume to be displaced. The purge gas volumes, which will eventually be required, cannot be calculated exactly because of the insufficient knowledge of the flow conditions in the furnace, especially with constructions which have a complicated design. The purge gas volume must, therefore, be established experimentally during the first purging cycle.

This method is safe as long as the temperature is maintained above the safety temperature level of 750 °C in the furnace chamber. But this method can also be applied in plants with a furnace chamber operated above 750 °C and integrated vestibules or cooling zones operated at a lower temperature, provided, however, that combustible protective gas is fed immediately into that part of the furnace which is operated above 750 °C and that the volume of flue gas produced by burning the protective gas is large enough to also purge the lower-temperature areas.

During regular operation of the furnace at a temperature above 750 °C there will be no explosion dangers. A short-term temperature drop below 750 °C, which may be required for process reasons, will not be dangerous, provided that the furnace door is not opened. However, if the door is opened to perform charging or discharging operations, a flame curtain must be provided to ensure that no explosive mixtures can arise in the furnace chamber.

The furnace must be provided with a suitable malfunction alarm triggering visual and audible alarms in the event of temperature declines below 750 °C, failure of the furnace heating system, and interruption of protective gas flow to the furnace. Temperature drops due the process reasons need not be signalled.

For lengthy shutdown periods the supply of gas to the furnace must be reliably interrupted. In addition, it must be ensured by adequate ventilation of the furnace chamber that non-explosive mixtures can be formed, e.g. from residues left in the charge.

Once the furnace is re-commissioned after lengthy downtime, it must be made absolutely sure that no gas or fuel gas (i. e from the pilot flames of the flame curtain) can enter into the furnace unless a steady pilot flame is established in the furnace. Gas is introduced after a pre-defined time. If no stable flame is established after a certain period of time, the gas valve closes automatically and the furnace will be flushed by air.

The starting sequence is started again automatically including flushing of the furnace. This will be repeated 3 times. An alarm will be triggered if no table flame is obtained. Upon failure of the gas supply, the pressure in the furnace chamber and connected zones may drop below the ambient pressure.

As a result, air will enter through openings unless it is burnt off with air. Failure of air supply must therefore be indicated by visual and audible alarms.

Upon failure of the supply of gas and in the case of defective air supply, an explosive mixture will form if combustible gas is used. This extremely dangerous condition may be avoided if, upon the interruption of the gas supply, air is fed – if possible advisable automatically – to the furnace chamber. Failure of gas flow to the furnace must be indicated by visual and audible alarms. If the O_2 concentration is recorded, deviations from the set point value must also be signalled.

For each furnace the following information must be recorded and entered into the furnace log:

- all tests
- all irregularities and malfunctions concerning the safety of the plant and action taken to cope with them
- structural changes and modifications, if any

The furnace log must be inspected at regular intervals by those responsible for the safe operation of the plant. The burner control is equipped with an automatic leakage test. This has to be checked at infrequent intervals.

The operating personnel must be given practical instructions on the safe operation of the furnace. This must be done under the guidance and supervision of a person, who is fully acquainted with the operation of the furnace, who knows the possible dangers in connection with the furnace and the generation of protective gas and who has experience in the action to be taken in the event of irregularities or malfunctions.

It is advisable to simulate sources of danger in order to familiarize the personnel with the action to be taken. In addition to the practical training, the operating staff must receive instructions on all types of protective gases used and their potential hazards. The instructions must be given prior to the use of the furnace and must be repeated at regular intervals but not less than once per year. It is recommended that a list of those attending be kept and that it is signed by the attendees. Shut-off, monitoring, and safety equipment must be checked for proper functioning at adequate intervals in accordance with the manufacturer's instructions.

Only if there is absolutely no other possibility, safety devices may be overridden for short periods, in order to remove or repair malfunctions or disturbances. However, this may only be done by competent experts and during any such work plant safety must be controlled manually.

Once the furnace must be inspected or for maintenance and repair work, the accident prevention regulations, the job instructions, and the operating instructions must be observed.

There are other small burners for pre-heating launders and molds. Usually these burners are mobile and must be placed on the equipment as required. Care must be taken when handling the burner equipment and placing it as required. The shut-off valve must be located in the reach of the personnel and it must also be assured that no obstructions are in the way of the shut-off valve.

The fixing equipment must be kept clean and in good operating condition. Since the pre-heat burners are mobile, a safe storage place should be designated. During storage the burners must be disconnected from the gas supply. Other burner equipment is used for crucible pre-heating. These burners are equipped with all safety equipment like a furnace. Therefore, the same safety measures are valid.

10.3.6 Charging

Charging of wet material is very dangerous and can result in severe explosion. This is particularly the case if the material, i. e scrap, is charged to liquid metal. This refers to loose scrap but also to molds which may collect cooling water in shrinkage cavities. The large ingots must be placed on the pre-heating bridge of the reverberatory furnace before they are pushed into the liquid metal.

This refers also to large pieces of scrap and stacks of ingots. Bulk materials charged to the rotary drum furnace are more of a problem. Usually this kind of scrap is stored in the open yard and may be exposed to rain and could be covered with snow. The material must be dry or at least the humidity must be reduced to an acceptable level by having the material stored for sufficient time underneath a roof. Material with low humidity can be placed into the empty rotary drum furnace provided if done very carefully. Oil is as bad as water. Oil-containing scrap must be stored to allow the oil to drain. This kind of scrap must not even be charged to the empty but hot furnace since the combustible oil will immediately create a flare out of the furnace door.

If oily or wet scrap cannot be dried prior to charging it to the furnace, it must be made sure that no ignition takes place or in the case of water the furnace is absolutely free of liquid metal. The furnace can operate at very low temperature of approx. 300-400 °C and while the temperature increases slowly the humidity is evaporated and the oil and other combustibles may burn. Charging is possible once the necessary precautions are taken but can be very, very hazardous if the material is wet. Disregarding these facts has caused too many severe accidents.

Since charging requires mechanical equipment and transport of bulky material, attention of the operators is required when moving in this area. Automatic charging machines must be equipped with optical and acoustic warning systems.

10.3.7 Liquid metal handling

Humidity and liquid metal do not go together. This must be considered if operating equipment and equipment components that will be in contact with liquid metal. Even traces of humidity may cause disturbances and hazards. Therefore, molds must be dried before being filled with liquid metal. Sometimes molds are coated. This coating needs drying as well. In order to avoid precipitating of humidity in the period between preparation and casting, the temperature of the mold must be heated up sufficiently. The molds in the ingot casting machine may come into contract with the coating water. Therefore, the casting machine must be equipped with a mold drying system shortly ahead of the pouring position.

10.3.8 Equipment

The need for safety precautions while operating all types of machines has always been a matter of major concern in accident prevention work because the severity of any injury is almost invariably high. The prevention of injuries can be accomplished, first and foremost, by proper guarding of the dangerous parts of the machines. Nevertheless, some machines must necessarily have a moving part exposed to allow it to do its job – for example, a pedestal grinding machine. In these cases safe operating procedures must be followed.

There are certain precautions which must be observed for all machines in the process plant or in the workshop. These are:

Machines must not be used unless the guards are in position and firmly secured. Personnel must report damaged or insecure guards to their supervisor immediately.

Operators shall not use a machine having an interlock guard if the guard does not function correctly, i. e if the machine can be operated without the guard in position. They are not permitted to wear loose clothing which could become entangled in the machine.

Never reach or climb over machinery.

For cutting machines, such as drills, shapers, milling machines, etc. it must be ensured that the job is firmly clamped down before starting the machine. Before starting operations, operators must ensure that nobody can be injured by the machine when it starts.

The logical time to install safety devices is when new machines are being built, while general construction work is being done or when alterations or repairs are being made. Best results can be accomplished with a minimum of expense and delay when the time plans and specifications are being prepared.

In order to make sure that the question of safety will not be overlooked, it is good to have all plans, specifications, and drawings checked for safety, making special provision for this in each set of specifications and in the remarks on each drawing.

Part IV
Plant design

1. Plant design

Petra Haag

1.1 Equipment arrangement

The most effective and economical equipment cannot be fully efficient if the arrangement of the plant sets limits to the operation. The plant concept should permit easy charging and furnace tending as well as uninterrupted metal transfer from the melting section to the casting furnaces and from there to the casting machine. And these requirements should be considered for a logical material flow within the production area.

It is also indispensable that the capacities and production rates of the various equipment are tuned to each other and match the plant production requirements. **Table 1.1** shows an example of the summary required to define the plant equipment. Basis of any evaluation are furnace capacities and the tuning of the cycles of melting and casting furnaces. Cycle diagrams (**Fig. 1.1**) are

Table 1.1: Production data for different annual production days

Net production	t/a	45,000	40,000	35,000
gross production	t/a	57,000	51,000	44,000
production time	d/a	350	300	250
production/day	t/d	143	142	143
availability	%	90	94	98
Melting furnace				
melting cycles/day	C/d	6	6	6
melt rate/cycle	t/C	20	20	20
net production/day	t/d	120	120	120
no. of melting furnaces		1	1	1
Casting furnace				
melting cycles/day	C/d	3	3	3
melt rate/cycle	t/C	4	4	4
melting time/cycle	h/C	1.5	1.5	1.5
casting time/cycle	h/C	4	4	4
auxiliary time/cycle	h/C	2	2	2
waiting time/cycle	h/C	0.5	0.5	0.5
holding time/cycle	h/C	6.5	6.5	6.5
bath capacity/furnace	t	25	25	25
no. of furnaces		2	2	2
total melt rate/day	t/d	23	22	23
Casting machine				
casting time/day	h/d	24	24	24
casting rate/time	t/h	143	142	143
casting rate/hour	t/h	6	6	6
ingot weight	kg	7	7	7
no. of ingots	M/h	850	844	850

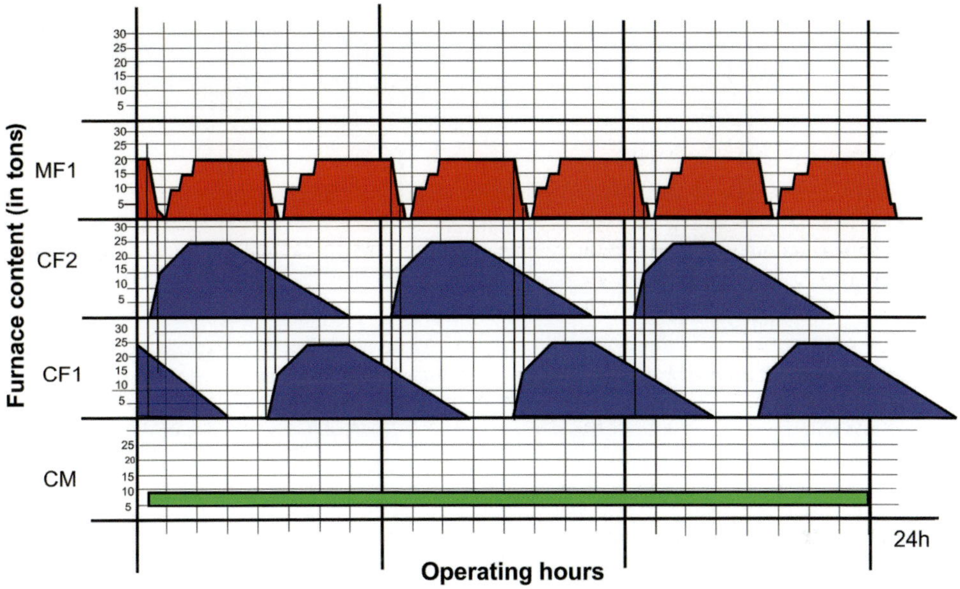

Fig. 1.1: Cycle diagram for melting and casting furnaces

very helpful for this task. First, the different cycles and their duration are defined in a diagram. It is good practice to indicate the bath content of the furnace. The times for charging, pre-heating, melting, holding and metal transfer and – as in the case of the melting furnace – for de-slagging are marked on the x axis representing the time axis. This should be able to include 24 hours of operation. We now have an indication of the content of metal in the furnace depending on the furnace cycle. Once we have completed the diagram for the melting furnace, we proceed with the casting furnace. The different cycles must then be linked in a way that the different cycles match each other. For example, discharge of metal from the melting furnace must match the charging of the casting furnace. The cycle of the melting furnace must be complete during the cycle time of the casting furnace. But in most cases the melting furnace is much faster than the casting furnace, i. e. the cycle time of the casting furnace is longer. It is also desirable for certain casting processes that they should commence without interruption for a lengthy period of time. This is not possible with one casting furnace only since preparing of the batch requires time which cannot be utilized for casting. This means casting has to be interrupted. With two casting furnaces operating in tandem, uninterrupted casting can be realized. While one furnace is casting, the metal batch in the other one is prepared. As soon as the first furnace is empty, the other one takes over. The melting furnace must match these requirements. The combination of the cycle diagrams shows the requirements for furnaces and their mode of operation.

The plant layout must permit the desired operation. It is, for instance, required for continuous casting that furnace tending and liquid metal transfer is possible while the other furnace is casting. Therefore the metal transfer launders must not cross each other. This is only possible with a very specific furnace arrangement.

The most common arrangement is to have two tiltable furnaces arranged opposite to each other whereby the tilting axis of each furnace is towards the center line of the system (**Fig. 1.2**). The

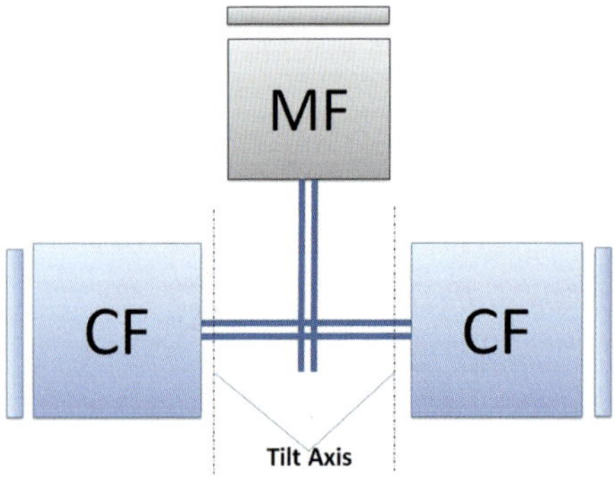

Fig. 1.2: Arrangement of melting and casting furnaces

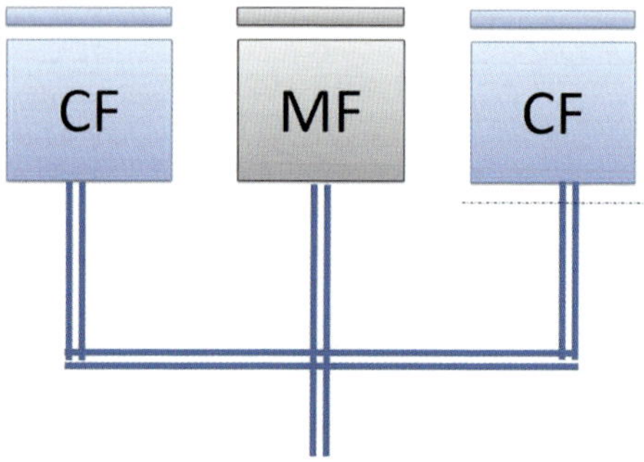

Fig. 1.3: Side-to-side arrangement of furnaces

discharge of the furnace feeds into a casting launder arranged in the center line between the two furnaces. The launder from the other furnace is closed by a simple ceramic gate. The tilting spout of each furnace is arranged in the tilting axis. This means that no level difference between furnace spout and launder is required permitting a turbulence-free flow of aluminum under a thin permanent oxide skin. This protects the liquid metal from oxidation and now oxides are entrapped in the metal flowing to the casting machine. Although a mechanical filter is arranged ahead of the casting machine, slight oxide contamination leads to an extended service life of the quite costly ceramic foam filter inserts. Since a limited collecting capacity of the foam filters is inevitable, a dual system with change-over possibility needs to be arranged ahead of the casting machine anyhow if casting is to continue over a lengthy period of time.

The metal flow is controlled by the tilting speed of the furnace. The level in the launder is used as control parameter; even slight fluctuations of the metal flow are monitored immediately and the tilting speed of the furnace is adjusted without notable time lag. The control loop permits to maintain the metal level within the launder within a range of +/- 2 mm. Although the control loop is quite simple, the elements for the furnace control have to be selected with care.

Tilting furnaces also allow maintaining a precise and constant metal temperature that is not possible with other systems. Metal is tapped from the surface whereby the input of energy to the furnace has immediate influence. Towards the furnace bottom, the temperature is gradually reducing due to the stable layer and the quite poor heat transfer within liquid metal.

While one furnace is casting, the next batch is melted and adjusted with regard to analysis and temperature. As soon as the first furnace is empty, the second furnace takes over without interruption of casting. Even a “flying” change of alloy is possible since each furnace is emptied at every casting cycle allowing proper cleaning of the furnace prior to the preparation of the new alloy.

This furnace arrangement is most efficient, very easy to operate and does not require complex pouring aid. This may be the reason that it is commonly used in the aluminum industry.

The arrangement, as described before, is certainly the very best solution. However, in existing plants the area for the new setup may be somewhat limited. In such a case, the two tiltable furnaces may be arranged side by side (**Fig. 1.3**). This layout results in somewhat longer casting launders that can be handled without major problems. There is in fact some advantage. Since the furnaces are arranged side by side, they can be operated from the same service floor for charging and furnace service. This arrangement is very convenient if the two tiltable reverberatory furnaces are melting/casting furnaces. They can operate to the casting machine without obstructing the operation of the other one. For continuous casting it must be assured that the melting and tending circuit of each furnace is within the casting time. This arrangement is used when clean scrap is processed. It is also convenient if one existing furnace shall be used. One of the furnaces could also be a stationary melting furnace designed as reverberatory furnace or twin chamber furnace feeding a tiltable casting furnace. If no interruption of casting is acceptable, metal needs to be transferred during the pouring operation of the casting furnace. However, the metal parameters have to be adjusted in the melting furnace prior to the transfer to the casting furnace. Alloy change during casting is also not possible since an undefined alloy with permanently changing characteristics is present during the transfer period.

For a new plant or if existing conditions permit, the melting furnace should be arranged above the casting furnace level since feeding of the casting furnace is only possible above the metal level. Therefore, the elevation of the melting furnace must be such that the bottom of the furnace bath is above the top level of the other furnace.

If this is not possible, a different method needs to be applied. The difference in elevation must be bridged by a metal pump. This could be a linear motor-type pump that allows lifting up to 1 m at a flow rate of max. 30 t/h. Problems and advantages of this system will briefly be discussed below. The oxidation and the foam generation during metal transfer are remarkable but can be handled.

Metal needs to be tapped from the bottom of the furnace. The temperature at the bottom is lower than on the bath surface due to the very stable formation of layers. Homogenization cannot take place by natural draft. The temperature difference between top and bottom is more than 50 °C with a bath depth of 800 mm. During tapping the bath depth gets lower and the temperature increases. This should be considered during casting.

The real problem with this arrangement is the controlled flow rate of metal to the casting machine. Stopper control is not very accurate and may cause clogging during a lengthy casting period. Therefore, a metal pump should be used to lift the metal up to above the top bath level at a controlled flow rate. This does not solve the problem of the different temperature levels. However, metal from the melting furnace can be transferred during casting so that continuous casting is possible.

Regarding metal quality there is formation of dross during pumping and metal transfer due to the strong agitation within the liquid metal and the unavoidable surface contact. Since the pump requires a certain heel, the furnace cannot be completely emptied. The remaining material has to be removed when the alloy changes. Therefore, a taphole and a collecting possibility must be considered. Metal parameters within the melting furnace must be adjusted carefully prior to transfer to the casting furnace and alloy change during casting is not possible.

A very vital item is the metal pump. Impeller systems do not provide enough pressure level for lifting the metal above bath level. A linear motor-type system can provide a pumping level of almost 1m at sufficient and controllable flow rate. A safe and reliable pumping system uses a vortex created in a special well arranged next to the furnace. This system is very safe but quite costly. A less expensive system uses a trough connected to the bottom of the furnace.

Both systems use the technology of the linear induction motor. The efficiency of the unit depends very much on the distance between inductor and metal. The indispensable refractory lining between inductor and liquid metal must be as thin as possible. Consequently, heat from liquid metal reaches the system and must be removed. For this reason the unit is equipped with a closed circuit cooling water system. The thin refractory layer is exposed to severe thermal and mechanical attack. Therefore, it is made of SiC tiles in case of a trough system and SiC tubes in case of a vortex system. Tiles or tubes have to be replaced at certain intervals. Another aspect is safety. In case of trouble, no dangerous situation occurs with the vortex system if one SiC tube breaks.

This is different with the trough system. If the SiC tile breaks, metal will leak uncontrolled from the furnace and may damage the equipment. Precautions are required to collect the leaking metal and to empty the furnace immediately and fast.

The furnace arrangement depends on the specific situation in the customer's plant. Safe and easy operation is ensured in tiltable furnaces. It is possible to have an absolutely turbulence-free metal transfer from the melting furnace to the casting furnace.

Metal pumping systems can be used. However, these methods require complex and maintenance intensive equipment. During metal transfer the metal level in the casting launder is automatically measured and the furnace tilting is controlled. In order to maintain the metal level in the launder with the metal level in the casting furnace during this operation period, the furnace tilts back. This control technique is proven, easy to operate, almost maintenance-free and cost-efficient.

The arrangement of casting machines requires a straight connection from the casting furnace to the casting equipment. Sufficient space should be allowed for the operators to move around the equipment and to take the necessary action during start-up and stop of casting. Space should also be allowed for in-line metal treatment equipment.

1.2 Plant layout

A typical plant layout for a secondary aluminium smelter is shown in **Fig. 1.4**. It shows the concept for a refining plant for the processing of a wide range of different scrap material. Therefore, the melting section comprises a 25 t rotary drum furnace and two 30 t reverberatory furnaces as casting furnaces. The final product will be ingots.

The plant consists of the scrapyard, the melting area, the casting area and the waste gas treatment.

Considering the metal quantity to be produced, a theoretical bath capacity of 25 t for each of the two casting furnaces will be required. This assumption assures that the metal quantity is available after one cycle of both tiltable rotary drum furnaces. This permits continuous operation of the casting plant. Furthermore, it must be considered that sufficient spare volume in the casting furnaces is available to be able to adjust the alloy by charging of additional corrective material. To do so, the furnaces should have sufficient spare capacity on top of the planned bath capacity of 25 t.

Raw material supplied to the plant passes a scale at the entrance gate to the plant and is carefully inspected (**Fig 1.5**). It is then directed to a designated storage area within the yard. The assumed metal composition is noted and registered together with its location in the yard. In cases, which do not permit an estimate of the analysis, a sample will be melted. Since this will require some time the material will be off-loaded at a designated storage area. The analysis of this material will be added to the record later.

Batches for the different production shifts of the melting plant will be composed and prepared within the yard facilities. This requires a batch calculation to arrive at the alloy as ordered by the customer. Each batch has the size of one complete furnace charge. The containers bearing the components for the charge are placed in the vicinity of the furnace in question to ensure fast loading of the furnace. The required alloying elements, such as pure aluminium, silicon or magnesium master alloys, are also placed next to the furnace in pre-calculated quantities.

The melting plant comprises a tiltable rotary drum furnace with a capacity of 25 t. For easy scrap handling and charging the batch charging area is designed as elevated section in the plant building.

The tiltable rotary drum furnaces will be loaded by means of a charging machine. Depending on the scrap to be processed, one, two or three chargings are required per furnace cycle. For batches

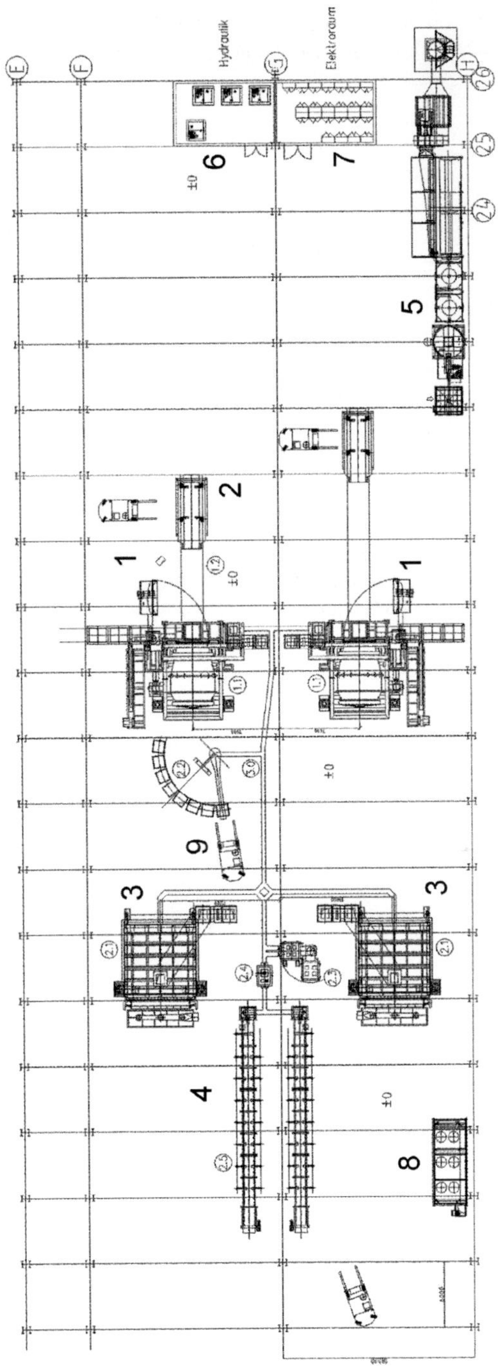

Fig. 1.4: Layout of melting plant

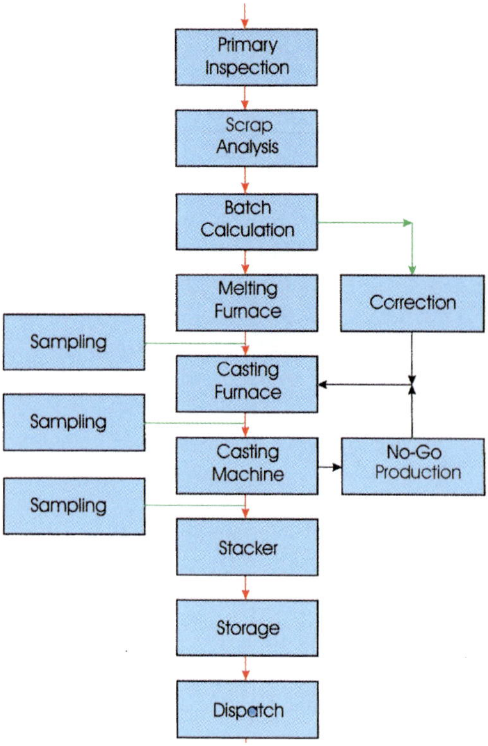

Fig. 1.5: Block diagram of an aluminium recycling plant

comprising small particles, melting takes place under a protective layer of salt. This is a mixture of NaCl and KCl with addition of a small quantity of fluorides (i. e. CaF_2). After melting, liquid metal will be transferred by means of a launder system to one of the casting furnaces. Thereafter, the salt slag will be discharged into slag pans arranged on a slag train.

It is also possible to cast liquid metal directly into sow molds by means of a casting station arranged next to the casting furnace.

Liquid metal can be directed to pre-heated transport crucibles via a launder system.

A sample will be taken for analysis prior to pouring or transfer of metal. If required, the additional alloying elements can be added to adjust the composition of the metal in the melting furnace already. However, it is more convenient and effective to do so in the casting furnaces.

The casting furnaces are arranged to permit continuous casting. They are designed as tiltable reverberatory furnaces with a preheating bridge next to the furnace door to permit charging of sows and stacks of ingots. To be able to melt sows, large size castings as well as bales or bundles of extrusion shapes, the furnace has a melting capacity of 3 t/h. Prior to casting, the batch in the furnace is adjusted to the final alloy composition and by introduction of energy the desired casting temperature is obtained. This requires that a sample is taken immediately before transfer of metal from the melting furnace. Sometimes it may even be necessary to add some alloying components, such as copper, in the melting furnace. A second sample is taken prior to casting. After the temperature is reached and the analysis is correct, the bath is skimmed and the furnace is ready for casting.

The metal is now transferred by means of a launder to the ingot casting machine. An in-line treatment system comprising de-gassing unit and metal filter is integrated in the launder system. The casting machine is designed as a water-cooled, chain-type machine. A stacking unit is arranged at the discharge of the casting machine. The completed stacks of ingots are discharged and transported to the storage area for dispatch. Depending on the customer, liquid metal can also be filled into a special crucible and delivered to the customer by road transport.

A waste gas scrubbing plant is arranged next to the production building. It comprises a bag house that is equipped with a reactor for additives. The additive storage bin is arranged at a location which allows supply of the additive by truck. Filter dust is collected in big bags. A storage area for production slag must be located in the vicinity of the plant as well. From here the slag is transported to the slag recycling plant.

Some auxiliary facilities are also required. A small quality control laboratory, essentially comprising a Quantometer and sample preparation equipment, is located next to the production area. The electrical equipment is located in a closed air-conditioned room. Attached to it is the small production office equipped with the process control computer.

For the closed circuit of the cooling water required for the casting machine, a water basin is required that is equipped with the necessary pumps and the cooling towers. A compressor station provides the required compressed air for the plant. If liquid metal is part of the product mix, a pre-heating station for crucibles is arranged in the crane range. Additionally, a repair station must be located in the vicinity of the production building.

Also some vehicles are required for preparation and charging of the raw materials. The vehicle pool comprises a front-end loader, standard forklift trucks and forklift trucks with rotating devices. A number of containers with a capacity of 1 or 2 tons are required for batch preparation as well as a number of dross bins must be available. A truck scale as well as smaller scales for batch preparation are also necessary.

Last but not least, tools for furnace tending are required, comprising skimming tools for manual operation and forklift attachments for skimming and stirring, different spoons for sampling, sample molds, small skimming tools for the casting machine and launders and thermocouples for manual temperature measurement and the like.

1.3 Plant personnel

An aluminium recycling plant can by no means be fully automated. Reliable and skilled personnel are required on different levels of the hierarchy (**Table 1.2**). The top management must have a background as metallurgical engineer or metallurgist and experience in the processing of aluminium. The production staff comprises the foremen and the operators of the equipment. The foremen should have a good technical background with some experience in metallurgical processing. The operators of the equipment are trained on the job. They need a good technical understanding and interest in the metallurgical processing of metals. In an operating plant this problem is solved by having the experienced operators teaching the newcomers. In future, it appears to be helpful if the plant operators undergo a training and apprenticeship as in other trades such as fitters or electricians. The plant equipment is becoming more and more complex and the operations are increasingly oriented towards high performance and trouble-free operation. Thus, the job at the furnace or the casting machine becomes quite demanding as far as technical knowledge is concerned. Furthermore, the operation of the various activities around the equipment is not always easy and requires also immediate action sometimes requiring less operating personnel. The different tasks of the personnel can be defined as follows:

The plant manager is responsible for the organization of the plant. Apart from this, his duty encompasses the entire production, the quality and the availability of the plant. The sales and administra-

Table 1.2: Plant personnel requirements

Unit / Staff	day shift	shift 1	shift 2	shift 3	Auxiliary shift t
Production management					
production manager	1				
foreman	1				
production planner	1				
secretary	1				
electrician		1	1	1	
fitter		1	1	1	
helper maintenance	2				
dispatch	2				
quality control	1	1	1	1	
yard helper		1	1	1	
Yard operation					
yard foreman	1				
forklift driver		1	1	1	
helper		1	1	1	
Rotary drum furnace					
furnace operator		1	1	1	
forklift driver		1	1	1	
helper		1	1	1	
Casting furnaces					
furnace operator		1	1	1	
forklift driver		to be handled by rotary drum furnace personnel			
helper		to be handled by rotary drum furnace personnel			
Casting machine		1	1	1	
caster		1	1	1	
stacker		1	1	1	
forklift driver		1	1	1	
helper		1	1	1	
total personnel/shift	10	15	15	15	0
total personnel	55				

tive manager, the quality control, the maintenance department and the operating staff will assist him.

The manager of quality control and quality assurance is responsible for the quality of the product and of the production as well as for the proper analysis and quality of the incoming scrap. He is assisted by the yard foreman and his crew, comprising generally one loader and one helper.

The maintenance foreman ensures a trouble-free operation of the plant. For each shift one responsible maintenance foreman will be on duty assisted by one electrician and one maintenance operator.

The shift foreman is responsible for production, safety and quality during his particular shift. The main duty comprises supervision of furnaces, casting machines and other production equipment. One man for charging the furnaces, one caster and two helpers for loading, stacking and transport assist the foreman.

The furnace operator reports to his foreman. His duties comprise the operation and monitoring of the furnace operation. He is assisted by a forklift driver for loading the charging machine or the furnace directly and for removing slag and skimmings. He is also responsible for supervising the furnace operation, skimming and the transfer into the casting furnace.

The caster is responsible for the operation of the casting machine. He reports to the foreman. He is assisted by a forklift driver for removing the stacked ingots.

Helpers are not assigned to a specific task. They will assist the operators when required.

The maintenance operator will report to the maintenance foreman. He is responsible for the maintenance of all equipment and mechanical repairs during his shift. In case of problems with the electrical equipment, he will call the electrician of the day shift.

The electrician will be required during day shift and reports to the maintenance manager. Since one electrician has to be stand-by, this duty cannot be performed by one man only.

The helpers are responsible for stacking the ingots leaving the casting machine according to the present methods and stacking the shape. Due to this heavy duty work in hot surroundings, a change of this personnel will take place every two hours. These people will then be replaced by the personnel for transport and vice-versa. The transport personnel is responsible for the required transport whenever a forklift operation is required.

The staffs for each shift are deployed according to the production schedule. Not all of the personnel is required for every shift. The yard personnel may work on day shift only and prepare the batches as required for the other shifts.

2. Plant implementation

Petra Haag

2.1 Feasibility and basic engineering

For the implementation of a new plant or any expansion or installation of new equipment, engineering work will be required (**Fig. 2.1**). All this work from first scratch to the commissioning of the plant requires skilled co-ordination which will be fairly straightforward for a small project but becomes very complex for the construction of a new plant. A market analysis shows the investor which type of product may be the most promising to operate an economical plant. Also the intended equipment and the available technologies have to be investigated. On this basis a very basic concept is developed. This conceptual engineering will form the basis for a detailed study. In order to obtain reliable data for the investments, feasibility engineering must be performed out with the target that the requirements of the plant are defined and investment cost is obtained. Basis will be the definition of processes required to reach the targets. This process engineering will also include the elaboration of production data, energy consumption and quantity of waste materials leaving the process (**Table 2.1**).

Once the plant concept is defined, usually competent suppliers of equipment are asked for quotations for the equipment required as part of the base data for the feasibility study. If the results are acceptable, engineering work will continue. If not, some data, concepts and technologies have to be investigated again until a satisfying concept is obtained. After the "go" decision is made, the project continues with the basic engineering.

The first step in the basic engineering of a project is the elaboration of the general layout. It is essential to plan the layout of the process units in a plant and the equipment within these units before any detail planning and cost estimate can commence. This phase of the project is most important as only a functional, well-planned arrangement of equipment and buildings can guarantee economical construction, good operation and efficient maintenance.

The location of all components is shown in their location relating to each other as well as to new and possibly existing parts of the buildings. Doors, gates, interior and exterior transport routes and flow of material are outlined. All important coordinates are measured on scale.

A very important document to be elaborated is the preliminary project schedule which is developed on the basis of the conceptual design. This will become a guide for the planning and recording of the project. All major activities, such as start and completion dates for engineering, design, procurement, manufacturing and erection of all major equipment, will be indicated. If planned and followed properly, the project schedule will later assure efficient coordination of engineering and construction activities, a precise estimate of manpower and a timely completion of the project.

The next step is the definition of equipment specifications. This document comprises a detailed specification of all plant components with descriptions and technical data including all individual values which are important to define guarantee data at the final engineering stage.

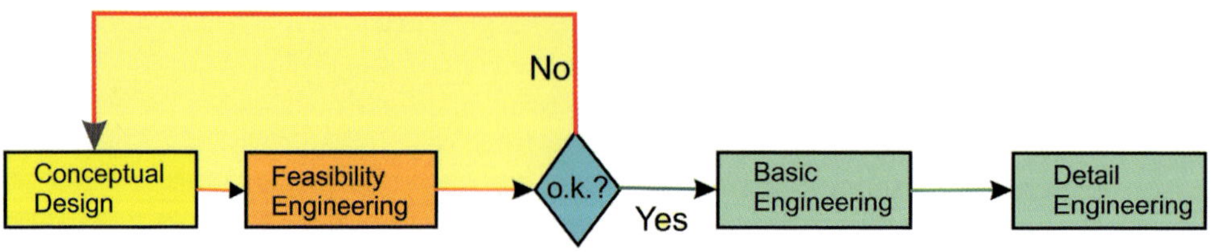

Fig. 2.1: Engineering phases

Table 2.1: Plant technical data and waste quantities

Melting furnace	Capacity	Specification	Value	Unit		
liquid metal quantities, max	25 t					
mean	20-23 t					
melt rate	5 t/h					
cycle time (test charge)	3.5 h					
operation mode	*gas/oxygen*					
gas quantity, max.	400 m^3/h		1,760,000	m^3/a		
oxygen quantity, max.	800 m^3/h		3,520,000	m^3/a		
burner rate	4000 kW					
flue gas quantity, max.	1200 m^3/h					
spec. energy consumption	400 kWh/t **feed**	*(gas + el. energy)*	503	kWh/t metal		
	gas/air					
gas quantity, max.	600 m^3/h		2,640,000	m^3/a		
oxygen quantity, max.	0 m^3/h					
burner rate	6000 kW					
flue gas quantity, max.	7000 m^3/h					
spec. energy consumption	600 kWh/t **feed**	*(gas + el. energy)*	754	kWh/t metal		
Recovery (typical - raw material mix)	**Feed**	**Recovery (sample) (%)**	**Melt loss (%)**	**Slag loss (%) v. S.**	**Total %**	**(t/a)**
swarf	12,000 t/a	98	5	5	8.6	1029
shredder	12,000 t/a	95	3	6	9.7	1159
bales	8,000 t/a	96	2	5	7.2	577
dross	12,000 t/a	55	2	8	51.9	6230
total feed	44,000 t/a					8995
total metal	35,005 t/a					

Table 2.1: Plant technical data and waste quantities (continued)

Salt consumption (typical - raw material mix)	Feed	Salt quantity (kg/t)	Salt quantity (t/a)	Slag (t/a)
swarf	12,000 t/a	180	2,160	3,789
shredder	12,000 t/a	150	1,800	3,319
bales	8,000 t/a	150	1,200	1,937
dross	12,000 t/a	75	900	7,370
total	44,000 t/a		6,060	16,415
Casting furnace				
liquid metal quantities, max	30 t			
mean	25 t			
melt rate	7 t/h			
operation mode	*gas/oxygen*			
gas quantity, max.	700 m^3/h		124,000	m^3/a
oxygen quantity, max.	1,400 m^3/h		248,000	m^3/a
burner rate	7,000 kW			
flue gas quantity, max.	2,100 m^3/h			
spec. energy consumption, melting	1,000 kWh/th Al/h			
spec. energy consumption, holding	9 kWh/th			

The detail engineering is performed to define every detail of the plant so that manufacturing and, finally, construction can be implemented. The construction activities terminate with the dry run of all equipment followed by test runs with material. During this phase all settings of equipment and instruments are made. As soon as trained production personnel is available, the final test is made to prove that the plant is able to produce as planned and production can start.

All actual engineering work starts with the basic engineering. Parallel to the engineering activities, the financing scheme for the project must be established, part of which is the feasibility study.

2.2 Feasibility study

Prior to any decision to install a new plant or go ahead with a major expansion, generally a feasibility study will be prepared. This study analyzes the market situation for the product and raw material, availability of resources and plant area as well as availability of personnel and many other materials and utilities. The target is to obtain the production cost which is then compared with the prices that can be achiered. The difference gives the revenue of the plant that shows at which time the initial investment is returned. For a profitable project the payback period should be in the range of 3 to 4 years.

The complete feasibility study includes a description of the project philosophy and plant capacity as well as a market analysis, plant description and recommended product mix. The following example is an extract of such a study performed for a project in the Middle East and limited to the cost aspects of the plant.

2.2.1 Basic data

Table 2.2 summarizes resources and prices as assumed in this calculation. This information is used in the following study.

2.2.2 Plant organization

The design is based on the work schedule which provides 1 hour for alloying, 45 minutes of producing the bottom plates and an average of 2 hours for casting. For a net production of 550-600 tons/month, it is recommended to work in 2 shifts per day of 9 hours per shift at 8 hours availability of the staff. The shift begins at 5 am and terminates at 11 pm. The 2-shift work is based on 22 days per month working time. A 3rd shift will be added as capacity increases. The weekly working time is 6 days. Friday will be the day off where the furnaces will only be kept warm. During the first period after the commissioning, the working time will be adjusted as required and as possible. For the first year of production an expatriate expert will ensure the smooth startup of the plant. The final staff comprises 17 people. As production increases to 9,000 tpa, a 3rd shift is planned and an increase of personnel accordingly.

The salaries and wages as per Table 2.2 are considered to include fringe benefits and other cost as well as overheads.

The data for the personnel required are based on an organization chart. **Table 2.3** shows the number of people required for training, startup and production. For the startup a certain number of production personnel is not required since only one shift per day is scheduled. This staff will be signed on as production increases and training will be obtained from the operational staff.

According to the time schedule, personnel has to be available ahead of production in order to obtain the required training and to assist during startup and commissioning. The expatriate production expert will be available during these periods and the first year of production for operation assistance.

Table 2.2: Basic input data for the feasibility study

Description	Value	Unit
Land	0.06	€/m^2
Buildings	59,000	€/m^2
Concrete	285,000	€/m^3
Vehicles	85,000	€
Office Equipment	21,000	€
Equipment	3,840,000	€
Shipment	2.2% of Contract	Value
Installation	17% of Contract	Value
Water (first filling of the system)	850	€/m^3
Manpower		
Expert (Consultant)	10,000	€/month
Plant Manager	2,500	€/month
Sales/Admin Manager	2,100	€/month
Commercial Staff	640	€/month
Quality Assurance	1,250	€/month
Foreman	850	€/month
Electrician	850	€/month
Maintenance	850	€/month
Loader	530	€/month
Caster	530	€/month
Helper	220	€/month
Energy		
Light fuel oil	0.10	€/l
Electrical	0.60	€/kWh
Raw Materials		
Clean Scrap	739	€/l
Painted Scrap	668	€/l
UBC	644	€/l
Silicon	1,562	€/l
Mis. Materials		
Water	11	€/m^3
Salt	53	€/l
Production	30	€/l
Commodities	5	€/l
Deposit Waste	40	€/l
Maintenance		
Vehicles		
Spares	7% of Equipment	Value
Marketing	12% of Sales	Value

Table 2.3: Required production personnel during startup and future production

	Startup	1st Year	Yearly
Expert (Consultant)	1.00	1.00	
Plant Manager	1.00	1.00	1.00
Sales/Admin Manager	1.00	1.00	1.00
Commercial Staff	2.00	2.00	2.00
Total Management	*4.00*	*4.00*	*4.00*
Quality Assurance	1.00	1.00	1.00
Foreman	2.00	3.00	4.00
Electrician	1.00	1.00	1.00
Maintenance	1.00	1.00	1.00
Loader	2.00	3.00	4.00
Caster	2.00	2.00	3.00
Helper	4.00	6.00	8.00
Driver		1.00	1.00
Office Assistant			1.00
Total Production	*13.00*	*18.00*	*24.00*
Total Plant	*17.00*	*22.00*	*28.00*

2.2.3 Land

Land will be rented from the government at a rate of \$ 0.06/m^2 per year. This cost will arise as soon as land allocation is finalized. Therefore, cost is estimated to start one year in advance.

2.2.4 Buildings

It is assumed that the production buildings will be constructed during the first stage. This refers also to the administration building and the plant service buildings. The total building space will be 4,500 m^2. Additionally, concrete is used for equipment foundation, storage areas and parking.

2.2.5 Fuel and electrical energy

The calculation is based on a rate of \$ 0.10/l for light fuel oil and \$ 0.60/KWh for electrical energy. An increase of 10 % per year is considered. The summary as per Table 2.2 includes production requirements as well as lighting and air-conditioning.

2.2.6 Raw material

Raw material will be scrap and silicon. In addition to the actual input of raw material, the melt loss is also considered in the summary of required raw material as per Table 2.1-B. In the first year production will be increased to 8,000 t/a, in the second year to 9,000 t/a and is maintained at this level for the remaining period. The product mix is regarded to be the input minimizing ratio. This means subsequent increase in raw material requirements accordingly and from the 3rd year onward a 3rd shift of production personnel is required.

2.2.7 Metal prices

The present situation on the metal market is characterized by increasing prices for aluminium. This refers to primary aluminium as well as to alloys. **Fig. 2.2** shows the development of prices

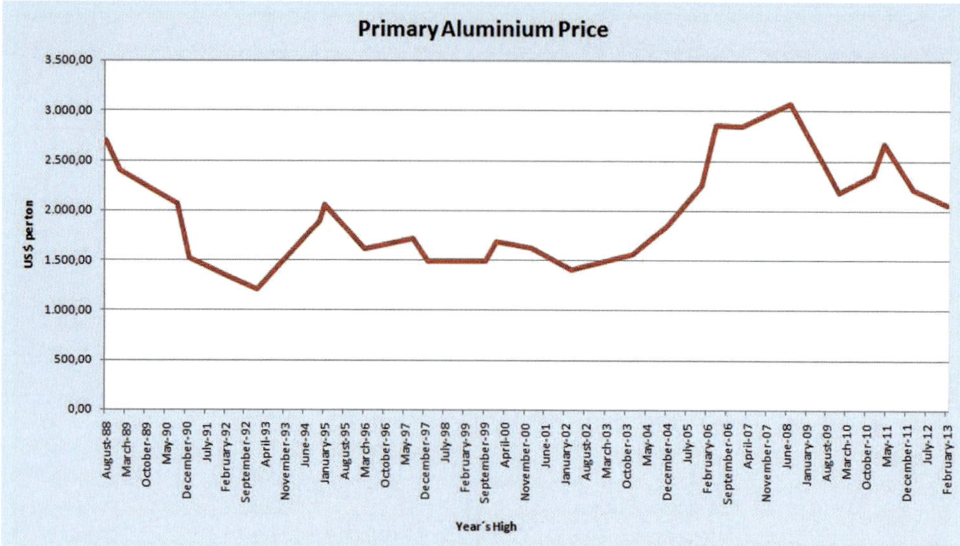

Fig. 2.2: Metal price development in $/ton aluminium

for aluminium. Although there is a fluctuation in prices the trend is indicated as climbing The main reason for this are the costs for energy and raw material. However, the differences in prices for remelt ingots and raw material remain more or less at the same level. This means that profit can be obtained from an operating plant, provided remelt cost stays within reasonable limits.

Basis for the calculation is a product metal price of $ 1,457.00/t for remelt ingots. For BC ingots the metal price will be 10 % lower which results in a price of $ 1,311.00/t. It has to be realized that all used data are average or medium prices which in certain cases may be lower or higher depending on the possible contracts obtained from users of this material. The calculation considers realistic prices for raw material, based on today's prices of scrap and realistic market prices for remelt ingots and BC ingots. An annual increase of both market prices for ingots and market price for scrap is assumed to be in the range of 2 %.

2.2.8 Maintenance

Maintenance cost is separated into labor and material. Due to the type of production, maintenance personnel is considered to be part of the operating staff. Material is estimated at 3 % of the equipment cost as annual requirement. This is regarded as a good average throughout the secondary aluminium industry.

2.2.9 Marketing cost

The marketing cost is assumed to be 12 % of the market price. This includes transport, 6 % customs duty and administration cost.

2.2.10 Summary

Table 2.4 summarizes all above-mentioned costs. It shows the startup and first year of production and gives an overview for the following years.

Table 2.4: Cost summary

Description	Startup Cost (in $)	Annnual Cost (in $)	Follow-up Cost (in $)
	(3 Months)	(1st Year)	annually
Land	26,730	21,384	21,384
Manpower			
Expert (Consultant)	30,000	120,000	0
Plant Manager	7,500	30,000	30,000
Sales/Admin Manager	6,300	25,200	25,200
Commercial Staff	3,840	15,360	15,360
Quality Assurance	3,750	15,000	15,000
Foreman	5,100	30,600	40,800
Electrician	2,550	10,200	10,200
Maintenance	2,550	10,200	10,200
Loader	3,180	19,080	25,440
Caster	3,180	12,720	19,080
Helper	2,640	15,840	21,120
Driver	0	6,360	6,360
Office Assistant	0	0	2,640
Energy			
Light fuel oil	24,000	1,152,000	1,152,000
Electrical	2,580	41,280	41,280
Raw Materials			
Clean Scrap	73,893	8,868,000	13,302,000
Painted Scrap	66,793	8,016,000	12,024,000
UBC	322,006	38,640,000	57,960,000
Silicon	72,633	5,810,640	8,715,960
Mis. Materials			
Water	1,584	25,344	38,016
Salt	83,475	1,335,600	2,003,400
Spares	268,800	268,800	268,800
Marketing	0	12,313,149	18,469,724
Total Cost	**1,013,085**	**76,802,757**	**114,217,964**

2.3 Calculations

2.3.1 Total investment

The investment is considered to be the amount paid for machines, buildings, furniture and car pool in the first year. This sum has been fixed and need not be verified. In the first run, the investment cost will occur during the start period only which leads to a breakeven point of 3.7 years. In a second calculation the investment is distributed linear over the complete investment period to simulate depreciations over this time. In the following, all values for the 2nd calculation are shown in parenthesis.

2.3.2 Floating assets

Before starting production a basic stock of raw materials has to be available. The value is calculated as the amount of material shown in the product mix, priced at market prices. As it is expected that the plant cannot operate at its full efficiency during the startup period, only 20 % of the normal amount of raw material is assumed to be required. Inventories of finished and unfinished products are calculated accordingly, taking into account that part of the production is recycled and the remaining part is held in stock during the startup period.

2.3.3 Startup cost

Startup costs are shipment, installation and the first filling of the system with water.

2.3.4 Miscellaneous

This item summarizes all costs not included above (e.g. marketing cost for the period).

2.3.5 Calculation rate of return

The results of the analysis are highly dependent on the rate of return used for the calculations. The profitability diminishes with increasing interest rates. As this rate has to be in the range between interest for invested capital and loans, it is assumed to be 8 %. A lower rate will lead to even better results and a higher rate to a later breakeven point.

All above-mentioned data are summarized in Table 2.4. All tables are essential for the calculation model which will be described in the next chapter.

2.3.6 Description of the calculation model

To get an idea of the feasibility of the investment, the following common methods of pre-investment analysis are used.

- Net present value method (NPV)
- Internal rate of return method (IRR)
- Payback method

Although the swift and popular "Payback Method" does not take into account the time value of the invested money, it is useful to explain the liquidity achievement of the investing company. NPV and IRR are more complicated to perform as both methods consider the present value of cash flows. They base on the used interest rate and assume immediate reinvestment of the generated cash.

2.3.7 Net present value

The difference of cash inflows and cash outflows (= surplus) of each year is discounted to its value at the beginning of the investment period. The sum of these so-called "present cash values" is the "Net present value (NPV)" of the investment. This, of course, leads to the conclusion that an investment is profitable if the NPV is not negative.

Adding the total sum invested to the NPV leads to the "Capitalized value of potential earnings (CVPE)". These 28.22 mil € can be regarded as the value to be earned with the investment.

Another question is how much money the investor has disposable at the end of the period. This "total accumulation of annuity (TAA)" is worked out adding the yearly surplus using the assumed interest rate as accumulation factor. Discounting the TAA to its value at the beginning of our investment period, again leads to the well-known NPV of the investment. The calculation shows a value of 20.27 mil €.

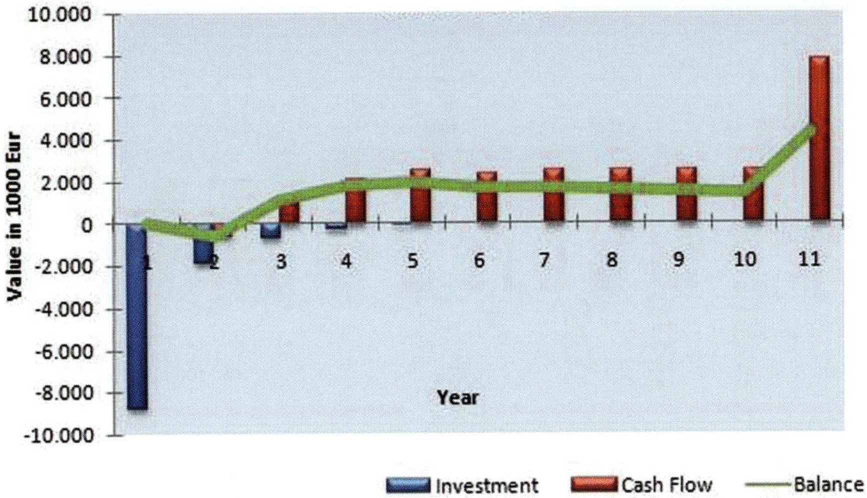

Fig. 2.3: Project cash-flow

As all the a.m. numbers are absolute, it might be of interest to get some ratios like the "Return on Investment (ROI)" to gain more information on the profitability of the project. The return on investment is defined as the quotient of CVPE and the investment. In this case the ROI is 175 % (**Fig. 2.3**).

2.3.8 Internal rate of return method

The internal rate of return is the rate at which the net present value of the investment equals 0. It represents the minimum necessary interest rate for the investment. The investment is regarded to be profitable if the internal rate of return is higher than the assumed rate. For this investment the rate is 34.13 % and definitely higher than the assumed 8 %.

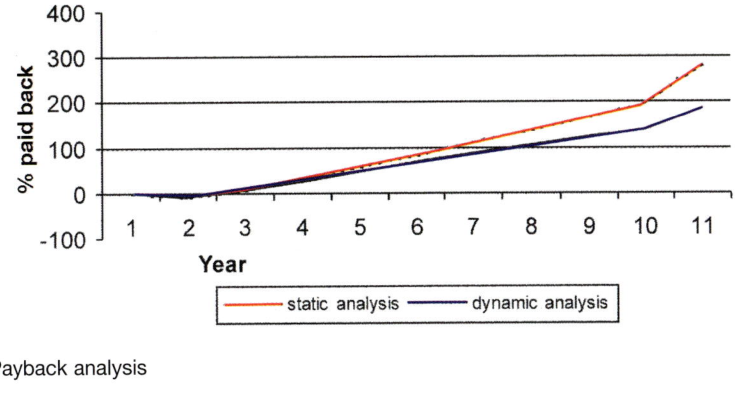

Fig. 2.4: Payback analysis

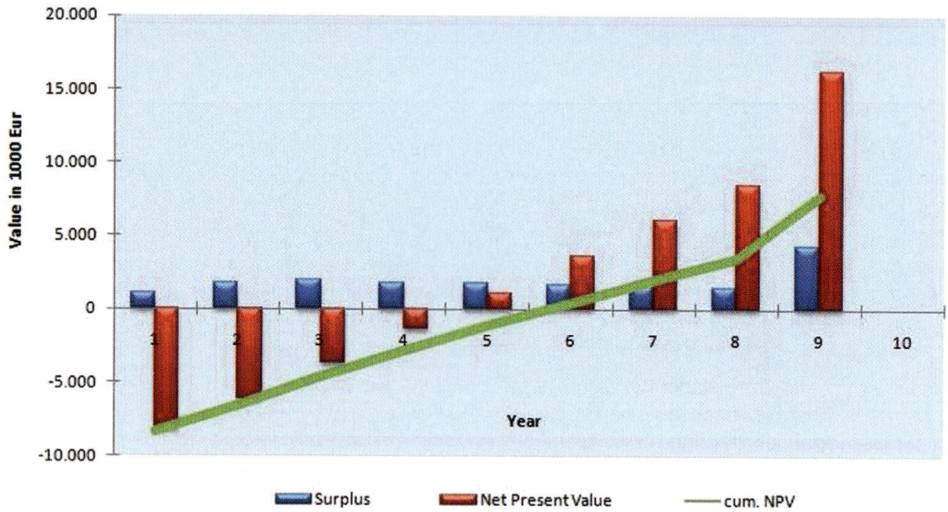

Fig. 2.5: Payback point

2.3.9 Payback method

The most interesting question seems to be the time in which the investment is recovered (paid back). This is the case as soon as the yearly surplus exceeds the interest payments. Basis for this is either the surplus (static) or the net present cash value (dynamic) of each year (**Fig. 2.4**). As the invested sum is distributed over the investment period of 10 years, this leads to a breakeven point of 5.9 years. The static breakeven point is 5 years, 7 months and the dynamic breakeven point 6 years, 5 months for the first run of calculations (**Fig. 2.5**).

2.3.10 Conclusion

The Pre-Investment Analysis indicates a breakeven point of 4.9 years. This start ROI is achieved even at less favorable assumptions for the development of metal prices. In European smelters this condition has been – and still is – an economical handicap. Due to their cost structure, with very high energy cost, the difference between raw material (scrap) and product does hardly cover the remelt cost. Quite the contrary is true for a project where energy cost can be kept at a low level. In our example melting cost amounts to $ 241.00/t. This is very well within the range between raw material cost and sales prices.

3. Detail engineering

Petra Haag

Hopefully, the project will go ahead if the results of the feasibility study permit and the financing of the project is settled. Contracts and agreements with government authorities and other parties involved have to be made.

For the technical implementation of the project, detail engineering will start.

Now the general layout and plant specification have to be detailed in drawings and specifications that allow procurement, fabrication and erection of the various equipment. This work is performed by experts, e.g. electrical design, furnace design or piping.

Arrangement plans are elaborated as layout and section drawings. The arrangement of the components is clearly defined as well as their location in relation to foundation and building. The drawings display connecting points for all plant supply and hydraulic piping. All main dimensions are shown. References and connecting points important for erection are specifically marked. Concealed equipment is only shown on the drawings in lining according to DIN, if necessarily required for clarification. The scale used should allow unmistakeable reading of the information. Usually a scale of 1:50 is applied.

Important piping and its connecting points are sketched on the basis of the general layout and the connections fixed. This refers to the connecting points to the plant supply as well as to the connections to various equipment.

The piping arrangement drawings include flue gas, gas, hydraulic, pressure and water piping.

Important cable routing and their joints are sketched on the basis of the general layout and the fixed connections. This relates to the connecting points to the plant supply and control room as well as to the connections to various equipment and its control stations.

The arrangement drawings for flue gas ducting are very important since they are usually of comparatively large size and must not present a hindrance for operation. These drawings define the exact routing of the ducting, the connecting points to the various plant components, branching, sections, fixings and shifting. The refractory lining of the ducting is also shown.

Basis for the design and construction of the foundations are detailed foundation drawings. These explain the exact shape of the foundations in sections and details such as edge protection, concrete covers and location and dimensions of the anchor points. All dimensions are clearly flagged. Connecting points for utilities, cables, water pipelines, hydraulic piping and gas piping for concealed piping are also shown in the drawing. Tubes and channels for underground cable and water pipelines are detailed as required.

Horizontal and vertical forces are detailed in load diagrams complete with the corresponding working points. If occurring, variation of the forces and oscillation amplitudes and frequency are given.

Based on the mechanical design, the civil design has to be executed in detail. This comprises a foundation design as well as the design of buildings and rooms to house auxiliary equipment and offices. This comprises structural steel buildings, concrete structure and masonry work.

Utility services have to be planned as well within the framework of civil engineering. The electrical engineering has to be executed in close cooperation with the detailed process engineering. The time schedule prepared during basic engineering is updated and extended to match the conditions of the project.

Parts of the process engineering are the operation and maintenance instructions which should be handed over to the production personnel as basis for their instructions to operate the plant and for training purposes. This documentation should be available ahead of production startup time.

Following the completion of construction and erection, the as-built drawings show the actual plant and conclude the documentation prepared by the supplier.

4. Construction

Petra Haag

All equipment manufactured and purchased has to be delivered to site. The transport may become quite a time-consuming task requiring substantial paperwork in case of an export project. The deliveries must be coordinated so that the equipment is supplied in the order of the construction procedure on site. Sufficient storage area for large parts and closed buildings for sensitive equipment, such as electrical cabinets and also refractory material, must be available on site. One important issue is that buildings and foundations are available. The conditions on site must permit trouble-free execution of the project activities. This refers specifically for the conditions in cold climate. It is essential to have sufficient temperature for the installation of refractory material. The sequence of construction should be structured according to the requirements of the plant, i. e. the sequence of commissioning. For instance, it does not make sense to startup the melting furnace if there is no possibility to discharge the molten metal. Thus, construction should commence according to the startup sequence and the planning must inevitably also consider the construction time. Furnaces always require installation of refractory lining, followed by curing of the refractory material. This is very time-consuming. The mechanical construction must also consider that efficient refractory material installation is possible. This is, for instance, in the case of the rotary drum furnace that the drum can be rotated once the refractory material installation work commences.

5. Project management

Petra Haag

To handle a project successfully an efficient project management is essential. All activities relating to a project are coordinated by the project manager including exchange of information to and from the customer. The actual duties of the project manager alter as the project advances. The different steps of a project have been described in other chapters and the challenging tasks of the project manager can be described simply as to complete the project according to the budget in the defined time and within the framework of the contract. This appears to be very simple but it includes a multitude of tasks to be fulfilled. This is shown in **Fig. 5.1**. The functions must not be performed by the project manager and his staff themselves but he has to coordinate all these

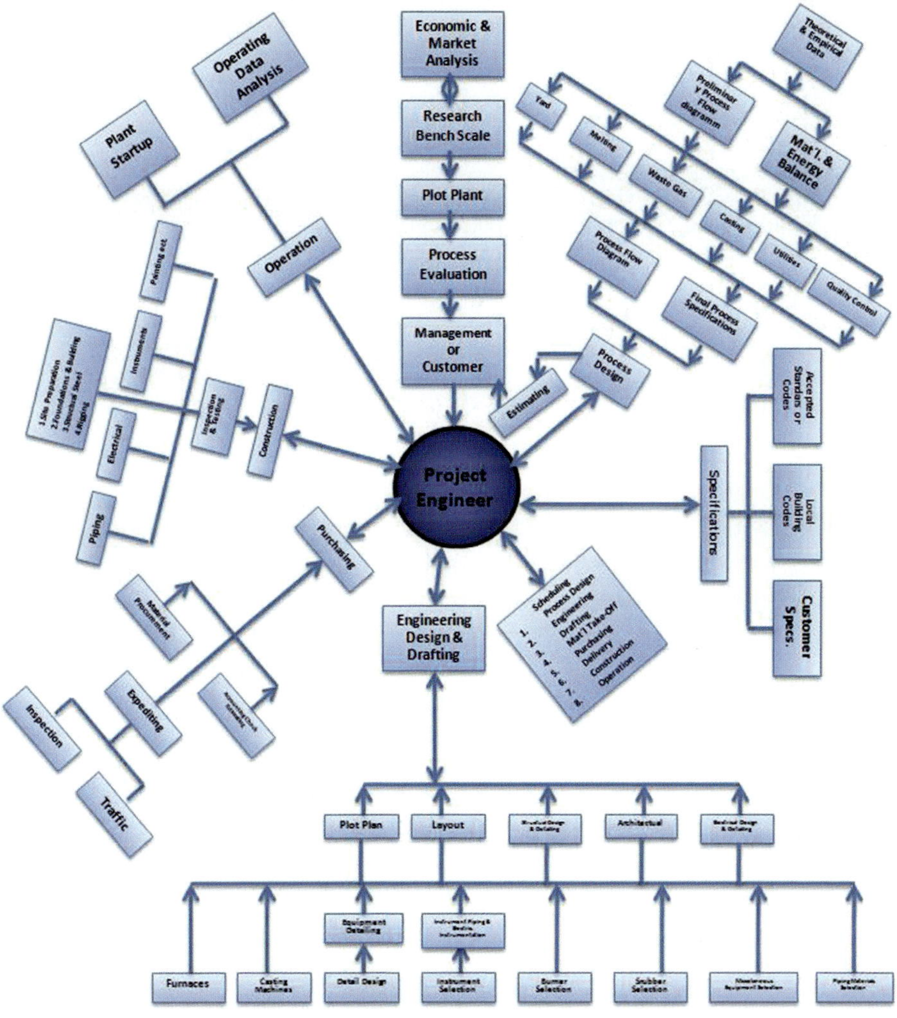

Fig. 5.1: Responsibilities of the project engineer

activities and make sure that the results are available within the given time and with the quality required. The duties of the project manager require skill and experience. He must have a technical background to be able to fulfil his task.

In the first phase of any project the project manager will be responsible for the relationship between the supplying company and customer. He has to find out the customer's requirements and preferences to enable the preparation of a detailed proposal matching the customer's wishes.

Site allocation is one of the early steps in the project.This involves the gathering of all data available on the conditions of the planned location of the plant. This again can be split into data on availability and price of resources such as energy, raw material, labor and process water (process related) and environmental conditions such as weather, site conditions, transport routes (for equipment in the beginning and marketing of products as the project commences) and general and environmental laws and regulations for the area.

Once all data is collected, the conceptual designing of the plant can go ahead. Cost can be estimated and a proposal will be presented to the customer.

Throughout this phase the project manager is responsible for the coordination of all planning activities and the liaison with the customer. He will channel all information, check plant layout and equipment design, crosscheck the cost estimates, schedule the time required for the project and submit the resulting offer to the customer. The project manager will discuss the proposal with the customer and supply him with additional explanations and information. This may include changes of design or plant layout, commercial conditions in a possible contract or the preparation of a feasibility study to calculate the risk and profitability of the project.

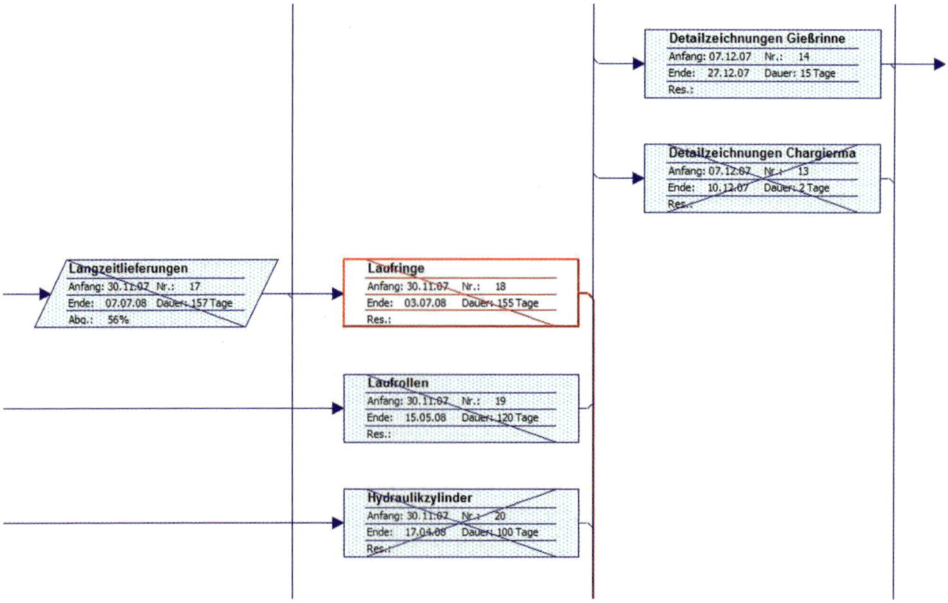

Fig. 5.2: Detail of a CPM (Critical Path Method) schedule showing the link between the different project activities, duration (Dauer) of an activity, planned start (Anfang), completion time (Ende) and buffer time (Res.), as well as % of completion (Abg.). If there is no buffer between start and completion the activities are on the "critical path" which means that any delay can negatively affect the objective to complete the project on time. Those activities are marked red

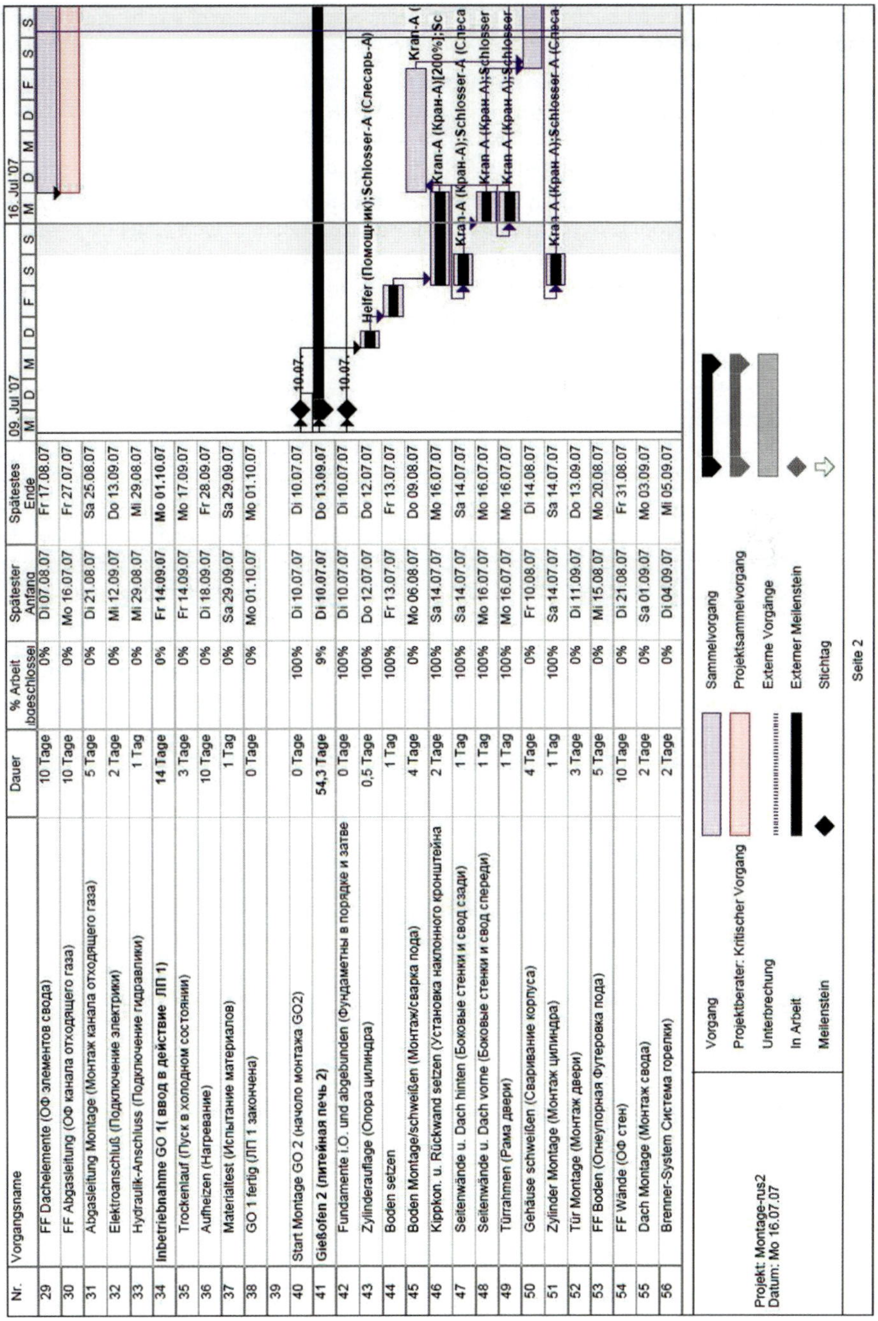

Nr.	Vorgangsname	Dauer	% Arbeit abgeschlossen	Spätester Anfang	Spätestes Ende
29	FF Dachelemente (ОФ элементов свода)	10 Tage	0%	Di 07.08.07	Fr 17.08.07
30	FF Abgasleitung (ОФ канала отходящего газа)	10 Tage	0%	Mo 16.07.07	Fr 27.07.07
31	Abgasleitung Montage (Монтаж канала отходящего газа)	5 Tage	0%	Di 21.08.07	Sa 25.08.07
32	Elektroanschluß (Подключение электрики)	2 Tage	0%	Mi 12.09.07	Do 13.09.07
33	Hydraulik-Anschluss (Подключение гидравлики)	1 Tag	0%	Mi 29.08.07	Mi 29.08.07
34	**Inbetriebnahme GO 1(ввод в действие ЛП 1)**	**14 Tage**	**0%**	**Fr 14.09.07**	**Mo 01.10.07**
35	Trockenlauf (Пуск в холодном состоянии)	3 Tage	0%	Fr 14.09.07	Mo 17.09.07
36	Aufheizen (Нагревание)	10 Tage	0%	Di 18.09.07	Fr 28.09.07
37	Materialtest (Испытание материалов)	1 Tag	0%	Sa 29.09.07	Sa 29.09.07
38	GO 1 fertig (ЛП 1 закончена)	0 Tage	0%	Mo 01.10.07	Mo 01.10.07
39					
40	Start Montage GO 2 (начало монтажа GO2)	0 Tage	100%	Di 10.07.07	Di 10.07.07
41	**Gießofen 2 (литейная печь 2)**	**54,3 Tage**	**9%**	**Di 10.07.07**	**Do 13.09.07**
42	Fundamente i.O. und abgebunden (Фундаметны в порядке и затве	0 Tage	100%	Di 10.07.07	Di 10.07.07
43	Zylinderauflage (Опора цилиндра)	0,5 Tage	100%	Do 12.07.07	Do 12.07.07
44	Boden setzen	1 Tag	100%	Fr 13.07.07	Fr 13.07.07
45	Boden Montage/schweißen (Монтаж/сварка пода)	4 Tage	0%	Mo 06.08.07	Do 09.08.07
46	Kippkon. u. Rückwand setzen (Установка наклонного кронштейна	2 Tage	100%	Sa 14.07.07	Mo 16.07.07
47	Seitenwände u. Dach hinten (Боковые стенки и свод сзади)	1 Tag	100%	Sa 14.07.07	Sa 14.07.07
48	Seitenwände u. Dach vorne (Боковые стенки и свод спереди)	1 Tag	100%	Mo 16.07.07	Mo 16.07.07
49	Türrahmen (Рама двери)	1 Tag	100%	Mo 16.07.07	Mo 16.07.07
50	Gehäuse schweißen (Сваривание корпуса)	4 Tage	0%	Fr 10.08.07	Di 14.08.07
51	Zylinder Montage (Монтаж цилиндра)	1 Tag	100%	Sa 14.07.07	Sa 14.07.07
52	Tür Montage (Монтаж двери)	3 Tage	0%	Di 11.09.07	Do 13.09.07
53	FF Boden (Огнеупорная Футеровка пода)	5 Tage	0%	Mi 15.08.07	Mo 20.08.07
54	FF Wände (ОФ стен)	10 Tage	0%	Di 21.08.07	Fr 31.08.07
55	Dach Montage (Монтаж свода)	2 Tage	0%	Sa 01.09.07	Mo 03.09.07
56	Brenner-System Система горелки)	2 Tage	0%	Di 04.09.07	Mi 05.09.07

Fig. 5.3: Detail of a bar chart developed from the CPM method showing activity (Vorgangsname), duration (Dauer), % completed (% Arbeit abgeschlossen), latest start (spätester Anfang) and latest completion (spätestes Ende)

Assuming the proposal was successful, the implementation phase of the project starts with the signing of the contract and everybody is happy now.

This phase starts with the detail design of the plant according to the contract and the detail design of the equipment. The project manager will find sub-suppliers for equipment, equipment parts, construction, etc. Subcontracts will be negotiated and awarded and the additional information will be integrated into the plant and equipment design.

Another one of the project manager's duties is to "transform" the cost and time estimates into "actual" data. His main tool to achieve this will be the project control with focus on cost and time management. At the beginning of the project a time schedule will be established. This covers all important project activity estimates such as performance time, required manpower, preliminary and following activities and the implied cost. Milestone dates are defined such as project start, shipment of items, commissioning, etc. Usually these milestones are connected to payments according to the contract. A frequent update of the activities reveals discrepancies to the estimates in time for the project manager to react in such way that the smooth performance of the project is ensured. Examples of a cost and time schedule can be seen in **Fig. 5.2**, the project time schedule showing the relationships between the activities including the critical path and **Fig. 5.3** with detailed information on each activity, such as start and finish date, delay, manpower planned and / or used. Closely connected to the time schedule is the cost plan, showing the estimated cost for each item (e. g. Steel structure casting furnace I, burners furnace I, ...) and activity (e.g. detail engineering hydraulics, shipment furnaces,....) and the related payments. A continuous updating of the data will allow the project manager a realistic overview and good chance to react on any difficulties that might arise due to deviations from the plan.

The manufacturing of the components and delivery of all plant equipment follow the detail design. The project manager certifies to the customer that the equipment conforms to the design, quality, price and time schedule as specified.

After the project manager has arranged the shipment of the equipment, the construction phase begins. Now the project manager has to make sure that construction staff and material is on site and the equipment can be installed according to the schedule. The coordination of all activities will be the main focus in this phase and will result in the timely start of commissioning of the plant.

Once the plant is in operation the project manager will conclude his activities with the completion of all necessary documentation. The plant will be handed over to the customer and the after-sales service starts. This is very important to maintain a good relationship with the customer. It is a very good marketing argument that a satisfied customer proudly shows the plant he has had constructed. The supplier should also inform the customer of new developments and assist him to obtain production know-how, contacts to his customers and required spare parts.

6. Final remarks

Christoph Schmitz

The plant is now ready for operation. If the equipment is designed as described in the previous chapters it should be able to operate economically and trouble-free. But a sound plant concept is one part of the success. The other parts are good operators and efficient plant management. Continuous improvement of the skill of operators and also adaptation of the equipment to new technologies will keep the entire operation at the highest possible level. This does not mean that any fancy new process must be introduced. If a new technology is selected, the plant management must assure that the present operation can be improved with regard to lower production cost, better environmental protection and improved quality.

Annex

Annex

Symbols

Symbol	Designation	Unit
A	area	m^2
a (b)	acceleration	m/s^2
C	radiation coefficient	$W/(m^2K^4)$
c	specific heat capacity	kJ/K
E (U)	internal (thermodynamic) energy	J (W)
F	force (force of weight)	N
f	factor	–
g	gravitational acceleration	
H	enthalpy	J (Ws)
h	specific enthalpy	J/kg
Hu	lower heating value	J/kg
I	electric current	A
k	heat transfer coefficient	$W/(m^2K)$
M, m	(material) mass	kg
N	material quantity	kmol
P	power	kW (kJ/s)
p	pressure	Pa (mbar)
Q	heat quantity	J
q	specific heat quantity	J/kg
R	electric resistance	Ω
R	general gas constant	J/(molK)
s	thickness (lining etc)	m
T	absolute temperature	K
U	electric tension	V
V	volume	m^3
V_n	gas volume at mean condition	m_n^3
v	specific gas volume	m^3/kg
w	gas velocity	m^3/s
α	heat transfer coefficient	$W/(h\ m^2K)$
β	burn – off factor	–
γ	density (solid and liquid matter)	Kg/dm^3
Δ	difference (i. e. temperature)	–
ε	radiation coefficient	$W/(m^2K^4)$

Symbols (continued)

Symbol	Designation	Unit
η	efficiency factor	–
ϑ	temperature	°C
λ	excess air factor	–
ρ	density (gas)	Kg/m^3
ϕ	wall factor	–

Unit conversion

Force	1 N = 0.102 kp
Pressure	1 N/m^2 = 1 Pa = 0.102 mm WG
	100 Pa = 1 h Pa = 1 mbar = 10.2 mm WG
Energy, heat values, mech. work	1 J = 1 Ws = 1 Nm 1 kWh = 3,600 kJ = 860 kcal
Mol volume	1 kmol = 22,4 mn^3
Temperature	K = °C + 273
Gas volume at 20 °C 100 °C 500 °C 800 °C 1,000 °C 1,100 °C 1,200 °C	1 m_n^3 = 1.07 m^3 1 m_n^3 = 1.36 m^3 1 m_n^3 = 2.83 m^3 1 m_n^3 = 3.39 m^3 1 m_n^3 = 4.66 m^3 1 m_n^3 = 5.03 m^3 1 m_n^3 = 5.40 m^3

Important material data

	Density [kg/dm³]	Melting point [°C]	Heat of fusion [kJ/kg]	Spec. heat capacity [kJ/(kg K)]
Aluminium, solid liquid	2.7 2.3	658	356	0.91
Copper	8.92	1,083	210	0.39
Magnesium	1.74	650	250	0.5
Silicon	2.4	1,410		0.74
Steel	7.7	1,350	-	0.50

Important material data (continued)

	Density (0 °C, 1 bar) [kg/m³]	Heating Value [kWh/m³]	Spec. heat capacity c_p [[kJ/(kg K)]	Mol mass [g/mol]	Gas constant [J/(kgK))
Air	1.29	-	0.92	28.97	287
Natural gas	0.72	10.2 -10.7	2.16	16.05	518
Propan	2.02	10.3	1.14	44.09	189
Nitrogen	1.25	-	1.04	28.02	297
Carbon oxide (CO_2)	1.25	-	1.04	28.01	297
Combustion products	1.45	-	1.15		
Light fuel oil	0.9 [kg/dm³]	11.4 [kWh/kg]			

Some data for quick estimate

R	universal gas constant	1.314	J/(kgK)	
q_m	theoretical energy required to heat up and melt aluminium from 20 °C solid to 720 °C liquid	316 1,138	kWh/t kJ/kg	This value is lower for certain alloys but is suitable for quick estimate
	actual values depend on the type of furnace and ranges from	600-1,800	kWh/t	
α_{ab}	heat absorption of a liquid aluminium bath with a temperature of 700 °C and a flue gas temperature of 1,100 °C	80-150	$W/(m^2\Delta\vartheta)$	$\Delta\vartheta$ = difference between flue gas temperature and bath temperature
V_{fl}	flue gas volume by combustion of 1 m³ of natural gas	11 m_n^3/m^3		The actual value stands for the furnace design. The upper furnace is valid for a furnace with bath circulation
	flue gas volume by combustion of 1 kg of light fuel oil	12 m_n^3/m^3		based on a λ = 10 %

Conversion of length units

Metric		Imperial
1 millimeter [mm]		0.03937 in
1 centimeter [cm]	10 mm	0.3937 in
1 meter [m]	100 cm	1.0936 yd
1 kilometer [km]	1,000 m	0.6214 mile

Imperial		Metric
1 inch [in]		2.54 cm
1 foot [ft]	12 in	0.3048 m
1 yard [yd]	3 ft	0.9144 m
1 mile	1,760 yd	1.6093 km
1 int nautical mile	2,025.4 yd	1.853 km

Conversion formulas of temperature units

	from Celsius	to Celsius
Fahrenheit	$[°F] = [°C] \times {}^{9}/_{5} + 32$	$[°C] = ([°F] - 32) \times {}^{5}/_{9}$
Kelvin	$[K] = [°C] + 273.15$	$[°C] = [K] - 273.15$

Literature

Literature

Part I

Altenpohl — Aluminium und Aluminiumlegierungen, Springer Verlag, Berlin 1965

Altenpohl, D. — Aluminium von innen betrachtet, Aluminium Verlag, Düsseldorf 1983

Aluminium — Aluminium Taschenbuch, 14th edition, Aluminium Verlag, Düsseldorf 1994

Gesing, A. — Review, state of the art and priorities – collection and sorting, International Aluminium Recycling, Workshop, Trondheim 2013

Ginsberg et al. — Die Tonerde, Walter de Gruyter & Co., Berlin 1964

Grjotheim, K.; Welch, B. J. — Aluminium Smelter Technology, Aluminium Verlag, Düsseldorf 1987

Kirchner, G. — The European and global dimension of aluminium recycling at present and in future, Erzmetall, Clausthal-Zellerfeld 2002

Klockmanns — Lehrbuch der Mineralogie, Ferdinand Enke Verlag, Stuttgart 1978

Krone — Aluminium Recycling, VDS, Düsseldorf 2000

Milbowen, P. — Aluminium recycling local to global supply chain, Aluminium International Today, Redbill

Münster, H. P. — Taschenbuch des Metallhandels, Hütting Verlag, Heidelberg 1997

Schneider, K. — Die Verhüttung von Aluminiumschrott, Metall Verlag, Berlin 1970

Schucht, S. — Ökologische Modernisierung und Strukturwandel in der deutschen Aluminium Industrie, Freie Universität Berlin, Berlin 1999

Voswinkel, C. et al. — Herstellung hochwertiger Gußteile für die Fahrzeugindustrie, Gießerei Verlag, Düsseldorf 1986

Part II

Alfaro, I. — Technische und wirtschaftliche Gesichtspunkte bei der Verarbeitung von Krätze, Aluminium Verlag, Düsseldorf 1986

Aluminium — Aluminium Taschenbuch, 14th edition, Aluminium Verlag, Düsseldorf 1994

Cziebos, H. — Hütte, Die Grundlagen der Ingenieurwissenschaft, Springer Verlag, Berlin 1991

Dubbel — Taschenbuch für den Maschinenbau, Springer Verlag, Berlin 1986

Escherle, A. — Application of pyrolisis in aluminium recycling, Erzmetall, Clausthal-Zellerfeld 2004

Fetcher, A. — Betriebserfahrungen und Wirtschaftlichkeit bei der Krätzebehandlung in Aluminiumgießereien, Aluminium Praxis, Düsseldorf 1981

Lechner, M. D. — Taschenbuch für Chemiker und Physiker. Vol. I, founded by Jean d'Ans, Ellen Lax. 4th edition, Springer Verlag, Berlin 1992

Maniruzzaman et al. — Computer simulation of floatation treatment systems, TMS Light Metals, Annual Meeting, Warrendale 1993

Mittag, C. — Die Hartzerkleinerung, Springer Verlag, Berlin 1953

Schubert, H. — Die Aufbereitung fester mineralischer Rohstoffe, VEB, Deutscher Verlag für Grundstoffindustrie, Leipzig 1975

Thome-Kozmienzky, K. J. — Optimierung der Abfallverbrennung, TK Verlag, 2004

Part III

Altenpohl, D. — Aluminium von innen betrachtet, Aluminium Verlag, Düsseldorf 1983

Altendorf, H.-J. — Potentials for an electronic control unit with fuel/air compound control in combustion processes, Gaswärme International, Band 54 (2005), Vulkan-Verlag, Essen 2005

Baker, P. W. et al. — Doss formation during metal transfer operations, TMS Conference, Sydney 1995

Bamji, P. J. — Submergence of light scrap using a linear induction melter, TMS Light Metals, Warrendale 1983

Baukal, C. E. — Oxygen-enhanced combustion, CRC Press, 1998

Biedenkopf, P. — Advanced tool for flexible and economical melting in the non ferrous industry, Aluminium International Today, Redbill 2004

Brunhuber	Gießerei-Lexikon, Gießerei Verlag, Düsseldorf 1986
Brunklaus, J.H. et al.	Industrieöfen, Bau und Betrieb, Vulkan-Verlag, Essen 1986
Buchholz; Rødseth	Investigation of heat transfer conditions in a reverberatory melting furnace by numerical modelling, TMS Light Metals, 2011
Chambers, P. et al.	Metall level sensing and control, TMS Conference, Melbourne 1993
Cramer, C.	Heizgasvermischung in Ofenanlagen, Abschlußbericht, Forschungsgemeinschaft Industrieofenbau, Aachen 1985
Cziebos, H.	Hütte. Die Grundlagen der Ingenieurwissenschaft, Springer Verlag, Berlin1991
de Graal, I.	Potential for increasing fuel efficiency for aluminium melting furnaces, Heat Processing, Vulkan-Verlag, Essen 2005
DGFS	Feuerfestbau, Deutsche Gesellschaft für Feuerfest und Schornsteinbau, Vulkan-Verlag, Essen 2003
Domagala, J.	Regenerative burner technology for aluminium melting furnaces industry
Donsbach, F. et al.	Sicheres und energiesparendes Schmelzen in Mittelfrequenztiegelöfen, Otto Geucher GmbH, Lammersdorf 2005
English, C.	The casthouse trough degasser, TMS Light Metals, Warrendale 1998
Epstein, S.G.	Summary of ten years of molten aluminium incident reporting, TMS Light Metals, Warrendale 1995
Essafi, R.	Optimierung der Gaszusammensetzung für das Schmelzen von Aluminium-Schrott in Drehtrommelöfen, Diplomarbeit, Clausthal-Zellerfeld 2001
Fehrenbach, I. et al.	Füllstandsmessung mit Laser, M&P,1991
Friesen, K. J. et al.	The coalescence behaviour of aluminium droplets under a molten flux cover, TMS, Warrendale 1997
Frischknecht, B. et al.	Sheet production of high quality products with mini mills, TMS Conference, Sydney 2001
German Federal Ministry	Government regulations: First General Administrative Regulation Pertaining the Federal Immission Control Act, (Technical instructions on air quality control (TA-Luft))
German Federal Ministry	Ministry for Environment, Nature Conservation and Nuclear Safety: TRK-Wert für chlorierte Dibenzodioxine und -furane, BArbBl 9/93 p. 71, Berlin 1993
Gluns, L. et al.	Technology and economies of oxy-fuel combustion, aluminium International Today, Redbill, 2004
Gripenberg, H. et al.	Six years experience from low-temperature oxyfuel in primary and remelting aluminium cast houses, TMS Light Metals, 2012
Godfrey, M. V.	Improving performance of gas fired crucible furnaces for Aluminium, Industrial Heating, 1989
Guthie, R. I. L. et al.	Impurity sources and control, TMS Light Metals, Warrendale, 3rd Australian Conference, 1983
Häberle, G.	Tabellenbuch Mechatronik, Europa Lehrmittel, Haan Gruiten
Hansen, W.	Heizöl. Handbuch für Industriefeuerungen, Springer Verlag, Berlin 1959
Heiligenstaed	Wärmetechnische Rechnungen für Industrieöfen, Stahl & Eisen Verlag, Düsseldorf 1941
Henning, B. et al.	Multimelter. The new generation of aluminium melting furnaces
Herkwich, G. et al.	Herkwich Horizontal D.C. Casting, TMS Conference, Sydney 1995
Hertwich	Homogenizing and sawing, Company publication, Hertwich Engineering GmbH, Braunau
Hielscher, U.	Qualitätsorientierte Schmelzprüfung in der Aluminiumgießerei, Gießerei, Düsseldorf 1987
Hoffmann, G.	Studie Erdgasverbrennung mit Sauerstoffangereicherter Luft, Ruhrgas, Essen 1985
Jasper GmbH	Company literature
Johansen, S. T. et al.	Particle floatation to bubbles in rotor stirred reactors, TMS Light Metals, Warrendale 1995
Kamiske, G. F.	Qualitätsmanagement von A-Z, Carl Hanser Verlag, München 1995

Kammer, C.	Grundlagen und Werkstoffe. Beeinflussung der Eigenschaften durch thermische und mechanische Behandlung. In: Aluminium Taschenbuch 1. 16th edition, Aluminium-Verlag, Düsseldorf 2002, ISBN 3-87017-274-6, pp. 278 ff.
Kreysa, E.	Induktionsöfen zum Speichern von Flüssigmetall in Aluminium Gießereien, Aluminium Verlag, Düsseldorf
Kromschroeder AG	Catalogue
Krug, H. P. et al.	A contribution to succession measurement after inline degasser
Lamtec	Combustion control, Company publication, Lamtec Meß- und Regeltechnik für Feuerungen GmbH & Co. KG
Liere-Netheler, W.	Direct firing of aluminium heat treatment systems, Heat Processing 1/2005, Vulkan-Verlag, Essen 2005
Marks	Standard handbook for mechanical engineers, 8th edition, McGraw-Hill Book Company, New York
Marquarotto, G.	Tomorrow's solution to improve casthouse managment and the operator's working conditions, Aluminium Verlag, Düsseldorf 2005
Meyer, H. J.	Use of regenerative heating technologies at aluminium melting and aluminium recycling furnaces, Heat Processing, Vulkan-Verlag, Essen 2004
Moser, R.	Energiesparmaßnahmen an flammenbeheizten Herdöfen in Aluminium Gießereien, Aluminium Verlag, Düsseldorf
Niehoff, T.	Oxyfuel. Solutions for energy and environmental conservation, TMS Light Metals, 2008
Niehoff, T.	Energy and emissions with oxyfuel, TMS 2010
Nilmani, M.	Low cost solutions to gas lancinging problems, TMS Conference, Melbourne 1993
Oldörp, R.	Umrüstung von Aluminiumschmelzöfen auf regenerative Beheizung, Gaswärme International, Vulkan-Verlag, Essen 2004
Pierce, D.C. et al.	The casthouse addressing scrap charging safety, TMS, Warrendale 1995
Piwinger, F.	Regelungstechnik für Praktiker, VDI-Verlag, Düsseldorf
Polmear, I. J.	Metallurgy of the light alloys. Wrought aluminium alloys, light alloys, 2nd edition, Chapman and Hall, New York 1989
Quesnel, S. et al.	Criteria for choosing refactories in aluminium holding and melting furnaces, TMS, Warrendale 1998
Ray, L. D.	Change term assessment of the ALSCAN hydrogen analyser, TMS Light Metals, Warrendale 1992
Rixen, K.	Regenerative burner technology for aluminium melting furnaces industry
Roben, F. J.	Trocknungs- und Vorwärmeanlage für Aluminium Schmelzöfen, Gießerei Verlag, Düsseldorf 1985
Roy, R. R. et al.	The role of salt flux in recycling of aluminium, TMS Light Metals, Warrendale 1998
Ruhrgas	Erdgasinformationen, issue 12, Ruhrgas AG, Essen 1983
Schaller, D. G.	The next generation of melting technology for aluminium melting, TMS Light Metals, Warrendale 1985
Schneider, W.	The new airsol-veil billet casting system, TMS Conference, Sydney 1995
Schuh, F.	Enzyklopädie der Naturwissenschaft und Technik, Zweiburgen Verlag, Weinheim 1979
Silny, A. et al.	Interfacial tension between aluminium, aluminium based alloys and chloride-fluoride melts, TMS, Warrendale 1996
Sindram, M.; Walter, D.	Trockene Rauchgasreinigung für Feuerungsanlagen und andere thermische oder chemische Prozesse, Optimierungskonzepte für die trockenen Rauchgasreinigung, Rheinkalk GmbH, Wülfrath
Sjödein, O.	ABB's electromagnetic sterring system improves work environment and saves energy, Aluminium World, London 2005
Slujeru, Y. et al.	Application of new hot top process for production of extrusion and forging billets, Showa Denko K. K., Tokyo 1992
Smalley, D.S.	A new coreless industrian furnace monitor, AFS-transactions, 1989
Taylor, M. B. et al.	The inert gas dross cooler, TMS, Warrendale 1995

Terrier, J.-C.	ALPUR TS, Pechiney's new global concept for aluminium treatment, TMS Light Metals, Warrendale 1998
Thomas, A.T.	Proceedings of 6th International Conference on Light Metals, Leoben, Austria, Aluminium-Verlag, Düsseldorf 1975
v. Stark, A.	Handbook of thermoprocessing technologies, Vulkan-Verlag, Essen 1998
VDI	VDI-Wärmeatlas, VDI-Verlag, Düsseldorf 1974
Vogg, H.	Abfallverbrennung Nr. 37 u. 37c, TÜV-Rheinland, 1992
Voswinkel, C. et al.	Herstellung hochwertiger Gußteile für die Fahrzeugindustrie, Gießerei Verlag, Düsseldorf 1986
Voswinkel, C. et al.	Herstellung hochwertiger Gußteile aus Aluminium für die Automobilindustrie, Gießerei, Düsseldorf 2004
Waite, P. D.	Improved metallurgical understanding of the ALCAN compact de-gasser, TMS Light Metals, Warrendale 1998
Wallach, P.	Personal proactive clothing, TMS Light Metals, Warrendale 1995
Weber, E.; Broche, W.	Apparate und Verfahren der industriellen Gasentreinigung, R. Oldenbourg Verlag, München 1973
Wiesner, A.	Schmelzen mit induktivem Wärmeübergang. Theorie und Praxis, Aluminium, Düsseldorf 1983
Winiam, N. et al.	New engineering solution upgrading an existing degassing system, TMS Light Metals, Warrendale 1993
Whipple, D.	Direct chaged melters. Bloom company literature; A control strategy for high production aluminium furnaces. Bloom company literature
Winter, H.	Prozessleittechnik in Chemieanlagen, Europa Lehrmittel, Haan Gruiten
Wohlschläger, G.	Direct firing of aluminium heat treatment systems, Heat Processing 1/2005, Vulkan-Verlag, Essen 2005

Part IV

Bainbridgh, L. F.	The casthouse financial aspects and model, TMS Conference, Melbourne 1993
Home, K. et al.	An integrated approach to the design and construction management of an aluminium smelter expansion, Aluminium World, London 2005
Marks	Standard handbook for mechanical engineers, 8th edition, McGraw-Hill Book Company, New York
Probst, H.-J.	Projektmanagement, Redline Wirtschaft, redline GmbH, Heidelberg
Reese, F. R.; Barrcew, M. H.	Project engineering, John Wiley & Sons, London 1957

Index

Index

D

F

G

J

L

M

N

O

P

S

U

V

W

X

Advertiser's Index